DIGITAL ELECTRONICS AND SYSTEMS

AUTHOR

Dr. Abhishek Bhatt

Professor

School of Data Science,

Symbiosis Skills and Professional University, Pune

REVIEWER

Dr. Dhiman Mallick

Assistant Professor

Indian Institute of Technology, Delhi

KHANNA BOOK PUBLISHING CO. (P) Ltd.

4C/4344, Ansari Road, Darya Ganj, New Delhi, India

All India Council for Technical Education

Nelson Mandela Marg, Vasant Kunj, New Delhi, India

Dear Readers,

To prevent the piracy, this book is secured with HIGH SECURITY HOLOGRAM on the front title cover. In case you don't find the hologram on the front cover title, please write us to at contact@khannabooks.com or whatsapp us at +91-99109 09320 and avail special gift voucher for yourself.

Specimen of Hologram on front Cover title:

Moreover, there is a SPECIAL DISCOUNT COUPON for you with EVERY HOLOGRAM.

How to avail this SPECIAL DISCOUNT:

Step 1: Scratch the hologram

Step 2: Under the scratch area, your "coupon code" is available

Step 3: Logon to www.khannabooks.com

Step 4: Use your "coupon code" in the shopping cart and get your copy at a special discount

Step 5: Enjoy your reading!

ISBN: 978-93-5538-887-2

Book Code: UG411EN

Digital Electronics and Systems

by Dr. Abhishek Bhatt

First Edition: 2026

Originally Published by:

All India Council for Technical Education, Ministry of Education, GOI

Exclusively Published & Distributed by:

Khanna Book Publishing Co. (P) Ltd.

4C/4344, Ansari Road, Daryaganj,
New Delhi (INDIA)

Visit us at: **www.khannabooks.com**

Write us at: **contact@khannabooks.com**

To view complete list of books,
Please scan the QR Code:

Printed in Bharat.

प्रो. टी. जी. सीताराम
अध्यक्ष
Prof. T. G. Sitharam
Chairman

आज़ादी का अमृत महोत्सव

अखिल भारतीय तकनीकी शिक्षा परिषद्
(भारत सरकार का एक सांविधिक निकाय)
(शिक्षा मंत्रालय, भारत सरकार)
नेल्सन मंडेला मार्ग, वसंत कुंज, नई दिल्ली–110070
दूरभाष : 011–26131498
ई–मेल : chairman@aicte-india.org

ALL INDIA COUNCIL FOR TECHNICAL EDUCATION
(A STATUTORY BODY OF THE GOVT. OF INDIA)
(Ministry of Education, Govt. of India)
Nelson Mandela Marg, Vasant Kunj, New Delhi-110070
Phone : 011-26131498
E-mail : chairman@aicte-india.org

FOREWORD

Engineers are the backbone of any modern society. They are the ones responsible for the marvels as well as the improved quality of life across the world. Engineers have driven humanity towards greater heights in a more evolved and unprecedented manner.

The All India Council for Technical Education (AICTE), have spared no efforts towards the strengthening of the technical education in the country. AICTE is always committed towards promoting quality Technical Education to make India a modern developed nation emphasizing on the overall welfare of mankind.

An array of initiatives has been taken by AICTE in last decade which have been accelerated now by the National Education Policy (NEP) 2020. The implementation of NEP under the visionary leadership of Hon'ble Prime Minister of India envisages the provision for education in regional languages to all, thereby ensuring that every graduate becomes competent enough and is in a position to contribute towards the national growth and development through innovation & entrepreneurship.

One of the spheres where AICTE had been relentlessly working since past couple of years is providing high quality original technical contents at Under Graduate & Diploma level prepared and translated by eminent educators in various Indian languages to its aspirants. For students pursuing 2nd year of their Engineering education, AICTE has identified 87 books, which shall be translated into 12 Indian languages - Hindi, Tamil, Gujarati, Odia, Bengali, Kannada, Urdu, Punjabi, Telugu, Marathi, Assamese & Malayalam. In addition to the English medium, books in different Indian Languages are going to support the students to understand the concepts in their respective mother tongue.

On behalf of AICTE, I express sincere gratitude to all distinguished authors, reviewers and translators from the renowned institutions of high repute for their admirable contribution in a record span of time.

AICTE is confident that these outcomes based original contents shall help aspirants to master the subject with comprehension and greater ease.

(Prof. T. G. Sitharam)

ACKNOWLEDGEMENT

The authors are grateful to the authorities of AICTE, particularly Prof. T. G. Sitharam, Chairman; Dr. Abhay Jere, Vice-Chairman; Prof. Rajiv Kumar, Member-Secretary, Dr. Ramesh Unnikrishnan, Advisor-II and Dr. Sunil Luthra, Director, Training and Learning or their planning to publish the books on ***Digital Electronics & Systems***.

I sincerely acknowledge the valuable contributions of the reviewer of the book Prof. (Dr.) Dhiman Mallick, Assistant Professor, Indian Institute of Technology, Delhi; Dr. M.S. Sutaone, Director, COEP Pune; Dr. S. P. Mahajan, Professor and Head, Dept. of E & TC, COEP, Pune; Dr. B. B. Ahuja, Ex- Director, COEP Pune; Dr. Swati Mujumdar, Pro Chancellor, SSPU, Pune; Dr. Gauri Shirurkar VC-SSPU, Pune. I am also thankful to my wife Vandana; my son Akshaj & Ajithesh, as Without their support, this work would not have come to fruition.

I would also like to acknowledge the reference books that have been instrumental in the preparation of this book: "Digital Logic and Computer Design" by M Morris Mano; “Fundamentals of Digital Circuits” by Anand Kumar; “Digital Electronics: Principles and Integrated Circuit” by Anil K. Maini; "Digital Logic: Applications and Designs" by John M Yarbrough; “Digital Fundamentals” by Floyd; “Digital Principles and Application” by Leach and Malvino. Additionally, I extend my gratitude to my students, over the years, who have made significant contributions that have enriched my experience and expertise in this subject. I am immensely grateful for their valuable input for making this book students’ friendly and giving a better shape in an artistic manner.

This book is an outcome of various suggestions of AICTE members, experts and authors who shared their opinion and thought to further develop the engineering education in our country. Acknowledgements are due to the contributors and different workers in this field whose published books, review articles, papers, photographs, footnotes, references and other valuable information enriched us at the time of writing the book.

Dr. Abhishek Bhatt

PREFACE

Welcome to the world of digital electronics! This book includes the topics recommended by AICTE, in a very systematic and orderly manner serves as your comprehensive guide to understanding the principles, concepts, and applications of digital electronics. Whether you are a student embarking on a journey of learning or a professional seeking to refresh your knowledge, this book is designed to provide you with a solid foundation in this fascinating field.

Digital electronics has revolutionized the way we live, work, and communicate. From the smartphones we rely on to the computers that power our daily tasks, digital electronics form the backbone of modern technology. Understanding the fundamentals of digital electronics is crucial for anyone involved in fields such as electrical engineering, computer engineering, robotics, and telecommunications.

In this book, we will embark on a journey that will take us through the intricacies of digital logic, circuit analysis, and design. We will explore the building blocks of digital systems, including logic gates, flip-flops, registers, and counters. We will delve into the world of combinatorial and sequential logic, understanding how to analyse, design, and optimize digital circuits.

Throughout the chapters, we will emphasize a hands-on approach to learning. We will provide practical examples, step-by-step explanations, and opportunities for you to apply your knowledge through exercises and projects. We will not only focus on the theoretical aspects of digital electronics but also explore its practical applications. From computer architecture and microprocessors to digital signal processing and communications systems, we will uncover how digital electronics plays a vital role in various domains.

As you progress through this book, we encourage you to actively engage with the material, ask questions, and seek deeper understanding. Digital electronics can sometimes be challenging, but with perseverance and practice, you will gain the necessary skills to analyse, design, and troubleshoot digital circuits with confidence.

We would like to express our gratitude to the countless researchers, educators, and engineers who have contributed to the field of digital electronics over the years. Their collective efforts have paved the way for the advancements we witness today. We hope that this book will serve as a tribute to their contributions and inspire you to further explore the exciting possibilities of digital electronics.

We sincerely hope that this book will be a valuable resource in your journey of learning and discovery. We invite you to immerse yourself in the world of digital electronics and embark on an adventure that will empower you to shape the future.

Happy reading! Happy Learning!

Dr. Abhishek Bhatt

OUTCOME BASED EDUCATION

For the implementation of an outcome based education the first requirement is to develop an outcome based curriculum and incorporate an outcome based assessment in the education system. By going through outcome based assessments evaluators will be able to evaluate whether the students have achieved the outlined standard, specific and measurable outcomes. With the proper incorporation of outcome based education there will be a definite commitment to achieve a minimum standard for all learners without giving up at any level. At the end of the programme running with the aid of outcome based education, a student will be able to arrive at the following outcomes:

PO1. Engineering knowledge: Apply the knowledge of mathematics, science, engineering fundamentals, and an engineering specialization to the solution of complex engineering problems.

PO2. Problem analysis: Identify, formulate, review research literature, and analyze complex engineering problems reaching substantiated conclusions using first principles of mathematics, natural sciences, and engineering sciences.

PO3. Design / development of solutions: Design solutions for complex engineering problems and design system components or processes that meet the specified needs with appropriate consideration for the public health and safety, and the cultural, societal, and environmental considerations.

PO4. Conduct investigations of complex problems: Use research-based knowledge and research methods including design of experiments, analysis and interpretation of data, and synthesis of the information to provide valid conclusions.

PO5. Modern tool usage: Create, select, and apply appropriate techniques, resources, and modern engineering and IT tools including prediction and modeling to complex engineering activities with an understanding of the limitations.

PO6. The engineer and society: Apply reasoning informed by the contextual knowledge to assess societal, health, safety, legal and cultural issues and the consequent responsibilities relevant to the professional engineering practice.

PO7. Environment and sustainability: Understand the impact of the professional engineering solutions in societal and environmental contexts, and demonstrate the knowledge of, and need for sustainable development.

PO8. Ethics: Apply ethical principles and commit to professional ethics and responsibilities and norms of the engineering practice.

PO9. Individual and team work: Function effectively as an individual, and as a member or leader in diverse teams, and in multidisciplinary settings.

PO10. Communication: Communicate effectively on complex engineering activities with the engineering community and with society at large, such as, being able to comprehend and write effective reports and design documentation, make effective presentations, and give and receive clear instructions.

PO11. Project management and finance: Demonstrate knowledge and understanding of the engineering and management principles and apply these to one's own work, as a member and leader in a team, to manage projects and in multidisciplinary environments.

PO12. Life-long learning: Recognize the need for, and have the preparation and ability to engage in independent and life-long learning in the broadest context of technological change.

COURSE OUTCOMES

After completion of the course the students will be able to:

CO-1: Understand the fundamentals of digital logic, including Boolean algebra, logic gates, truth tables, and logic simplification techniques.

CO-2: Analyze and design combinational and sequential digital circuits using various components such as logic gates, multiplexers, decoders, flip-flops, counters, and registers.

CO-3: Become familiar with various digital components and their characteristics, including different types of logic families (TTL, CMOS), integrated circuits (ICs), and programmable logic devices (PLDs) like field-programmable gate arrays (FPGAs).

CO-4: Understand the applications and use cases of digital electronics in various fields, such as shift registers, counters, memory devices etc.

CO-5: Enhance their problem-solving and critical thinking abilities by analyzing complex digital circuits, designing solutions like A/D and D/A convertors.

CO-6: Learn about emerging digital technologies: Depending on the course, students might also explore recent trends in digital design, such as low-power design techniques, hardware security, or emerging nanotechnologies.

Course Outcomes	Expected Mapping with Programme Outcomes *(1- Weak Correlation; 2- Medium correlation; 3- Strong Correlation)*											
	PO-1	PO-2	PO-3	PO-4	PO-5	PO-6	PO-7	PO-8	PO-9	PO-10	PO-11	PO-12
CO-1	3	2	2	2	1	-	-	-	-	-	-	1
CO-2	3	1	2	1	-	-	-	-	-	-	-	-
CO-3	3	3	2	1	-	-	-	-	-	-	-	1
CO-4	3	3	3	2	1	-	-	-	-	-	-	-
CO-5	3	3	2	1	1	-	-	-	-	-	-	1
CO-6	3	3	3	1	2	2	-	-	-	-	1	3

GUIDELINES FOR TEACHERS

To implement Outcome Based Education (OBE) knowledge level and skill set of the students should be enhanced. Teachers should take a major responsibility for the proper implementation of OBE. Some of the responsibilities (not limited to) for the teachers in OBE system may be as follows:

- Within reasonable constraint, they should manoeuvre time to the best advantage of all students.
- They should assess the students only upon certain defined criterion without considering any other potential ineligibility to discriminate them.
- They should try to grow the learning abilities of the students to a certain level before they leave the institute.
- They should try to ensure that all the students are equipped with the quality knowledge as well as competence after they finish their education.
- They should always encourage the students to develop their ultimate performance capabilities.
- They should facilitate and encourage group work and team work to consolidate newer approach.
- They should follow Blooms taxonomy in every part of the assessment.

Bloom's Taxonomy

Level	Teacher should Check	Student should be able to	Possible Mode of Assessment
Create	Students ability to create	Design or Create	Mini project
Evaluate	Students ability to justify	Argue or Defend	Assignment
Analyse	Students ability to distinguish	Differentiate or Distinguish	Project/Lab Methodology
Apply	Students ability to use information	Operate or Demonstrate	Technical Presentation/ Demonstration
Understand	Students ability to explain the ideas	Explain or Classify	Presentation/Seminar
Remember	Students ability to recall (or remember)	Define or Recall	Quiz

GUIDELINES FOR STUDENTS

Students should take equal responsibility for implementing the OBE. Some of the responsibilities (not limited to) for the students in OBE system are as follows:

- Students should be well aware of each UO before the start of a unit in each and every course.
- Students should be well aware of each CO before the start of the course.
- Students should be well aware of each PO before the start of the programme.
- Students should think critically and reasonably with proper reflection and action.
- Learning of the students should be connected and integrated with practical and real life consequences.
- Students should be well aware of their competency at every level of OBE.

ABBREVIATIONS AND SYMBOLS

List of Abbreviations

Abbreviations	Full form	Abbreviations	Full form
A	Ampere. — A unit of electric current	DTL	Diode-Transistor Logic
AMOLED	Active Matrix Organic LED.	ECL	Emitter coupled logic
ALU	Arithmetic Logic Unit	EPROM	Erasable Programmable Read-Only Memory
ANSI	American National Standards Institute	EEPROM	Electrically Erasable Programmable Read-only Memory
ARM	Advanced Reduced instruction set computing Machine	ESD	Electrostatic Discharge
ASCII	American Standard Code for Information Interchange	FET	Field-Effect Transistor.
ASIC	Application Specific Integrated Circuit	FIFO	First-In, First-Out (a queue)
ADC	Analog-To-Digital Converter	FILO	First-In, Last-Out (a stack)
BJT	Bipolar Junction Transistor	FPGA	Field-Programmable Gate Array
bps	Bit per second	GND	Ground
Bps	Byte per second	GPIO	General-Purpose Input/Output
BW	Bandwidth	HAL	Hardware Abstraction Layer
CCD	Charge-Coupled Device	HDMI	High-Definition Multimedia Interface
CMOS	Complementary Metal-Oxide Semiconductor	HDL	Hardware description language
CPU	Central Processing Unit	IC	Integrated Circuit
DAC	Digital-To-Analog Converter	IGBT	Insulated-Gate Bipolar Transistor
dBm	Unit level for power ratio expressed in decibels (dB) with reference to one mill watt (mW)	DIP	Dual-Inline Package
DEMUX	Demultiplexer,	DMM	Digital Multimeter
DRAM	Dynamic Random-Access Memory	PCB	Printed Circuit Board.
DSP	Digital Signal Processor	PCIe	PCI Express
ISA	Instruction Set Architecture	PROM	Programmable Read-only Memory

Abbreviations	Full form	Abbreviations	Full form
IEEE	Institute of Electronic and Electrical Engineers of USA	RAM	Random Access Memory.
JFET	Junction Field-Effect Transistor.	ROM	Read-only Memory
LCD	Liquid Crystal Display	RTC	Real-time Clock
LED	Light-Emitting Diode	RTL	Resistor-Transistor Logic
MOSFET	Metal-Oxide-Semiconductor Field-Effect Transistor	SRAM	Static Random-Access Memory
MUX	Multiplexer,	SSD	Solid-State Drive
NVMe	NVM Express	USB	Universal Serial Bus.
NVRAM	Non-Volatile RAM	VLSI	Very-Large-Scale Integration
OLED	Organic LED	VHSIC	Very High Speed Integrated Circuit
Op-amp	Operational amplifier.	VHDL	VHSIC Hardware Description Language

List of Symbols

Symbol	Symbol Name in Maths	Symbols Meaning	Example
^	caret / circumflex	and	x ^ y
·	and	and	x · y
+	plus	or	x + y
&	ampersand	and	x & y
|	vertical line	or	x | y
∨	reversed caret	or	x ∨ y
$\bar{X}$	bar	not – negation	$\bar{x}$
x’	single-quote	not – negation	x’
!	Exclamation mark	not – negation	! x
¬	not	not – negation	¬ x
~	tilde	negation	~ x
⊕	circled plus / oplus	exclusive or – xor	x ⊕ y

Symbol	Symbol Name in Maths	Symbols Meaning	Example
$\Leftrightarrow$	equivalent	if and only if (iff)	p: this year has 366 days q: this is a leap year $p \Leftrightarrow q$
$\Rightarrow$	implies	Implication	p: a number is a multiple of 4 q: the number is even $p \Rightarrow q$
$\in$	Belong to/is an element of	Set membership	$A = \{1, 2, 3\}$ $2 \in A$
$\notin$	Not element of	Negation of set membership	$A=\{1, 2, 3\}$ $0 \notin A$
$\forall$	for all	Universal Quantifier	$2n$ is even $\forall\, n \in N$ where N is a set of Natural Numbers
$\leftrightarrow$	equivalent	if and only if (iff)	p: x is an even number q: x is divisible by 2 $p \leftrightarrow q$
$\nexists$	there does not exist	Negation of existential quantifier	b is not divisible by a, then $\nexists\, n \in N$ such that $b = na$
$\exists$	there exists	Existential quantifier	b is divisible by a, then $\exists\, n \in N$ such that $b = na$
$\because$	because / since	Because shorthand	$a = b, b = c$ $\Rightarrow a = c\ (\because a = b)$
$\therefore$	therefore	Therefore shorthand (Logical consequence)	$x + 6 = 10$ $\therefore x = 4$

LIST OF FIGURES

Unit 5 Sequential Circuit and System

Title of Figure	***Page No.***

Title of Figure	*Page No.*

Title of Figure	***Page No.***

LIST OF TABLES

CONTENTS

Appendix

1 Fundamentals: Digital Systems

UNIT SPECIFICS

Through this unit we have discussed the following aspects:

- *To distinguish between the BCD, Binary, Octal, Binary, and Decimal number systems.*
- *To Convert numbers between different number systems i.e., BCD, Binary, Octal, Binary, and Decimal.*
- *To analyze various complement methods used with different radix.*
- *To identify and understand various Binary Code along with Error Detection and Correction Codes.*
- *To perform various arithmetic operations using Binary numbers.*

RATIONALE

We utilise the number system daily to complete our work and go about our everyday routines. We have a system with distinct symbols and predetermined values in place for this. This system turns into what we refer to as a number system, aiding in all of our calculations. The fundamental characteristics of a number system include distinct symbols, consistency, comparable value outputs, and ease of replication. The decimal system, which has a base power of 10, continues to be the most fundamental system used by humans, although machines cannot utilise the same system.

Machines understand numbers differently from humans, necessitating the need for a completely different system. Every time we enter text into a device, the characters are converted into numbers that only the computer can comprehend. When a number is made up of digits and several values, a device needs a positional number system. Each digit has a distinct location and a symbol of its own. The value of each digit depends on the digit itself, where it appears in the number, and its base.

PRE-REQUISITES

Mathematics: Basic Calculation (Class XII)

UNIT OUTCOMES

List of outcomes of this unit is as follows:

U1-O1: Describe basic Number System

U1-O2: Describe the Decimal, Octal, Binary, and Hexadecimal Number system

U1-O3: Explain Conversion of Number system while describing their behaviour.

U1-O4: Realize the role of various codes like weighted, non-weighted, Gray etc.

U1-O5: Apply error correction and detection in codes.

Unit-1 Outcomes	EXPECTED MAPPING WITH COURSE OUTCOMES (1- Weak Correlation; 2- Medium correlation; 3- Strong Correlation)					
	CO-1	CO-2	CO-3	CO-4	CO-5	CO-6
U1-O1	3	3	3	-	3	1
U1-O2	1	1	2	2	1	-
U1-O3	2	1	3	1	2	1
U1-O4	-	-	3	1	2	2
U1-O5	3	3	3	-	3	1

➢ Scan the below QR codes for more information on the topic written below the QR code.

Introduction to Number Systems

Fixed-Point Representation

Binary Codes

Error Detection and Correction Codes

Unit Outcome

Digital circuits typically manage the data using binary signals. The information may include special characters, alphabetic values, or numerical values. Therefore, it is necessary to convert the information into an appropriate format before it can be analysed, processed, or stored using digital systems. To do this, a coding method is used to convert each number, letter, or special character into a distinct set of 0s and 1s. Furthermore, data transmission and reception error detection and correction must be used for many purposes, such as mathematical computations. Multiple codes may be used in any digital system for various operations, and it can be essential to transform data between different codes. Code converter circuits are needed for this function and will be covered later.

1.1. Introduction to Number Systems

The most crucial component in digital electronics is the number system, which is utilised to represent the data. The four most used bases for representing numbers are decimal, binary, octal, and hexadecimal (Radix). Comprehension, analysing, and characterising digital systems like microprocessors, logic circuits, computers, etc. requires a thorough understanding of these number systems.

Since the weight of each digit depends on its relative position inside the number, a number system is described as a system of position weighted systems. It is the mathematical notation for consistently employing digits or other symbols to represent the numbers in a particular set.

Any number system has an ordered group of symbols known as digits, and there are rules for carrying out arithmetic operations such as addition, multiplication, and so on. A number made up of these digits often has two parts: integer and fractional, which are distinguished by a radix coding called encoding. Various codes with varying functions may exist, and they may be expressed as follows:

$$(N)_b = d_{n-1}d_{n-1} \dots d_i \dots d_1 d_2 d_0 \,.\, d_{-1} d_{-2} \dots d_{-f \dots} d_{-m}$$

where

N = a number

b = a number

n = number of digits in integer portion

m = number of digits in fractional portion

d_{n-1} = most significant digit (MSD)

d_{-m} = least significant digit (LSD)

The base (or) radix of a number system, can be defined as the number of different symbols (digits or characters) used in that number system. if the base of number system is ‘r’, the no. of different symbols used in the system is ‘r’ (the different symbols are ‘0’ to ‘r-1’). The largest value of digits in base ‘r’ system is ‘r-1’.

For example, if a number system has been represented using the digit from 0 – 9, then the base of the system is 10 (as the digits are from 0 to 9, i.e., 10. In this so-called decimal number system, the base (radix) value of Decimal number system is '10'. Hence, depending on radix, there are four main types of number system are as enumerated below (Refer Fig. 1.1)

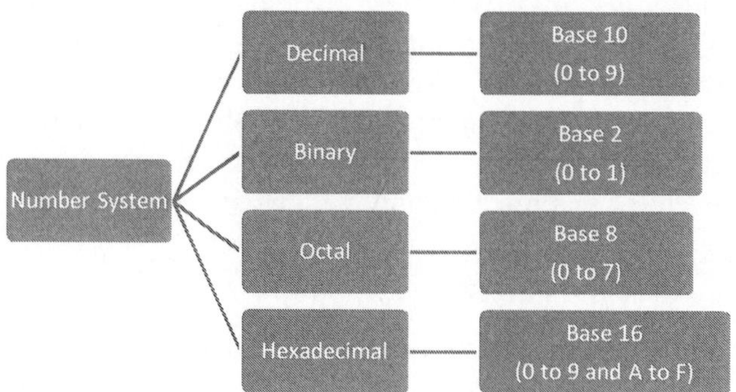

Fig. 1.1: Classification of the Number system

Fig 1.1, shows the four basic types of the number system. However, Table 1.1 gives some unique characteristics of such number systems, that can be derived using its Base (radix) value

Table 1.1: Characteristics of a Number system

Term	General	Binary	Decimal	Octal	Hexadecimal
Base (Radix)	***r***	2	10	8	16
Total Possible symbols	***r***	2	10	8	16
Symbols (From-to)	***0,1,2…….. (r-1)***	0, 1	0,1, 2….9	0,1,2…..7	0,1,2,3,4,5,6,7,8,9, A, B, C, D, E, F
Highest Possible Digit	***(r-1)***	1	9	7	F (15)

Any positional number system can be expressed as sum of products of place value and digit value for base '*r*' system, the place (or) weights of different digits in mixed number systems is

$$\ldots\ldots\ldots\ldots r^3 r^2 r^1 r^0 . r^{-1} r^{-2} r^{-3} \ldots\ldots\ldots\ldots$$

In decimal number system is

$$\ldots\ldots\ldots\ldots 10^3 10^2 10^1 10^0 . 10^{-1} 10^{-2} 10^{-3} \ldots\ldots\ldots\ldots$$

For Examples

$Q.1: (256.72)_{10}$

$Sol: (2\times10^2+5\times10^1+6\times10^0+7\times10^{-1}+2\times10^{-2})$

$Q.2: (234.14)_5$

$Sol.: (2\times5^2+3\times5^1+4\times5^1+1\times5^{-1}+4\times5^{-2})$

$Q.3: (346.71)_8$

$Sol: (3\times8^2+4\times8^1+6\times8^0+7\times8^{-1}+1\times8^{-2})$

1.2 Binary number system

The smallest unit of information is called a bit, which is an acronym for "binary digit." Either "0" or "1" apply. A byte is a group of eight bits. The word length or word size of a system refers to how many bits it can process in a single clock cycle. Although it may differ from computer to computer, this word length is set for a single machine and is determined by its system architecture. The word's size could be one byte, two bytes, four bytes, or even more.

The two discrete signal levels LOW and HIGH can also stand as substitutes for the binary digits "0" or "1," respectively. A bit is a binary digit that can either be '0' or '1'. Alternate names for the two states (or levels) include ON and OFF and TRUE and FALSE. Consequently, the analysis and design of digital systems may be done using the binary number system. George Boole created the Boolean algebra in 1854. In his essay "An Investigation of the Laws of Thought," George Boole, who was studying the mathematical theory of LOGIC, developed the idea of the binary number system. Additionally, in 1938, Claude Shannon organised Boole's work; later, in his book Symbolic Analysis of Relay and Switching Circuits, similar logic ideas were modified for use in the construction of digital hardware.

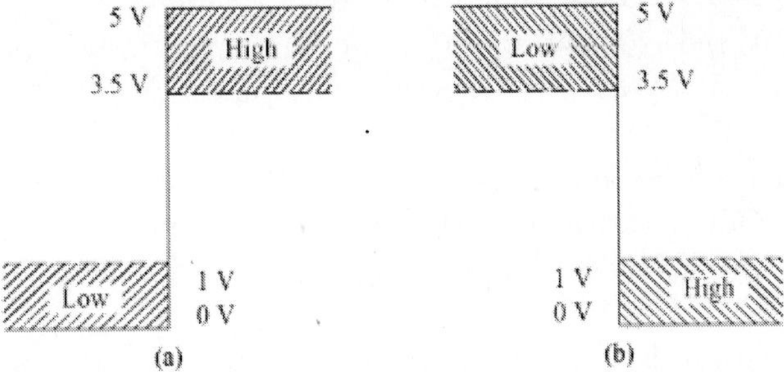

Fig. 1.2: Digital Signal Representation (a) Positive Logic (b) Negative Logic

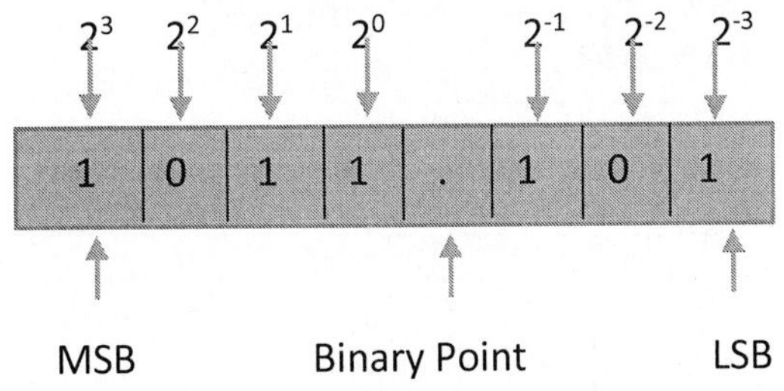

Fig. 1.3: Binary Value positions as the power of 2

In Fig. 1.3, shown above, the most important Bit is the left most bit with the highest weight known as Most Significant Bit (MSB). Least Significant Bit (LSB) refers to the right most bit that is the lightest in weight. The number system with base (or radix) two (2) is known as binary number system (Refer Fig.1.2, and Table 1.2). To represent the voltage levels of a digital circuit, a binary number system is used. A digital circuit has just two voltage levels: logic High and Logic low. The high voltage is +5V and the low voltage is +0V. The binary numbers represent the logic low as a 0 and the logic high as a 1.

- The base value of binary number system is '2', *(i.e., r =2)*
- The different symbols used in Binary number system are 0 to r-1. (i.e. 0 to 0). Those are 0,1. The maximum digit in binary system = r-1 =2-1 =1
- The numbers in binary number system can be represented as

$$\ldots\ldots 2^3 2^2 2^1 2^0 . 2^{-1} 2^{-2} 2^{-3} \ldots\ldots$$

- The most significant bit (MSB) is the bit on the left, and the least significant bit is the bit on the right (LSB). A group of four bits is referred to as a nibble in the binary number system, and a group of eight bits is referred to as a byte.

***Advantages of Binary Number System*:**

In any digital system, arithmetic and Logical operations are the backbones. Using Binary System following are the advantages;

- George Boole laid the foundation to represent mathematics of logic by using the binary notation of '0' and '1'. The idea of Boolean algebra was shown to be quite helpful in resolving many logical issues. Given that logic's mathematical underpinnings (often referred to as Boolean algebra) had been simplified to binary notation.
- The use of 0s and 1s allows for the convenient representation of all types of data. Thus, for usage in computer systems, the binary number system clearly outperformed the other forms of number systems.
- Devices used in electronics for hardware designing generally used to work in switching mode i.e., ON and OFF (for transistors the two modes are saturation and cut-off). It is very convenient and efficient to operate the devices using Binary logic in these two distinctly different modes (ON and OFF).
- When the relevant data are represented as 0s and 1s, it is very convenient to do all arithmetic operations like addition, subtraction, multiplication, division, etc.

Table 1.2: 4- bit Binary Number and their equivalent Decimal Number

Binary Number				**Decimal Number**
2^3	2^2	2^1	2^0	
0	0	0	0	0
0	0	0	1	1
0	0	1	0	2
0	0	1	1	3
0	1	0	0	4
0	1	0	1	5
0	1	1	0	6
0	1	1	1	7
1	0	0	0	8
1	0	0	1	9
1	0	1	0	10
1	0	1	1	11
1	1	0	0	12
1	1	0	1	13
1	1	1	0	14
1	1	1	1	15

George Boole
1815 - 1864

1.2.1 Decimal to Binary Conversion

The supplied decimal value has been divided by 2 recursively and the remainders have been noted until we have obtained 0 or 1 as the quotient, which is the binary equivalent of the given decimal number. For a better understanding, consider the following example.

Example 1.1. Convert $(87)_{10}$ to a Binary number $(X)_2$.

Solution: A decimal number can be converted to a binary number by successively dividing the number by 2 as follows:

$87 \div 2 = 43$	remainder 1	LSB(Least Significant Bit)
$43 \div 2 = 21$	remainder 1	
$21 \div 2 = 10$	remainder 0	
$10 \div 2 = 5$	remainder 0	
$5 \div 2 = 2$	remainder 1	
$2 \div 2 = 1$	remainder 0	
$1 \div 2 = 0$	remainder 1	MSB(Most Significant Bit)

Therefore $(87)_{10} = (1010011)_2$

Note that the first remainder becomes the most significant bit (MSB). The last remainder becomes the least significant bit (LSB). Converting Fractional Decimal Number into Binary Number, the conversion is affected by continuous multiplication by 2 and keeping track of the integers.

Example 1.2: Using decimal to binary formula, convert $(29)_{10}$ into a binary number.

Solution: Using decimal to binary formula, we have,

$29 \div 2 = 14$	remainder 1	LSB(Least Significant Bit)
$14 \div 2 = 7$	remainder 0	
$7 \div 2 = 3$	remainder 1	
$3 \div 2 = 1$	remainder 1	
$1 \div 2 = 0$	remainder 1	MSB(Most Significant Bit)

Therefore $(29)_{10} = (11101)_2$

Example 1.3: Convert $(145)_{10}$ into a binary number.

Solution: Using decimal to binary formula, we have,

$145 \div 2 = 72$	remainder 1	LSB(Least Significant Bit)
$72 \div 2 = 36$	remainder 0	
$36 \div 2 = 18$	remainder 0	
$18 \div 2 = 9$	remainder 0	
$9 \div 2 = 4$	remainder 1	
$4 \div 2 = 2$	remainder 0	
$2 \div 2 = 1$	remainder 0	
$1 \div 2 = 0$	remainder 1	MSB(Most Significant Bit)

Therefore $(145)_{10} = (10010001)_2$

Decimal (Fraction) to Binary Conversion

- Multiply the number by the base (=2) and
- Take the integer (either 0 or 1) as a coefficient then
- Take the resultant fraction and repeat the division.

Example 1.4: Convert $(0.625)_{10}$ to Binary Number
Solution: Multiply the fraction part by 2, and record the values of Integer, Fraction and Coefficient Value

	Integer		Fraction	Coefficient	
$0.625*2=$	1	.	25	$a_{-1}=1$	MSB
$0.25*2=$	0	.	50	$a_{-2}=0$	
$0.50*2=$	1	.	00	$a_{-3}=1$	LSB
$0.00*2=$	0	.	00	$a_{-4}=0$	As Value goes Zero so no further Calculation required

Hence $(0.625)_{10}=(0.a_{-1}a_{-2}a_{-3})_2=(0.101)_2$

1.2.2 Binary to Decimal Conversion

The weights of all the locations in a binary number that contain a 1 are added up to convert a binary number to a decimal number.

$$(Decimal\ Number)_{10}=[d_0(d_0\times 2^0)+d_1(d_1\times 2^1)+d_2(d_2\times 2^2)+.............+d_{n-1}(d_{n-1}\times 2^{n-1})]$$

Where, $d_0, d_1, d_2, ... d_{n-1}$ are individual digits of the binary number starting from the right-most position

Example 1.5: Convert $(1010111)_2$ to decimal number.
Solution:

MSB						LSB	
2^6	2^5	2^4	2^3	2^2	2^1	2^0	**Bit Weights**
1	0	1	0	1	1	1	**Binary Value**
1*2^6	0*2^5	1*2^4	0*2^3	1*2^2	1*2^1	1*2^0	**Multiply**
64	0	16	0	4	2	1	**Calculate Value**
		64+16+4+2+1= $(87)_{10}$					**Final Decimal Value**

Example 1.6: Convert *$(1001.0101)_2$* to its Decimal equivalent.
Solution: Here first integer part is converted to decimal than the fractional part will be converted.
Part A: Integer Part

$$(1001)_2=1\times 2^0+0\times 2^1+0\times 2^2+1\times 2^3=1+0+0+8=9$$

Part B: Fractional Part

$$(.0101)_2=0\times 2^{-1}+1\times 2^{-2}+0\times 2^{-3}+1\times 2^{-4}=0+0.25+0+0.0625=0.3215$$

Therefor the decimal equivalent of *$(1001.0101)_2$* is *$(9.3215)_{10}$*

Example 1.7: Determine the decimal equivalent of the following Binary Numbers
(a) 110101 (b) 101101 (c) 11111111 (d) 00000000

Solution:

$$(a)\qquad (110101)_2 = 1\times 2^5 + 1\times 2^4 + 0\times 2^3 + 1\times 2^2 + 0\times 2^1 + 1\times 2^0$$
$$= 32+16+0+4+0+1=(53)_{10}$$

$$(b)\qquad (101101)_2 = 1\times 2^5 + 0\times 2^4 + 1\times 2^3 + 1\times 2^2 + 0\times 2^1 + 1\times 2^0$$
$$= 32+0+8+4+0+1=(45)_{10}$$

$$(c)\qquad (11111111)_2 = 1\times 2^7 + 1\times 2^6 + 1\times 2^5 + 1\times 2^4 + 1\times 2^3 + 1\times 2^2 + 1\times 2^1 + 1\times 2^0$$
$$= 128+64+32+16+8+4+2+1=(255)_{10}$$

$$(d)\qquad (00000000)_2 = 0\times 2^7 + 0\times 2^6 + 0\times 2^5 + 0\times 2^4 + 0\times 2^3 + 0\times 2^2 + 0\times 2^1 + 0\times 2^0$$
$$= 0+0+0+0+0+0+0+0=(0)_{10}$$

Example 1.8: Determine the decimal Equivalent for the following Binary Numbers.
(a) $(101101.10101)_2$ (b) $(1100.1011)_2$ (c) $(1001.0101)_2$ (d) $(0.10101)_2$

Solution:

$$(a)\ (101101.10101)_2 = 1\times 2^5 + 0\times 2^4 + 1\times 2^3 + 1\times 2^2 + 0\times 2^1 + 1\times 2^0$$
$$+\ 1\times 2^{-1} + 0\times 2^{-2} + 1\times 2^{-3} + 0\times 2^{-4} + 1\times 2^{-5}$$
$$= 32+0+8+4+0+1+\frac{1}{2}+0+\frac{1}{8}+0+\frac{1}{32} = (45.65625)_{10}$$

$$(b)\ (1100.1011)_2 = 1\times 2^3 + 1\times 2^2 + 0\times 2^1 + 0\times 2^0$$
$$+\ 1\times 2^{-1} + 0\times 2^{-2} + 1\times 2^{-3} + 1\times 2^{-4}$$
$$= 8+4+0+0+\frac{1}{2}+0+\frac{1}{8}+\frac{1}{16} = (12.6875)_{10}$$

$$(c)\ (1001.0101)_2 = 1\times 2^3 + 0\times 2^2 + 0\times 2^1 + 1\times 2^0$$
$$+\ 0\times 2^{-1} + 1\times 2^{-2} + 0\times 2^{-3} + 1\times 2^{-4}$$
$$= 8+0+0+1+0+\frac{1}{4}+0+\frac{1}{16} = (9.3125)_{10}$$

$$(d)\ (0.10101)_2 = 1\times 2^{-1} + 0\times 2^{-2} + 1\times 2^{-3} + 0\times 2^{-4} + 1\times 2^{-5}$$
$$= \frac{1}{2}+0+\frac{1}{8}+0+\frac{1}{16} = (0.65625)_{10}$$

1.2.3 Signed Binary Number System

A number with a sign is said to be signed. The plus (+) sign is used to indicate a positive number in the context of the digital system, and the absence of any sign denotes a positive value. However, a negative number is indicated by the Minus (-) sign. When representing a signed binary number, a second bit is employed as the sign bit. The most important element will be this sign bit. At MSB a positive number was represented by a "0" bit and a negative number by a "1" bit.

The representation of signed numbers is in two ways

1. Sign-magnitude form

2. Complement form
    a. 1's Complement
    b. 2's Complement

Reduction in the software is a benefit of subtracting using the complement approach. the use of a single digital circuit for addition, subtraction, and multiplication. Sign-magnitude form. The figure below represents the categorization to represent binary numbers with ***signed magnitude*** (Refer Fig. 1.4)

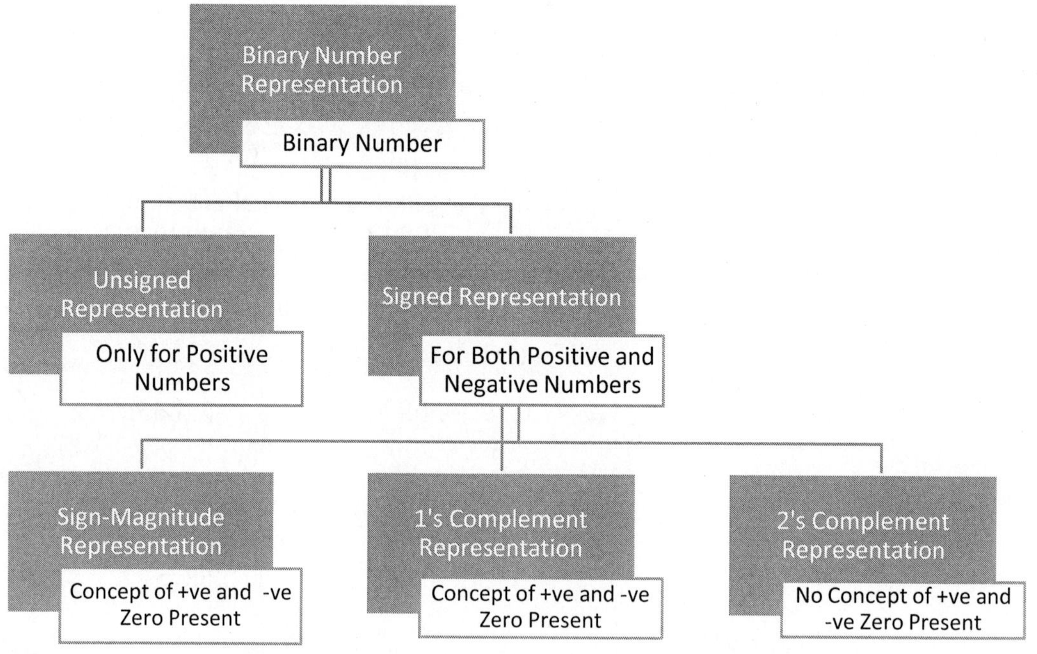

Fig. 1.4: Signed and unsigned Binary magnitude format

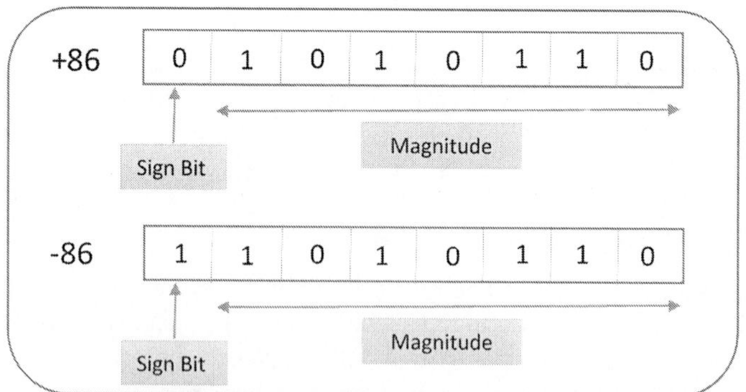

Fig. 1.5: Signed Magnitude representation of Binary numbers

MSB represents the 'sign' to represent the positive and negative magnitude using sign bit. While, MSB is '0' it represents a positive value and '1' denoting a negative value. Except the MSB the remaining bits represents the magnitude. ***For example,*** +86 would be 01010110 and that for -86 would be 11010110 (Refer Fig. 1.5). Decimal values between $-(2^n-1)$ and $+(2^n-1)$ can be represented using an n-bit binary representation i.e., for an eight-bit sequence *(n=8)* the range is from +127 to -127 while using the sign bit magnitude format.

Example 1.9: Find the decimal equivalent of the following Binary Numbers assuming sign-magnitude representation of the binary numbers.

(a) $(101100)_2$ (b) $(001000)_2$ (c) $(0111)_2$ (b) $(1111)_2$

Solution:

(a) Sign bit is 1, which means answer is Negative
Magnitude =01100= $(12)_{10}$, Hence, $(101100)_2=(-12)_{10}$

(b) Sign bit is 0, which means answer is Positive
Magnitude =01000= $(8)_{10}$, Hence, $(001000)_2=(+8)_{10}$

(c) Sign bit is 0, which means answer is Positive
Magnitude =111= $(7)_{10}$, Hence, $(0111)_2=(+7)_{10}$

(d) Sign bit is 1, which means answer is negative
Magnitude =111= $(7)_{10}$, Hence, $(1111)_2=(-7)_{10}$

1's Complement form

In general, the 1's complement [also termed as (r-1)'s complement, as for binary r=2] of a binary number is obtained by complementing all its bits, i.e., by replacing all 0s with 1s and 1s with 0s (Refer Fig. 1.6).

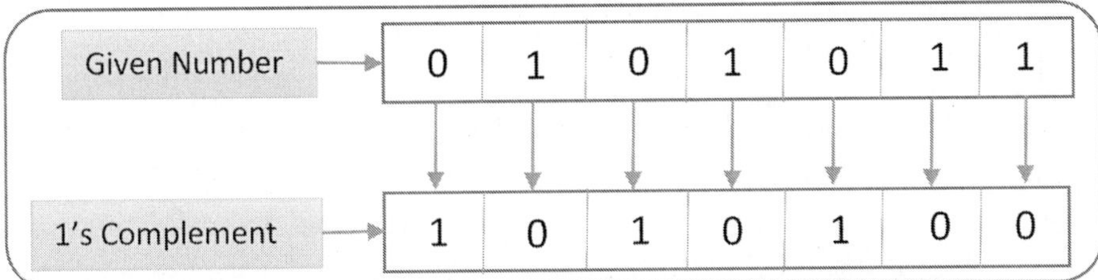

Fig. 1.6: Convert a binary number to its 1's complement form

The positive numbers remain the same in the format of the 1's complement. On the other hand, negative numbers are created by adding one to their positive counterparts. Decimal values in the range of -127 to +127 can be represented using the eight-bit version of the 1's complement format. *If the number is positive, a sign bit of "0" is added in front of MSB, and the magnitude is represented in true binary*. However, if the integer is negative, the magnitude is displayed in one's complement form, and a sign bit of 1 is inserted before MSB. To get one's complement, transpose the true binary representation's bits (1 to 0, 0 to 1).

Example 1.10: Represent the following numbers in 1's complement form

(a) +51 and -51
(b) +7 and -7
(c) +8 and -8
(d) +15 and -15

Solution: in 1's complement form

(a) $(+51)_{10} = (0110011)_2$ and $(-51)_{10}$ =1's complement of $(+51)_{10} = (1001100)_2$

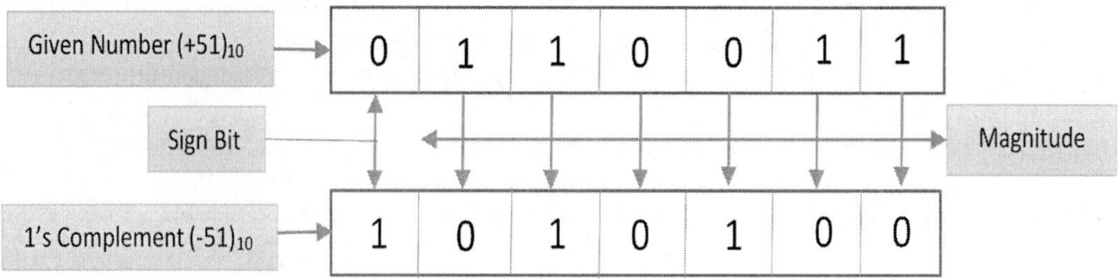

Similarly,

(b) $(+7)_{10} = (0111)_2$ and $(-7)_{10}$ =1's complement of $(+7)_{10} = (1000)_2$
(c) $(+8)_{10} = (01000)_2$ and $(-8)_{10}$ =1's complement of $(+8)_{10} = (10111)_2$
(d) $(+15)_{10} = (01111)_2$ and $(-15)_{10}$ =1's complement of $(+15)_{10} = (10000)_2$

2's Complement form

The Least Significant Bit (LSB) of the 1's complement of the number is multiplied by one to produce the 2's complement of the binary number.

2's Complement form = 1's Complement form + 1

Given Number	1	1	0	1	0	1	1
1's Complement	0	0	1	0	1	0	0
Add 1							1
2's Complement	0	0	1	0	1	0	1

1's Complement	1	1	0	1	0	1	0
Add 1							1
2's Complement	1	1	0	1	0	1	1

Fig. 1.7: Convert a binary number to its 2's complement form

(NOTE: Two's complement of 0 is 0.)

If the given number is positive, a sign bit of "0" is added in front of MSB to reflect the magnitude in true binary form. ***If the given number is negative***, a sign bit of one is added in front of MSB, and the magnitude is displayed in its 2's complement form. (Refer Fig. 1.7).

Similar to how 1's complement indicates the sign, 2's complement does the same, using a '0' for the plus sign and a '1' for the minus sign, with the remaining bits used to represent magnitude. Negative magnitudes are represented by the 2's complement of their positive equivalent, whilst positive magnitudes are represented in the same way as in the case of sign bit or 1's complement representation.

For example, +43 would be represented as 0101011, and -43 would be represented as 0010100 using 1's complement. But 2's complement form of -43 is 0010101 Here one must note that, if the 2's complement of the magnitude of +43 gives a magnitude of -43, then the reverse process is also true, i.e., the 2's complement of the magnitude of -43 gives a magnitude of +43 (Refer Fig. 1.7). The range of numbers represented in 2's complement form is $-(2^{n-1})$ *to* $+(2^{n-1}-1)$, i.e., -128 to +127, for an n-bit word which includes sign bit (Refer Table 1.3).

Table 1.3: Sign-magnitude, representation using four bits 1's and 2's complement.

Decimal Number	**Binary Number**		
	Sign-Magnitude	*1's Complement*	*2's Complement*
0	0000	0000	0000
1	0001	0001	0001
2	0010	0010	0010
3	0011	0011	0011
4	0100	0100	0100
5	0101	0101	0101
6	0110	0110	0110
7	0111	0111	0111
-8	-------	-------	1000
-7	1111	1000	1001
-6	1110	1001	1010
-5	1101	1010	1011
-4	1100	1011	1100
-3	1011	1100	1101
-2	1010	1101	1110
-1	1001	1110	1111
-0	1000	1111	-------

Example 1.11: Represent $(-51)_{10}$ using 2's complement.
Solution:

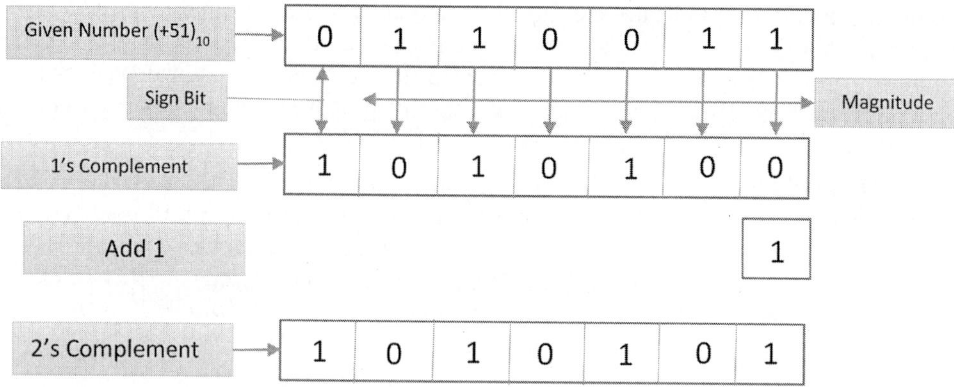

Example 1.12: Find the 2's Complement of the given Binary Numbers
(a) 01001110 (b) 00110101 (c)110110
Solution:

(a)		(b)		(c)	
Number :	01001110	Number :	00110101	Number :	110110
I's Complement :	10110001	I's Complement :	11001010	I's Complement :	001001
Add 1 :	1	Add 1 :	1	Add 1 :	1
2's Complement:	10110010	2's Complement :	11001011	2's Complement :	001010

Example 1.13: Represent $(-17)_{10}$ in
(a) Sign Magnitude (b) 1's Complement (c) 2's Complement
Solution:
The minimum number of bits required to represent (+17)10 in signed number format is six.

Hence, $(+17)_{10}=(010001)_2$

Therefore, $(-17)_{10}$ can be represented by,
(a) Sign Magnitude $(110001)_2$
(b) 1's Complement $(101110)_2$
(c) 2's Complement $(101111)_2$

1.2.4 Binary Arithmetic Operations

The fundamentals of binary addition and subtraction are comparable to those that apply to the decimal number system, which we are all familiar with. There are only two digits in the binary number system—0 and 1—and any number may be represented by these two digits. The operations of addition, subtraction, multiplication, and division are referred to as binary arithmetic. The least significant bit, or the bit on the right-most side, is where binary arithmetic operations begin. In the following sections, we will go over each procedure in order.

Binary addition:

Basic rules of binary addition are as follows:

- 0 + 0 = 0

- 0 + 1 = 1
- 1 + 0 = 1
- 1 + 1 = 0 With a carry of '1' to the next more significant bit.
- 1 + 1 + 1 = 1 with a carry of '1' to the next more significant bit.

An example will help us to understand the addition process. Let us take two binary numbers 10001001 and 10010101. The below example of binary arithmetic clearly explains the binary addition operation, the carried 1 is shown on the upper side of the operands (Refer 1.8).

1							**1**		←	**Carry**
	1	0	0	0	1	0	0	1	←	Number A
	1	0	0	1	0	1	0	1	←	Number B
1	0	0	0	1	1	1	1	0	←	SUM

Fig. 1.8: Example of binary arithmetic

Example 1.14: Solve the Following

(a) $(111.101)_2 + (0.111)_2$ (b) $(1101.101)_2 + (111.011)_2$ (c) $(1011.101)_2 + (1110.1011)_2$

Solution

(a)	(b)	(c)
111.101	1101.101	1011.101
000.111	111.011	1110.1011
$(1000.100)_2$	$(10101.000)_2$	$(11010.0101)_2$

Binary Subtraction:

There are four simple rules of binary subtraction

- 0 – 0 = 0
- 0 – 1 = 1, borrow 1 from the next more significant bit
- 1 – 0 = 1
- 1 – 1 = 0

Evaluate 1.15: $(110110)_2 - (10110)_2$

Solution:

The number of bits in the subtrahend are 5 while that of minuend are 6. We make the number of bits in the subtrahend equal to that of minuend by taking a `0' in the sixth place of the subtrahend.

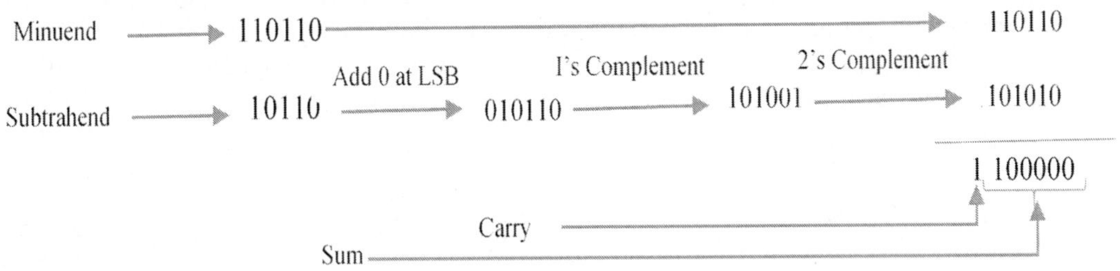

Minuend → 110110 → 110110
Add 0 at LSB
1's Complement
2's Complement
Subtrahend → 10110 → 010110 → 101001 → 101010
1 100000
Carry
Sum

After dropping the carry-over, we get the result of subtraction to be 100000.

Example 1.16: Solve the following

(a) $(1011)_2 - (0110)_2$ (b) $(101001)_2 - (011010)_2$ (c) $(111111)_2 - (010101)_2$

Solution:

(a)	(b)	(c)
1011	101001	111111
0110	011010	010101
$(0101)_2$	$(01111)_2$	$(101010)_2$

Binary multiplication

The following is a list of the basic rules of multiplication: Although binary multiplication may appear to be more challenging than binary addition or subtraction, it is essentially a straightforward process. Here are some rules to be followed,

- $0\times0=0$
- $1\times0=0$
- $0\times1=0$
- $1\times1=1$ (there is no carry or borrow for this)

Example 1.17: Solve the following

(a) $(100.01)_2$ X $(10.1)_2$ (b) $(101011)_2$ X $(11)_2$

Solution:

(a)	(b)
1 0 0 0 1	1 0 1 0 1 1
X 1 0 1	X 1 1
1 0 0 0 1	1 0 1 0 1 1
0 0 0 0 0	1 0 1 0 1 1
1 0 0 0 1	1 0 0 0 0 0 0 1
1 0 1 0 1 0 1	

The repeated left-shift and add technique, also known as the repeated edition algorithm, is used to multiply binary integers in a manner comparable to that of decimal numbers. However, the "repeated add and right-shift" algorithm is used in microprocessors and microcomputers. Because using the "repeated left-shift and add" approach to perform binary multiplication is generally significantly more convenient. Additionally, mixed binary integers can be multiplied in binary by executing multiplication without taking the binary point into account. The binary point is then positioned after n bits, beginning from the LSB, where n equals the total number of bits in the fractional parts of the multiplicand and multiplier.

Binary division:

Binary division is the process of repeatedly subtracting, whereas binary multiplication is the process of repeated addition. An example will help explain the operation more clearly. Multiplication and subtraction are the other two binary arithmetic operations that make up binary division. The process for obtaining binary division is the same as that for decimal division.

Example 1.18: Solve

(a) $(11010)_2 \div (101)_2$. (b) (1110101)2 $\div (1001)_2$

Solution

```
101 ) 11010 ( 101 → Quotient
      101
      ---
       110
       101
      ---
         1  → Remainder
```

Here '101' is the quotient and '1' is the remainder.

```
                   1101        ← Quotient
Divisor → 1001 ) 1110101       ← Dividend
                 1001
                 ----
                  1011
                  1001
                  ----
                 001001
                   1001
                   ----
                   0000
```

Here '1101' is the quotient and '0' is the remainder.

1.2.5 One's and Two's Complements Arithmetic

In digital computers, complements are employed for logical operations and to make the subtraction operation simpler. There are two different types of complements for any radix-r system (radix r indicates the base of the number system) (Refer Table 1.4).

- For Example, the 9's complement of $(2496)_{10}$ would be $(7503)_{10}$, i.e., by subtracting each digit from 9. However, the 10's complement is obtained by adding 1 to the 9's complement. Hence, the 10's complement of $(2496)_{10}$ is $(7504)_{10}$.
- Hen dealing with octal number, the 7's complement of an octal number is calculated by reducing each octal digit from 7. For example, the 7's complement of $(562)_8$ would be $(215)_8$. The 8's complement is calculated by adding 1 to 7's complement. Hence, the 8's complement of $(562)_8$ would be $(216)_8$.
- The complement of 15 in a hexadecimal number system is obtained by subtracting each hex digit from 15. In the case of $(3BF)_{16}$ the complement of 15 would be $(C40)_{16}$. However, 16's complement can be calculated by adding '1' to the 15's complement. Hence, the 16's complement of $(3BF)_{16}$ would be $(C41)_{16}$.

Table 1.4: Summary for r's and (r-1)'s complement

S.N.	Complement	Examples	Description
1	Radix Complement	2's Complement (Binary) 8's Complement (Octal) 10's Complement (Decimal) 16's Complement (Hexadecimal)	The radix complement is referred to as the r's complement
2	Diminished Radix Complement	1's Complement (Binary) 7's Complement (Octal) 9's Complement (Decimal) 15's Complement (Hexadecimal)	Diminished Radix Complement is referred to as the (r-1)'s complement

Procedure for subtraction using r's Complement:

The r's complement of a positive number N in base r with an integer part of n digits can be expressed as $r^n - N$ for $N \neq 0$ and 0 for $N=0$.

Example 1.19: Find the *r's* complement of

(a) $(52520)_{10}$ (b) $(0.3267)_{10}$ (c) $(101100)_2$ (d) $(0.0110)_2$

Solution:

(a) 10's complement of $(52520)_{10}$ is 10^5-52520= 47480 (as there are 5 integer digits in the number so $n=5$)

(b) 10's complement of $(0.3267)_{10}$ is 10^0-0.3267= 0.6733 (as NO integer digits in the number so $n=0$)

(c) 2's complement of $(101100)_2$ is 2^6- 101100= $(100000\text{-}101100)_2$=010100 (as $n=6$)

(d) 2's complement of $(0.0110)_2$ is 2^0- 0.0110= $(1\text{-}0.0110)_2$=0.1010 (as $n=0$)

The definition given above can be used to determine the r's complement for a number, which exists for every base r (r greater than but not equal to 1). The examples are considered for base 2 (Binary) and base 10 (Decimal) only as they are of most interest to us.

Let N_1 and N_2 be two numbers of the base *'r'* number system than to perform N_1-N_2 using *r's* complement

(a) If $N_1 > N_2$

Step-1: Add the r's complement of the subtrahend (N_2) to the minuend.

Step-2: if there is a carryout, Ignore it.

(b) If $N_1 < N_2$

Step-1: Add the *r's* complement of the subtrahend (N_2) to the minuend.

Step-2: if there is no carryout.

Step-3: Result is in *r's* complement form and hence takes the *r's* complement of the result.

Example 1.20: Solve the following using *r's* complement

(a) $(72532)_{10}$- $(3250)_{10}$ (b) $(3250)_{10}$ - $(72532)_{10}$ (c) $(1010100)_2$ – $(1000100)_2$ (d) $(1000100)_2$ - $(1010100)_2$

Solution:

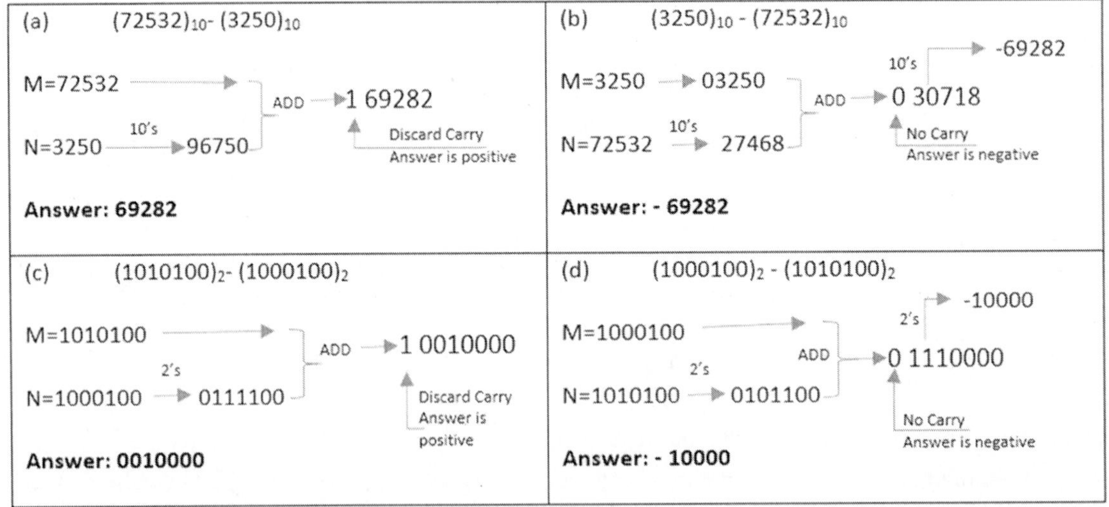

Procedure for subtraction using (r-1)'s Complement:

Given a positive number N in the base r with the integer part of n digits, fractional part of m digits the *(r-1)'s* complement of N can be defined as $r^n - r^{-m} - N$.

Example 1.21: Find the *(r-1)'s* complement of

(a) $(52520)_{10}$
(b) $(0.3267)_{10}$
(c) (25.369)10
(d) $(101100)_2$
(e) $(0.0110)_2$

Solution:

(a) 9's complement of $(52520)_{10}$ is 10^5-10^0-52520 = 99999- 52520= 47479

Note: As there are 5 integer digits in the number so n=5, no fractional part so m=0

(b) 9's complement of $(0.3267)_{10}$ is 10^0- 10^{-4} -0.3267= 0.9999-0.3267= 0.6733

Note: As NO integer digits in the number so n=0, and fractional part is upto 4 places so, m=4

(c) 9's complement of $(25.369)_{10}$ is 10^2-10^{-3}-25.369 = 99.999-25.369=74.36; *Note: n=2, m=3*

(d) 1's complement of $(101100)_2$ is 2^6- 2^0-101100= 111111-101100= 010011; *Note: n=6, m=0*

(e) 1's complement of $(0.0110)_2$ is 2^0- 2^{-4} - 0.0110= 0.1111- 0.0110 = 0.1001; *Note: n=0, m=4*

From the above examples, we see that the 9's complement of a decimal number is formed by subtracting each digit from 9. The 1's complement of a binary number is even simpler to form. The 1's are changed to 0s and 0s to 1s.

Let N_1 and N_2 be two numbers of the base *'r'* number system than to perform N_1-N_2 using *(r-1)'s* Complement.

(a) If $N_1 > N_2$

Step-1: Add the*(r-1)'s* complement of the subtrahend (N_2) to the minuend.

Step-2: if there is a carryout, this will be called end-around carry (EAC), so bring this carry around and add it to LSB.

(b) If $N_1 < N_2$

Step-1: Add the *(r-1)'s* complement of the subtrahend (N_2) to the minuend.

Step-2: if there is no carry out, result is in *(r-1)'s* complement form

Step-3 Hence take the *(r-1)'s* complement of the result.

Example 1.22: Solve the following using *(r-1)'s* complement.

(a) $(72532)_{10}$- $(3250)_{10}$

(b) $(3250)_{10}$ - $(72532)_{10}$

(c) $(1010100)_2$ – $(1000100)_2$

(d) $(1000100)_2$ - $(1010100)_2$

Solution:

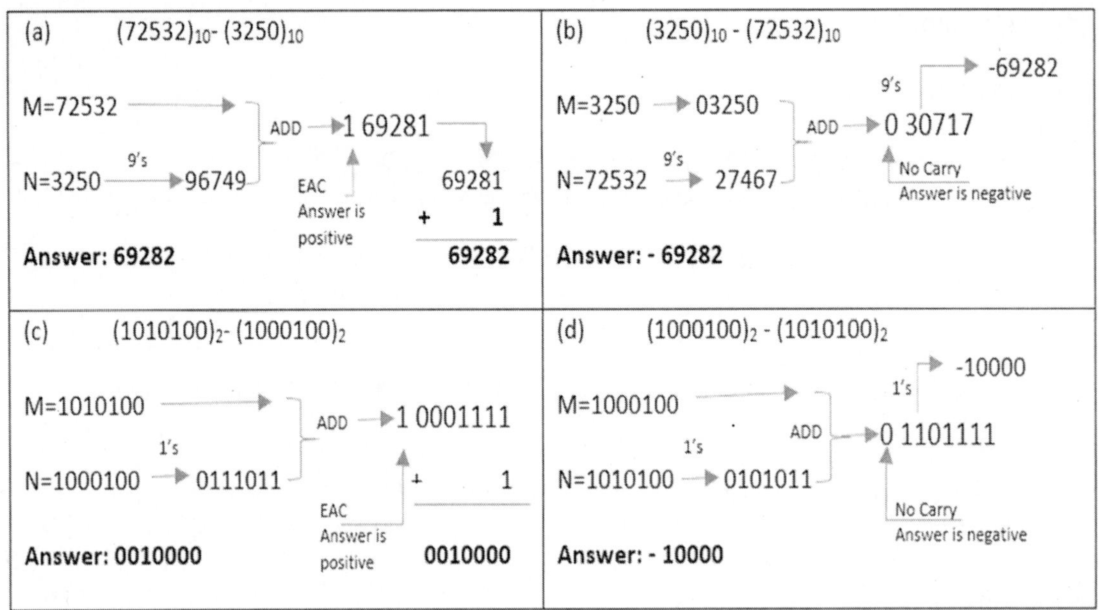

Addition Using the 2 's Complement Method:

The most popular method in binary arithmetic is called the 2's complement method. By adhering to the fundamental rules of binary addition, it is capable of processing both positive and negative binary numbers. The first phase will result in the 2's complement of the negative number, which will then be added. Whether a carry is produced after addition must be taken into consideration. To illustrate this in detail, let us consider the following four different cases (Refer Fig. 1.9):

Case-1: Both the numbers are positive.

Case-II: Larger of the two numbers is positive.

Case-III: Larger of the two numbers is negative.

Case-IV: Both the numbers are negative.

As a result, when all circumstances are considered, it is generally evident that if there is an overflow, the sum is erroneous, as shown by the example below.

It is worth mentioning here that arithmetic operations using *n* no. of bits are constrained to a range from, -2^{n-1} to $+(2^{n-1}-1)$, while using the 2's complement notation.

From the above examples, as 8 bits are used *(n=8)* the range of output value after addition must lies between -128 to +127.

Case-I: When both the numbers are positive.

44 + 45 = 89:	104 + 45 = 149:	127 + 1 = 128:
1 1 1	1 1 1	1 1 1 1 1 1 1
0 0 1 0 1 1 0 0	0 1 1 0 1 0 0 0	0 1 1 1 1 1 1 1
+ 0 0 1 0 1 1 0 1	+ 0 0 1 0 1 1 0 1	+ 0 0 0 0 0 0 0 1
0 1 0 1 1 0 0 1	1 0 0 1 0 1 0 1	1 0 0 0 0 0 0 0
No overflow nor carryout. Sum is Correct	Overflow, no carryout. Sum is not correct.	Overflow, no carryout. Sum is not correct.

Case-II: Larger of the two numbers is positive.

-39 + 92 = 53:	10 + -3 = 7:	-1 + 1 = 0:
1 1 1 1	1 1 1 1 1	1 1 1 1 1 1 1 1 1
1 1 0 1 1 0 0 1	0 0 0 0 1 0 1 0	1 1 1 1 1 1 1 1
+ 0 1 0 1 1 1 0 0	+ 1 1 1 1 1 1 0 1	+ 0 0 0 0 0 0 0 1
0 0 1 1 0 1 0 1	0 0 0 0 0 1 1 1	0 0 0 0 0 0 0 0
Carryout without overflow. Sum is correct.	Carryout without overflow. Sum is correct.	Carryout without overflow. Sum is correct.

Case-III: Larger of the two numbers is negative.

-75 + 59 = -16:	-10 + 3 = -7:	- 92 + 39= -53:
1 1 1 1 1 1	1 1	1 1
1 0 1 1 0 1 0 1	1 1 1 1 0 1 1 0	1 0 1 0 0 1 0 0
+ 0 0 1 1 1 0 1 1	+ 0 0 0 0 0 0 1 1	+ 0 0 1 0 0 1 1 1
1 1 1 1 0 0 0 0	1 1 1 1 1 0 0 1	1 1 0 0 1 0 1 1
No overflow nor carryout. Sum is correct.	No overflow nor carryout. Sum is correct	No overflow nor carryout. Sum is correct.

Case-IV: Both the numbers are negative.

-19 + -7 = -26:	-103 + -69 = -172:	-39 + - 92 = -131:
1 1 1 1 1 1	1 1 1 1 1 1	1
1 1 1 0 1 1 0 1	1 0 0 1 1 0 0 1	1 1 0 1 1 0 0 1
+ 1 1 1 1 1 0 0 1	+ 1 0 1 1 1 0 1 1	+ 1 0 1 0 0 1 0 0
1 1 1 0 0 1 1 0	0 1 0 1 0 1 0 0	0 1 1 1 0 1 0 1
Carryout without overflow. Sum is correct.	Overflow, with incidental carryout. Sum is not correct.	Overflow, with incidental carryout. Sum is not correct.

Fig.1.9: Binary arithmetic using 2's complement method

Comparison between 1's and 2's complements

A comparison between 1's and 2's complement reveals using the Table 1.5 below.

Table 1.5: Comparison between 1's and 2's complement

Parameters	1's complement representation	2's complement representation
Process of Generation	Easy to use, as 1's complement of a given binary number can be attained by merely inverting all bits from 0's to I's and vice versa.	A bit complicated as compared to 1's complement. As to attain the 2's complement of a given binary number, first it needs to find its 1's complement by inverting all bits and then adding 1 to the LSB.
Example	Given No. / Binary / 1's Complement: $(9)_{10}$ / $(1001)_2$ / 0110	**Given No.** / **Binary** / **1's Complement** / **2's Complement**: $(9)_{10}$ / $(1001)_2$ / $(0110)_2$ / $(0111)_2$
Number Representation	If we want to represent the sign binary number, we can use the 1's. If we have a number 0, then it will not be possible to use it in the form of ambiguous representation.	If we want to represent the sign binary number, we can also use the 2's. If we have a number 0, then it will possible to use it as an unambiguous representation of all given numbers.

Parameters	1's complement representation	2's complement representation
Range of Values	If there is an *n-bit* Binary representation, the 1's complement will range from $-(2^{(n-1)}-1)$ as the lowest negative number to $(2^{(n-1)}-1)$ as the largest positive number (i.e., -127 to +127).	If there is an *n-bit* Binary representation, the 2's complement will range from $-(2^{(n-1)})$ as the lowest negative number, to $(2^{(n-1)}-1)$ as the largest positive number (i.e., -128 to +127).
Representation of '0'	The 1's complement introduces the concept of +0 and -0. The +0 is 00000000, and -0 will be represented by 11111111.	In 2's complement, both -0 or +0 can be represented as 0000000 (+0) only. It means that number 0 is always considered as a positive in the 2's complement. As like if add 1 to -0, i.e., 11111111 it gives 00000000 (+0). Hence, in general, 2's complement will be used in binary arithmetic.
Sign Extension	In the 1's complement, sign extension is used to convert the given sign into another sign for any signed integer.	Same as like 1's complement, in 2's complement also it needs to convert a given sign into another sign for any signed integer.
End-Around Carry-Bit	If carry has been generated after performing an arithmetic operation (addition) using 1's complement, this carry is called EAC and needs to add the ***end-around carry*** bit to LSB.	The 2's complement ignores this type of addition of EAC, i.e., here carry has been discarded.
Ease of operation	Due to the occurrence of EAC, 1's complement arithmetic operation is difficult as compared to the 2's complement arithmetic operation.	As 2's complement has to do nothing with EAC, hence it is simpler than 1's complement arithmetic operation.

1.3 Octal Number System

The bae value of octal number system is '8'. (i.e. r= 8) the number of different symbols used in octal system are 0 to *r-1*. (i.e., =0 to 7). Those are 0,1,2,3,4,5,6,7 (Refer Table 1.6).

The maximum digit in the octal number system = r-1

=8-1

=7

The numbers in the octal number system can be represented as

$$\ldots\ldots\ldots.8^3 8^2 8^1 8^0 . 8^{-1} 8^{-2} 8^{-3} \ldots\ldots.$$

Table 1.6: Octal numbers with its and decimal equivalent

Decimal	Binary	Octal	Decimal	Binary	Octal
0	000	0	**4**	100	4
1	001	1	**5**	101	5
2	010	2	**6**	110	6
3	011	3	**7**	111	7

1.3.1 Decimal to octal conversion:

Similar to how decimal (Base 10) to binary (Base 2) conversion is done, converting a decimal (Base 10) number to its octal (Base 8) equivalent is also possible. The sole distinction is that "8" does progressive division (for the integer part) and progressive multiplication (for the fractional part). It is important to keep in mind that an octal integer is defined by the radix "8." The following example serves as a demonstration of the procedure.

Example 1.23: convert $(266)_{10}$ to its octal equivalent.

Solution: A decimal number can be changed to an octal number by dividing it by 8 in steps as shown below:

$266 \div 8 = 33$ remainder 2 LSD (right-most digit)
$33 \div 8 = 4$ remainder 1
$4 \div 8 = 0$ remainder 4 MSB (left-most digit).

Therefore $266_{10} = 412_8$

Example 23: Find the Octal equivalent of $(73.75)_{10}$

Solution

Integer Part					Fractional Part			
Divide by 8	Quotient	Remainder		↑	Multiply by 8	Output	Carry	↓
73 ÷ 8	9	1	LSD		0.75 X 8	6.00	6	
9 ÷ 8	1	1			0.00 X 8	0.00	0	
1 ÷ 8	0	1	MSD					

Therefore $(73.75)_{10} = (111.6)_8$

1.3.2 Octal to Decimal conversion:

The octal number system's "8" radix is employed. In order to represent numerals, eight different symbols (0–7) will be employed. In general, each number in octal may include both an integer and a fractional component, or both. However, a (octal) point will be used to separate the two halves (.). We define the weights in power of '8' and can represent an octal integer in decimal form as

$$\ldots\ldots\ldots\ldots 8^3 8^2 8^1 8^0 . 8^{-1} 8^{-2} 8^{-3} \ldots\ldots\ldots\ldots$$

Example 1.24: Convert the following octal numbers to its decimal number equivalent

(a) (347)8 (b) $(712.63)_8$ (c) $(1205.6)_8$

Solution:

(a) $(347)_8 = 3 \text{ X } 8^2 + 4 \text{ X } 8^1 + 7 \text{ X } 8^0 = 192 + 32 + 7 = (231)_{10}$

(b) $(712.63)_8 = 7 \text{ X } 8^2 + 1 \text{ X } 8^1 + 2 \text{ X } 8^0 + 6 \text{ X } 8^{-1} + 1 \text{ X } 8^{-2} = (458.796875)_{10}$

(c) $(1205.6)_8 = 1 \text{ X } 8^3 + 2 \text{ X } 8^2 + 0 \text{ X } 8^1 + 5 \text{ X } 8^0 + 6 \text{ X } 8^{-1} = (680.75)_{10}$

1.3.3 Binary to Octal Conversion

By taking groups of three bits, beginning with the LSB, and swapping them out with an octal Value digit, a binary number can be transformed into an octal number.

Example 1.25: convert $(10111111010110.0011)_2$ to an octal number

Solution:

1. Given No.: 10111111010110.0011
2. Make a group of 3 bits (Starts from decimal point in both the directions)

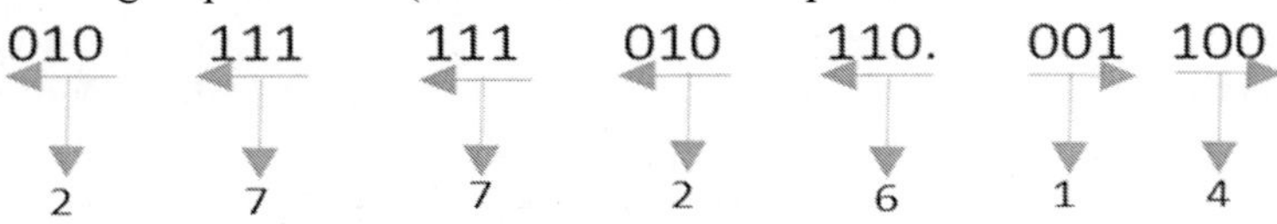

Answer= $(27726.14)_8$

1.3.4 Octal to Binary Conversion:

Each octal digit is represented as its 3-bit equivalent binary number when converting an octal number to a binary number. The octal number can be indirectly converted to binary by first being converted to digital form, after which it can be further converted to binary form.

Example 1.26: convert $(32766.14)_8$ to its binary equivalent

Solution:

Method 1:

(a) Given No.: $(27726.14)_8$

(b) Represent each digit by a group of 3 bits (Starts from the decimal point in both directions)

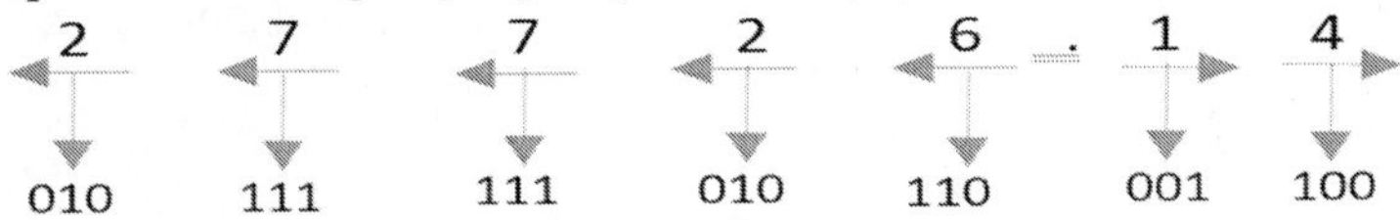

Answer= $(10111111010110.0011)_2$

Method 2:

$(27726.14)_8$ —Decimal→ $(12246.1875)_1$ —Binary→ $(10111111010110.0011)_2$

The octal arithmetic rules are comparable to binary to decimal arithmetic. Octal arithmetic procedures involving octal representations of integers are typically not of relevance to us. Long binary data strings are typically entered into digital systems using this number system. This makes it considerably simpler to input binary data into a digital system. By transforming octal numbers to binary numbers and then applying the binary addition rule, arithmetic operations can be carried out. Octal numbers must first be translated to binary in order to execute multiplication and division in the octal number system.

Example 1.27: Evaluate

(a) $(712.63)_8 + (121.37)_8$ (b) $(676.12)_8 + (343.46)_8$

Solution:

(a) $(712.63)_8 + (121.37)_8$

```
  712.63
  121.37
 --------
(1034.22)8
```

(b) $(676.12)_8 + (343.46)_8$

```
  676.12
  343.46
 --------
(1241.60)8
```

1.4 Hexadecimal Number System

Large binary numbers that are stored and processed inside of a computer can be represented in a compressed manner using the Hexadecimal number system. The Hexadecimal Number System's base value is "16,", i.e., r = 16. The number of different symbols used in the Hexadecimal number system is from 0 to *r-1*, i.e., from '0' to '15'. Those are 0,1,2,3,4,5,6,7,8,9, A, B, C, D, E, and F.

The maximum digit in the Hexadecimal number system = r-1
=16-1
=15(F)

$$\ldots\ldots\ldots\ldots 16^3 16^2 16^1 16^0 . 16^{-1} 16^{-2} 16^{-3} \ldots\ldots\ldots\ldots$$

Handling and remembering binary numbers are not very convenient, and also decimal number can't be used in computers. Hence, hexadecimal number system, gives an edge over binary and decimal number system. As by using a four digit (0000 to FFFF) 65536 different addresses can be expressed. Additionally, the hexadecimal representation of the memory's contents makes handling it much easier.

1.4.1 Decimal to Hexadecimal conversion

As with other conversions from decimal, the method of going from decimal to hexadecimal is comparable. The progressive division and multiplication factor in this situation is 16, as the hexadecimal number system's base is 16. The following examples assist to further illustrate how the conversion process works.

Example 1.28: Determine the Hexadecimal equivalent of (82.25)10

Solution:

Integer Part					Fractional Part		
Divide by 16	Quotient	Remainder		↑	Multiply by 16	Output	Carry
82 ÷ 16	5	2	LSD		0.25 X 16	4.00	4
5 ÷ 16	0	5	MSD		0.00 X 16	0.00	0

Therefore (82.25)10 = (52.4)16

1.4.2 Hexadecimal to decimal conversion:

The power of 16 can be used to convert a hexadecimal number to a decimal value. This could be better appreciated by using the example below.

Example 1.29: Find the Decimal number equivalent of $(1E0.2A)_{16}$

Solution:

Hex Value	**1**	**E**	**0**	**.**	**2**	**A**
Multiply by	16^2	16^1	16^0	.	16^{-1}	16^{-2}
After multiplication	$1 X 16^2$	$14 X 16^1$	$0 X 16^0$	.	$2 X 16^{-1}$	$10 X 16^{-2}$
Decimal output	$1 X 16^2+14 X 16^1+0 X 16^0+2 X 16^{-1}+10 X 16^{-2} = (480.164)_{10}$					

Therefore $(1E0.2A)_{16} = (480.164)_{10}$

1.4.3 Hexadecimal to binary conversion

Simply replace each hexadecimal digit with its 4-bit binary equivalent to convert a hexadecimal number to its binary equivalent. The corresponding values are displayed in the Table 1.7 below.

Table 1.7: convert a Hexadecimal number to its binary equivalent number

Decimal	Binary	Hexadecimal	Octal	Decimal	Binary	Hexadecimal	Octal
0	0000	0	0	8	1000	8	10
1	0001	1	1	9	1001	9	11
2	0010	2	2	10	1010	A	12
3	0011	3	3	11	1011	B	13
4	0100	4	4	12	1100	C	14
5	0101	5	5	13	1101	D	15
6	0110	6	6	14	1110	E	16
7	0111	7	7	15	1111	F	17

Example 1.30: Convert the following to its binary equivalent.

(a) $(2F9A.13)_{16}$ (b) into Binary

Solution:

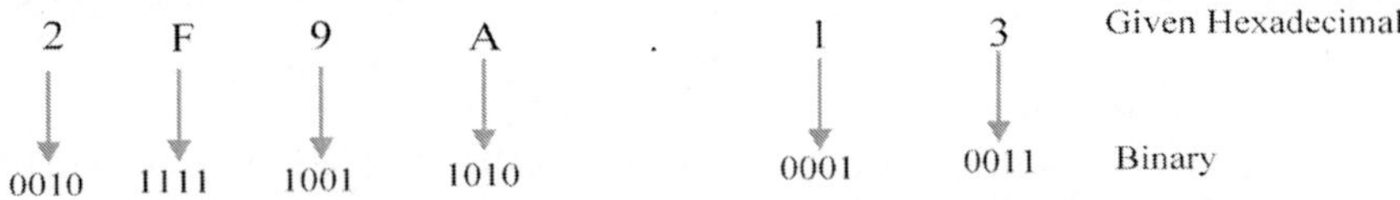

Therefore $(2F9A.13)_{16}$ equivalent of Binary is $(10111110011010.00010011)_2$

1.4.4 Binary to hexadecimal conversion

By creating groups of four bits for the integer part of binary numbers starting at the LSB and advancing towards the MSB, and then replacing each group of four bits with its hexadecimal representation, binary numbers can be transformed into their equivalent hexadecimal numbers. The aforementioned process is repeated for the fractional part, starting at the bit next to the binary point and going to the right. The following example will help you better understand the conversion.

Example 1.31: Convert the following to $(10111110011010.00010011)_2$ to its Hexadecimal equivalent.

(a) $(10111110011010.00010011)_2$ (b) $(10111111010110)_2$

Solution:

(a)

The given binary number is 10111110011010.00010011. First Start making Group of 4 bits

0010 1111 1001 1010 . 0001 0011

Using the Table ...... Hex equivalent value for each Binary group can be written

Therefore $(10111110011010.00010011)_2 = (2F9A.13)_{16}$

(b)

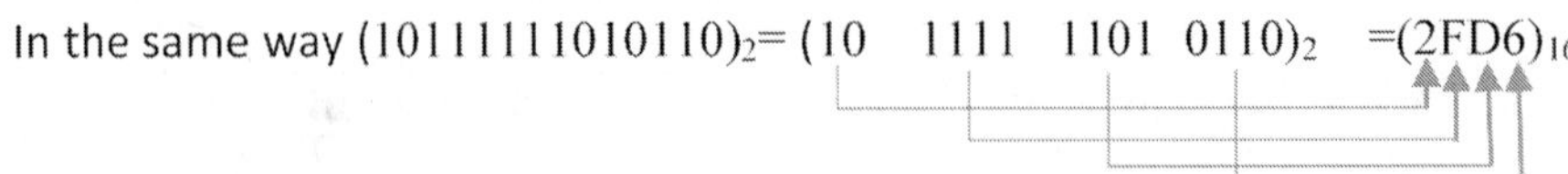

Therefore $(10111111010110)_2 = (2FD6)_{16}$

1.4.5 Hexadecimal to octal conversion

There are several methods for converting a hexadecimal number to an octal number. The number can either be converted to binary first, then to an octal number, or it can be converted from hexadecimal to decimal first, then from decimal to an octal number.

Example 1.32: Convert the following Hexadecimal numbers into Octal Numbers

(a) $(A72E)_{16}$ (b) (0.BF85)16

Solution:

(a)

Given Hex No.	A		7		2		E
Binary Equivalent	1010		0111		0010		1110
Combine all Values			1010011100101110				
Make a group of 3 bits each	001	010	011	100	101	110	
Octal Equivalent	1	2	3	4	5	6	

Therefore, the octal equivalent of $(A72E)_{16}$ is $(123456)_8$

(b)

Given Hex No.	B		F		8		5
Binary Equivalent	1011		1111		1000		0101
Combine all Values			1011111110000101				
Make a group of 3 bits each	001	011	111	110	000	101	
Octal Equivalent	1	3	7	6	0	5	

Therefore, octal equivalent of $(0.BF85)_{16}$ is $(0.127605)_8$

1.4.6 Octal to hexadecimal conversion

There are several ways to convert an octal number to a hexadecimal number. One can either convert the octal number to decimal first and then further convert the decimal number to hexadecimal, or one can convert the octal number to binary first and then further convert it to hexadecimal.

Example 1.33: Convert following Octal Numbers to Hexadecimal numbers

(a) (532)8 (b) $(6327.4051)_8$

Solution:

(a)

Given Hex No.	5	3	2
Binary Equivalent	101	011	010
Combine all Values		101011010	
Make a group of 3 bits each	0001	0101	1010
Octal Equivalent	1	5	A

Therefore, octal equivalent of $(532)_8$ is $(15A)_{16}$

(b)

Given Hex No.	6	3	2	7	.	4	0	5	1
Binary Equivalent	110	111	010	111		100	000	101	001
Combine all Values		110111010111					100000101001		
Make a group of 4 bits each	1101	1101	0111			1000	0010	1001	
Hexadecimal Equivalent	D	D	7			8	2	9	

Therefore, an octal equivalent of $(6327.4051)_8$ is $(DD7.829)_{16}$

1.4.7 Hexadecimal arithmetic operations

To perform the arithmetic operation using hexadecimal number, first it needs to convert hexadecimal values to their binary equivalent. Since, the digital circuit can handle the information only in binary form. However, entering data using hex values is simpler for a person to do. Moreover, the rules for arithmetic operations are similar to the rules that used for decimal, octal and binary arithmetic. Arithmetic operation while using hexadecimal numbers will be easier to realize with the following examples.

Example 1.34: Solve the following

(a) $(EAB.C)_{16} + (143.3)_{16}$ (b) $(7F)_{16}$ + $(BA)_{16}$

Solution:

	(a) $(EAB.C)_{16} + (143.3)_{16}$					**(b) $(7F)_{16}$ + $(BA)_{16}$**		
	Using direct addition					By converting to Binary		
	E	A	B	.	C	7F	0111 1111	$(127)_{10}$
+	1	4	3	.	3	BA	1011 1010	$(186)_{10}$
	$(FEE.F)_{16}$					(139)16	(1 0011 1001)*	$(313)_{10}$

**whenever the sum of a group is greater than 9 (1001) then add 6 (0110) to that group this idea will be discussed later when we discuss BCD addition.Note: The positioned weights method can be used to convert any number with base "r" to its decimal equivalent. By multiplying each digit of the number with base "r" by its position weight and adding the product term, the decimal number can be calculated using this procedure. if the number is $D_3D_2D_1D_0.D_{-1}D_{-2}$ with base r, then its decimal equivalent is*

$$\ldots\ldots\ldots\ r^3\ r^2\ r^1\ r^0\ .\ r^{-1}\ r^{-2}\ldots\ldots$$

$$\ldots\ldots\ldots\ D_3 D_2\ D_1\ D_0\ .\ \ D_{-1}\ D_{-2}\ \ldots\ldots$$

$$= (D_3 x\ r^3)+ (D_2\ x\ r^2)+ (D_1 x\ r^1) + (D_0\ x\ r^0) + (D_{-1} x\ r^{-1}) + (D_{-2}\ x\ r^{-2})$$

1.5 Approaches to store real number

According to Fig. 1.10, below there are two main methods used in contemporary computers to store real numbers (i.e., numbers with a fractional component). Both Fixed Point Notation and Floating-Point Notation fall under this category. While floating point numbers allow for a variable number of digits after the decimal point, fixed point notation has a fixed number of digits after the decimal point. These buildings are described below –

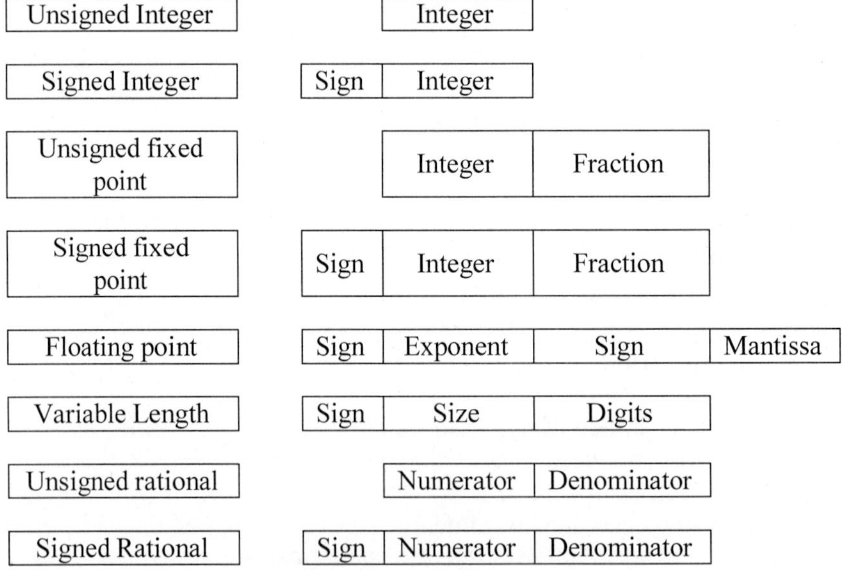

Fig. 1.10: Major approaches to store real numbers

1.5.1 Fixed-Point Representation:

The integer and fractional parts of this encoding each have a set number of bits. For instance, if the least value that can be saved is 0000.0001 and the maximum value is 9999.9999. A fixed-point number representation consists of three fields: the sign field, integer field, and fractional field.

These figures can be represented using:

- Signed representation: range from $-(2^{(k-1)}-1)$ to $(2^{(k-1)}-1)$, for k bits.
- 1's complement representation: range from $-(2^{(k-1)}-1)$ to $(2^{(k-1)}-1)$, for k bits.
- 2's complementation representation: range from $-(2^{(k-1)})$ to $(2^{(k-1)}-1)$, for k bits.

Because it is more straightforward for mathematical operations and has an obvious feature, the complementation representation of 2 is preferred in computer systems.

Example 1.35: Assume number is using a 32-bit format which reserves 1 bit for the sign, 15 bits for the integer part, and 16 bits for the fractional part.

Then, -43.625 is represented as following:

1	000000000101011	1010000000000000
Sign Bit	Integer Part	Fractional Part

Where, 0 is used to represent + and 1 is used to represent. 000000000101011 is 15-bit binary value for decimal 43 and 1010000000000000 is 16 bit binary value for fractional 0.625.

Note that 8-bit exponent field is used to store integer exponents $-126 \leq n \leq 127$.

The smallest normalized positive number that fits into 32 bits is $(1.00000000000000000000000)_2 x 2^{-126} = 2^{-126} \approx 1.18 x 10^{-38}$, and the largest normalized positive number that fits into 32 bits is $(1.11111111111111111111111)_2 x 2^{127} = (2^{24}-1) x 2^{104} \approx 3.40 x 10^{38}$. These numbers are represented as follows,

Smallest	0	10000000	00000000000000000000000
	Sign Bit	Exponent Part	Mantissa Part

Largest	0	01111111	11111111111111111111111
	Sign Bit	Exponent Part	Mantissa Part

The precision of a floating-point format is the number of positions reserved for binary digits plus one (for the hidden bit). In the examples considered here the precision is 23+1=24. The gap between 1 and the next normalized floating-point number is known as machine epsilon. the gap is $(1+2^{-23})-1=2^{-23}$for above example, but this is same as the smallest positive floating-point number because of non-uniform spacing unlike in the fixed-point scenario.

(Note that non-terminating binary numbers can be represented in floating point representation, e.g., 1/3 = $(0.010101\ ...)_2$ cannot be a floating-point number as its binary representation is non-terminating.)

1.5.2 Floating-Point Number representation

A convenient way to represent a number contains integer or fractional part or both, is by using Floating-point notation. Floating point representation not only makes the arithmetic operations relatively much easier, Additionally, with the same number of digits, it expands the range of numbers (from lowest to largest). Typically, the following equation can be used to express floating-point numbers:

$$N = m \times b^e$$

Where,
N=Number given
b = ***numeration****, the base of the number system*
m= ***mantissa,*** *the fractional part also called significand*
e = ***exponent,*** *the integer part*

For a binary number system, *b=2*, hence the equation becomes

$$N = m \times 2^e$$

Similarly, for Hexadecimal Number system

$$N = m \times 16^e$$

Similarly, for Decimal number system

$$N = m \times 10^e$$

Mantissa (m), or the *significand* represents the fractional part. Represent in the form of $(\pm d.dddd \ldots dd)$ with total number of *p-digits.* However, each digit *d* is going to be varied from *0* to *b-1* inclusive (i.e., for binary the value of *d* may be 0 or 1). Moreover, if the leading digit of m is nonzero, a number is said to be normalized.

For example, decimal numbers 0.0003754 and 3754 will be represented in floating point notation as 3.754×10^{-4} and 3.754×10^{3}respectively. A hex number $257.ABF$ will be represented as $2.57ABF \times 16^2$. in the case of normalized binary numbers, the leading digit, which is the most significant bit is always '1' and thus does not need to store explicitly.

Additionally, the most significant bit is set to "1" immediately to the right of the radix point when a given mixed binary integer is expressed as a floating-point number. The exponent and mantissa can both be positive or negative numbers.

For example, a mixed binary number such as (110.1011)2 will be represented in floating point notation as $0.1101011 \times 2^3 = .1101011e + 0011$. Here .1101011 is the mantissa and $e + 0011$ implies that the exponentis +3. Another example, (0.000111)2 will be written as $.111e - 0011$, with .111 being the mantissa and $e - 0011$ implying an exponent of -3. Also (-0.00000101)2 may be written as $-0.101 \times 2^{-5} = -.101e - 0101$ where $-.101$ is the mantissa and $e - 0101$ indicates the exponent of -5. If we wanted to represent the mantissa using eight bits, then .1101011 and .111 would be represented as .11010110 and .11100000 respectively.

As the mantissa is kept in an unsigned integer, if the mantissa is stored in n bits, it can represent a decimal value between '0' and 2^n-1. If *M* is the largest number such that 10^M-1 is less than or equal to 2^n-1, then *M* is the precision expressed as decimal digits of precision. For example, if the mantissa is expressed in 20 bits, then decimal digits of precision can be found to be about 6, as $2^{20}-1$ equals *1048575*, which is a little over 10^6-1. We will briefly describe the commonly used formats for binary floating-point number representation.

All sorts of information are represented in digital computers using the binary number system. Binary bits are used to represent alphanumeric characters (i.e., 0 and 1). Designing and storing digital representations is simpler, and their accuracy and precision are higher.

For example, the binary number system, octal number system, decimal number system, hexadecimal number system, and other number representation systems are all examples of digital number representation approaches. The most relevant and well-liked number representation scheme for digital computer systems, however, is the binary one.

1.5.3 IEEE Floating point Number Formats

The IEEE-754 standard, whose full name is IEEE Standard for Binary Floating-point Arithmetic (ANSI/IEEE STD 754-1985), is the most widely used format for encoding floating-point integers. Binary Floating-point Arithmetic for Microprocessor Systems, IEC 60559:1989 is another name for it. IEEE-754r is an ongoing revision to IEEE-754. A different related standard, EEE 854-1987, expands on IEEE-754 to incorporate binary and decimal arithmetic. Below is a quick summary of the key elements of the IEEE-754 standard and an introduction to other related standards.

ANSI/IEEE-754 Format

The most popular representation for real numbers on computers, including Intel-based personal computers, Macintoshes, and the majority of UNIX platforms, is IEEE-754 floating point. Four formats for representing floating-point numbers are specified. Single-precision, double-precision, single-extended precision, and double-extended precision formats are among them. The characteristics of the four formats included in the IEEE-754 standard are listed in Table 1-8. The single-precision and double-precision formats are the two that are most frequently used out of the four that were discussed. It is uncommon to use the single-extended and double-extended precision forms.

Table 1.8: Characteristic parameters of IEEE-754 formats.

Precision	Sign (bits)	Exponent (bits)	Mantissa (bits)	Total length (bits)	Decimal digits of precision
Single	1	8	23	32	>6
Single-extended	1	≥11	≥32	≥44	>9
Double	1	11	52	64	>15
Double-extended	1	≥15	≥64	≥80	>19

Fig. 1.11 shows the basic constituent parts of the single- and double-precision formats. The sign, the exponent, and the mantissa are the three fundamental parts of floating-point numbers as they are expressed using various formats, as illustrated in the image. A "0" stands for a positive number, while a "1" stands for a negative number. In order to acquire the stored exponent, a bias equal to 2n-1 - 1 is added to the actual exponent. This allows the n-bit exponent field to represent both positive and negative exponent values.

For the single-precision format, this is equal to 127 for an eight-bit exponent, while for the double-precision format, it is equal to 1023 for an 11-bit exponent. The inclusion of bias permits the employment of an exponent within the ranges of -1023 to +1024, which corresponds to a range of 0-2047 in the second instance, and -127 to +128, which corresponds to a range of 0-255 in the first case. The 2's complement form is always used to indicate a negative exponent. The single-precision format provides a range of 10-38 to 10-38+38, or 2-127 to 2+127. The figures are 2-1023 to 2+1023, which in the case of the double-precision format translates to 10-308 to 10+308.

For the representation of special values, the extreme exponent values are set aside. For instance, in the single-precision format, the skewed exponent value for an exponent value of -127 is zero, represented by an exponent field with only zeros. The value of the floating-point number is precisely zero in the case of a biased exponent of zero if the mantissa is also zero. The mantissa represents a denormalized number that does not have the assumed leading bit of "1" if it is nonzero. An all-1s exponent field represents a biased exponent of +255, which corresponds to an actual exponent of +128. Infinity is represented as a number if the mantissa is zero. Positive and negative infinity are separated using the sign bit. The number is a "NaN" if the mantissa is nonzero (Not a Number). A value that doesn't represent a real number is represented by the value NaN. The range of exponent values that an eight-bit exponent can represent is from -126 to +127.

The sign of the mantissa is indicated by the MSB of byte 1 in Fig.1.11(a). An eight-bit exponent is represented by the last seven bits of byte | and the MSB of byte 2. A 23-bit mantissa is produced using the final seven bits of byte 2, as well as the 16 bits from bytes 3 and 4.

A change has been made to the mantissa m. The normalised mantissa's left-most bit is always "1." This "1" is always implied but not explicitly stated. Similar justification is available for the double-precision format depicted in Fig. 1.11(b).

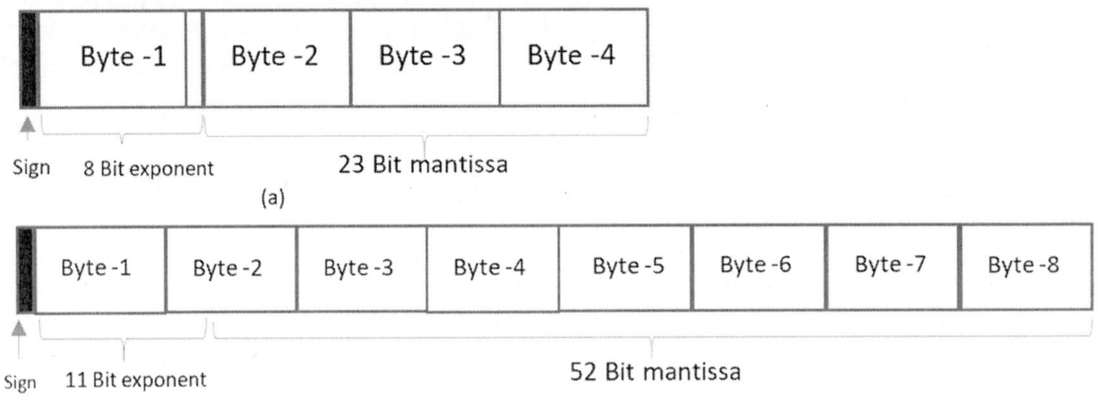

Fig. 1.11: basic constituent parts of the single- and double-precision formats

Step-by-step transformation of $(23)_{10}$, into an equivalent floating-point number in single-precision IEEE format is as follows:

$$(23)_{10} = (10111)_2 = 1.0111e + 0100$$

0	00000100	0	0111000 00000000 00000000
Sign	8 Bit exponent	Sign	23 Bit mantissa

$(+23)_{10}$ = 01000001 10111000 00000000 00000000

$(-23)_{10}$ = 11000001 10111000 00000000 00000000

IEEE-754r Format

An ongoing upgrade of the IEEE-754 standard is called IEEE-754r, as was previously indicated. The addition of the 128-bit format and decimal format is the most visible improvement to the standard, and the revision's primary goal is to expand the standard wherever it has proven required. Since most business data is stored in decimal form and binary floating point cannot accurately represent decimal fractions, the standard's extension to include decimal floating-point representation has become important. Results achieved by using binary floating point to represent decimal data are likely to differ from those obtained by using decimal arithmetic.

Many of the definitions have been updated during the editing process for uniformity and clarity. The 128-bit "quad-precision" format is a brand-new format that has been added to the list of binary formats. Three new decimal forms that match binary format lengths have also been described. Some examples of these are decimal forms with a seven-, sixteen-, and thirty-four-digit mantissa that can be normalised or denormalized. The formats combine a portion of the exponent and mantissa into a combination field in order to achieve maximum range (determined by the number of exponent bits) and precision (determined by the number of mantissa bits), and then compress the remaining mantissa using tightly packed decimal encoding. But a thorough explanation of the modification is outside the purview of this work.

IEEE-854 Standard

The primary goal of the IEEE-854 standard was to establish a floating-point arithmetic standard free from the radix and word length restrictions of the more well-known IEEE-754 standard. The IEEE standard for radix-independent floating-point arithmetic is IEEE-854 for this reason. Although the standard exclusively addresses binary and decimal floating-point arithmetic, it offers sufficient recommendations for those considering the use of any other radix value, such as 16 in the hexadecimal number system, to implement floating point. Four formats are listed in this standard as well, including the single, single-extended, double, and double-extended precision forms.

Fig. 1.12 represents IEEE (Institute of Electrical and Electronics Engineers) has standardized Floating-Point.

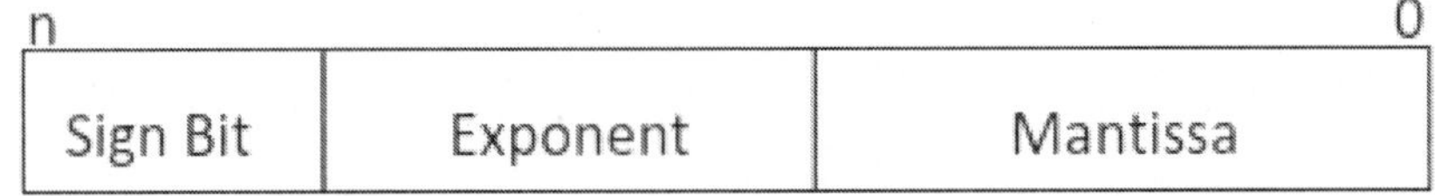

Fig. 1.12: IEEE (Institute of Electrical and Electronics Engineers) has standardized Floating-Point Representation

So, actual number is $(-1)^s(1+m)x2^{(e\text{-Bias})}$, where s is the sign bit, m is the mantissa, e is the exponent value, and *Bias* is the bias number. The sign bit is 0 for positive number and 1 for negative number. Exponents are represented by or two's complement representation.

According to IEEE 754 standard, the floating-point number is represented in following ways:

- Half Precision (16 bit): 1 sign bit, 5 bit exponent, and 10-bit mantissa
- Single Precision (32 bit): 1 sign bit, 8 bit exponent, and 23-bit mantissa
- Double Precision (64 bit): 1 sign bit, 11 bit exponent, and 52-bit mantissa
- Quadruple Precision (128 bit): 1 sign bit, 15 bit exponent, and 112-bit mantissa

1.5.4 Special Value Representation

The IEEE 754 standard contains some unusual values that depend on various mantissa and exponent values.

- All the exponent bits 0 with all mantissa bits 0 represents 0. If sign bit is 0, then +0, else -0.
- All the exponent bits 1 with all mantissa bits 0 represent infinity. If sign bit is 0, then $+\infty$, else $-\infty$.
- All the exponent bits 0 and mantissa bits non-zero represents denormalized number.
- All the exponent bits 1 and mantissa bits non-zero represent error.

Example 1.36: Determine the floating-point representation of $(-142)_{10}$ using the IEEE single precision format.

Solution:

Step 1	**Determine Binary equivalent of $(142)_{10}$**	**$(142)_{10} = (10001110)_2$**
Step 2	Equivalent floating-point number	$(10001110)_2 = 1.0001110 \text{ X } 2^7 = 1.000111e + 0111$
Step 3	Find Mantissa	0001110 00000000 00000000
Step 4	Find the Exponent	00000111
Step 5	Find the biased exponent	00000111+01111111=10000110
Step 6	Find the sign of Mantissa	1 (Represent negative)
Step 7	Final value of $(-142)_{10}$	11000011 00001110 00000000 00000000

1.5.5 Floating-Point Arithmetic

It is required to do a few tests prior to executing arithmetic operations on floating-point integers, including determining the signs of the two mantissas and examining any potential exponent misalignment. For instance, the addition and subtraction procedures require that two integers' exponents be made equal if they are not already equal. The smaller of the two numbers' mantissa is then shifted to the right, and the exponent is increased after each shift until the two exponents are equal. The alignment of the binary points and equivalence of the exponents make addition and subtraction operations simple. Naturally, a magnitude check is also necessary when subtracting two numbers to discover which is smaller.

Addition and Subtraction

If N_1 and N_2 are two floating-point numbers given by

$$N_1 = m_1 \times 2^e$$

$$N_2 = m_2 \times 2^e$$

Then

$$N_1 + N_2 = m_1 \times 2^e + m_2 \times 2^e = (m_1 + m_2)2^e$$

And

$$N_1 - N_2 = m_1 \times 2^e - m_2 \times 2^e = (m_1 - m_2)2^e$$

N1 > N2 is a presumption made during the subtraction procedure. It could be necessary to post-normalize the result after the addition or subtraction operation.

Multiplication and Division

The mantissas of the two numbers are multiplied and their exponents are added when two floating-point numbers are being multiplied. When performing a division operation, the dividend mantissa is divided by the divisor mantissa, resulting in the mantissa of the quotient, and the exponent of the quotient is obtained by subtracting the two exponents (i.e., dividend exponent minus divisor exponent).

If

$$N_1 = m_1 \times 2^{e1} \ and \ N_2 = m_2 \times 2^{e2}$$

Then

$$N_1 \times N_2 = (m_1 \times m_2) \times 2^{(e1+e2)}$$

And

$$N_1 / N_2 = (m_1 / m_2) \times 2^{(e1-e2)}$$

As with addition and subtraction operations, post-normalization may once again be necessary after multiplication or division.

Example 1.36: Solve the following using floating-point numbers. Verify the answers by performing equivalent decimal addition.

(a) $(39)_{10} + (19)_{10}$ (b) $(1E)_{16} + (F3)_{16}$ (c) $(21)_8 - (17)_8$

Solution:

(a) $(39)_{10} = (100111)_2 = 0.100111 \text{ X } 2^6$
$(19)_{10} = (10011)_2 = 0.10011 \text{ X } 2^5 = 0.010011 \text{ X } 2^6$
$(39)_{10} + (19)_{10} = 0.100111 \text{ X } 2^6 + 0.010011 \text{ X } 2^6 = (0.100111 + 0.010011) \text{ X } 2^6$
$= 0.111010 \text{ X } 2^6 = (111010)_2 = (58)_{10}$ ***Hence Verified***

(b) $(1E)_{16} = (00011110)_2 = 0.\ 00011110 \text{ X } 2^8$
$(F3)_{16} = (11110011)_2 = 0.\ 11110011 \text{ X } 2^8$
$(1E)_{16} + (F3)_{16} = 0.\ 00011110 \text{ X } 2^8 + 0.\ 11110011 \text{ X } 2^8 = (0.\ 00011110 + 0.\ 11110011) \text{ X } 2^8$
$= 0.111010 \text{ X } 2^6 = (000100010001) = (111)_{16}$ ***Hence Verified***

(c) $(21)_8 = (010001)_2 = 0.\ 010001 \text{ X } 2^6$
$(17)_8 = (001111)_2 = 0.\ 001111 \text{ X } 2^6$
$(21)_8 + (17)_8 = 0.\ 010001 \text{ X } 2^6 - 0.\ 001111 \text{ X } 2^6 = (0.\ 010001 - 0.\ 001111) \text{ X } 2^6$
$= 0.000010 \text{ X } 2^6 = (02)_8$ ***Hence Verified***

1.6 Digital signals and Digital systems

A digital system is a network of digital modules that manipulates discrete informational elements that are internally represented in binary form. A vast range of consumer and industrial devices, including automated industrial machinery, pocket calculators, microprocessors, digital computers, digital watches, TV games, and signal processing, among others, now use digital systems.

1.6.1 Characteristics of Digital systems

- Digital systems manipulate discrete elements of information.
- Discrete elements are nothing but the digits such as 10 decimal digits or 26 letters of alphabets and so on.
- Digital systems use physical quantities called signals to represent discrete elements.
- In digital systems, the signals have two discrete values and are therefore said to be binary.
- A signal in digital system represents one binary digit called a bit. The bit has a value either 0 or 1.

Numerous technologies produced information-carrying signals from analogue signals. Both in terms of values and timing, these signals are continuous. Digital signals, in contrast to analogue signals, are discrete in terms of value and time (Refer fig. 1.13). These signals, which are made up of various voltage values, are represented by binary numbers.

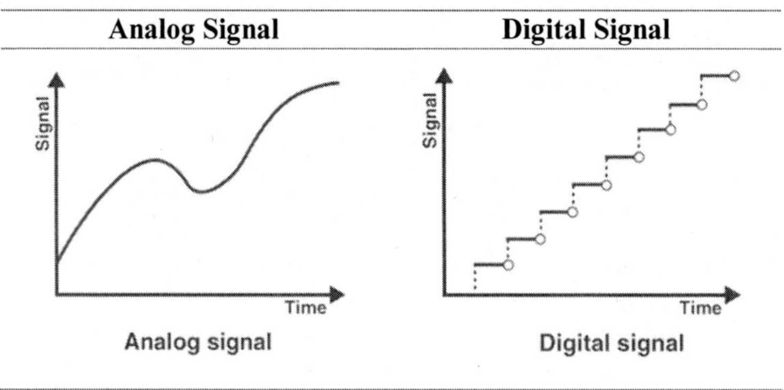

Fig. 1.13: Representation of Analog and Digital Signal

Quantizing the quantities is necessary in order to imitate a physical process in a digital computer. Real-time continuous signals that are used to represent the process' variables are quantized by an analog-to-digital conversion device. Numerical techniques are used to simulate a physical system in a digital computer whose behaviour is specified by mathematical equations. The digital computer manipulates the variables in their natural form when the problem to be solved is fundamentally discrete, as in commercial applications. Fig. 1.14 displays a block diagram of the digital computer. Programs are kept in the memory together with input, output, and intermediate data.

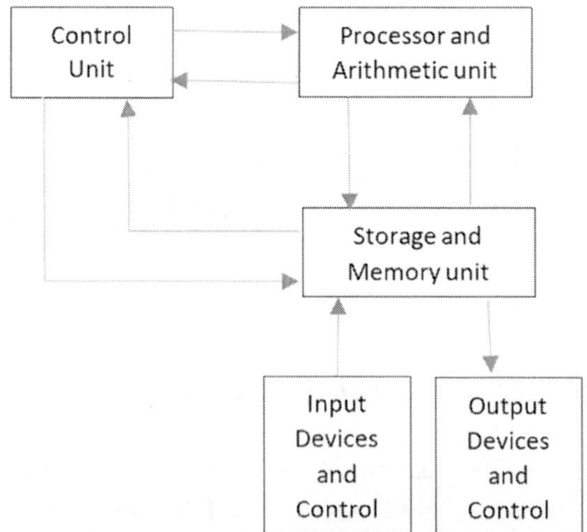

Fig. 1.14: A block diagram ‘of the digital computer

The processor unit carries out mathematical operations as well as other data processing activities as directed by a software. The information transfer between the various units is monitored by the control unit. The program is stored in memory, and the control unit pulls the instructions out one at a time. The control unit notifies the processor to carry out the action provided by each instruction. Data and programmes are both kept in memory. The processor modifies the data in accordance with the program's instructions under the supervision of the control unit.

Using an input device such a punch-card reader or teletypewriter, the user's programme and prepared data are sent into the memory unit. The computations' output is transferred to an output device, like a printer, and the printed results are displayed to the user. Special digital systems that are driven by electromechanical components and controlled by electronic digital circuits make up the input and output devices.

A digital computer works with discrete information components, and these components are represented in binary form. The binary number system can be used to express calculations' operands. Data processing is done by means of binary logic components using binary signals. Other discrete elements, such as decimal digits, are represented in binary codes. Binary storage elements store quantities.

1.6.2 Difference between Analog and Digital signals

Table 1.9: Difference between Analog and Digital Signals

Analog	Digital
An analog signal is a continuous signal that represents physical measurements:	Digital signals are time separated signals which are generated using digital modulation.
It is denoted by sine waves	It is denoted by square waves
It uses a continuous range of values that help you to represent information.	Digital signal uses discrete 0 and 1 to represent information.
Temperature sensors, FM radio signals, Photocells, Light sensor, Resistive touch screen are examples of Analog signals.	Computers, CDs, DVDs are some examples of Digital signal.
The analog signal bandwidth is low	The digital signal bandwidth is high.
Analog signals are deteriorated by noise throughout transmission as well as write/read cycle.	Relatively a noise-immune system without deterioration during the transmission process and write/read cycle.
Analog hardware never offers flexible implementation.	Digital hardware offers flexibility in implementation.
It is suited for audio and video transmission.	It is suited for Computing and digital electronics.
Processing can be done in real-time and consumes lesser bandwidth compared to a digital signal.	It never gives a guarantee that digital signal processing can be performed in real time.
Analog instruments usually have s scale which is cramped at lower end and gives considerable observational errors.	Digital instruments never cause any kind of observational errors.
Analog signal doesn't offer any fixed range.	Digital signal has a finite number, i.e., 0 and 1.

In a digital system, discrete pieces of information are represented by physical quantities known as signals. The most typical types of electrical signals are voltages and currents. The signals in all current modern digital systems are referred to as binary since they only have two discrete values. Because many-valued electrical circuits are less reliable than binary signals, the designer of digital systems is limited to using binary signals. In other words, it is possible to create a circuit with ten states using a discrete voltage value for each state, but the operation of such a circuit would be extremely unreliable. Contrarily, a transistor circuit with an on/off switch has two potential signal values and can be built to be very dependable. Digital systems that are confined to accept discrete values are further constrained to take binary values because to this physical restriction of components and the tendency of human logic to be binary.

Differential equations, whose solutions as a function of time reflect the whole mathematical behaviour of the process, can mathematically explain many physical systems. A physical system is directly simulated by a computer. A specific portion of the process being studied has an analogue in the computer's Fach section. Analog computers use continuous signals to represent their variables, often changing electric voltages throughout time. The process variables and the signal variables are thought to behave analogously. As a result, analogue voltage measurements can be used in place of process variables. Because "analogue computer" has evolved to refer to a computer that manipulates continuous variables, the phrase analogue signal is occasionally used instead of continuous signal.

1.7 Binary Codes

Digital computers will use binary codes to better represent and process the numbers. Binary code is used to represent the data in either binary or decimal form. Since humans can comprehend and retain values in decimal format, data must be specified in decimal format for improved human comprehension. These user-provided decimal input data will be internally stored in a computer system utilising a particular sort of coding called a binary code.

Furthermore, four binary storage elements are needed to store each decimal digit using binary mean. Since a four-bit binary value is used to represent each decimal digit (0–9). These additional arithmetic and logical operations will be carried out using this binary representation. Finally, binary numbers are transformed back into decimal values for human usage and further comprehension.

The fundamental number system conversion procedure was already covered in earlier parts. However, in coding, a certain set of symbols is used to represent numbers, letters, or words. The collection of symbols is referred to as a code. A set of binary bits is used to represent, store, and transmit digital data. Binary code is another name for this category. The number and alphabetic letter both serve as representations of the binary code.

For example,

Decimal	Binary Number	Binary Coded Decimal (BCD)
$(395)_{10}$	$(110001011)_2$	$(0011\ 1001\ 0101)_{BCD}$

The decimal number $(395)_{10}$, when converted to binary, is equal to $(110001011)_2$ and consists of nine binary digits, this process is called change of radix or simply conversion. However, when the same number, is represented internally using a code (for example here it is BCD code), each decimal digit will occupy four bits. Hence, to represent $(395)_{10}$, a total of 12 bits, i.e., 001110010101 will be used. The first four bits (0011) represent a 3, the next four (1001) used to represent 9, and the last four (0101) gives 5.

Hence, there is a difference between conversion and coding. Bits obtained through coding are mixtures of 1s and 0s that are organised in accordance with the guidelines established for a particular type of code. It is crucial to understand that in a digital system, a string of 1s and 0s sometimes represents a binary number and other times another discrete quantity of information as defined by a particular binary code.

1.8 Classification of Binary Codes:

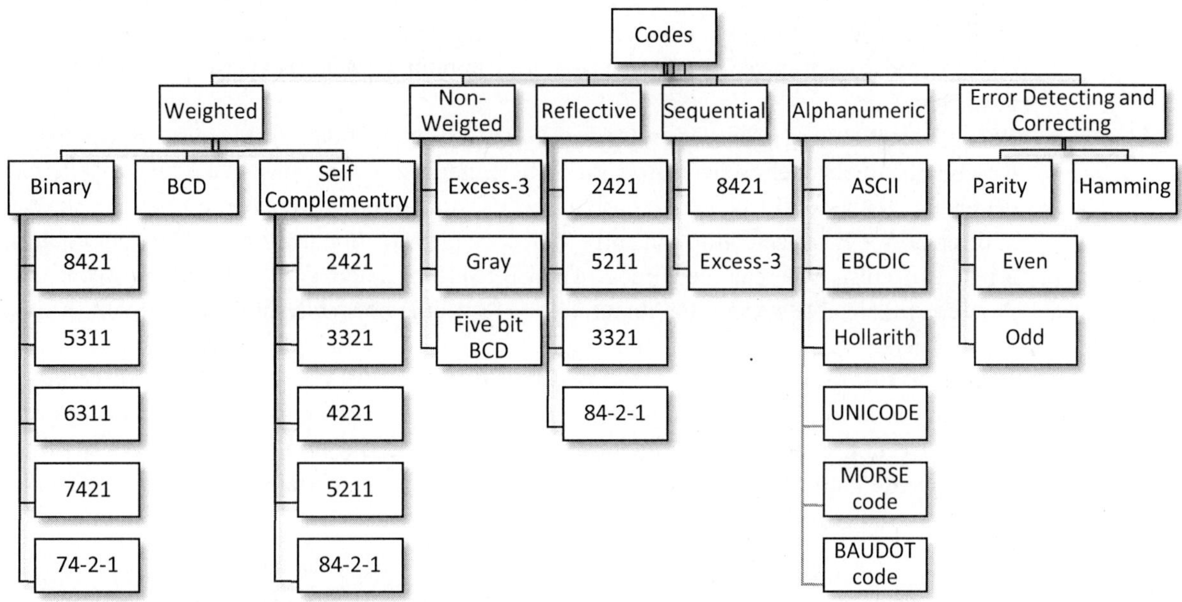

Fig. 1.15: Classification of Binary codes

Classification of binary codes can be represented as shown in Fig. 1.15.

- **Weighted Code-** The position-weighting principle is observed by the weighted codes. Each position of the number represents a specific weight.
  Ex.: BCD, 8421, 2421 etc.
- **Non-Weighted codes-** The term "non-weighted codes" refers to codes that do not assign a fixed value to each digit location inside the number or any weight to any of the digit positions.
  Ex.: Excess -3 code, gray code etc.
- **Self-complementing codes (Reflective code)** – If the code word is the 9's complement of N, or "9-N," then the code is said to be self-complementing. can be acquired using the letter N as a code. by reversing the order of the 0s and 1s. In a self-complementing code, the code for 9 is the complement of the code for 0, the code for 8 is the complement of the code for 1, and so forth. Self-complementing codes have an advantage that their logical complement is the same as the arithmetic complement. In order for a code to be self-complementing, all of the weights must add up to 9.

Ex.: 2421, 5211,642-3, 84-2-1 etc.

- **Sequential codes**- Sequential codes are those in which two successive binary digits differ by just one digit.
  Ex.: 8421 and Excess-3 codes.
- **Alphanumeric codes** are symbols that represent alphanumeric data, such as alphabetic letters and decimal numbers, as a series of 0s and 1s.
  Ex.: ASCII code, EBCDIC code.
- **Gray codes**, often known as cyclic codes, are those where each code word differs from the one before it just by one-bit position. Additionally, known as unit distance codes. The unit distance codes have special advantages that they minimize transitional errors of flashing.
  Exp. Gray Code.
- **Error Detection and Correction Codes:** During transfer of data from source to destination may be due to channel properties error could be induced in the dataset i.e., the data bits may get change (either 0 to 1 or 1 to 0) during transmission. In order to make reception of the data to be error free Error detection and correction code plays an important role. The error could be a result of noise, whereas the source of noise could be internal or external. However, in a physically realizable system it is impossible to avoid the noise interference, but with the help of error detection and correction codes original message could be recovered from noisy data. Error detection and correction is a two-step process, in first step consist of error detection, codes are used to identify or detect the error, whether it is present or not. Further, by using error correction code the error bit position will be find out and the particular bit will be complemented in order to recover the original message. The classification of error detection and correction codes is represented in Fig. 1.16.

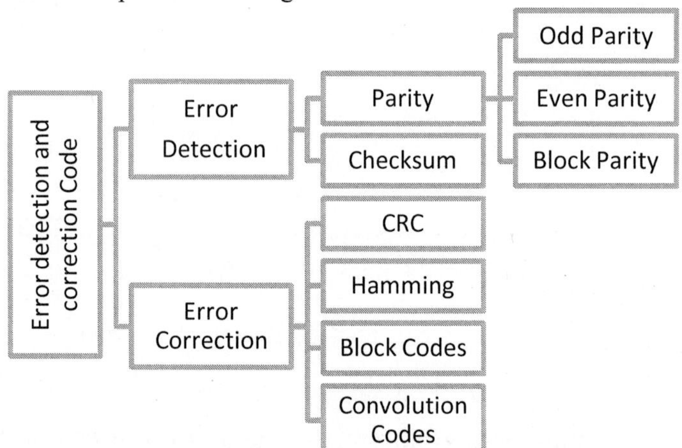

Fig. 1.16: Classification of error detection and correction codes

1.9 BCD Code *or* 8421 code *or* Natural BCD Code

Binary coded decimal, sometimes known as BCD, is a means to represent decimal digits. A 4-bit binary format is typically employed. Since BCD values directly correlate with decimal numbers and representing a given decimal number in an analogous BCD code is a far less laborious activity,

BCD codes are the most widely used codes. As a result, converting a number from BCD to decimal and vice versa is quite simple.

The BCD code is actually known as the 8421 BCD code, where the numbers 8, 4, 2, and 1 stand in for the weights of the various bits in the four-bit groups, starting with the MSB and moving towards the LSB. The four-bit binary equivalent of each decimal digit in the integer and fractional sections of a decimal number is used to represent the BCD equivalent of that value.

Table 1.10: Representation of decimal values in the form of BCD Codes

		BCD Codes		
Decimal	**Binary**	**8421 BCD**	**4221 BCD**	**5421 BCD**
0	0000	0000	0000	0000
1	0001	0001	0001	0001
2	0010	0010	0010	0010
3	0011	0011	0011	0011
4	0100	0100	1000	0100
5	0101	0101	0111	1000
6	0110	0110	1100	1001
7	0111	0111	1101	1010
8	1000	1000	1110	1011
9	1001	1001	1111	1100

As an example, to find the BCD equivalent of $(49.16)_{10}$, just every digit needs to be converted into its equivalent 4-bit binary value. Such as 4,9,1 and 5 in binary is equivalent to 0100, 1001, 0001, and 0101. Hence BCD equivalent of $(49.16)_{10}$, will be written as $(0010\ 0011.0001\ 0101)_{BCD}$.

The BCD code are called weighted code, which means that there is a weight assigned to each bit in a four-bit group. Combinedly a four-bit group is going to represent a decimal digit. Hence, for all the weighted codes each bit in the four-bit group will be assigned a weight, finally weights of all position of 1's is going to be added to represent a given decimal digit. Some of the codes that belongs to such weighted code category, other than BCD code are 4221, 8421, 84-2-1, and 5421 etc. codes.

The Table below summarizes some of the weighted codes along with their binary equivalent value. The representation of a particular decimal value may be different for different code. Such as, $(98.16)_{10}$ will be written as $(1111\ 1110.0001\ 1100)_{4221}$ in 4221 BCD code and $(1100\ 1011.0001\ 1001)_{5421}$ in 5421 BCD code. Since the 8421 code is the most popular of all the BCD codes, it is simply referred to as the BCD code (Refer Table 1.10).

1.9.1 BCD-to-Binary Conversion

There are two steps involved in converting a BCD number to its decimal equivalent. In first step A given BCD number can be converted into an equivalent Decimal number (Step-1 in Table 1.11), by arranging the BCD number as a group of 4-bits. After converting the BCD number to decimal equivalent, it will be further converted to binary by using simple method of decimal to binary conversion (Step-2 of Table 1.11).

Example 1.37: Find the binary equivalent of the BCD number $(0010\ 1001.0111\ 0101)_{BCD}$

Solution:

Table 1.11: Converting a BCD number to a decimal equivalent

	The given BCD No.	$(0010\ 1001.0111\ 0101)_{BCD}$
Step 1	Convert it to decimal (using Group of 4 Bits)	$(29.75)_{10}$
Step 2	Binary equivalent of the decimal No.	$(11101.11)_2$

1.9.2 Binary-to-BCD conversion

The process of going from BCD to binary to decimal is known as the reverse conversion. The procedure for converting from binary to BCD is the same as the procedure for converting from BCD to binary when done backwards. By first figuring out its decimal counterpart and then writing the matching BCD equivalent, a given binary number can be transformed into its equivalent BCD number.

Example 1.38: find the BCD equivalent of the binary number 10101011.101

Solution:

Step 1	The given Binary No.	$(10101011.101)_2$
Step 2	Convert it to decimal equivalent	$(171.625)_{10}$
Step 3	BCD equivalent of the decimal No.	$(0001\ 0111\ 0001.\ 0110\ 0010\ 0101)_2$

1.9.3 Higher-Density BCD Encoding

The number of bits necessary to represent a particular decimal number in normal BCD encoding is always more than the amount needed to express the same number in straight binary encoding. A three-digit decimal number, for instance, needs 12 bits to be represented in traditional BCD format. However, because 210 > 103, just 10 bits would be required to encode these three decimal digits individually. The Chen-Ho encoding and the tightly packed decimal are two examples of such encoding systems. The latter has the benefit of encoding two digits in the ideal seven bits and one digit in four bits, similar to ordinary BCD, in subsets of the encoding.

1.9.4 Packed and Unpacked BCD Numbers

In the case of unpacked BCD numbers, the machine stores each four-bit BCD group that corresponds to a decimal digit in a separate register. If the registers are eight bits or wider in this scenario, the register space is wasted.

Two BCD digits are kept in a single eight-bit register for packed BCD numerals. In order to combine two BCD digits into one eight-bit register, the top register number must be moved four times to the left, and the upper and lower register numbers must then be added. The storage of the decimal digits '5' and '7' serves as an example of the process.

- Decimal digit 5 is initially stored in the eight-bit register as: 0000 0101.
- Decimal digit 7 is initially stored in the eight-bit register as: 0000 0111.
- After shifting to the left 4 times, the digit 5 register reads: 0101 0000.
- The addition of the contents of the digit 5 and digit 7 registers now reads: 0101 0111.

Example 1.39: How many bits would be required to encode decimal numbers 0 to 9999 in straight binary and BCD

Codes? What would be the BCD equivalent of decimal 27 in 16-bit representation?

Solution:

Step 1	Total number of decimals to be represented (0-9999)	$10000 = 10^4 = 2^{13.29}$
Step 2	the number of bits required for straight binary encoding	14
Step 3	The number of bits required for BCD encoding	16
Step 4	BCD equivalent of 27 in 16-bit representation	0000000000100111

1.10 Excess-3 Codes

Another significant BCD code is the excess-3 code. It is particularly important for mathematical operations since it solves the problem of adding two decimal digits whose sum is greater than nine when using the 8421 BCD code. The excess-3 code greatly simplifies arithmetic operations and has no such restriction. Table 1.12 lists the excess-3 code for the decimal numbers 0-9.

Table 1.12: Excess-3 code for the decimal numbers 0-9

Decimal	**Binary**	**Excess-3**	Decimal	**Binary**	**Excess-3**
0	0000	0011	5	0101	1000
1	0001	0100	6	0110	1001
2	0010	0101	7	0111	1010
3	0011	0110	8	1000	1011
4	0100	0111	9	1001	1100

Example 1.40: Find the excess-3 code for the decimal number

(a) $(597)_{10}$ (b) $(23.57)_{10}$

Solution:

(a)

Step 1	Given Decimal Number	$(597)_{10}$
Step 2	Its corresponding Four-bit binary equivalent	0101 1001 0111
Step 3	Add 3 (0011) to individual group	0101 1001 0111 + 0011 0011 0011 1000 1100 1010
Step 4	Excess-3 code for $(597)_{10}$ is therefore given by	1000 1100 1010

(b)

Step 1	Given Decimal Number	$(23.57)_{10}$
Step 2	Its corresponding Four-bit binary equivalent	0010 0011 . 0101 0111
Step 3	Add 3 (0011) to individual group	0010 0011 . 0101 0111 + 0011 0011 0011 0011 0101 0110 1000 1010
Step 4	Excess-3 code for $(23.57)_{10}$ is therefore given by	0101 0110 1000 1010

Example 1.41: Find the decimal number equivalent for the given excess-3 code 110010100011.01110101

Solution:

Step 1	Given Excess-3 Number	110010100011.01110101
Step 2	Its corresponding Four-bit group binary	1100 1010 0011 . 0111 0101
Step 3	Subtract3 (0011) to individual group	1100 1010 0011 . 0111 0101 - 0011 0011 0011 . 0011 0011 1001 0111 0000 . 0100 0010
Step 4	Decimal for (1001 0111 0000.0100 0010) excess-3 value is therefore given by	$(970.42)_{10}$

The complement of the excess-3 code of a given decimal number provides the excess-3 code for the decimal number's complement, which is a crucial property that makes this code appealing for performing arithmetic operations. The excess-3 code can be used successfully for both addition and subtraction of decimal integers since adding the 9's complement of a decimal value B to a decimal number of A results in A - B.

1.10.1 BCD Addition and Subtraction in Excess-3 Code:

In the section below, we'll demonstrate how to add and subtract BCD values using the excess-3 code.

Addition:

The excess-3 code can be very effectively used to perform the addition of BCD numbers. The steps to be followed for excess-3 addition of BCD numbers are as follows:

1) The given BCD numbers are written in excess-3 form by adding '0011' to each of the four-bit groups,
2) The two numbers are then added using the basic laws of binary addition.
3) Add '0011' to all those four-bit groups that produce a carry, and subtract '0011' from all those four-bit groups that do not produce a carry during addition.
4) The result thus obtained is in excess-3 form.

Subtraction:

Subtraction of BCD numbers using the excess-3 code is similar to the addition process discussed above, the steps to be followed for excess-3 subtraction of BCD numbers are as follows:

1) Express both minuend and subtrahend in excess-3 code.
2) Perform subtraction following the basic laws of binary subtraction.
3) Subtract '0011' from each invalid BCD four-bit group in the answer.
4) Subtract '0011' from each BCD four-bit group in the answer if the subtraction operation of the relevant four-bit groups required a borrow from the next higher adjacent four-bit group.
5) Add '0011' to the remaining four-bit groups, if any, in the result.
6) This gives the result in excess-3 code.

The process of addition and subtraction can be best illustrated with the help of following examples.

Example 1.42: Add $(0011\ 0101\ 0110)_{10}$ and $(0101\ 0111\ 1001\ )_{10}$ using the excess-3 addition method and verify the result using equivalent decimal addition.

Solution

The excess-3 equivalents of 0011 0101 0110 and 0101 0111 1001 are 0110 1000 1001 and 1000 1010 1100 respectively. The addition of the two excess-3 numbers is given as follows:

```
0110    1000    1001
1000    1010    1100
--------------------
1111    0011    0101
```

The outcome of the aforementioned addition is 1100 0110 1000 after subtracting 0011 from the groups that did not yield a carry and adding 0011 to the ones that did. As a result, the excess-3 code for the actual result is 1100 0110 1000. The outcome is represented by the BCD code 1001 0011 0101, which is 935 in BCD. This is the correct answer as the addition of the given BCD numbers 0011 0101 0110 = $(356)_{10}$ and 0101 0111 1001 = $(579)_{10}$ yields $(935)_{10}$ only.

Example 1.43: Perform $(185)_{10}$-$(8)_{10}$using the excess-3 code.

Solution:

$(185)_{10}$ = $(0001\ 1000\ 0101)_{BCD}$, the excess-3 equivalent of $(0001\ 1000\ O101)_{BCD}$ = $(0100\ 1011\ 1000)_{XS\text{-}3}$,
$(8)_{10}$ = $(008)_{10}$ = $(0000\ 0000\ 1000)_{BCD}$, the excess-3 equivalent of $(0000\ 0000\ 1000)_{BCD}$ = $(0011\ 0011\ 1011)_{XS\text{-}3}$

Subtraction is performed as follows:

```
  0100   1011   1000
− 0011   0011   1011
--------------------
  0001   0111   1101
```

The least significant column of four-bit groups required a borrow during the subtraction operation, whereas the other two columns did not. Additionally, an erroneous BCD code group was created using the least significant column. This now represents the result of subtraction stated in excess-3 code after subtracting 0011 from the result of this column and adding 0011 to the results of the other two columns. The result in BCD code is therefore 0001 0111 0111. The decimal equivalent of 0001 0111 0111 is $(177)_{10}$, which is the correct result.

1.11 Gray Codes:

Frank Gray of Bell Labs created the Gray code, sometimes called **Cyclic Code, Reflected Binary Code (RBC), Reflected Binary (RB)**, or Grey code, which was patented in 1953. Two subsequent values in this unweighted binary code only differ by one bit. The maximum error that can occur in a system using the binary Gray code to encode data is far lower as a result of this feature than the worst-case error experienced in the case of pure binary encoding.

Table 1.13: Gray codes for different bit sequences

1-bit sequence			2-bit sequence			3-bit sequence			4-bit sequence		
Decimal	Binary	Gray	Decimal	Binary	Gray	Decimal	Binary	Gray	Decimal	Binary	Gray
0	0	0	0	00	00	0	000	000	0	0000	0000
1	1	1	1	01	01	1	001	001	1	0001	0001
			2	10	11	2	010	011	2	0010	0011
			3	11	10	3	011	010	3	0011	0010
						4	100	110	4	0100	0110
						5	101	111	5	0101	0111
						6	110	101	6	0110	0101
						7	111	100	7	0111	0100
									8	1000	1100
									9	1001	1101
									10	1010	1111
									11	1011	1110
									12	1100	1010
									13	1101	1011
									14	1110	1001
									15	1111	1000

The binary and Gray code counterparts of the decimal numerals 0 through 15 are shown in Table 1.13. The last entry rolls over to the first entry in the four-bit Gray code numbers mentioned in Table 2.3, meaning that the last and first entries also only differ by one bit. This is referred to as the Gray code's cyclic property. The term "Gray code" was first used to refer to a particular binary code for non-negative integers and was dubbed the binary-reflected Gray code or just the Gray code, even though there can be more than one Gray code for a given word length. (Refer Table 1.13).

There are various ways by which Gray codes with a given number of bits can be remembered. One such way is to remember that the least significant bit follows a repetitive pattern of '2' (11, 00, 11, ...), the next higher adjacent bit follows a pattern of '4' (1111, 0000, 1111, ...) and so on. By prefixing a "0" to the Gray code for n-I bits to obtain the first 2n-1 numbers, and then a "1" to the reflected Gray code for n-1 bits to gain the remaining 2n-1 numbers, we can also construct the n-bit Gray code recursively. The code that Gray reflects is simply the code written backwards. Table 2.3 provides an illustration of the reflect-and-prefix method's use in the creation of higher-bit Gray codes. Between the rows of bits that represent the Gray codes, there are columns of bits that show the intermediate process of writing the code, followed by the identical code written in reverse.

1.11.1 Binary-Gray code conversion

Let's examine the algorithm used to convert binary data to grayscale. View the translation of "11101" in binary to its Gray equivalent. (Refer Fig. 1.17).

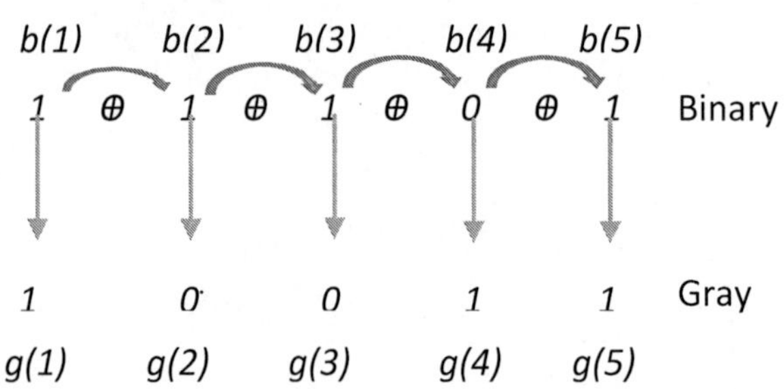

How operation executed

- *g(1) = b(1)*
- *g(2) = b(1) XOR b(2)*
- *g(3) = b(2) XOR b(3)*
- *g(4) = b(3) XOR b(4)*
- *g(5) = b(4) XOR b(5)*

Note: The XOR operation produces a 1 if the bits are different, and produces a 0 if the bits are equal.

Fig. 1.17: Binary to Gray code conversion

Gray's most significant bit (MSB) is directly derived from binary's MSB. The remaining grey bits are created by performing an XOR operation on the binary bit that came before (b(i-1)) and the current binary bit (b(i)). For the scenario depicted in the graphic above:

1.11.2 Gray to binary conversion

Let's now comprehend the algorithm used to convert grayscale data to binary. See the binary equivalent of the colour "10011" Gray converted. (Refer Fig. 1.18).

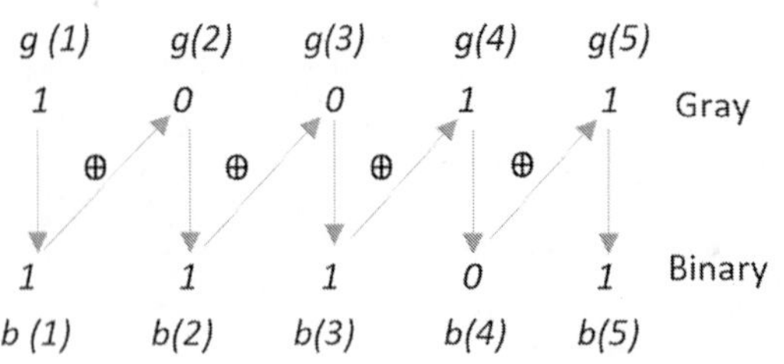

How operation executed

- *b(1) = g(1)*
- *b(2) = b(1) XOR g(2)*
- *b(3) = b(2) XOR g(3)*
- *b(4) = b(3) XOR g(4)*
- *b(5) = b(4) XOR g(5)*

Note: The XOR operation produces a 1 if the bits are different, and produces a 0 if the bits are equal.

Fig.1.18: Gray to Binary Conversion

In binary, the most significant bit (MSB) is directly derived from the MSB in grey. The remaining binary bits are created by performing an XOR operation on the current Gray bit (g(i)) and the preceding binary bit (b(i-1)). For the scenario depicted in the graphic above:

Example 1.44: Find the gray code for the given binary code below

(a) $(1101)_2$ (b) $(11010010)_2$ (c) $(1001011)_2$ (d) $(10101010)_2$

Solution:

(a) $(1011)_G$ (b) $(10111011)_G$ (c) $(1101110)_G$ (d) $(11111111)_G$

Example 1.45: Find the Binary equivalent for the Gray code values given below

(b) $(1001)_G$ (b) $(10110101)_G$ (c) $(10101010)_G$ (d) $(10000010)_G$

Solution:

(b) $(1110)_2$ (b) $(11011001)_2$ (c) $(11001100)_2$ (d) $(11111100)_2$

1.11.3 Applications of Gray Codes

1. Because it reduces the likelihood of errors, the Gray code is employed in the transmission of digital signals.
2. In angle-measuring equipment, the Gray code is chosen over the simple binary code. The likelihood of an angle being misread is virtually eliminated when using the Gray code, compared to when using straight binary to express the angle. The Gray code's cyclic characteristic is advantageous in this application.
3. The axes of Karnaugh maps, a graphical method for minimising Boolean equations, are labelled using the Gray code.
4. Gray codes are used to address programme memory in computers in a way that uses the least amount of electricity. This is because improvements in the programme counter result in fewer address lines changing state.
5. Given that code mutations typically result in incremental changes, grey codes are also highly helpful in genetic algorithms. Occasionally, though, a one-bit change might result in a significant leap, so resulting to new properties.

Example 46: Given the sequence of three-bit Gray code as (000, 001, 011, 010, 110, 111, 101, 100), write the next three numbers in the four-bit Gray code sequence after 0101.

Solution:

The first eight of the 16 Gray code numbers of the four-bit Gray code can be written by appending '0' to the eight three-bit Gray code numbers. The remaining eight can be determined by appending '1' to the eight three-bit numbers written in reverse order. Following this procedure, we can write the next three numbers after 0101 as 0100, 1100 and 1101

Extra Note: The first three in the list, 8421, 2421, and 5211, are weighted binary codes, whereas the latter two are not. You can see that the 2421 and 5211 codes are what are known as reflective codes because the code for the decimal 9 is the complement of the code for the decimal 0, the code for the decimal 8 is the complement of the code for the decimal 1, the code for the decimal 7 is the complement of the code for the decimal 2, the code for the decimal 6 is the complement of the code for the decimal 3, and the code for the decimal 5 is the complement of the code for the decimal 4. Reflective code also exists for excess-3. A reflective code is not the 8421 code.

1.12 The reflective Code:

A code is said to be reflective when code for 9 is complement for the code for 0, and so is for 8 and 1 codes, 7 and 2, 6 and 3, 5 and 4. Codes 2421 (2+4+2+1=9), 5211(5+2+1+1=9), 84-2-1 (8+4+(-2) +(-1) =9) and excess-3 are reflective, whereas the 8421 code is not.

Only discrete forms of data can be processed by digital computers. Numerous physical systems produce data continuously. Before being applied to a digital system, these data must be transformed into digital or discrete form. An analog-to-digital converter is used to transform continuous or analogue information into digital form. When converting analogue data to digital data, it can be useful to apply the reflected code provided in Table 1.14.

Table 1.14: Comparison of reflected code with its equivalent decimal and excess-3

Decimal	84-2-1	2421	5211	Excess-3
0	0000	0000	0000	0011
1	0111	0001	0001	0100
2	0110	0010	0011	0101
3	0101	0011	0101	0110
4	0100	0100	0111	0111
5	1011	1011	1000	1000
6	1010	1100	1010	1001
7	1001	1101	1100	1010
8	1000	1110	1110	1011
9	1111	1111	1111	1100

Since a number in the reflected code only changes by one bit as it moves from one number to the next, it has an advantage over pure binary numbers. When the analogue data are represented by a continuous change in shaft position, the reflected code is often applied. Segments of the shaft have been divided up, and each segment has been given a number. When detection is detected in the line separating any two segments, there is less uncertainty if neighboring segments are made to match to adjacent reflected-code numbers. There is other alternative such codes in addition to the reflected code displayed in Table 1 14.

*Note: The **Gray Code**, often called **reflected binary code**, is a series of binary number systems. The initial N/2 values are compared to the last N/2 values in reverse order, which is why this code is known as reflected binary code. Two successive values in this code are separated by one bit of binary digits.*

1.13 Alphanumeric Codes

Binary codes are used to represent alphanumeric data and are also known as character codes. The codes encode alphanumeric data in a way that a computer can read, understand, and process. This data includes letters of the alphabet, integers, mathematical symbols, and punctuation. These codes allow us to connect input-output devices to the computer, such as keyboards, printers, visual display units, etc. The 12-bit Hollerith code was one of the more well-known alphanumeric codes used in the early development of computers, when punched cards were the primary means of data input and output. Alphanumeric data on punched cards was then encoded using the Hollerith code.

The punched card medium, however, has completely disappeared from the scene, making the code obsolete. The ASCII and EBCDIC codes are two frequently used alphanumeric codes. Unlike the latter, which is mostly utilised with bigger systems, the former is common with microcomputers and is present on almost all personal computers and workstations.

The amount of characters that can be encoded by conventional character encodings like ASCII, EBCDIC, and their variants is limited. Since no single encoding actually has enough characters to support all of the languages spoken in the European Union, multilingual computer processing is not possible using these encodings. The most comprehensive character encoding technique that enables text of various formats and languages to be encoded for use by computers is called Unicode, and it was created jointly by the Unicode Consortium and the International Standards Organization (ISO). The pages that follow provide descriptions of various codes.

1.13.1 ASCII code

The ASCII (American Standard Code for Information Interchange), pronounced 'ask-ee', is strictly a seven-bit code based on the English alphabet. ASCII codes are used to represent alphanumeric data in computers, communications equipment and other related devices. The code was first published as a standard in 1967. It was subsequently updated and published as ANSI X3.4-1968, then as ANSI X3.4-1977 and finally as ANSI X3.4-1986. Since it is a seven-bit code, it can at the most represent 128 characters. It currently defines 95 printable characters including 26 upper-case letters (A to Z), 26 lower-case letters (a to z), 10 numerals (0 to 9) and 33 special characters including mathematical symbols, punctuation marks and space character.

It also specifies codes for 33 obsolete, nonprinting control characters that have an impact on how text is processed. All other characters have been rendered obsolete, with the exception of "carriage return" and/or "line feed," by modern mark-up languages and communication protocols, the transition from text-based to graphical devices, and the abolition of teleprinters, punch cards, and paper cassettes. Ascii-8, often known as US ASCII-8 or ASCII-8, is an eight-bit variant of the ASCII code. A maximum of 256 characters can be represented using the eight-bit version.

Appendix-A, lists the ASCII codes for all 128 characters. When the ASCII code was introduced, many computers dealt with eight-bit groups (or bytes) as the smallest unit of information. The eighth bit was commonly used as a parity bit for error detection on communication lines and other device-specific functions, Machines that did not use the parity bit typically set the eighth bit to '0'.

As a result of the growing use of computers, numerous ASCII code variations have developed throughout time to make it easier for languages other than English that employ a Roman-based alphabet to be expressed. All ASCII printable characters in some of these variations correspond to their seven-bit ASCII code equivalents. One such example is the eight-bit standard ISO/IEC 8859, which was created as a real expansion of ASCII while maintaining the original character mapping. This made it possible to represent a larger number of languages. The most popular character encodings in use today remain to be ISO-8859-1, Windows-1252, and the original seven-bit ASCII, despite the standard's incompatibilities and restrictions.

1.13.2 EBCDIC code

Another frequently used alphanumeric code, the EBCDIC (Extended Binary Coded Decimal Interchange Code), is most popular with larger systems and is pronouced "eb-si-dik." IBM developed the code to expand the binary-coded decimal system at the time. All IBM mainframe computer peripherals and operating systems employ the EBCDIC coding system, and their operating systems offer ASCII and Unicode modes to enable conversion between other encodings.

It should be noted that EBCDIC does not offer any technical advantages over the ASCII code, ISO-8859, or Unicode. In the past, it was significant because punch cards allowed for relatively simpler data entry into larger machines. Since punch cards are no longer utilised on mainframes, the code is only used for backwards compatibility on modern mainframe machines,

Table 1.15: Sample codes for EBCDIC and its HEX equivalent

Char	EBCDIC	HEX	Char	EBCDIC	HEX	Char	EBCDIC	HEX
A	1100 0001	C1	P	1101 0111	D7	4	1111 0100	F4
B	1100 0010	C2	Q	1101 1000	D8	5	1111 0101	F5
C	1100 0011	C3	R	1101 1001	D9	6	1111 0110	F6
D	1100 0100	C4	S	1110 0010	E2	7	1111 0111	F7
E	1100 0101	C5	T	1110 0011	E3	8	1111 1000	F8
F	1100 0110	C6	U	1110 0100	E4	9	1111 1001	F9
G	1100 0111	C7	V	1110 0101	E5	blank	...	...
H	1100 1000	C8	W	1110 0110	E6	.	...	...
I	1100 1001	C9	X	1110 0111	E7	(	...	...
J	1101 0001	D1	Y	1110 1000	E8	+	...	...
K	1101 0010	D2	Z	1110 1001	E9	$	...	...
L	1101 0011	D3	0	1111 0000	F0	*	...	...
M	1101 0100	D4	1	1111 0001	F1	)	...	...
N	1101 0101	D5	2	1111 0010	F2	–	...	...
O	1101 0110	D6	3	1111 0011	F3	/		

Since it uses an eight-bit coding, it can store up to 256 characters. Characters in EBCDIC are listed in Table 1-15 in both binary and hexadecimal formats. In EBCDIC, a single byte is split into two groups of four bits called nibbles. The first four-bit group, known as the "zone," denotes the type of character, while the second group, known as the "digit," designates the particular character.

This code is used by each and every IBM computer and accessory. Because it uses an 8-bit coding, 256 characters can fit in it. To familiarise yourself with the EBCDIC code, see some of the characters below. (A more thorough version is provided in **Appendix-A**, though.

1.13.3 Unicode

As briefly discussed in the previous sections, there are not enough characters in encodings like ASCII, EBCDIC, and their variants to allow for the encoding of all forms, scripts, and languages of alphanumeric data. These encodings prevent computer processing in several languages as a result. These encodings also have compatibility issues. The same number may be used by two separate encodings for two distinct characters, or various numbers may be used for the same characters.

For example, code 4E (in hex) represents the upper-case letter 'N' in ASCII code and the plus sign '+' in the EBCDIC code, Unicode, developed jointly by the Unicode Consortium and the International Organization for Standardization (ISO), is the most complete character encoding scheme that allows text of all forms and languages to be encoded for use by computers. It not only enables the users to handle practically any language and script but also supports a comprehensive set of mathematical and technical symbols, greatly simplifying any scientific information exchange. The Unicode standard has been adopted by such industry leaders as HP, IBM, Microsoft, Apple, Oracle, Unisys, Sun, Sybase, SAP and many more.

Unicode and ISO-10646 Standards

Before we go on to highlight some of Unicode's key features, it should be noted that ISO-10646 is another standard that shares many of Unicode's goals and methods of operation. While ISO-10646 is a project of the International Organization for Standardization, Unicode is the creation of the Unicode Consortium, a group of multilingual software producers (originally primarily headquartered in the US). Despite the fact that each body releases its own standards independently, they have agreed to keep the Unicode and ISO-10646 code tables compatible and closely coordinate any additional expansions.

The ***code table*** *established by ISO-10646 and Unicode gives each character a unique number regardless of the platform, software, or language used. The table includes the necessary characters to represent almost all known languages and scripts. The list also contains the Japanese, Chinese, and Korean characters in addition to the Greek, Latin, Cyrillic, Arabic, Arabian, and Georgian scripts. Devanagari, Bengali, Gurmukhi, Gujarati, Oriya, Telugu, Tamil, Kannada, Thai, Tibetan, Ethiopic, Sinhala, Canadian Syllabics, Mongolian, Myanmar, and more scripts are also included in the list. Uncovered scripts will soon be included. A significant number of artistic, typographic, mathematical, and scientific symbols are also covered in the code table.*

In Unicode, various letters are represented by hexadecimal numbers prefixed by the letter 'U+'. For example, 'A' and 'e' in basic Latin are respectively represented by U+0041 and U+0065. The first 256 code numbers in Unicode are compatible with the seven-bit ASCII-code and its eight-bit variant ISO-8859-1. Unicode characters U+0000 to U+007F (128 characters) are identical to those in the ASCII code, and the Unicode characters in the range U+0000 to U+OOFF (256 characters) are identical to ISO-8859-1 ***(Refer Annexure-A)***.

Use of Combining Characters

Combining characters, which are accents or other diacritical markings added to the preceding character· rather than entire characters on their own, are given code numbers by Unicode. This enables any accent to be applied to any character. Even though Unicode supports character combining, it also gives distinct codes to precomposed characters—commonly used accented characters. This is carried out to provide compatibility with earlier encodings. As an example, the character *'ï'* can be represented as the precomposed character U+O0E4. It can also be represented in Unicode as U+0061 (Latin lower-case letter 'a') followed by U+00A8 (combining character '..').

Unicode and ISO-10646 Comparison

Although the code tables for Unicode and ISO-10646 are the same, Unicode has many more features than ISO-10646. While the ISO-10646 standard only offers a complete character set, the Unicode standard offers a number of additional features that are related to it, including character properties, text normalisation algorithms, and bidirectional text handling to ensure that mixed texts with both right-to-left and left-to-right scripts are displayed correctly.

1.14 Seven-segment Display Code

Seven-segment displays (Refer to Fig. 1.19) are widely used and can be seen practically anywhere, including on gas pumps, digital clocks, pocket calculators, and electronic test equipment. Our display requirement is always satisfied by a single seven-segment display or a stack of similar displays. There are seven-segment displays with LED and LCD technology. Additionally, there are common cathode-type LED displays where the individual diodes are interconnected as shown in Fig. 1.19 (c), and common anode-type LED displays where the arrangement of various diodes, designated a, b, c, d, e, f, and g, is as shown in Fig. 1.19 (b). Each display unit typically has a Dot Point (DP).

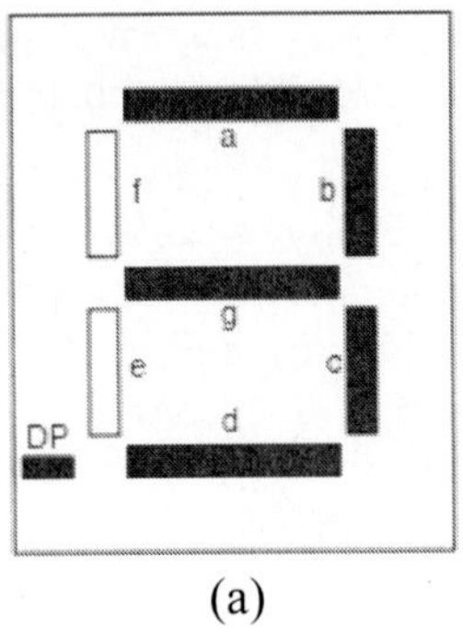

(a)

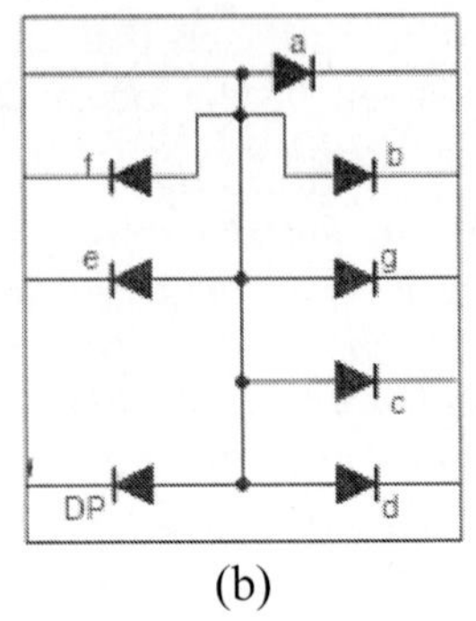

(b)

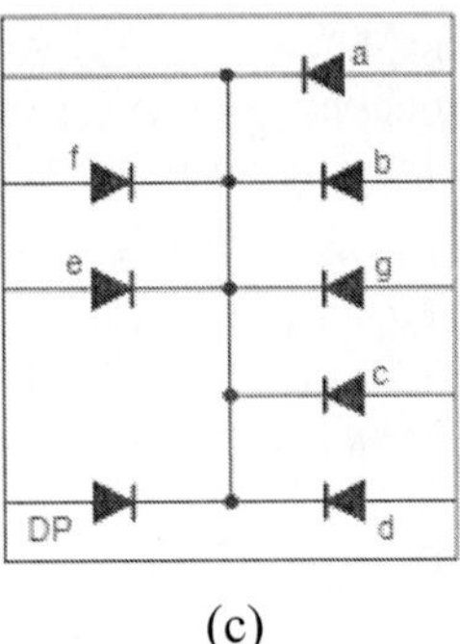

(c)

Fig.1.19: Seven-segment displays

Table 1.16: Decimal values and its equivalent Seven Segment display Values

Decimal/ Alphabet	Common cathode type ('1' mean ON)								Common Anode type ('0' mean ON)							
	a	b	c	d	e	f	g	DP	a	b	c	d	e	f	g	DP
0	1	1	1	1	1	1	0		0	0	0	0	0	0	1	
1	0	1	1	0	0	0	0		1	0	0	1	1	1	1	
2	1	1	0	1	1	0	1		0	0	1	0	0	1	0	
3	1	1	1	1	0	0	1		0	0	0	0	1	1	0	
4	0	1	1	0	0	1	1		1	0	0	1	1	0	0	
5	1	0	1	1	0	1	1		0	1	0	0	1	0	0	
6	0	0	1	1	1	1	1		1	1	0	0	0	0	0	
7	1	1	1	0	0	0	0		0	0	0	1	1	1	1	
8	1	1	1	1	1	1	1		0	0	0	0	0	0	0	
9	1	1	1	0	0	1	1		0	0	0	1	1	0	0	
a	1	1	1	1	1	0	1		0	0	0	0	0	1	0	
b	0	0	1	1	1	1	1		1	1	0	0	0	0	0	
c	0	0	0	1	1	0	1		1	1	1	0	0	1	0	
d	0	1	1	1	1	0	1		1	0	0	0	0	1	0	
e	1	1	0	1	1	1	1		0	0	1	0	0	0	0	
f	1	0	0	0	1	1	1		0	1	1	1	0	0	0	

The display pattern in figure "2.1" might have the DP either to the left (as shown) or to the right. Table 2.8 provides the binary code for displaying various numeric and alphabetic characters for both the common cathode and the common anode type displays. This type of display may be used to display numerals from 0 to 9 and letters from A to F. In a common cathode-type display, a "I" illuminates a segment, while in a common anode-type display, a "0" illuminates a segment (Refer Table 1 16).

1.15 Error Detection and Correction Codes

Error can be treated as a condition when the information received at the output doesn't match with the given input information transmitted i.e., there is a variation in received output to actual output. It happens due to the reason that during transmission, digital signals encountered with various type of noise (Source could be internal or external) that will further introduce errors, while transmitting binary data (bits) from one system to other (Refer Fig. 1.20). In general, the data transfer in digital systems will be in the form of 'Bits' so the occurrence of error means that a bit '0' may change to '1' or a bit '1' may change to '0'. Hence, due to this the data gets corrupted during transmission (from source to receiver). As already said, this occurs due to either by external noise component or some other physical imperfections. Finally, the problem is that, the received output data is not as per the desired output data. This mismatched data is called "Error". The data errors will lead to change overall impression of the output that causes loss of important / secured data. Even one bit of change in data may affect the whole system's performance.

The problem of error detection and repair is of enormous practical importance in any application involving digital systems, whether it be a digital computer or a digital communication set-up. During the bit stream's passage from the transmitter to the receiver, errors inevitably slip in due to noise or other impediments. Digital systems are sensitive to faults and have a tendency to malfunction if the bit error rate is higher than a specific threshold level, therefore any such error can be devastating if it is not identified and fixed.

Check bits (also known as redundant bits) are extra bits that are added to the information-carrying bit stream in order to detect errors and repair them. The resulting bit sequence will have a special property that aids in mistake localization and detection. Due to the fact that they do not convey any information, the extra bits that were added to the bit stream (original data) are also known as superfluous bits. However, including redundant bits makes it possible to achieve error-free or trustworthy information transmission from source to destination. On the other hand, as a tradeoff, it makes inefficient utilization of bandwidth. In this section, some common error detection and correction codes will be discussed with examples.

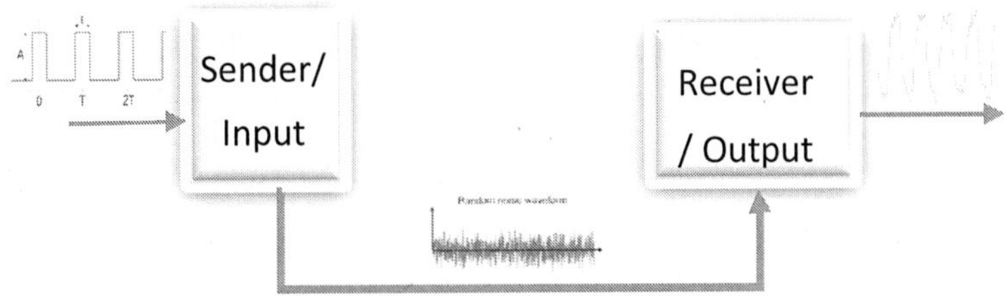

Fig.1.20: Error introduced while transmission from sender to receiver

1.15.1 Types of Errors

In a given data sequence, if at any point "1" is converted to "0" or "0" is changed to "1," as was already mentioned in the preceding section, it is referred to as a "error." Consequently, there are generally three sorts of errors that might happen during data transmission from the transmitter to the receiver (Refer Fig. 1.21). They are enumerated as

- Single bit errors (Error in only one bit)
- Multiple bit errors (Error in two or more than two bits)
- Burst errors (Errors occurred in Bunch/sequence of data)

Single Bit Data Errors

In the given data sequence when there is only one bit has changed its value from '0' to '1' or '1' to '0', is called as "Single bit error". However, possibility of occurrence for single bit error is very rare in a digital system using serial communication. Moreover, there is a fairly good chance that this type of error occurs in parallel communication system. As there is a high probability that single line to be noisy while transmitting data parallelly, as data is going to be transferred bit wise in single line, as shown in fig. below (Refer Fig. 1.21).

Multiple Bit Data Errors

In the given data sequence when two or more than two bits get changed simultaneously, from 1 to 0 or 0 to 1, the system is said to be encountered with 'Multiple bit error'. The phenomenon is very severed, as it changed the whole meaning of data drastically. Serial and parallel both type of data transmission system suffered from this phenomenon.

Burst Errors

In the given data sequence when two or more than two bits get changed simultaneously as like multiple errors but in case of 'Burst Error' they changed in a sequence (Bunch), i.e., from starting point (First error bit) to end point (Last error bit) the bits get changed from 1 to 0 or 0 to 1 as shown in Fig. 1.21. The set of bits that get changed has caused error in data sequence, and this error is called as "Burst error". It is calculated in from the first bit change to last bit change. There is a high probability of burst errors to be occurred while dealing with serial communication and such errors are very difficult to resolve. The error has been identified form fourth bit to 6th bit. As the bits from 4th and 6th position has got changed from 1 to 0 or 0 to 1. Hence, set of bits from 4th to 6th are called "Burst error.

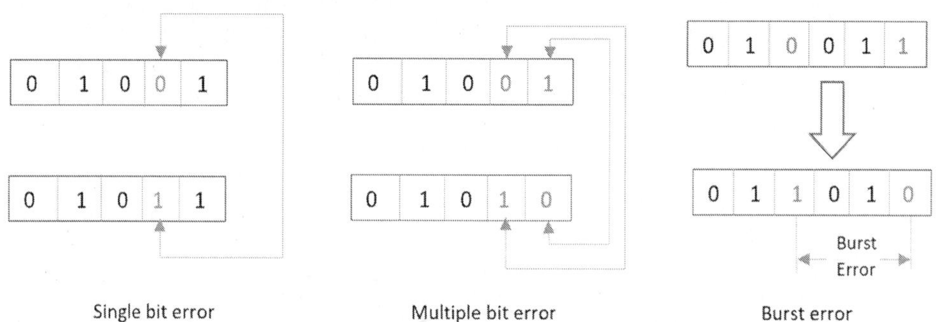

Fig. 1.21: Types of errors (Based on number of bit change) that may occur in data transmission

1.15.2 Error-Detecting codes

Even and odd Parity check Codes:

Parity as the name suggest is used to identify the number of I's present into a data sequence. The data is meant to have an even parity if there are an even number of 1s. However, the data is meant to have an odd parity if there are an odd number of 1s present. To find one bit errors, parity codes are employed. The data sequence will be extended by one more bit. Depending on the type of parity, an additional bit, known as the parity bit, is utilised to be added to the data to make the number of 1s either even or odd (Refer Fig. 22). There are two types of parity checking which are

- Even Parity – Here the total number of 1's in the message is made even.
- Odd Parity – Here the total number of 1's in the message is made odd.

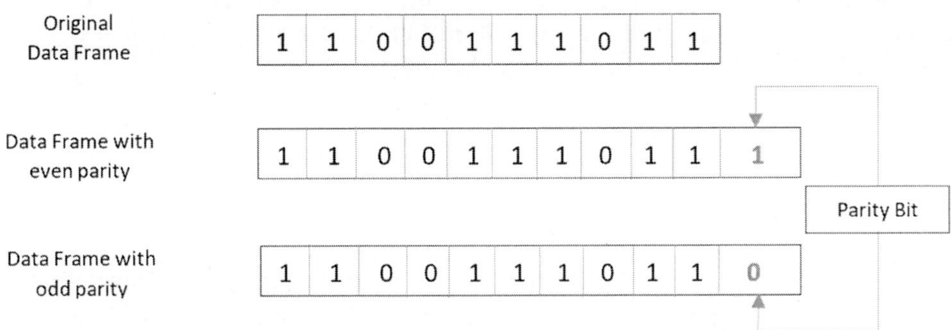

Fig. 1.22: Even and odd parity

Error Detection by Parity Check

Table 1.17: Four bit binary and its equivalent ODD/EVEN parity codes

4 Bit Message	Message with odd parity	Message with even parity
0000	0000**1**	0000**0**
0001	0001**0**	0001**1**
0010	0010**0**	0010**1**
0011	0011**1**	0011**0**
0100	0100**0**	0100**1**
0101	0101**1**	0101**0**
0110	0110**1**	0110**0**
0111	0111**0**	0111**1**
1000	1000**0**	1000**1**
1001	1001**1**	1001**0**
1010	1010**1**	1010**0**
1011	1011**0**	1011**1**
1100	1100**1**	1100**0**
1101	1101**0**	1101**1**
1110	1110**0**	1110**1**
1111	1111**1**	1111**0**

Sender's End – The sender counts the number of ones in a frame while it is being created and then adds the parity bit, the value of which is calculated as follows -

- In the case of even parity, the parity bit value is 0 if the number of 1s is even. The parity bit value is 1 if a sequence of ones is odd.
- When there is an odd number of ones, the parity bit value is 0. The parity bit value is 1 if the number of ones is even.

Receiver's End – The receiver counts the number of 1s in each frame it receives. If the count of 1s is even during an even parity check, the frame is approved; otherwise, it is discarded. If the count of 1s is odd during an odd parity check, the frame is accepted; otherwise, it is refused. Some of the 4-bit messages to be transmitted are shown in the following table along with their parity bits. The bits in red color (Bold) are the parity bits (Refer Table 1.17).

If the received data differs from the sent data on the receiving end, an error has occurred. There is a mistake if the supplied date has an even parity code and the received data has an odd parity code (Refer Fig. 1.23).

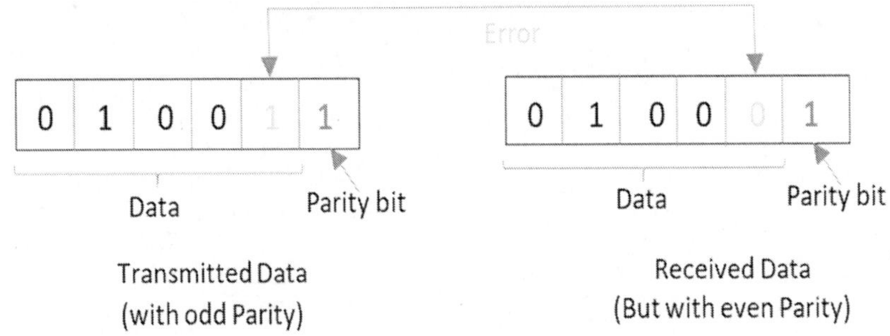

Fig.1.23: Parity codes used for error detection

Therefore, neither even nor odd parity codes are used to fix the sent data; rather, they are solely used to detect errors. Even parity is frequently employed and has all but become standard.

Example 1.46: check the following received data for even parity scheme

(a)˙ 10101010 (b) 11110110 (c)10111001

Solution:

a) No. of 1's in the word is even is 4 so there is no error
b) No. of 1's in the word is even is 6 so there is no error
c) No. of 1's in the word is odd is 5 so there is error

Example 1.47: check the following received data for odd parity scheme

a) 10110111 b) 10011010 c)11101010

Solution:

a) No. of 1's in the word is 6 i.e., even so word has error
b) No. of 1's in the word is 4 i.e., even so word has error
c) No. of 1's in the word is 5 i.e., odd so there is no error

Note: The parity check fails if two bits get changed, as after that parity remains unchanged but the data received has the errors. Also, through parity check one could not correct the error. Hence, parity check codes can only be used for only one-bit error detection.

Block Parity

A block of data with rows and columns is considered to exist when multiple binary words are sent and received at once. Let's take a look at a 4 x 8 block, which is formed of four 8-bit words that need to be communicated, as an example. For both rows and columns, parity bits are assigned.

Even parity is given to each eight-bit word and for each column in Fig. 1.24. The parity bit in this instance is referred to as block parity. From the transmitter side, the data block is sent. The first row of the received block of data(b) has no errors since the even parity is preserved. Odd parity replaces even parity in the second row. Consequently, the second row contains an error. The third and fourth rows are kept in parity, therefore there is no error.

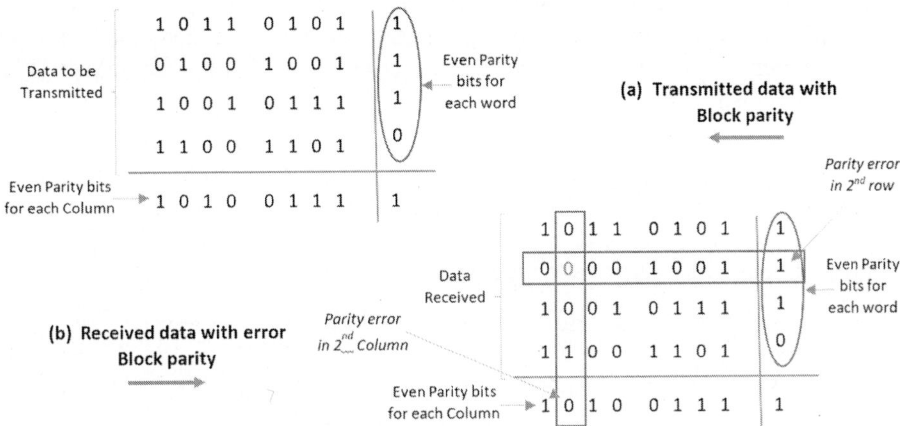

Fig. 1.24: Block Parity Codes

Let's examine the column-wise parity in order to identify the incorrect bit in the second row. If you do this, you'll see that the even parity has been switched to the odd parity in the second column. Therefore, the second column contains an error. There is no change in the even parity and hence no inaccuracy in any other columns in any other columns. The bit that lies at the point where the second row and second column intersect is an incorrect bit. The incorrect bit is highlighted in red in the image above. We can change bit 0 to bit 1 because the fault is only one bit. As a result, block parity allows for error detection and rectification using parity codes.

Parity error in 1st and 2nd Column

Even Parity bits for each word

Data Received

1	0	1	1	0	1	0	1	1
0	1	0	0	1	0	0	1	1
1	0	0	1	0	1	1	1	1
0	0	0	0	1	1	0	1	0

Even Parity bits for each Column

1	0	1	0	0	1	1	1	1

Fig. 1.25: Handling of errors using Block Parity codes

It has been seen from Fig. 1.25 that there is no parity bit mistake for each row. However, the even parity has been switched to odd parity for the first and second columns, which is incorrect. The first and second columns of the result show the error. Only the error can be detected in this scenario. There is no information exposing the row number where the mistakes occurred, hence it is impossible to determine which row has the error because there is no parity change for the row.

Checksums:

Simple parity cannot find two mistakes within a single word. Use a form of 2-dimensional parity to get around this. Each word that is communicated is added to the total of words that have already been transmitted, with the total being stored at the transmitter end. The check sum is the total that is added up at the end of transmission. sent to the recipient up until that point. The transmitted sum and the receiver's sum can be compared. If the two sums match, there were no errors at the receiving end. In teleprocessing systems, the receiving location can request a resend of all the data if there is a problem.

For checksum as shown in Fig. 1.26, The checksum generator at the sender side divides the data into equal subunits of n bits in length. Typically, this bit has a 16-bit length. Then, using the one's complement method, these subunits are combined together. These components make up the sum. After that, the resulting bit is completed. The original data unit has this checksum applied to the end of it before being sent to the recipient.

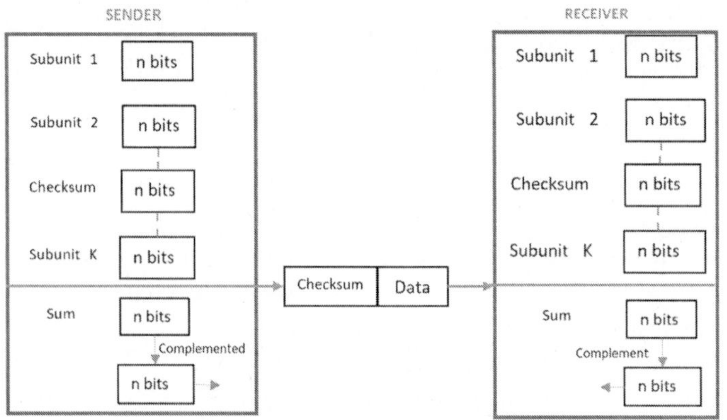

Fig.1.26: Error detection using Checksum

For example, If the data unit to be transmitted is 10101001 00111001, the following procedure is used at Sender site and Receiver site.

Solution:

<table>
<tr><th colspan="2">Sender End</th><th colspan="2" rowspan="3">Data Transmitted</th><th colspan="2">Receiver End</th></tr>
<tr><td>10101001</td><td>Subunit 1</td><td>10101001</td><td>Subunit 1</td></tr>
<tr><td>00111001</td><td>Subunit 2</td><td>00111001</td><td>Subunit 2</td></tr>
<tr><td>11100010</td><td>Sum (Using 1's Complement)</td><td>Data</td><td>Checksum</td><td>00011101</td><td>Checksum</td></tr>
<tr><td rowspan="2">00011101</td><td rowspan="2">Checksum (Complement of sum)</td><td rowspan="2">1010001
00111001</td><td rowspan="2">00011101</td><td>11111111</td><td>Sum</td></tr>
<tr><td>00000000</td><td>Sum's Complement</td></tr>
</table>

After receiving the data and the checksum, the receiver sends it to the checksum checker. This data unit is divided into several equal-length subunits by the checksum checker, who then adds all of the subunits. One of the subunits in these subunits is the checksum. After that, the resulting bit is completed. The data is error-free if the complemented result is zero. If the outcome is non-zero, the data has an error and is rejected by the receiver.

The sum's complement is *00000000.* Hence, there is no error into the data received. Utilizing a checksum has the main benefit of detecting both faults involving an odd number of bits and errors involving an even number of bits. However, if one or more bits of one subunit are damaged and their corresponding bit or bits of another subunit, i.e., subunit two, are also damaged, the problem error cannot be recognised. Since the total of those columns stays constant.

***But what happened if*,** the data transmitted along with checksum is 10101001 00111001 00011101. However, the data received at destination is **0**0101001 **1**0111001 00011101.

Sender End		**Data Transmitted**		**Receiver End**	
10101001	Subunit 1			00101001	Subunit 1
00111001	Subunit 2			10111001	Subunit 2
11100010	Sum (Using 1's Complement)	Data	Checksum	**00011101**	Checksum
00011101	Checksum (Complement of sum)	1010001 00111001	**00011101**	11111111	Sum
				00000000	Sum's Complement

Despite data corruption, the problem is not noticed. When a message is conveyed, it could be distorted or scrambled by noise.

1.15.3 Error-Correcting codes

If a code word can always be inferred from an incorrect word, it is said to be an error-correcting code. We can transmit some data along with error-detecting code to distinguish the original message from the corrupt message we received. Error-correcting codes are the name for this kind of code. While error-correcting codes use the same methodology as error-detecting codes, they also identify the precise location of the faulty bit.

In error-correcting codes, parity check has a straightforward method for spotting mistakes and a complex process for pinpointing the location of the damaged bit. To recover the original message, the corrupt bit's value is reversed (from 0 to 1 or 1 to 0). "Error Correction Codes" are the codes that are utilised for both error detection and correction. The error correction techniques are of two types. They are,

- Single bit error correction
- Burst error correction

"Single bit error correction" is the procedure or technique used to fix single bit errors. Burst error correction is a technique for identifying and fixing burst defects in a data series. A code must have a minimum distance of three to qualify as a single bit error correcting code. The fewest number of bits by which any two code words must differ is known as the minimum distance for that code. A code with a minimum distance of three can identify (but not correct) two-bit faults in addition to single-bit errors. It is essential to be able to identify and locate incorrect digits in order to do error repair. if the error's exact location has been identified. The message can then be fixed by complimenting the incorrect digit.

There are a variety of error-correcting codes available based on the mathematical concepts that are used with them. However, historically, these codes have been classified into **Linear block codes** and **Convolution codes**.

1. Linear Block Codes: In block codes, in fixed-size blocks of bits, the message is contained. In this, the redundant bits are added for correcting and detecting errors.
    a. Cyclic Redundancy Check
    b. Longitudinal Redundancy Check

c. Hamming Code

2. Convolution Codes: The message consists of data streams of random length, and parity symbols are generated by the sliding application of the Boolean function to the data stream.

Cyclic Redundancy Check

Every cyclic shift of a codeword produces a new code word in a cyclic code, which is a linear (n, k) block code. Here, k denotes the message's initial length (the number of information bits), and n denotes the message's final length (the number of check bits) The implementation of these codes is simple when shift registers include feedback connections. They are frequently employed for error detection because of this. At low redundancy levels, cyclic redundancy check (CRC) algorithms offer a respectably high level of security.

CRC Code Generation

The cyclic code for a given data word is generated as follows.

- The data word is first appended by a number of 0's equal to the number of check bits to be added.
- This new data bit sequence is then divided by a special binary word whose length equals n+ 1 (n being the number of check bits to be added).
- The remainder obtained as a result of modulo-2 division is then added to the dividend bit sequence to get the cyclic code.
- The code word so generated is completely divisible by the divisor used in the generation of the code.
- Thus, when the received code word is again divided by the same divisor, an error-free reception should lead to an all '0' remainder.
- A nonzero remainder is indicative of the presence of errors.

Example 1.48: A bit stream 1101011011 is transmitted using the standard CRC method. The generator polynomial is x^4+x+1. What is the actual bit string transmitted?

Solution-

- The generator polynomial $G(x) = x^4 + x + 1$ is encoded as 10011.
- Clearly, the generator polynomial consists of 5 bits.
- So, a string of 4 zeroes is appended to the bit stream to be transmitted.
- The resulting bit stream is 1101011011**0000**.

Now,

- The code word to be transmitted is obtained by replacing the last 4 zeroes of 1101011011**0000** with the CRC.
- Thus, the code word transmitted to the receiver = 1101011011**1110**.

Notes:

- For a burst error of length $n - 1$, the probability of error detecting is 100 %.
- A burst error of length equal to $n + 1$, the prob. of error detecting reduces to $1 - (1/2)^{n-1}$.
- A burst error of length greater than $n - 1$, the probability of error detecting is $1 - (1/2)^n$.

Longitudinal Redundancy Check

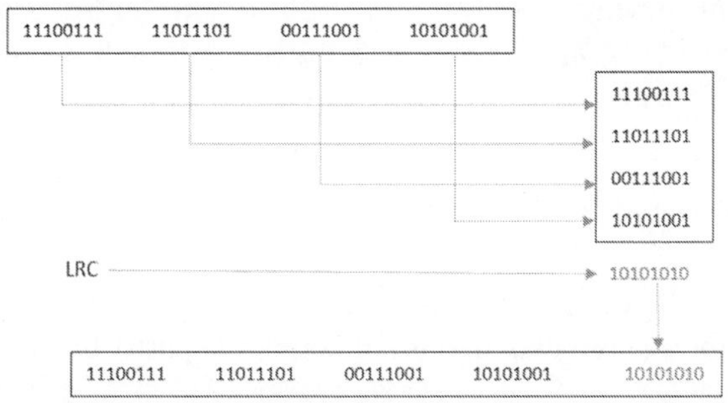

Fig. 1.27: Original data message including LRC

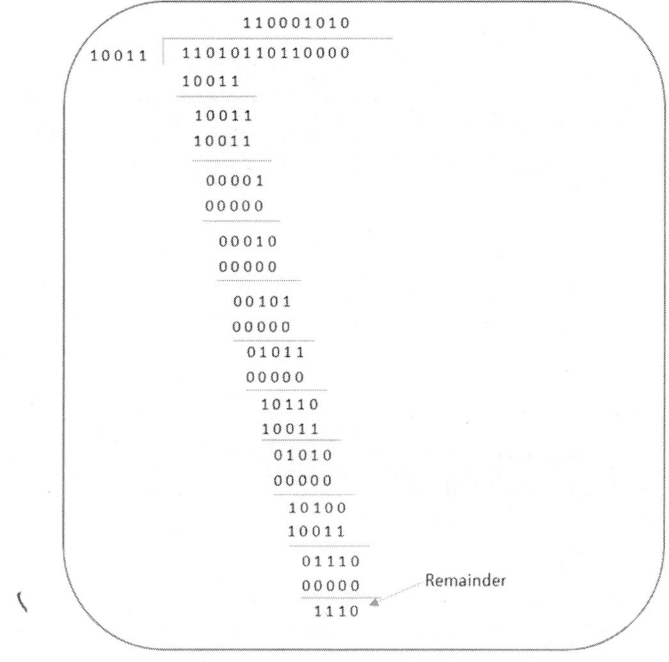

From here, CRC = 1110.

In the longitudinal redundancy approach, a BLOCK of bits is organised into rows and columns in a table format, and the parity bit for each column is calculated independently (Refer Fig. 1.27). Along with our initial data bits, a set of these parity bits is also transmitted.

As we calculate the parity of each column separately, the longitudinal redundancy check is a bit-by-bit parity computation. This technique can quickly identify burst and single-bit mistakes, but it cannot identify 2-bit errors that occur in the same vertical slice (Refer Fig. 1.27).

1.15.4 Hamming Code: A collection of error-correction codes called Hamming code can be used to find and fix mistakes that may happen as data is sent from the sender to the receiver or is stored there. **R.W. Hamming** created this technique for error correction. To make sure that no bits were lost during the data transfer, redundant bits are extra binary bits that are created and added to the information-carrying bits. The number of redundant bits can be calculated using the following formula:

$$2^r \geq m + r + 1$$

$$where,\ r = Redundant\ bits\ \&\ m = Message\ bits$$

Suppose the number of data bits is 7, then the number of redundant bits can be calculated using $2^4 \geq 7 + 4 + 1$. Thus, the number of redundant bits= 4

***General Algorithm of Hamming code* –**

The Hamming Code is nothing more than the use of additional parity bits to enable error detection.

Table 1.18: General Algorithm of Hamming code and its redundant bits calculation

Position	R8	R4	R2	R1
0	0	0	0	0
1	0	0	0	1
2	0	0	1	0
3	0	0	1	1
4	0	1	0	0
5	0	1	0	1
6	0	1	1	0
7	0	1	1	1
8	1	0	0	0
9	1	0	0	1
10	1	0	1	0
11	1	0	1	1

- R1 (0001)--- bit at 2^0 position is '1'---{1,3,5,7,9,11}
- R2 (0010)--- bit at 2^1 position is '1'---{2,3,6,7,10,11}
- R4 (0100)--- bit at 2^2 position is '1'---{4,5,6,7}
- R4 (1000)--- bit at 2^3 position is '1'---{8,9,10,11}

1) Write the bit positions starting from 1 in binary form (1, 10, 11, 100, etc).
2) All the bit positions that are a power of 2 are marked as parity bits (1, 2, 4, 8, etc).
3) All the other bit positions are marked as data bits.
4) Each data bit is included in a unique set of parity bits, as determined its bit position in binary form.
   a) Parity bit 1 covers all the bits positions whose binary representation includes a 1 in the least significant position (1, 3, 5, 7, 9, 11, etc).
   b) Parity bit 2 covers all the bits positions whose binary representation includes a 1 in the second position from the least significant bit (2, 3, 6, 7, 10, 11, etc).
   c) Parity bit 4 covers all the bits positions whose binary representation includes a 1 in the third position from
   the least significant bit (4–7, 12–15, 20–23, etc).

d) Parity bit 8 covers all the bits positions whose binary representation includes a 1 in the fourth position from the least significant bit bits (8–15, 24–31, 40–47, etc).
e) In general, each parity bit covers all bits where the bitwise AND of the parity position and the bit position is non-zero.

5) Since we check for even parity set a parity bit to 1 if the total number of ones in the positions it checks is odd.
6) Set a parity bit to 0 if the total number of ones in the positions it checks is even.

Determining the position of redundant bits These redundancy bits are placed at the positions which correspond to the power of 2 (Refer Fig. 1.28). As per the above discussion:

1. The number of data bits = 7 say for example ($D_7D_6D_5D_4D_3D_2D_1$)
2. The number of redundant bits = 4 say for example ($R_4R_3R_2R_1$)
3. The total number of bits = 11 say for example (D_7 to D_1 and R_4 to R_1)
4. The redundant bits are placed at positions corresponding to power of 2, i.e, 1, 2, 4, and 8

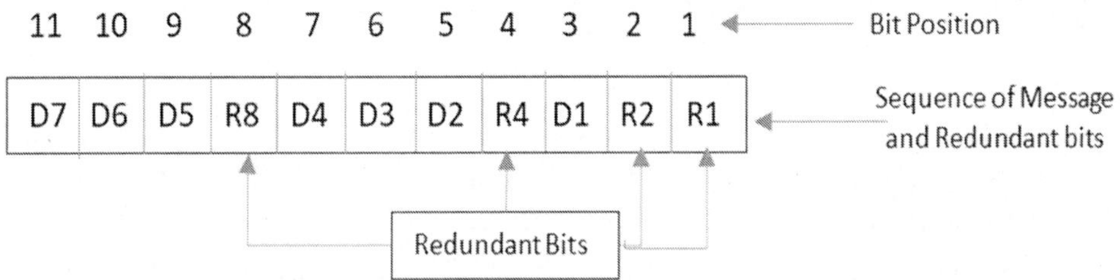

Fig. 1.28: Calculation for positioning redundant bits in Hamming code

Example 1.49: Suppose the data to be transmitted is 1011001, the bits will be placed as follows:

Solution:

Step-1: Position the data bits

		11	10	9	8	7	6	5	4	3
	1	0	1	R8	1	0	0	R4	1	R2

Step-2: Determine the Parity bits

- R1—Bit position 1,3,5,7,9,11—Considering the even parity for R1—R1,1,0,1,1,1—Hence, R1=0
- R2—Bit position 2,3,6,7,10,11—Considering the even parity for R2—R2,1,0,1,0,1—Hence, R2=1
- R4—Bit position 4,5,6,7—Considering the even parity for R4—R4,0,0,1—Hence, R4=1
- R8—Bit position 8,9,10,11—Considering the even parity for R8—R8,1,0,1—Hence, R8=0

Step-3: Thus, the data transferred is:10101001110

11	10	9	8	7	6	5	4	3	2	1
1	0	1	0	1	0	0	1	1	1	0

Step-4: **Error detection and correction –**

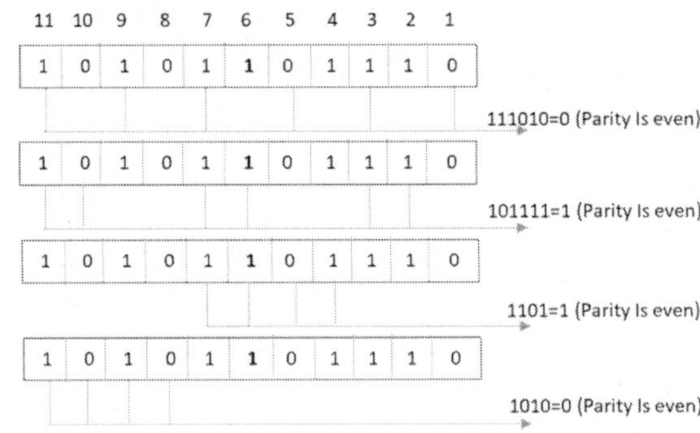

Fig. 1.29: Error detection and correction using Hamming Code

Suppose in the above example the 6th bit is changed from 0 to 1 during data transmission, hence, the data transferred is 10101101110. Then it gives new parity values in the binary number (Refer Fig. 1.29)

Step-5: a combination of $R_8R_4R_2R_1$ bits gives the binary number as 0110 whose decimal representation is 6. Thus, the bit 6 contains an error. To correct the error the 6th bit has to be complemented i.e., changed from 1 to 0. Hence, finally the error has been detected and corrected as well.

Note: Any length of communication can have single-bit errors fixed using the Hamming code. Despite being able to identify two-bit errors, the Hamming code is unable to pinpoint their exact positions. However, as demonstrated above, the number of parity bits needed to be communicated with the message depends on how lengthy the message is. The number of parity bits n required to encode m message bits is the smallest integer that satisfies the condition $(2^n - n) > m$

Example 1.50: By writing the parity code (even) and threefold repetition code for all possible four-bit straight binary numbers, prove that the Hamming distance in the two cases is at least 2 in the case of the parity code and 3 in the case of the repetition code.

Solution: The generation of codes is shown Below

Binary Number	Parity Code (Even Parity)	Three-time repetition code	Binary Number	Parity Code (Even Parity)	Three-time repetition code
0000	00000	000000000000	**1000**	10001	100010001000
0001	00011	000100010001	**1001**	10010	100110011001
0010	00101	001000100010	**1010**	10100	101010101010
0011	00110	001100110011	**1011**	10111	101110111011
0100	01001	010001000100	**1100**	11000	110011001100
0101	01010	010101010101	**1101**	11011	110111011101
0110	01100	011001100110	**1110**	11101	111011101110
0111	01111	011101110111	**1111**	11110	111111111111

An examination of the parity code numbers reveals that the number of bit disagreements between any pair of code words is not less than 2. It is either 2 or 4. It is 4, for example, between 00000 and 10111, 00000 and 11011, 00000 and 11101, 00000 and 11110 and 00000 and 01111. However, in the case of the threefold repetition code, it is either 3, 6, 9 or 12 and therefore not less than 3 under any circumstances.

Convolution Codes

We have so far talked about the preference for systematic unchanged code in the context of linear codes. In this case, if the data in total of n bits is transferred, k bits will serve as message bits and n+k bits will serve as parity bits. The parity bits are removed from the entire data set during encoding, and the message bits are then encoded. The parity bits are now added once more, and the entire data is encoded once more.

The following Fig. 1.30, quotes an example for blocks of data and stream of data, used for transmission of information.

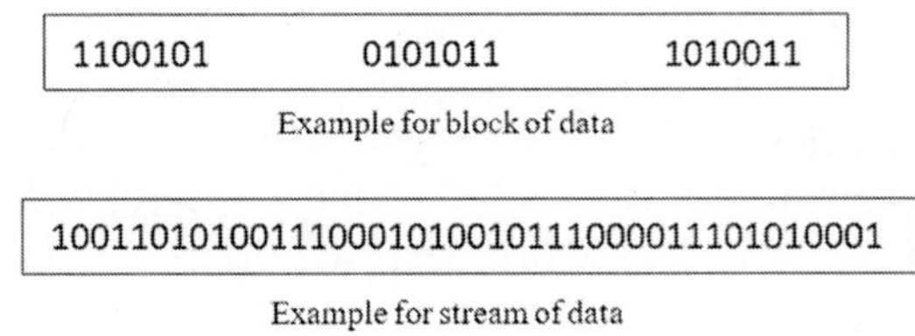

Fig. 1.30: Convolution Codes

The entire procedure is time-consuming, which has disadvantages. When the system is busy, the allocation of buffer is the key issue. Convolution codes eliminate this flaw. when the entire data stream is given symbols before being transferred. A buffer is not required because the data is a stream of bits.

1.16 Binary storage and registers.

The discrete information components in a digital computer must exist physically in a data storage medium. Additionally, the information storage media must have binary storage elements for storing individual bits when discrete informational items are represented in binary form. A binary cell is an object with two stable states and the ability to store one bit of data. Excitation impulses are received at the cell's input, which causes it to enter one of the two states. A physical quantity that distinguishes between the two states is the cell's output. When a cell is in one stable condition, the information stored in it is a "1," and when it is in the other stable state, it is a "0," Electronic flip-flop circuits, ferrite cores used in memories, and card positions with or without holes are a few examples of binary cells.

Registers

A collection of binary cells is called a register. Since a cell only holds one bit of data, any discrete amount of data with n bits can be stored in a register with n ceils. An n-tuple number of 1s and '0s make up a register's state, with each bit corresponding to a register cell's slate. The interpretation given to the data recorded in a register determines its substance. Consider, for example, the following 16-cell register:

Value of Bit Stored	**1**	**1**	**0**	**0**	**0**	**0**	**1**	**1**	**1**	**1**	**0**	**0**	**1**	**0**	**0**	**1**
Bit Position	1	2	3	4	5	6	7	8	9	10	11	12	13	14	15	16

Physically, the register is made up of 16 binary cells, each of which may store either a 1 or a 0. Let's say the register's bit arrangement is as displayed. The 16-tuple number represents the status of the register 110000111 1001001, Clearly, a register with cells can be in one of 2^n possible states, Now, if one assumes that the content of the register represents a binary integer, then obviously the register can store any binary number from 0 to 2^{16}—1.

In the specific illustration, the binary equivalent of the number 50121 is what is stored in the register. The content of the register is any two meaningful characters if it is assumed that the register stores alphanumeric characters of an eight-bit code (unassigned bit combinations do not represent meaningful information). In the EBCDIC code, the aforementioned example represents the two characters C (left eight bits) and I. (right eight bits). The content of the register is a four-digit decimal number, on the other hand, if one understands the contents of the register as four decimal digits represented by a four-bit code.

The decimal value 9096 is used in the aforementioned example of the excess-3 code. Since no decimal digit is allocated to the bit combination 1100, the contents of the register are meaningless in BCD. It is evident from this example that a register can hold one or more distinct informational components and that different types of informational elements may interpret the same bit configuration in different ways. The computer must be programmed to process the data in accordance with the type of information stored, and the user must input meaningful data into registers.

Register Transfer

The registers in a digital computer give it its personality. Simply said, the memory unit is a collection of thousands of registers used to store digital data. The processor unit is made up of a number of registers that contain the operands used to perform operations. Every input or output device must have at least one register to store the data transferred to or from it. An infer-register transfer operation, a fundamental operation in digital systems, entails transferring the data held in one register into another. The control unit uses registers to keep track of various computer sequences.

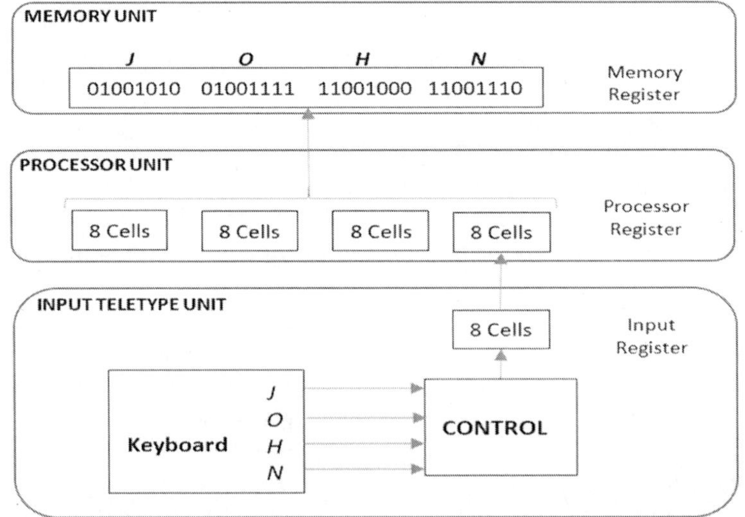

Fig. 1.31: Transfer of Information among registers

Fig. 1.31 depicts the movement of data between registers and shows visually how binary data is transferred from a teletype keyboard into a register in the memory unit. It is assumed that the input teletype machine has a keyboard, a control circuit, and an input register. The control writes an equivalent eight-bit alphanumeric character code into the input register every time a key is depressed. We'll presume that an odd-parity eighth bit of the ASCII code was used.

After each transfer, the input register is cleared to allow the control to insert a fresh eight-bit code when the keyboard is pressed again. The information from the input register is transferred into the eight least significant cells of a processor register. The previous character is shifted to the next eight cells to the left before each eight-bit character is transmitted to the processor register. The processor register is filled and its contents are transferred into a memory register when a transfer of four characters is finished. The transfer of the characters JOHN after the four appropriate keys were pressed resulted in the material being saved in the memory register seen in Figure 1.31

A computer must be equipped with (1) devices that contain the data to be processed and (2) circuit parts that modify individual bits of information, as shown in Fig. 1.32, to handle discrete quantities of binary data. A register is the thing that holds data the most frequently. Digital logic circuits are used to manipulate binary variables. The technique of adding two 10-bit binary values is shown in Fig. 1.32.

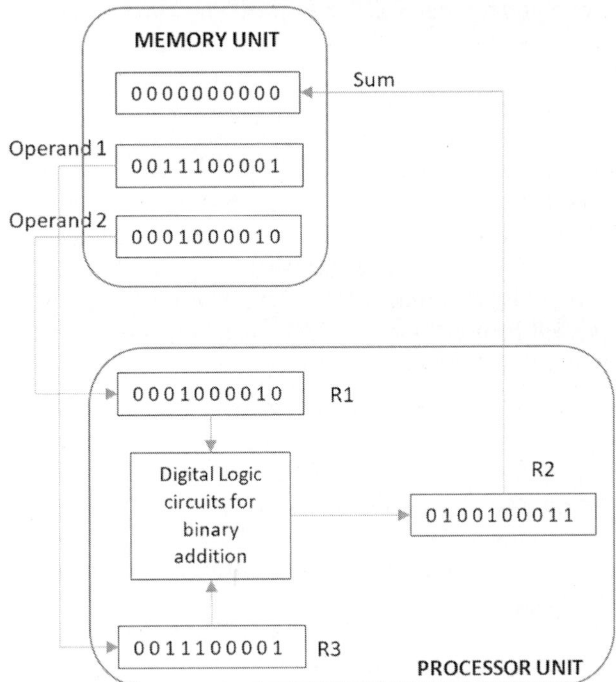

Fig.1.32: Example of Binary Information Processing

Only three of the memory unit's thousands of registers, which ordinarily number in the thousands, are visible in the diagram. Three registers, R1, R2, and R3, as well as digital logic circuits that modify the bits in R1 and R2 and transfer a binary number equal to their arithmetic total into R3 make up the portion of the processor unit that is being displayed.

Memory registers serve as information storage devices and are unable to handle the two operands. The data kept in memory can, however, be moved to processor registers. A memory register can receive results from processor registers and store them there until another use. The diagram depicts the transfer of data from two memory registers into R1 and R2 for two operands. The total is generated by the digital logic circuitry and sent to register R3 for storage. Now, R3's content can be transferred back to a memory register.

The information is processed by digital logic circuitry. The next part provides an introduction to digital logic circuits and their operational capabilities. Future chapters will address the topic of registers and registration transfer activities once more.

Unit summary

Various binary systems are shown in this chapter that can be used to represent information in digital components. The binary number system is described, and examples of binary codes are provided to demonstrate how decimal and alphanumeric data are represented.

The preceding chapter on number systems is expanded upon in this one. The preceding chapter covered the binary, hexadecimal, and octal number systems after introducing some fundamental ideas that are shared by all number systems and giving an overview of the well-known decimal number system. Octal and hexadecimal number systems are frequently employed to represent groups of binary digits, despite the fact that the binary system of representation is the one that is most frequently used in digital systems, including computers. When utilised to represent bigger decimal numbers, the binary coding scheme, known as the straight binary code and covered in the previous chapter, becomes increasingly difficult to handle. Numerous binary codes have developed throughout time to address this weakness as well as fulfil many other unique functions. This chapter will go through some of the more well-known binary codes, such as those that can effectively represent numeric and alphabetic data and those that can carry out specialised tasks like error detection and correction.

The next logical step after discussing various approaches to representing numeric and alphanumeric data in the previous two chapters is to learn the guidelines for manipulating data. Arithmetic and logic operations are two categories of operations that are carried out on binary data. Addition, subtraction, multiplication, and division are the four basic operations of mathematics. The fundamental operations in logic are AND, OR, and NOT. The rules pertaining to arithmetic operations are addressed in this chapter, but those pertaining to logic operations will be covered in the following chapter.

Solved Examples

1. Define the binary number system?
   The binary system has a base 2, and it consists of only two digits 0 and 1.
2. What is the size of bit, nibble, byte, word and double word in terms of number of bits?
   Bit = 1 bits, Nibble = 4 bits, Byte = 8 bits, word = 16 bits, double word = 32 bits.
3. What is meant by a bit?
   Bits are the binary digits like 0 and 1.
4. What is the best Example of Digital system?
   Digital Computer.

5. How many types of number system are there?
There are four types of number system: Decimal, Binary, Octal, Hexadecimal
6. Which code is called as minimum change code and why?
Gray code is called as minimum change code because it has a very special feature that only one bit will change, each time the decimal number is incremented.
7. What are the advantages of digital systems over analog systems? What is the chief limitation to the use of digital techniques?
Digital systems have a number of advantages over analog systems. Digital systems are more versatile, easier to store daft, easier to design, less affected by noise, more accurate and precise than analog systems. The only major drawback of digital techniques is that "the real world is not digital, it is analog".
8. What are the applications of the octal number system.
The applications of the octal number system are as follows:
For the efficient use of microprocessors.
For the efficient use of digital circuits.
It is used to enter binary data and display of information.
9. What is the use of resistors in digital electronics?
Resistors are used to control the current flow in digital electronics.
10. What is an analog signal? Give two examples of analog signals.
A signal which can assume any value in a given range is known as an analog signal. A sinusoidal signal and amplitude modulated signal are analog signals.
11. What is a digital signal? Give two examples of digital signals.
A signal which can assume only two possible values is known as a digital signal. Voltage levels 0 V and 5 V, and the presence or absence of pulse are digital signals.
12. Differentiate between analog and digital signals.
An analog signal is continuous and can assume any value in a given range, whereas a digital signal can have only two discrete values.
13. What are the two kinds of electronic circuits?
The two kinds of electronic circuits are- analog and digital.
14. What are analog circuits? Give a few examples.
Analog circuits, are those electronic circuits in which voltages and currents vary continuously through the given range They can take infinite values within the specified range. Signal generators, radio frequency transmitters and receivers, power supplies, electric motors, and speed controllers are some examples of analog devices, or we can say analog circuits are electronic circuits meant to process analog signals.
15. What are digital circuits?
Digital circuits are those electronic circuits in which the voltage levels can assume only a finite number of distinct values.
16. Why digital circuits are called switching circuits?
Digital circuits are often called switching circuits because the voltage levels in a digital circuit are assumed to be switched from one level to another instantaneously, that is, the transition time is assumed to be zero.
17. How are switching circuits classified?
Switching circuits are classified as (a) combinational switching circuits and (b) Sequential switching circuits.

18. Why are digital circuits called logic circuits?
    Digital circuits are also called logic circuits because each type of digital circuit obeys a certain set of logic rules.
19. What do you mean by circuit logic?
    The manner in which a logic circuit responds to input is referred to as the circuit's logic.
20. What is a hybrid system?
    Systems in which both analog and digital techniques are applied in the same system are called hybrid systems.
21. Name the three stages of digital system design?
    The three stages of digital system design are system design, logic design, and circuit design.
22. What do you mean by system design? Give an example.
    System design involves breaking the overall system into subsystems and specifying the characteristics of each such system. For example, the system design of a digital computer involves specifying the number and type of memory units, arithmetic units, and input-output devices, as well as specifying the interconnection and control of those subsystems.
23. What is circuit design?
    Circuit design involves specifying the interconnection of specific Components such as resistors, transistors, and diodes to form a gate, flip-flop, or any other logic building block.
24. What is the advantage of fixed point representation compared to floating point representation?
    Complexity and the cost of algorithm is less in fixed point representation, so it is suitable for time domain filtering.
25. What is the advantage of floating point representation compared to fixed point representation?
    Quantization error is small and dynamic range is high for floating point representation so it is suitable for frequency domain algorithm.
26. How the resolution can be increased in floating point representation of numbers?
    More the number of bits used in fraction part better will be the resolution.

Review Questions:

1. Determine the binary numbers represented by the following decimal numbers.
   (i) 25.5 (ii) 10.625 (iii) 0.6875
2. Find 2's complement of
   (i) (110110)2 (ii) (0110011)2
3. List the advantages of hexadecimal number system.
4. How will you detect overflow in signed-magnitude and Z's complement integer additions.
5. Perform the following subtractions using 2's complement method.
   (i) 01000 – 01001 (ii) 01100 – 00011 (iii) 0011.1001 – 0001.1110
6. Convert the decimal number 82.67 to its binary, hexadecimal and octal equivalents.
7. Add 20 and (-15) using 2's complement.
8. Add 648 and 487 in BCD code
9. Solve the following equations for X (i) 23.610 = X2 (ii) 65.53510 = X16
10. Perform the following additions using 2's complement
    (i) -20 to +20 (ii) +25 to -15 (iii) 23 – 48 (iv) – 48 – 23
11. Convert the
    (i) decimal number 430 to Excess-3 code
    (ii) binary number 10110 to Gray code.
12. Convert 2222 in Hexadecimal number.

13. Subtract –27 from 68 using 2's complements.
14. Divide $(101110)_2$ by $(101)_2$.
15. Perform following subtraction
    (i) 11001-10110 using 1's complement
    (ii) 11011-11001 using 2's complement
16. Convert the decimal number 45678 to its hexadecimal equivalent number.
17. Convert $(177.25)_{10}$ to octal.
18. Perform the following subtraction using 1's complement
    (i) 11001 – 10110 (ii) 11011 - 11001

Multiple Choice Questions (MCQs)

1. What is Digital Electronics?
   a) Field of electronics involving the study of digital signal
   b) Engineering of devices that digital signal
   c) Engineering of devices that produce digital signal
   d) All pf the mentioned
2. Which of the following is correct for Digital Circuits?
   a) Less susceptible to noise or degradation in quality
   b) Use transistors to create logic gates to perform Boolean logic
   c) Easier to perform error detection and correction with digital signal
   d) All of the mentioned
3. What is a Circuit?
   a) Open-loop through which electrons can pass
   b) Closed-loop through which electrons can pass
   c) Closed-loop through which Neutrons can pass
   d) None of the mentioned
4. Which of the following is an example of a digital Electronic?
   a) Computers
   b) Information appliances
   c) Digital cameras
   d) All of the mentioned
5. Which of the following is a type of digital logic circuit?
   a) Combinational logic circuits
   b) Sequential logic circuits
   c) Both a & b
   d) None of the mentioned
6. What is the minimum distance required for single error detection according to Hamming's analysis in Digital Electronics?
   a) 1 b) 2 c) 3 d) 4
7. Which of these error-detecting codes enables to find double errors in Digital Electronic devices?
   a) Parity method
   b) Check sum method

c) Bit generation method
d) Odd-Even method

8. What minimum distance is required for a single error correction according to Hamming's analysis in Digital Electronics?
a) 1 b) 2 c) 3 d) 4
9. Any signed negative binary number is recognised by its ________
a) MSB b) LSB c) Byte d) Nibble
10. The parameter through which 16 distinct values can be represented is known as ________
a) Bit b) Byte c) Word d) Nibble
11. If the decimal number is a fraction, then its binary equivalent is obtained by ________ the number continuously by 2.
a) Dividing b) Multiplying c) Adding d) Subtracting
12. The representation of octal number (532.2)8 in decimal is ________
a) (346.25)10
b) (532.864)10
c) (340.67)10
d) (531.668)10
13. The decimal equivalent of the binary number (1011.011)2 is ________
a) (11.375)10
b) (10.123)10
c) (11.175)10
d) (9.23)10
14. An important drawback of binary system is ________
a) It requires very large string of 1's and 0's to represent a decimal number
b) It requires sparingly small string of 1's and 0's to represent a decimal number
c) It requires large string of 1's and small string of 0's to represent a decimal number
d) It requires small string of 1's and large string of 0's to represent a decimal number
15. The decimal equivalent of the octal number $(645)_8$ is ______
a) $(450)_{10}$ b) $(451)_{10}$ c) $(421)_{10}$ d) $(501)_{10}$
16. The largest two-digit hexadecimal number is ________
a) $(FE)_{16}$ b) $(FD)_{16}$ c) $(FF)_{16}$ d) $(EF)_{16}$
17. The quantity of double word is ________
a) 16 bits b) 32 bits c) 4 bits d) 8 bits
18. 1's complement of 1011101 is ___________
a) 0101110 b) 1001101 c) 0100010 d) 1100101
19. 2's complement of 11001011 is ___________
a) 01010111 b) 11010100 c) 00110101 d) 11100010
20. On subtracting (010110)2 from (1011001)2 using 2's complement, we get ___________
a) 0111001 b) 1100101 c) 0110110 d) 1000011
21. Binary coded decimal is a combination of __________
a) Two binary digits
b) Three binary digits

c) Four binary digits
d) Five binary digits

22. Code is a symbolic representation of __________ information.
    a) Continuous b) Discrete c) Analog d) Both continuous and discrete
23. When numbers, letters or words are represented by a special group of symbols, this process is called __________
    a) Decoding b) Encoding c) Digitizing d) Inverting
24. The excess-3 code for 597 is given by __________
    a) 100011001010 b) 100010100111 c) 010110010111 d) 010110101101
25. Which of the following codes is a sequential code?
    a) 8421 code b) 2421 code c) 5421 code d) 2441 code
26. Which of these code pairs correctly represent Digital Electronics reflective codes?
    a) 2421 and 5211
    b) 2421 and 8421
    c) 5211 and 8421
    d) 5421 and 2421
27. Which of the following options correctly represent the characteristic of Excess – 3 code?
    a) It is a reflexive as well as a sequential code
    b) It is a reflexive code but not a sequential code
    c) It is a sequential code but not a reflexive code
    d) It is neither a reflexive code nor a sequential code

Reference and Suggested Readings

- *A. Anand Kumar, Fundamentals of digital circuits Second Edition, PHI Learning Private Limited, ISBN (978-81-203-3679-7)*
- *Floyd, T. L. (2005) Digital Fundamentals, Prentice-Hall Inc., USA.*
- *Morris Mano, M. and Kime, C. R. (2003) Logic and Computer Design Fundamentals, Prentice-Hall Inc., USA.*
- *MacWilliams, F. J. and Sloane, N. J. A. (2006) The Theory of Error-Correcting Codes, North-Holland Mathematical Library, Elsevier Ltd, Oxford, UK.*
- *Rafiquzzaman, M. (2005) Fundamentals of Digital Logic and Microcomputer Design, Wiley-Interscience, New York, USA.*
- *Tocci, R. J. and N. S. Widmer. 2004. Digital Systems Principles and Applications, 9th ed. Upper Saddle River, NJ: Prentice Hall.*
- *Tokheim, R. L. (1994) Schaum's Outline Series of Digital Principles, McGraw-Hill Companies Inc., USA.*
- *Yarbrough, John M. Digital logic: Applications and design. Eagan: West Publishing Company, 1997.*

2 Logic Gates and Boolean Algebra

UNIT SPECIFICS

Through this unit we have discussed the following aspects:

- *To Differentiate between basic, universal and special purpose Logic Gates.*
- *To understand various IEEE/ANSI standards and application relevant packages.*
- *To analyze various rules and postulates of Boolean Algebra.*
- *To identify and implement SOP and POS for Boolean expression and their reduction using* $\sum$ *and* $\prod$ *Nomenclature.*
- *To perform Quine–McCluskey Tabular Method for Multi-Output Functions*

RATIONALE

The building blocks of a digital circuit are logic gates, which execute numerous logical operations that are required by any digital circuit. These can take two or more inputs but only produce one output. The mix of inputs applied across a logic gate determines its output. Logic gates use Boolean algebra to execute logical processes. Logic gates are found in nearly every digital gadget we use on a regular basis. Logic gates are used in the architecture of our telephones, laptops, tablets, and memory devices.

Boolean algebra is a type of logical algebra in which symbols represent logic levels. The digits (or symbols) 1 and 0 are related to the logic levels in this algebra; in electrical circuits, logic 1 will represent a closed switch, a high voltage, or a device's "on" state. An open switch, low voltage, or "off" state of the device will be represented by logic 0.

At any one time, a digital device will be in one of these two binary situations. A light bulb can be used to demonstrate the operation of a logic gate. When logic 0 is supplied to the switch, it is turned off, and the bulb does not light up. The switch is in an ON state when logic 1 is applied, and the bulb would light up. In integrated circuits (IC), logic gates are widely employed.

PRE-REQUISITES

Basic Mathematics, Basic Switching Theory

UNIT OUTCOMES

List of outcomes of this unit is as follows:

U2-O1: *Describe binary logic gates and their categories.*

U2-O2: *Describe IEEE/ANSI standards and their relevant packages.*

U2-O3: *Understand rules, dual and postulates of Boolean algebra.*

U2-O4: *Implement SOP and POS for Boolean expression.*

U2-O5: *Apply Quine–McCluskey Tabular Method for Multi-Output Functions.*

Unit-2 Outcomes	***EXPECTED MAPPING WITH COURSE OUTCOMES*** *(1- Weak Correlation; 2- Medium correlation; 3- Strong Correlation)*					
	CO-1	***CO-2***	***CO-3***	***CO-4***	***CO-5***	***CO-6***
U2-O1	3	3	3	-	3	1
U2-O2	1	1	2	2	1	-
U2-O3	2	1	3	1	2	1
U2-O4	-	-	3	1	2	2
U2-O5	3	3	3	-	3	1

➢ Scan the below QR codes for more information on the topic written below the QR code.

Logic Gates

IEEE/ANSI Standard Symbols

Boolean Algebra

Quine–McCluskey Tabular Method

Unit outcome

The basic postulates and theorems of Boolean algebra are found in this unit. The correlation between a Boolean expression and its equivalent interconnection of gates is emphasized. All possible logic operations for two variables are investigated and from that, the most useful logic gates are derived. The characteristics of digital gates available in integrated circuit from are presented early in this chapter but a more detailed analysis describing the applications of gates in the upcoming chapter.

2.1 Binary Logic:

Binary logic deals with a variable that takes on two discrete values and with operations that assume logical meaning. the two values the variables take may be called by different names (e.g., true and false, yes and no, etc.), but for our purpose, it is convenient to think in terms of bits and assign the values of ***'1'*** and ***'0'***.

2.1.1 Introduction

Binary logic is used to describe, the manipulation and processing of binary information. It is particularly suited for the analysis and design of digital systems. For example, digital logic circuits perform binary arithmetic whose behavior is most conveniently expressed by means of binary variables and logical operations. The binary logic to be introduced in this section is equivalent to an algebra called ***Boolean algebra***. The formal presentation of two-valued Boolean algebra is covered in more detail in this chapter. The purpose of this section is to introduce Boolean algebra in a heuristic manner and relate it to digital logic circuits and binary signals.

Definition of Binary Logic: Binary logic consists of binary variables that perform logical operations. To represent such variables letters of the alphabet have been used, such as A, B, C, x, y, z, etc. However, each variable will have two and only two distinct possible values '1' and '0'. There are three basic logical operations 'AND', 'OR', and 'NOT'.

- AND: AND operation of binary variables may be represented by a dot (.) or by the absence of an operator. For example,

$$X.Y = Z \text{ or } XY = Z, \qquad \textit{read as} \quad \textit{"X AND Y is equal to Z"}$$

This AND logic can be interpreted to mean that z = 1 if and only if x = 1 and y= 1; otherwise, z =0. (Remember that x, y, and z are binary variables and can be equal either to 0 or 1, nothing else.)

- OR: OR operation is represented by a plus sign (+).
  For example

$$X + Y = Z\,, \qquad \textit{read as} \quad \textit{"X OR Y is equal to Z"}$$

This OR logic can be interpreted to mean that z = 1 if x = 1 or if y = 1 or if both x = 1 and y = 1. However, if both x = 0 and y = 0, then only z= 0.

- NOT: NOT operation is represented by a prime (sometimes by a bar).
  For example,

$$X' = Z \text{ or } \bar{X} = Z \quad \textit{read as "X NOT is equal to Z"}$$

This NOT logic can be interpreted to mean that z is what x is not. in other words. If x= 1, then z = 0. However, if x = 0, then z= 1.

Binary logic resembles binary arithmetic, and the operation AND as well as OR have some similarities to arithmetic multiplication and addition, respectively. As the symbols used for AND as well as

OR are the same as those used for arithmetic multiplication and addition, However, binary logic should not be confused with binary arithmetic. One should realize that with AND, and OR logical operations are going to be performed, that is why it is usually called Logical AND, and Logical OR. Moreover, an arithmetic variable designates a number that may consist of many digits, however, logic variable is always either a 1 or 0. For example,

1+1=1 (In case of Logical OR)

1+1=2 (In case of Arithmetic Sum)

For each combination of the values of x and y, there is a value of z specified by the definition of the logical operation. These definitions may be listed in a compact form by using ***truth tables***. *A truth table is a table of all possible combination of the variables, showing the relation between the values that the variables may take and the result of the operation.*

For example, the truth tables for the operations AND, and OR with variables x and y is obtained by listing all possible values that the variables may have when combined by pairs. The result of the operation for each combination is then listed in a separate row. The truth tables for AND, OR, and NOT are listed in Table 2.1. Such truth table clearly demonstrate the definition of the operations for a particular logical operation.

Table 2.1: Truth tables of logical operations

AND			**OR**			**NOT**	
x	y	$x.y$	x	y	$x+y$	x	$\bar{x}$
0	0	0	0	0	0	0	1
0	1	0	0	1	1	1	0
1	0	0	1	0	1		
1	1	1	1	1	1		

2.1.2 Switching Circuits and Binary Signals:

The use of binary variables and the application of binary logic are demonstrated by the simple switching circuits of Fig. 2.1. Let the manual switches A and B represent two binary variables with values equal to 0 when the switch is open and 1 when the switch is closed. Similarly, let the lamp L represent a third binary variable equal to 1 when the light is on and 0 when the light is off. For the switches in series, the light turns on if A and B are closed. for the switches in parallel, the light turns on if A or B is closed, it is obvious that the two circuits can be expressed by means of binary logic with the AND, and OR operations, respectively:

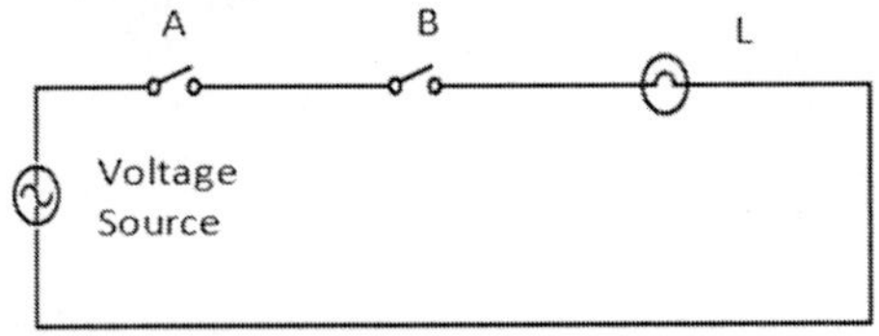

(a) Switches in series- logical AND

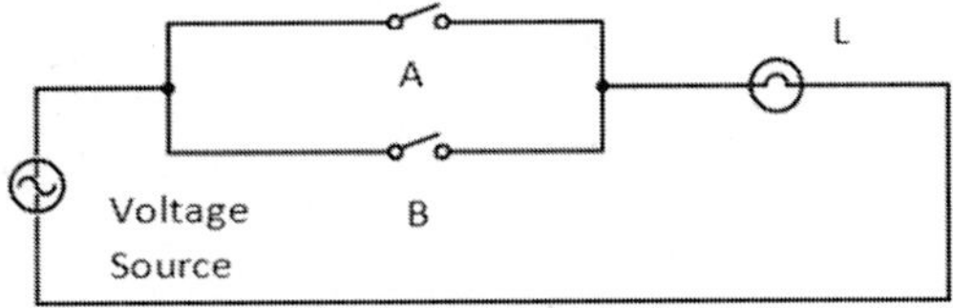

(b) Switches in Parallel-Logical OR Gate

Fig. 2.1: Switching circuits that demonstrate binary logic

The output for the above circuits can be given by following equations

$L = A \cdot B$ .... for the circuit of Fig. 2.1(a)

$L = A + B$............ for the circuit of Fig. 2.1(b)

Electronics digital circuits are sometimes called ***switching circuits*** because they ***behave like a switch***, with the active element such as a transistor either conducting/Saturation (switch closed) or not conducting/Cut-off (switch open). Instead of changing the switch manually, an electronic switching circuits uses binary signals to control the conduction or non-conduction state of the active element. The same circuit (Fig. 2.1) can be designed with the help of diodes and transistors as shown in figure below (Refer Fig. 2.2).

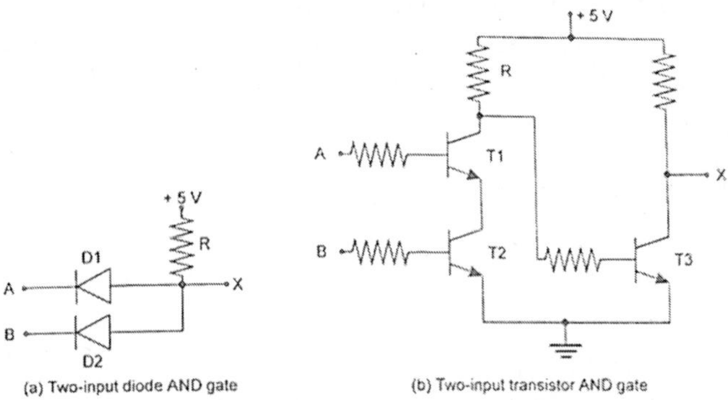

Fig. 2.2: Two Input Discrete AND gate using (a) Diodes (b) Transistors.

Electrical signals such as voltages or current exist throughout a digital system in either one of two recognizable values (except during transition). Voltage – operated circuits, for example, respond to two separate voltages levels which represent a binary variable equal to logic- 1 or logic- 0. For example, a particular digital system may define logic -1 as a signal with a nominal value of 3 volts, and logic -0 as a signal with a nominal value of 0 volt. As shown in Fig. 2.3 each voltage level has an acceptable deviation from the nominal. The intermediate region between the allowed regions is crossed only during state transitions. The input terminals of digital circuits accept binary signals within the allowable tolerances and respond at the output terminal with binary signals that fall within the specified tolerances.

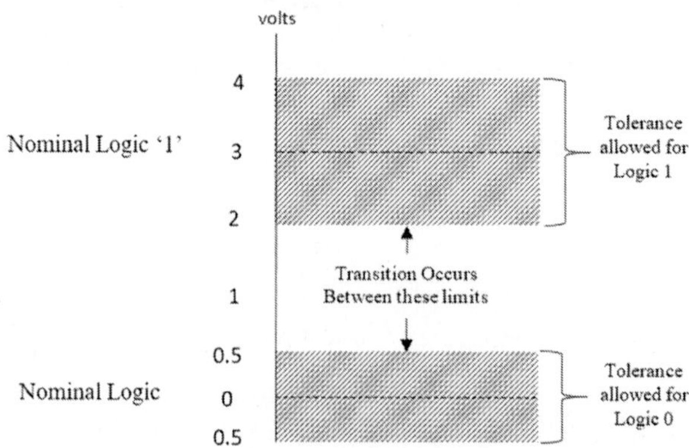

Fig. 2.3: Example of binary signals

2.1.3 Integrated Circuits:

Digital circuits are invariably constructed with integrated circuits. An integrated circuits (IC) is a small silicon semiconductor crystal, called a chip, that consist of transistors, diodes, registers, and capacitors. The various components are interconnected inside the chip to form an electronics circuit. The chip is mounted on a metal or plastic package, and connections are welded to external pins to form the IC. integrated circuits differ from other electronics circuits composed of detachable components in that individual components in the IC cannot be separated or disconnected and the circuit inside the package is accessible only through the external pins.

Integrated circuits (ICs) come in two types of packages, the flat package and the dual-in-line (DIP) package. As shown in Fig. 2.4. The dual-in-line (DIP) package is the most widely used type because of the low price and easy installation on circuit boards. The envelope of the IC package is made of plastic or ceramic. Most package have standard sizes, and has a numeric designation printed on the surface of the package for identification. Each vender publishes a data book or catalog that provides the necessary information concerning the various products.

For example, four AND gates are enclosed inside a 14-pin dual-in-line package with dimensions of 20 x 8 x 3 millimeters. An entire microprocessor is enclosed within a 40- pin dual-in-line package with dimensions of 50 x 15 x 4 millimeters.

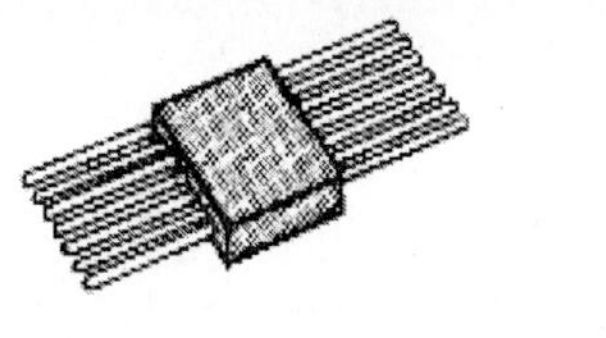

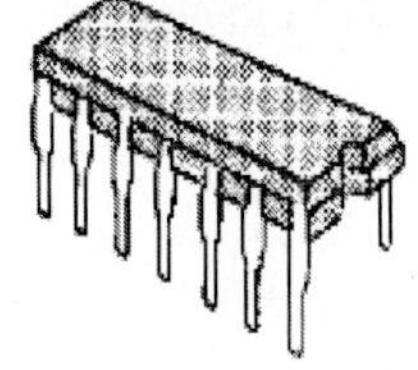

Flat Package Dual in Line

Fig. 2.4: Integrated – circuits packages

The cost of ICs is very low, which makes them economical to use. Their reduced power consumption makes the digital system more economical to operate. They have a high reliability against failure, so the digital system needs less repairs. The operating speed is higher, which make them suitable for high – speed operations. The use of ICs reduces the number of external wiring connections because many of the connections are internal to the package. Because of these advantages, digital systems are always constructed with integrated circuits.

Integrated circuits are classified in two general categories, linear and digital. Linear ICs operate with continuous signals to provide electronic functions such as amplifiers and voltages comparators. Digital integrated circuits operate with binary signals and are made up of interconnected ***digital gates***. Here we are concerned only with digital integrated circuits.

2.2 Logic Gates

Logic gates are ***electronic circuits*** that can be used to implement the most ***elementary logic expressions***, also known as ***Boolean expressions***. The logic gate is the most basic building block of combinational logic. There are three basic logic gates, namely *the OR gate*, *the AND gate* and *the NOT gate*. Other logic gates that are derived from these basic gates are the NAND gate, the NOR gate, the EXCLUSIVE-OR gate, and the EXCLUSIVE-NOR gate.

This chapter deals with logic gates and some related devices such as buffers, drivers, etc., as regards their basic functions. The treatment of the subject matter is mainly with the help of respective truth tables and Boolean expressions. The chapter is adequately illustrated with the help of solved examples. Towards the end, the chapter contains application-relevant information in terms of popular type numbers of logic gates from different logic families and their functional description to help application engineers in choosing the right device for their application.

Electronics digital circuits are also called ***logic circuits*** because, with the proper input they establish logical manipulation paths. Any desired information for computing or control can be operated upon by passing binary signals through various combinations of logic circuits, each signal representing a variable and carrying one bit of information.

Logic circuits that perform the logical operations such as AND, OR, and NOT are shown with their symbols in Fig. 2.5, these circuits, called gates are block of hardware that produce a logic -1 or logic -0 output signal if input logic requirements are satisfied. Note that these circuits are called gates or digital circuits or switching circuits or logic circuits all four names are widely used, but we shall refer to the circuits as AND, OR, and NOT gates. The NOT gate is sometimes called an inverter circuit since it inverts a binary signal.

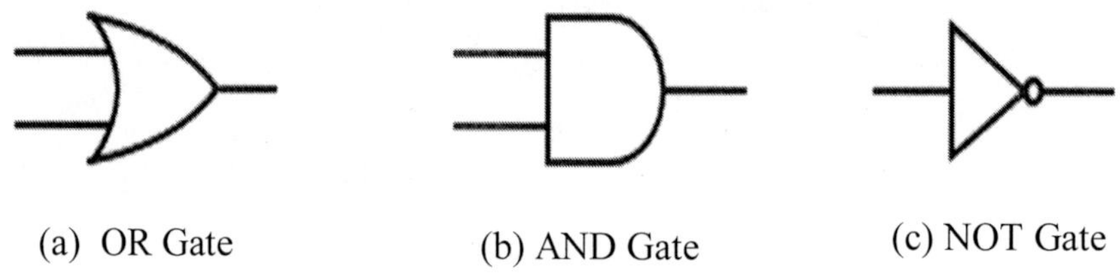

(a) OR Gate (b) AND Gate (c) NOT Gate

Fig. 2.5: Logical symbols for AND, OR, and NOT gate

2.2.1 Positive and Negative Logic

The binary variables, as we know, can have either of the two states, i.e., the logic '0' state or the logic '1' state. These logic states in digital systems such as computers, for instance, are represented by two different voltage levels or two different current levels. If the more positive of the two voltage or current levels represent a logic '1' and the less positive of the two levels represents a logic '0', then the logic system is referred to as a ***Positive Logic*** system. If the more positive of the two voltage or current levels represent a logic '0' and the less positive of the two levels represents a logic '1', then the logic system is referred to as a ***Negative Logic*** system.

For example, If the two voltage levels are 0 V and +5 V, then in the positive logic system the 0 V represents a logic '0' and the +5 V represents a logic '1'. Whereas, In the negative logic system, 0 V represents a logic '1' and +5 V represents a logic '0'.

It is interesting to note, as we will discover in the latter part of the chapter, that a ***positive OR is a negative AND****. That is, OR gate hardware in the positive logic system behaves like an AND gate in the negative logic system. The reverse is also true. Similarly, a positive NOR is a negative NAND, and vice versa.*

2.2.2 Truth Table and Logic gates

A truth table lists all possible combinations of input binary variables and the corresponding outputs of a logic system. The logic system output can be found from the logic expression, often referred to as the ***Boolean expression***, that relates the output with the inputs of that very logic system. When the number of input binary variables is only one, then there are only two possible inputs, i.e., '0' and '1'. If the number of inputs is two, there can be four possible input combinations, i.e., 00, 01, 10 and 11. Fig. 2.6 (b) shows the truth table of the two-input logic system represented by Fig. 2.6 (a). The logic system of Fig. 2.6 (a) is such that Y=0 only when both A=0 and B=0. For all other possible input combinations, output Y=1. Similarly, for three input binary variables, there could be eight different possible combinations, I.e., 000, 001, 010, 011, 100, 101, 110, and 111.

This statement can be generalized to say that, if a logic circuit has n binary inputs, its truth table will have 2^npossible input combinations, or in other words 2^n rows. Fig. 2.7 shows the truth table of a three-input logic circuit, and it has 8 (=2^3) rows. Incidentally, as we will see later in the chapter, this is the truth table of a three-input AND gate. It may be mentioned here that the truth table of a three-input AND gate as given in Fig. 2.7 is drawn following the positive logic system, moreover, in all further discussion throughout the book, we will use a positive logic system unless otherwise specified.

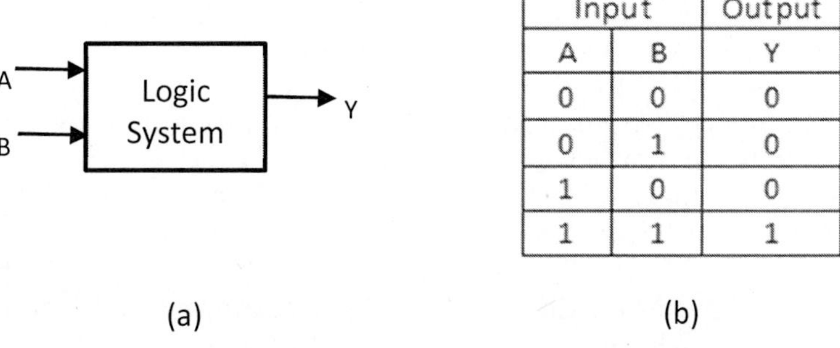

Input		Output
A	B	Y
0	0	0
0	1	0
1	0	0
1	1	1

Fig. 2.6: Two input logic system and sample truth table.

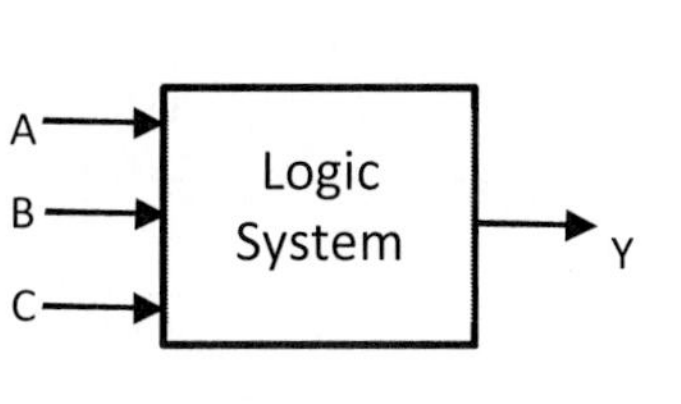

Input			Output
A	B	C	Y
0	0	0	0
0	0	1	0
0	1	0	0
0	1	1	0
1	0	0	0
1	0	1	0
1	1	0	0
1	1	1	1

Fig. 2.7: Three input logic system and sample truth table.

2.3 Categories of Logic Gates:

Logic gates can be categorized into three different categories as given

- Basic Gates
- Special Purpose Gates
- Universal Gates

2.3.1 Basic Gates

The logic gate is the most basic building block of any digital system, including computers. Each one of the basic logic gates is a piece of hardware or an electronic circuit that can be used to implement some basic logic expression. While laws of Boolean algebra could be used to do manipulation with binary variables and simplify logic expressions, these are actually implemented in a digital system with the help of electronic circuits called logic gates. The three basic logic gates are the OR gate, the AND gate, and the NOT gate.

The 'OR' Gate

An OR gate performs an ***ORing operation*** on two or more than two logic variables. The OR operation on two independent logic variables A and B is written as Y=A+B and reads as 'Y equals A OR B' and not as A plus B. An OR gate is a logic circuit with two or more inputs and one output.

Statement: *In an OR gate the output of an OR gate is LOW only and only when all of its inputs are LOW, whereas the output will be HIGH if any of the input is HIGH.*

This statement when interpreted for a positive logic system means the following. The output of an OR gate is a logic '0' only when all of its inputs are at logic '0'. For all other possible input combinations, the output is a logic '1'. Fig. 2.8 shows the circuit symbol and the truth table of a two-input OR gate. The operation of a two-input OR gate is explained by the logic expression

$$Y = A + B$$

As an illustration, if we have four logic variables and we want to know the logical output of $Y = A + B + C$, then it would be the output of a three-input OR gate with A, B, and C as its inputs.

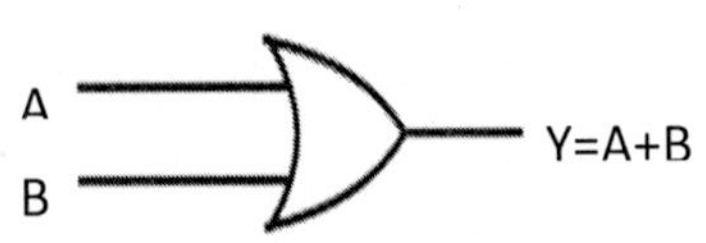

(a) Logic Symbol for two Input OR gate

Input		Output
A	B	Y=A+B
0	0	0
0	1	1
1	0	1
1	1	1

(b) Truth Table for two Input OR gate

Fig. 2.8: Two – input OR gate.

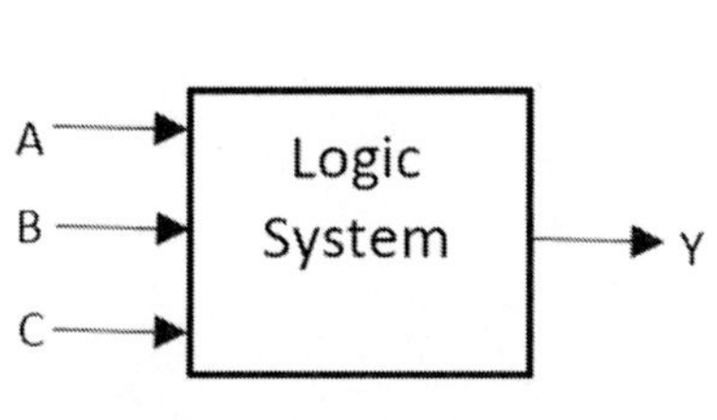

Input			Output
A	B	C	Y
0	0	0	0
0	0	1	0
0	1	0	0
0	1	1	0
1	0	0	0
1	0	1	0
1	1	0	0
1	1	1	1

Fig. 2.9: (a) Three-input OR gate, (B) the truth table of a three-input OR gate.

Fig. 2.9 shows the circuit symbol of three-input OR gates and it also shows the truth table of a three-input OR gate. Logic expressions explaining the functioning of three-input OR gates is given below

$$Y = A + B + C$$

Example 2.1: Implement a four-input OR gate using two-input OR gates only?

Solution: There are two different possible combinations to implement 4 input OR gate using 2 input OR gates. A, B, C, and D, are logic inputs and Y_3 is the logical output. Both the approaches are shown in Figure Below

(a) Fig. 2.10 (a) shows one possible arrangement the output can be calculated as.

$$Y_1 = A + B; \quad Y_2 = Y_1 + C = A + B + C$$
$$Y_3 = Y_2 + D = A + B + C + D$$

(b) Fig. 2.10 (b) shows another possible arrangement and the output can be calculated as,

$$Y_1 = A + B, Y_2 = C + D,$$
$$Y_3 = Y_1 + Y_2 = A + B + C + D$$

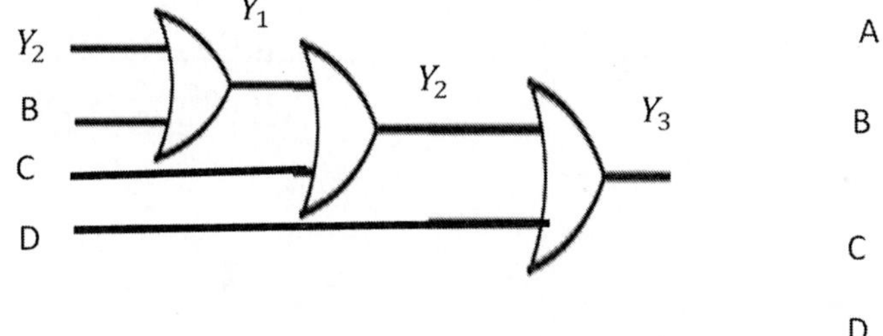

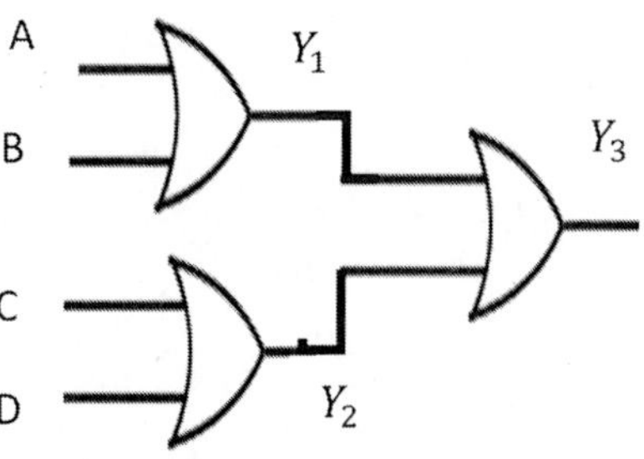

Fig. 2.10: Example 2.1

Example 2.2: For the OR gate Draw plot the output waveform (Refer Fig. 2.11(a)).

Solution: Fig. 2.11(b) shows the output waveform. It can be drawn by following the truth table of the OR gate

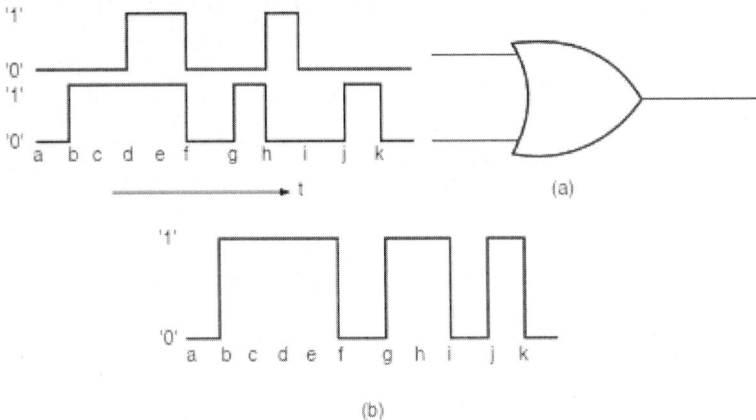

Fig. 2.11: Example 2.2

The 'AND' Gate

An AND gate is a kind of logic circuit that may have two or more inputs but only one output. The output of an AND gate is HIGH only when all of its inputs are in the HIGH state. In all other cases, the output is LOW. When interpreted for a positive logic system, this means that the output of the AND gate is a logic '1' only when all of its inputs are in the logic '1' state. In all other cases, the output is logic '0'. The logic symbol and truth table of a two-input AND gate are shown in Fig. 2.13 (a) and (b) respectively. Fig. 2.13 (a) and (b) show the logic symbols of three-input and four-input AND gates respectively. Fig. 2.13 (c) gives the truth table of a four-input AND gate.

Statement: *In an AND gate the output of an AND gate is HIGH only and only when all of its inputs are HIGH, whereas the output will be LOW if any of the input is LOW.*

The AND operation on two independent logic variables A and B is written as Y=A.B and reads as Y equals ***A AND B*** and ***not as A multiplied by B***. Here, A and B are input logic variables and Y is the output. An AND gate performs an ANDing operation. The same AND logic can be designed for three and four input AND gate (Refer Fig. 2.13)

Logic Symbol

A

B

Y

Expression

$Y = A.B$

Truth Table

Input		**Output**
A	B	Y
0	0	0
0	1	0
1	0	0
1	1	1

Fig. 2.12: Two-input AND gate.

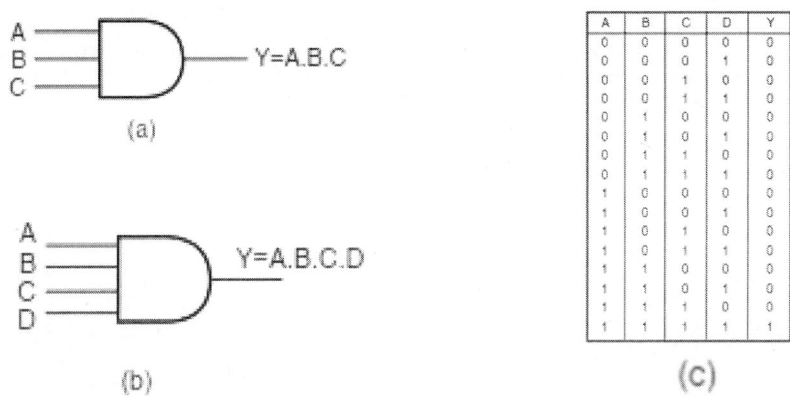

A	B	C	D	Y
0	0	0	0	0
0	0	0	1	0
0	0	1	0	0
0	0	1	1	0
0	1	0	0	0
0	1	0	1	0
0	1	1	0	0
0	1	1	1	0
1	0	0	0	0
1	0	0	1	0
1	0	1	0	0
1	0	1	1	0
1	1	0	0	0
1	1	0	1	0
1	1	1	0	0
1	1	1	1	1

Fig.2.13: (a) Three-input AND gate, (b) four-input AND gate and (c) the truth table of a four-input AND gate.

If we interpret the basic definition of OR as well as AND gate for a negative logic system, we have an interesting observation. We find that an OR gate in a positive logic system is an AND gate in a negative logic system. Also, a positive AND is a negative OR.

Example 2.3: Show the logic arrangement for implementing a four-input AND gate using two-input AND gates only.

Solution: Fig. 2.14 shows the hardware implementation of a four-input AND gate using two-input AND gates. Same as Example 1, two approaches can be used whereas, only the first one is presented here (Refer Fig. 2.14), the second approach is similar as discussed in example 1.

The output of AND gate can be calculated as

$$Y_1 = A.B; \quad Y_2 = Y_1.C = A.B.C$$
$$Y_3 = Y_2.D = A.B.C.D$$

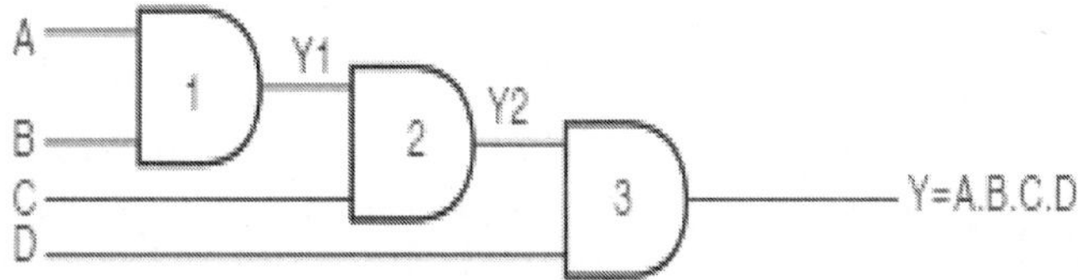

Fig.2.14: Implementation of a four -input AND gate using two-input AND gates.

The 'NOT' Gate

A NOT gate is a one-input, one-output logic circuit whose output is always the complement of the input. That is, a LOW input produces a HIGH output, and vice versa. When interpreted for a positive logic system, a logic '0' at the input produces a logic '1' at the output, and vice versa. It is also known as a 'complementing circuit' or an 'inverting circuit'. Fig. 2.15 shows the circuit symbol and the truth table.

The NOT operation on a logic variable X is denoted as X' or $\bar{X}$. That is, if X is the input to a NOT circuit, then its output Y is given by

$$Y = \bar{X} \text{ or } X' \quad \text{and reads as } Y \text{ equals NOT } X$$
$$\text{Thus, if } X = 0, Y = 1 \text{ and if } X = 1, Y = 0$$

Logic Symbol	Expression	Truth Table
X — Y=$\bar{X}$ (a) X — Y=$\bar{X}$ (b)	$Y = \bar{X}$	X | Y 0 | 1 1 | 0

Fig.2.15: (a) Circuit symbol of a NOT circuit and (b) the truth table

Example 2.4: For the logic circuit arrangements of Figs 2.15 (a) and (b), draw the Remark about what you will observe at the output.

Fig. 2.16: Example 2.4.

Solution:

Circuit Arrangement	Remark
	In the case of the OR gate arrangement of Fig. 2.15(a), the output will be permanently in logic '1' state as the two inputs can never be in logic '0' state together owing to the presence of the inverter.
	In the case of the AND gate arrangement of Fig. 2.15(b), the output will be permanently in logic '0' state as the two inputs can never be in logic '1' state together owing to the presence of the inverter.

2.3.2 Special Purpose Gates

<u>The 'EXCLUSIVE-OR' (XOR or EX-OR) Gate</u>

The EXCLUSIVE-OR gate, commonly written as EX-OR gate, is a two-input, one-output gate. Fig. 2.17 (a) and (b) respectively show the logic symbol and Expression for the EX-OR gate, whereas Fig. 2.17 (c) represents the truth table of a two-input EX-OR gate. As can be seen from the truth table, the output of an EX-OR gate is a logic '1' when the inputs are unlike and a logic '0' when the inputs are like.

Although EX-OR gates are available in integrated circuit form only as two-input gates, unlike other gates which are available in multiple inputs also, multiple-input EX-OR logic functions can be implemented using two-input gates. The output of a multiple-input EX-OR logic function is a logic '1' when the number of 1s in the input sequence is odd and a logic '0' otherwise. Further, an all 0s input sequence also produces a logic '0' at the output. The output of a two-input EX-OR gate is expressed by

$$Y = A \oplus B = \bar{A}B + A\bar{B}$$

Logic Symbol	Expression	Truth Table		
		Input		**Output**
		A	B	Y=A□B
A, B — Y=A⊕B	$Y = (A \oplus B) = \overline{A}B + A\overline{B}$	**0**	0	0
		0	1	1
(a)		**1**	0	1
		1	1	0

Fig. 2.17: (a) Circuit symbol of a NOT circuit and (b) the truth table.

Statement: *If odd nos. of Inputs are High, output is high otherwise output is low.*

Since an X-OR gate produces an output 1 only when the inputs are not equal, it is called an anti-coincidence gate or inequality detector. The output of an X- OR gate is the modulo sum of its two inputs. The name Exclusive-OR is derived from the fact that its output is a 1, only when exclusively one of its inputs is a 1 (it excludes the conduction when both the inputs are 1).

Examples 2.5: Implement a NOT circuit using a two-input EX-OR gate?

Solution: Refer to the truth table of a two-input EX-OR gate reproduced in Fig. 2.17 (c). It is clear from the truth table that, if one of the inputs of the gate is permanently tied to logic '1' level, then the other input and output perform the function of a NOT circuit. Fig. 2.18 shows the implementation.

Fig. 2.18: (a) Implementation of a NOT circuit using an EX-OR gate.

<u>The 'EXCLUSIVE-NOR' (X-NOR or EX-NOR) Gate</u>

EXCLUSIVE-NOR (commonly written as EX-NOR) means NOT of EX-OR, i.e., the logic gate that we get by complementing the output of an EX-OR gate.

Statement: *<u>For an EX-NOR Gate the output is high if even no. (Like 2,4,6....) of inputs are high otherwise (0, 1, 3....) the output is low.</u>*

Fig. 2.19 shows its circuit symbol along with its truth table. The truth table of an EX-NOR gate is obtained from the truth table of an EX-OR gate by complementing the output entries.

Logic Symbol	Expression	Truth Table
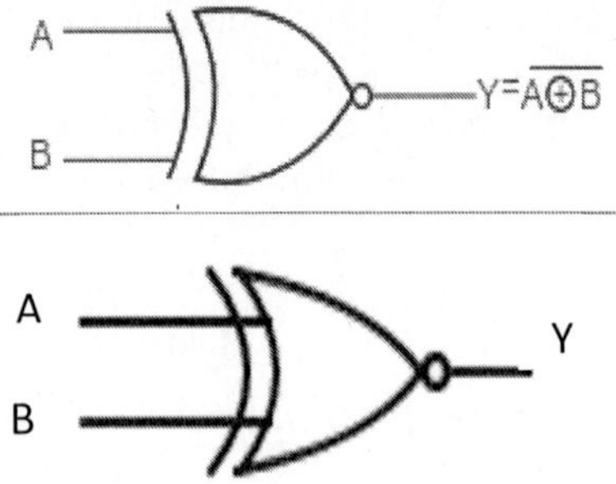	$X = A \odot B = AB + \bar{A}\bar{B}$ $\overline{A \oplus B} = \overline{A\bar{B} + \bar{A}B}$	
(a)	(b)	(c)

Input		Output
A	B	$Y = A \odot B$
0	0	1
0	1	0
1	0	0
1	1	1

Fig. 2.19: (a) Circuit symbol of a two-input EXCLUSIVE-NOR gate, (b) expression, and (c) the truth table.

Statement: for an EX-NOR Gate the output is high if even no. (Like 2,4,6….) of inputs are high otherwise (0, 1, 3….) the output is low.

2.3.3 Universal Gates

Though logic circuits of any complexity can be realized by using only the three basic gates (AND, OR and NOT), however, there are two universal gates (NAND and NOR), each of which can also realize logic circuits single-handedly. The NAND and NOR gates are also, therefore, called universal building blocks. Both NAND and NOR gates can perform all three basic logic functions (AND, OR, and NOT). Therefore, any logic expression can be converted and implemented using AND logic or NOR logic.

The 'NAND' Gate

NAND stands for NOT along with AND. Hence, an AND gate followed by a NOT circuit makes it a NAND gate. Figure 2.19(b) shows the circuit symbol of a two-input NAND gate.

Logic Symbol	Expression	Truth Table
	$Y = \overline{A.B}$ (Read as A anded with B whole Bar)	
(b)	(c)	(d)

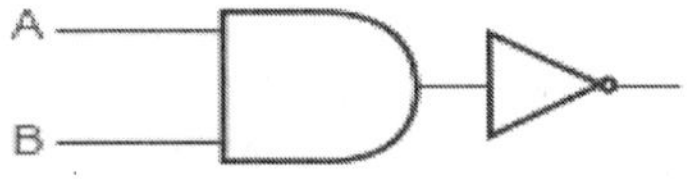

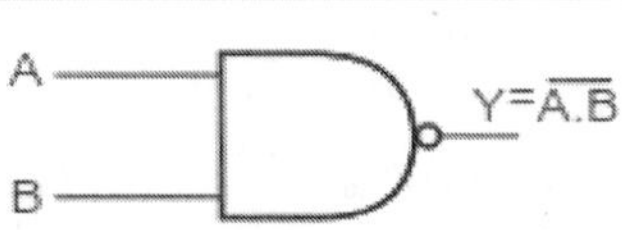

Input		Output
A	B	$Y = \overline{A.B}$
0	0	1
0	1	1
1	0	1
1	1	0

Fig. 2.20: Two-input NAND (a) Implementation using an AND gate and a NOT circuit, (b) Circuit symbol (c) Expression, and (d) Truth table

The truth table of a NAND gate is obtained from the truth table of an AND gate by complementing the output entries (Refer Fig. 2.10). The output of a NAND gate is a logic '0' when all its inputs are a logic '1'. For all other input combinations, the output is a logic '1'.

Statement: *when any of the inputs is low, the output is high. However, if all the inputs are high, the output is Low.*

Concept of Bubbled OR

The NAND gate can perform the function of the OR Gate (Refer Fig. 2.21). A NAND function can also be realized by first inverting the inputs and then ORing the inverted inputs. Hence, a NAND gate is a combination of two NOT gates and an OR Gate. The Corresponding Output expression is

$$X = \overline{AB} = \bar{A} + \bar{B}$$

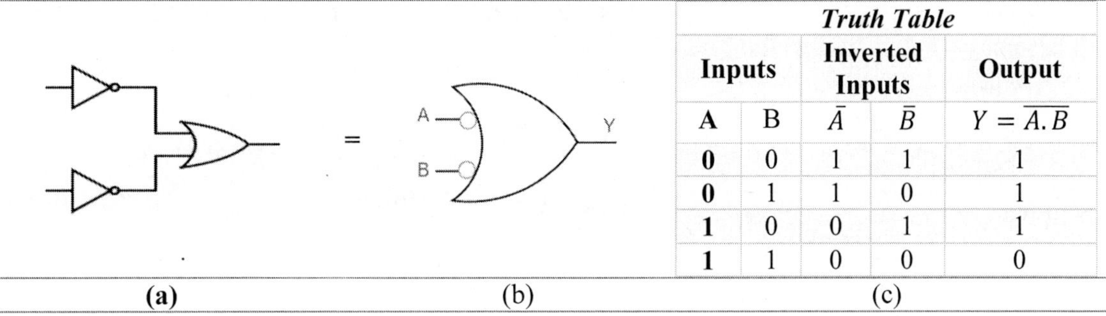

Truth Table

Inputs		Inverted Inputs		Output
A	B	$\bar{A}$	$\bar{B}$	$Y = \overline{A.B}$
0	0	1	1	1
0	1	1	0	1
1	0	0	1	1
1	1	0	0	0

(a) (b) (c)

Fig. 2.21: Bubbled OR gate.

The 'NOR' Gate

Logic Symbol | **Expression** | **Truth Table**

A B Y

(a)

A B Y=$\overline{A+B}$

$Y = \overline{(A+B)}$

(Read as A ORed with B whole bar)

A	B	Y
0	0	1
0	1	0
1	0	0
1	1	0

(b) (c) (d)

Fig.2.22: Two-input NOR (a) Implementation using an OR gate and a NOT circuit, (b) Circuit symbol (c) Truth table

NOR stands for NOT along with OR. An OR gate followed by a NOT circuit makes it a NOR gate (Refer Fig. 2.22). The truth table of a NOR gate is obtained from the truth table of an OR gate by complementing the output entries.

Statement: *when any of the inputs is high, the output is low. However, if all the inputs are low, the output is high.*

The output of a NOR gate is a logic '1' when all its inputs are logic '0'. For all other input combinations, the output is a logic '0'. The output of a two-input NOR gate is logically expressed as

$$Y = \overline{(A + B)}$$

Concept of Bubbled AND Gate

Looking at the truth table of a two-input NOR gate, we see that the output X is 1 only when both A and B are equal to 0, i.e., only when both $\bar{A}$ and $\bar{B}$ are equal to 1. That means, a NOR gate is equivalent to an AND gate with inverted inputs and the corresponding output expression is, X = $\bar{A}\bar{B}$. Hence, a NOR function can also be realized by first inverting the inputs and then ANDing those inverted inputs. Thus, a NOR gate is a combination of two NOT gates and an AND gate (see Fig. 2.23). Hence,

$$Y = \overline{A + B} + \bar{A}\bar{B} =.$$

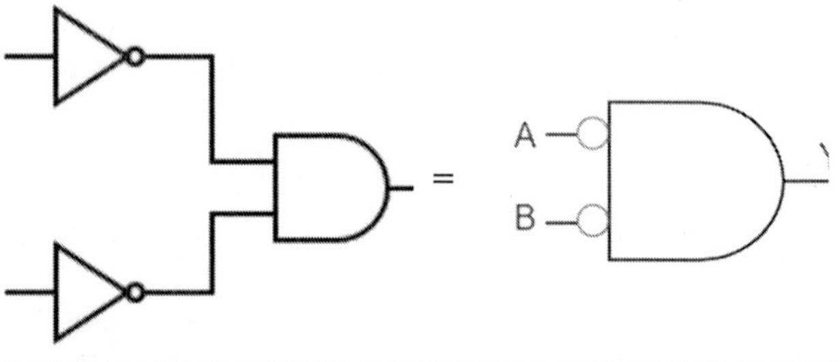

Truth Table				
Inputs		**Inverted Inputs**		**Output**
A	B	$\bar{A}$	$\bar{B}$	$Y = \overline{A + B}$
0	0	1	1	1
0	1	1	0	0
1	0	0	1	0
1	1	0	0	0

(a) (b) (c)

Fig. 2.23: Bubbled AND Gate.

The AND gate with inverted inputs (Fig. 2.22a) is called a bubbled AND gate. So, a NOR gate is equivalent to a bubbled AND gate whose truth table is shown in Fig.2.22b. A bubbled AND gate is also called a negative AND gate. Since its output assumes the HIGH state only when all its inputs are in LOW states, a NOR gate is also called an active-LOW AND gate.

NOR gate realization using transistors:

A discrete two-input NOR gate is shown in Fig. 2.24 (a).

- When A = 0 V and B = 0 V, both transistors T1 and T2 are OFF; so, no current flows through R and, therefore, no voltage drop occurs across R. Hence, the output voltage X = + 5 V (logic 1).
- When either A = + 5 V or B = +5 V or when both A and B are equal to + 5 V, the corresponding transistor T1 or T2 or both T1 and T2 are ON. Therefore, X is at Vc(sat) with respect to ground and equal to 0V (logic 0).

The truth table is shown in Fig. (b). The IC 7402 contains four two-input NOR gates; the IC 7427 contains three three-input NOR gates, and the IC 7425 contains two four-input NOR gates.

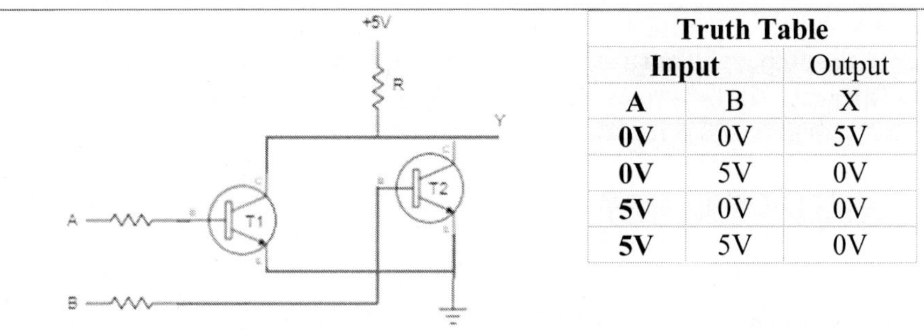

Truth Table		
Input		Output
A	B	X
0V	0V	5V
0V	5V	0V
5V	0V	0V
5V	5V	0V

Fig. 2.24: Discrete two-input NOR gate.

Why they Called Universal gates

OR, AND, and NOT gates are the three basic logic gates as they together can be used to construct the logic circuit for any given Boolean expression. NOR and NAND gates have the property that they individually can be used to hardware-implement a logic circuit corresponding to any given Boolean expression. That is, it is possible to use either only NAND gates or only NOR gates to implement any Boolean expression. It is for this reason that NAND and NOR gates are universal gates.

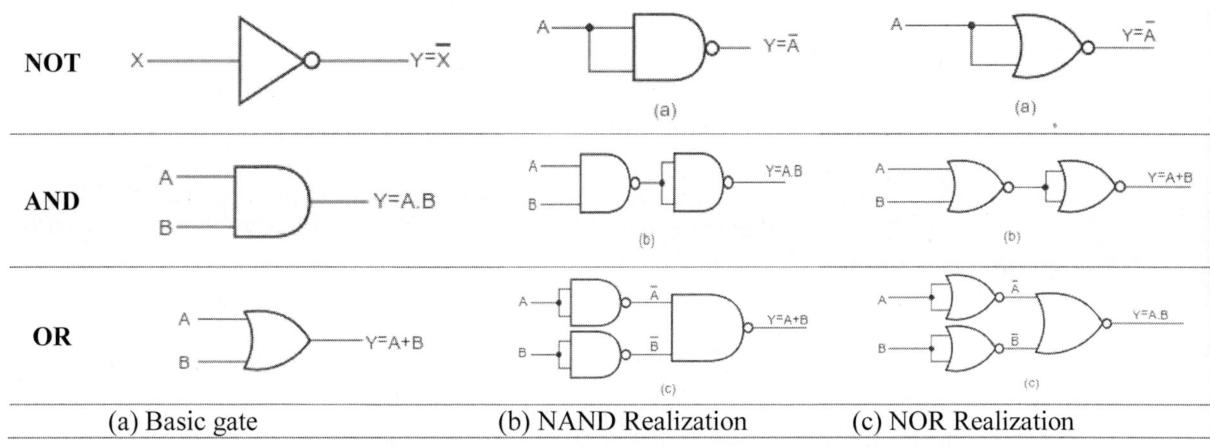

(a) Basic gate (b) NAND Realization (c) NOR Realization

Fig.2.25: Implementation of basic logic gates using only either NAND or NOR gates.

As an illustration, Fig. 2.25 (b) shows how two-input NAND gates can be used to construct a NOT gate AND gate and a two-input OR gate. Fig. 2.25 (c) shows the same using NOR gates. Understanding the conversion of NAND to OR and NOR to AND requires the use of DeMorgan's theorem, which is discussed in upcoming section on Boolean algebra.

2.3.4 INHIBIT Gate

There are many situations in digital circuit design where the passage of a logic signal needs to be *either enabled or inhibited* depending upon certain other control inputs. INHIBIT here means that the gate produces a certain fixed logic level at the output irrespective of changes in the input logic level.

As an illustration, if one of the inputs of a four-input NOR gate is permanently tied to the logic '1' level, then the output will always be at the logic '0' level irrespective of the logic status of other inputs. This gate will behave as a NOR gate only when this control input is at logic '0' level. This is an example of the INHIBIT function. The INHIBIT function is available in integrated circuit form for an AND gate, which is basically an AND gate with one of its inputs negated by an inverter. The negated input acts to inhibit the gate. In other words, the gate will behave like an AND gate only when the negated input is driven to a logic '0'. Fig. 2.26 shows the circuit symbol and truth table of a four-input INHIBIT gate.

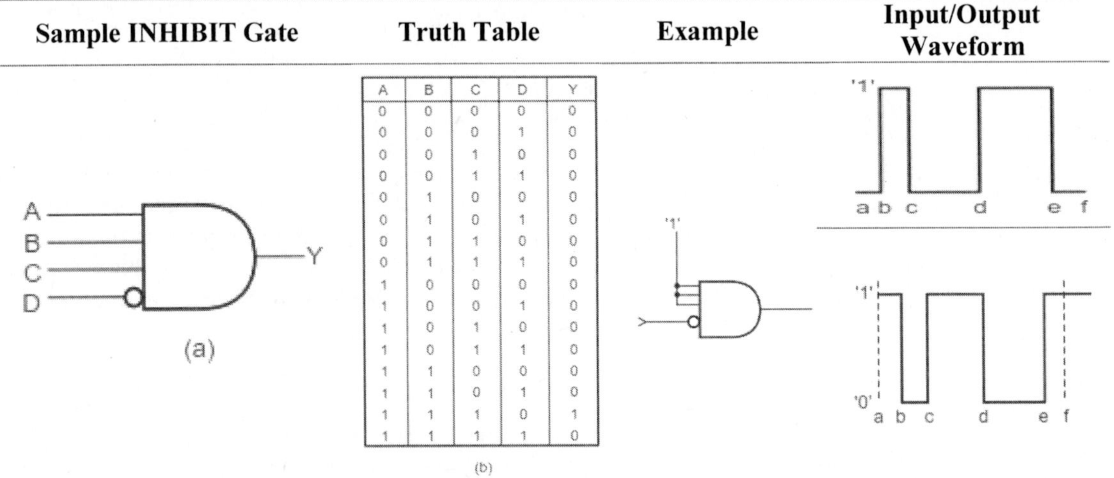

A	B	C	D	Y
0	0	0	0	0
0	0	0	1	0
0	0	1	0	0
0	0	1	1	0
0	1	0	0	0
0	1	0	1	0
0	1	1	0	0
0	1	1	1	0
1	0	0	0	0
1	0	0	1	0
1	0	1	0	0
1	0	1	1	0
1	1	0	0	0
1	1	0	1	0
1	1	1	0	1
1	1	1	1	0

Fig.2.26: INHIBIT gate.

Observation: *Since all other inputs of the gate have been permanently tied to logic '1' level, a logic '0' at the INHIBIT input would produce a logic '1' at the output and a logic '1' at the INHIBIT input would produce a logic '0' at the output.* The output waveform is therefore the inversion of the input waveform and is shown in Fig. 2.26.

Example 2.6: Construct the Truth Table of EX-OR and EX-NOR and Implement them using basic gates.

Solution: The truth table for EX-OR constructed below shows that can be represent by the equation below

$$A \oplus B = A\bar{B} + \bar{A}B$$

The corresponding logic diagram and truth table is shown in Fig. 2.27:

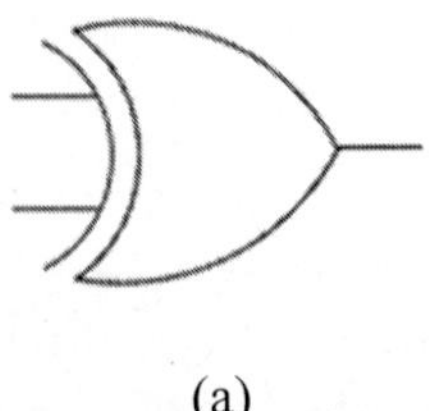

(a)

A	B	Y=A⊕B
0	0	0
0	1	1
1	0	1
1	1	0

(b)

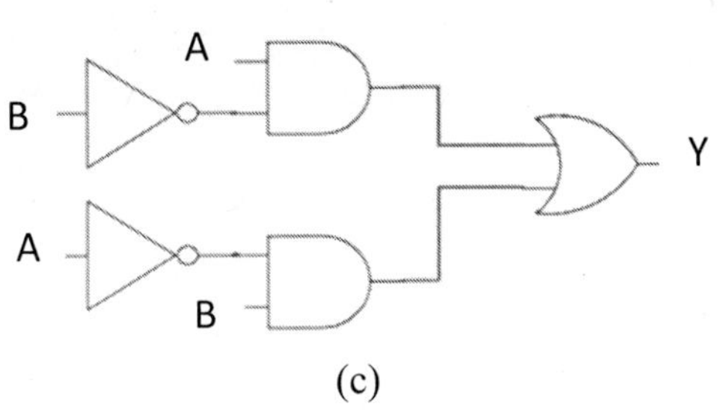

A	B	$\bar{A}$	$\bar{B}$	$\bar{A}B$	$A\bar{B}$	$\bar{A}B + A\bar{B}$
0	0	1	1	0	0	0
0	1	1	0	1	0	1
1	0	0	1	0	1	1
1	1	0	0	0	0	0

(c) (d)

Fig. 2.27: Two input EX-OR gate (a) Logic Symbol (b) Truth Table (c) EX-OR gate using Basic Gates (d) Truth Table

The truth table for EX-NOR constructed below shows that can be represent by the equation below

$$A \odot B = AB + \bar{A}\bar{B}$$

The corresponding logic diagram and truth table is shown in Fig. 2.28

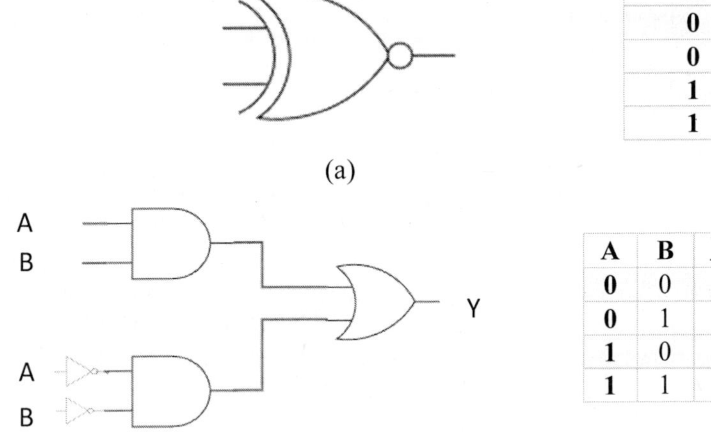

A	B	Y=A□B
0	0	1
0	1	0
1	0	0
1	1	1

(a) (b)

A	B	$\bar{A}$	$\bar{B}$	AB	$\bar{A}\bar{B}$	$AB + \bar{A}\bar{B}$
0	0	1	1	0	1	1
0	1	1	0	0	0	0
1	0	0	1	0	0	0
1	1	0	0	1	0	1

(c) (d)

Fig. 2.28: Two input EX-OR gate (a) Logic Symbol (b) Truth Table (c) EX-OR gate using Basic Gates (d) Truth Table

2.3.5 Gates with Open Collector/Drain Outputs

These are gates where we need to connect an external resistor, called the *pull-up resistor*, between the output and the DC power supply to make the logic gate perform the intended logic function. Depending on the logic family used to construct the logic gate, they are referred to as gates with open collector output (in the case of the TTL logic family) or open drain output (in the case of the MOS logic family). Moreover, Logic families are discussed in detail in upcoming section.

The advantage of using open collector/open drain gates lies in their capability of providing an ANDing operation when outputs of several gates are tied together through a common pull-up resistor, without having to use an AND gate for the purpose. This connection is also referred to as WIRE-AND connection. 2.29(a) shows such a connection for open collector NAND gates.

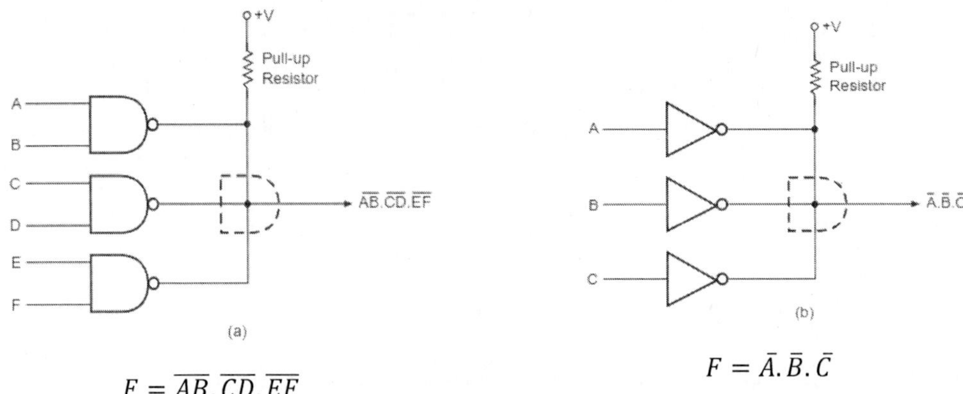

$F = \overline{AB}.\overline{CD}.\overline{EF}$ $\qquad F = \bar{A}.\bar{B}.\bar{C}$

Fig.2.29: WIRE-AND connection with open collector/drain devices.

Fig. 2.29 (b) shows a similar arrangement for NOT gates. The disadvantage is that they are relatively slower and noisier. Open collector/drain devices are therefore not recommended for applications where speed is an important consideration.

2.4 Tristate Logic Gates

Tristate logic gates have three possible output states, i.e. the logic '1' state, the logic '0' state and a high-impedance state. The high-impedance state is controlled by an external ENABLE input. The ENABLE input decides whether the gate is active or in the high-impedance state. When active, it can be '0' or '1' depending upon input conditions. One of the main advantages of these gates is that their inputs and outputs can be connected in parallel to a common bus line. Fig. 2.30 (a) shows the circuit symbol of a tristate NAND gate with active HIGH ENABLE input, along with its truth table. The one shown in Fig. 2.30 (b) has active LOW ENABLE input. When tristate devices are paralleled, only one of them is enabled at a time. Fig. 2.30 (c) shows paralleling of tristate inverters having active HIGH ENABLE inputs.

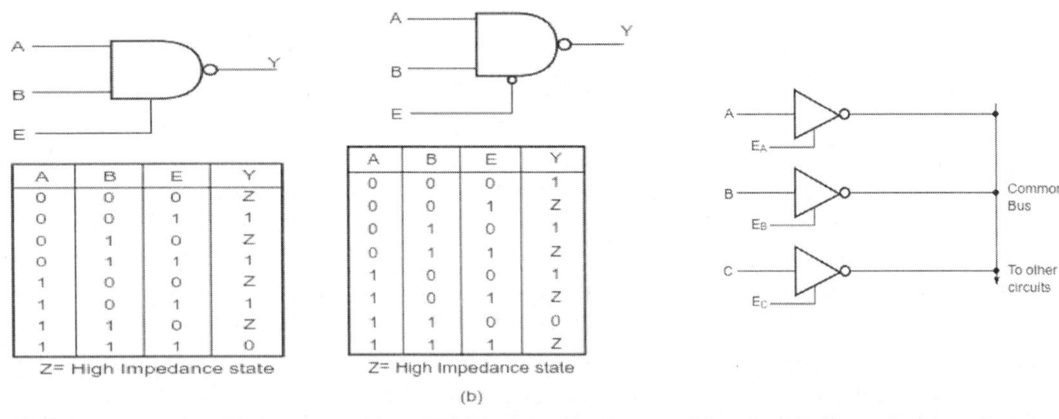

A	B	E	Y
0	0	0	Z
0	0	1	1
0	1	0	Z
0	1	1	1
1	0	0	Z
1	0	1	1
1	1	0	Z
1	1	1	0

Z= High Impedance state

A	B	E	Y
0	0	0	1
0	0	1	Z
0	1	0	1
0	1	1	Z
1	0	0	1
1	0	1	Z
1	1	0	0
1	1	1	Z

Z= High Impedance state

(a) Tri-state Active High (b) Tri-state Active Low (c) Paralleling of tristate inverters

Fig.2.30: Tristate devices

2.5 Schmitt Gates (Trigger)

The logic gates discussed so far have a single-input threshold voltage level. This threshold is the same for both LOW-to-HIGH and HIGH-to-LOW output transitions. This threshold voltage lies somewhere between the highest LOW voltage level and the lowest HIGH voltage level guaranteed by the manufacturer of the device. These logic gates can produce an erratic output when fed with a slow varying input. Fig. 2.31 shows the response of an inverter circuit when fed with a slow varying input both in the case of an ideal signal [Refer Fig. 2.31(a)] and in the case of a practical signal having a small amount of AC noise superimposed on it [Refer Fig. 2.31(b)].

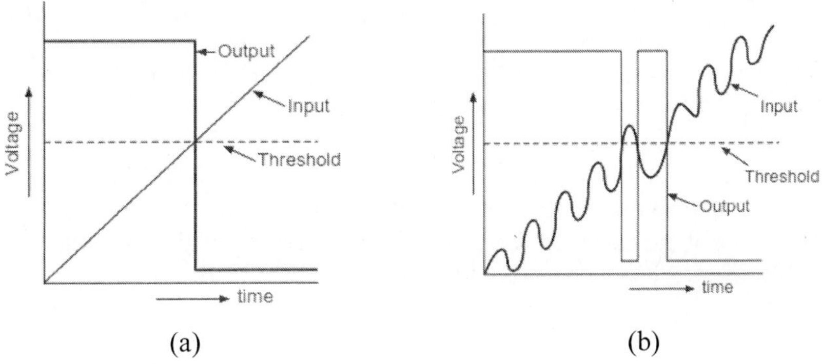

Fig.2.31: Response of conventional inverters to slow varying input.

A possible solution to this problem lies in having two different threshold voltage levels, one for LOW-to-HIGH transition and the other for HIGH-to-LOW transition, by introducing some positive feedback in the internal gate circuitry, a phenomenon called hysteresis.

There are some logic gate varieties, mainly in NAND gates and inverters, that are available with built-in hysteresis. These are called Schmitt gates, which interpret varying input voltages according to two threshold voltages, one for LOW-to-HIGH and the other for HIGH-to-LOW output transition.

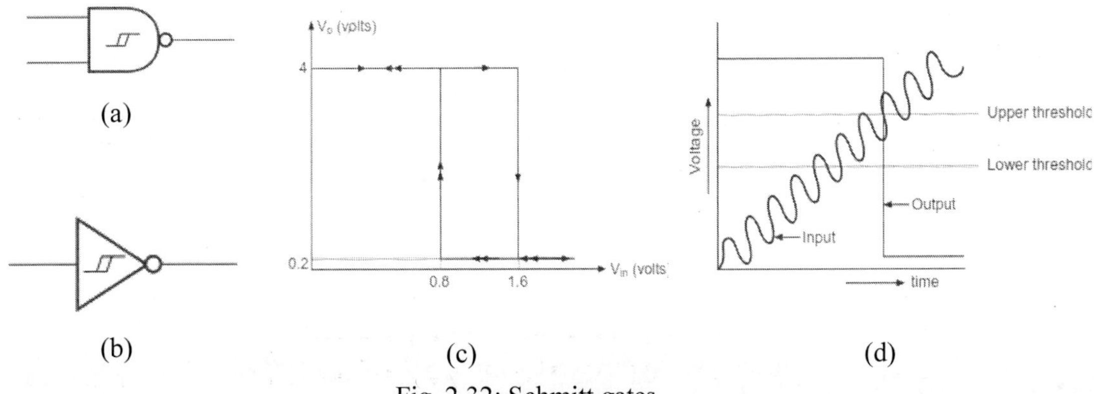

Fig. 2.32: Schmitt gates.

Fig. 2.32 (a) and (b) respectively show circuit symbols of Schmitt NAND and Schmitt inverter. Schmitt gates are distinguished from conventional gates by the small 'hysteresis' symbol reminiscent of the B-H loop for a ferromagnetic material. Fig. 2.32 (c) shows typical transfer characteristics for such a device. The difference between the two threshold levels is the hysteresis. These characteristics have been reproduced from the data sheet of IC 74LS132, which is a quad two-input Schmitt NAND belonging to the low-power Schottky TTL family. Fig. 3.32 (d) shows the response of a Schmitt inverter to a slow varying noisy input signal. We will learn more about different logic families in upcoming section. It may be mentioned here that hysteresis increases noise immunity and is used in applications where noise is expected on input signal lines.

2.6 AND-OR-INVERT Gates

AND-OR and OR-AND gates can be usefully employed to implement sum-of-products and product of-sums Boolean expressions respectively. Fig. 2.33 (a) and (b) respectively show the symbols of AND-OR-INVERT and OR-AND-INVERT gates. Another method for designating the gates shown in Fig. 2.33 is to call them two-wide, two-input AND-OR-INVERT or OR-AND-INVERT gates as the case may be. The gate is two-wide as there are two gates at the input, and two-input as each of the gates has two inputs. Other varieties such as two-wide, four-input AND-OR-INVERT (Fig. 2.34) and four-wide, two-input AND-OR-INVERT (Fig. 2.35) are also available in IC form.

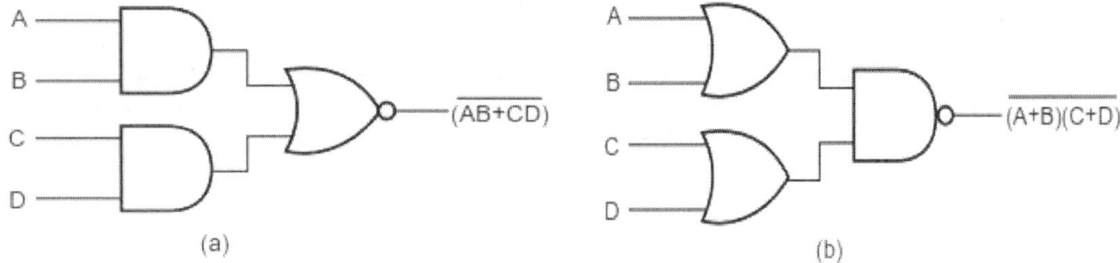

Fig. 2.33: (a) AND-OR-INVERT (b) OR-AND-INVERT

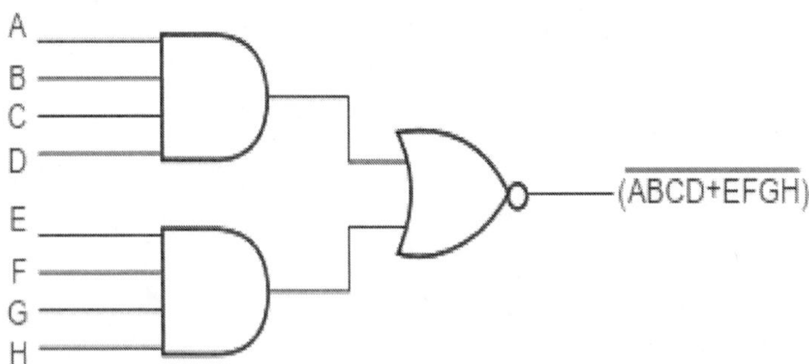

Fig. 2.34: Two-wide Four-Point AND-OR-INVERT

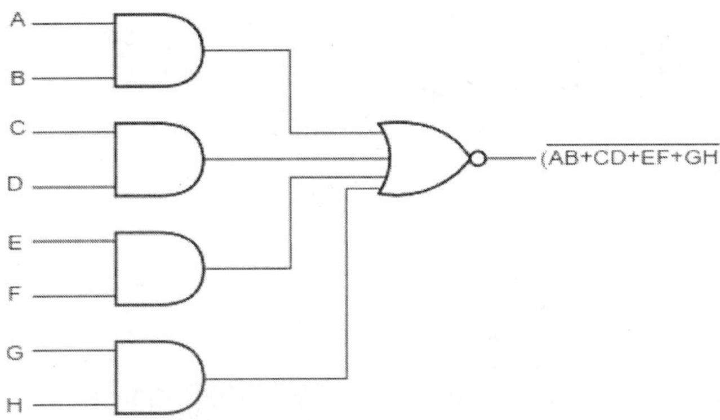

Fig. 2.35: Four-wide Two-Point AND-OR-INVERT

2.7 Buffers and Transceivers

Logic gates, discussed in the previous pages, have a limited load-driving capability. A buffer has a larger load-driving capability than a logic gate. It could be an inverting or noninverting buffer with a single input, a NAND buffer, a NOR buffer, an OR buffer or an AND buffer. 'Driver' is another name for a buffer. A driver is sometimes used to designate a circuit that has even larger drive capability than a buffer. Buffers are usually tristate devices to facilitate their use in bus-oriented systems.

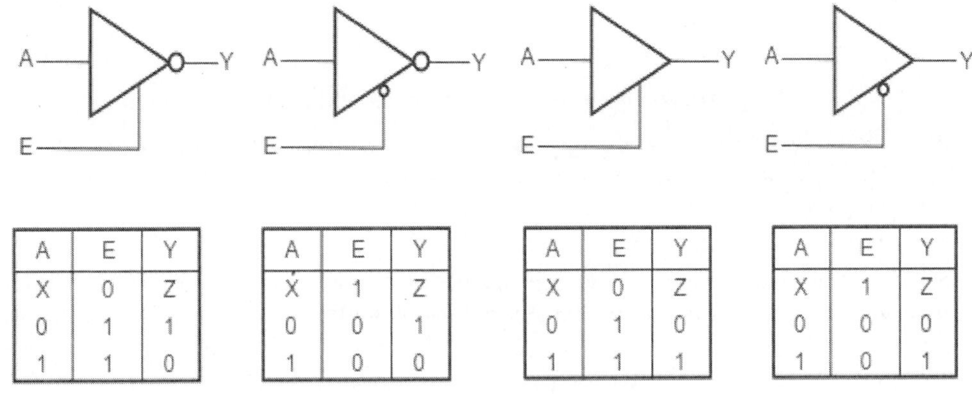

A	E	Y
X	0	Z
0	1	1
1	1	0

A	E	Y
X	1	Z
0	0	1
1	0	0

A	E	Y
X	0	Z
0	1	0
1	1	1

A	E	Y
X	1	Z
0	0	0
1	0	1

Z = High Impedance State

Fig. 2.36: Inverting tristate and noninverting tristate buffers.

Fig. 2.36 shows the symbols and functional tables of inverting and noninverting buffers of the tristate type. A transceiver is a bidirectional buffer with additional direction control and ENABLE inputs. It allows flow of data in both directions, depending upon the logic status of the control inputs. Transceivers, like buffers, are tristate devices to make them compatible with bus-oriented systems.

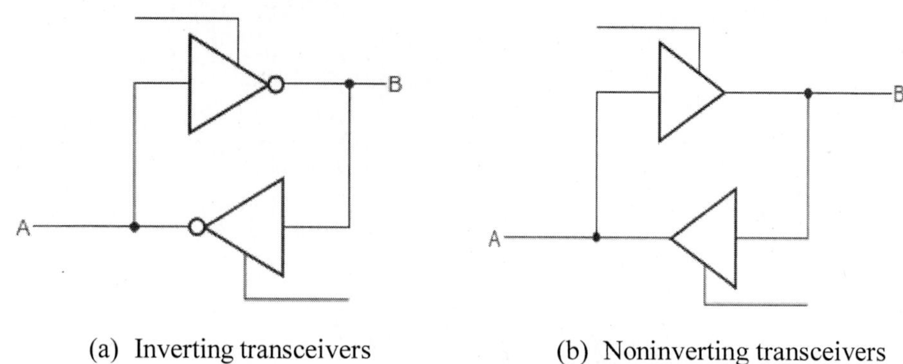

(a) Inverting transceivers (b) Noninverting transceivers

Fig. 2.37: Inverting and noninverting transceivers.

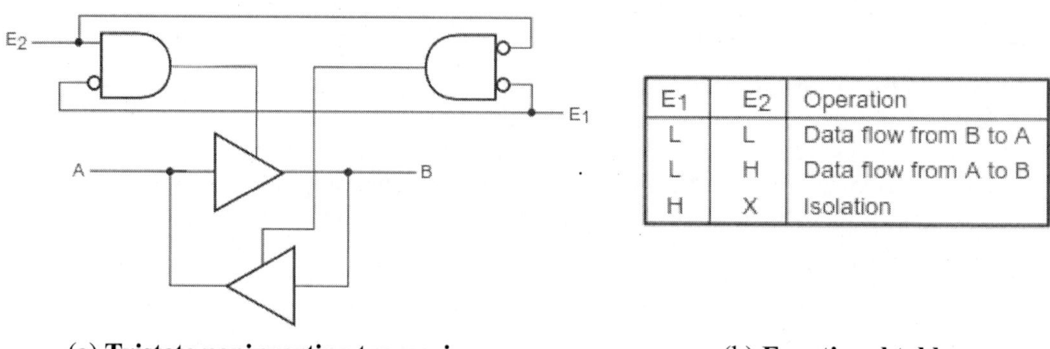

E_1	E_2	Operation
L	L	Data flow from B to A
L	H	Data flow from A to B
H	X	Isolation

(a) **Tristate noninverting transceiver** (b) **Functional table**

Fig. 2.38: Tristate noninverting transceiver and its Functional table

Fig. 2.37 (a) and (b) respectively show the circuit symbols of inverting and noninverting transceivers. Fig. 2.38 shows a typical logic circuit arrangement of a tristate noninverting transceiver with its functional table.

Some of the common applications of inverting and noninverting buffers are as follows. Buffers are used to drive circuits that need more drive current. Noninverting buffers are also used to increase the fan-out of a given logic gate. This means that the buffer can be used to increase the number of logic gate inputs to which the output of a given logic gate can be connected. Yet another application of a noninverting buffer is its use as a delay line. It delays the signal by an amount equal to the propagation delay of the device. More than one device can be connected in cascade to get larger delays.

2.8 IEEE/ANSI Standard Symbols

The symbols used thus far in the chapter for representing different types of gates are the ones that are better known to all of us and have been in use for many years. Each logic gate has a symbol with a distinct shape. However, for more complex logic devices, e.g., sequential logic devices like flip-flops, counters, registers or arithmetic circuits, such as adders, subtractors, etc., these symbols do not carry any useful information. A new set of standard symbols was introduced in

1984 under IEEE/ANSI Standard 91–1984. The logic symbols given under this standard are being increasingly used now and have even started appearing in the literature published by manufacturers of digital integrated circuits.

The utility of this new standard will be more evident in the following paragraphs as we go through its salient features and illustrate them with practical examples.

2.8.1 IEEE/ANSI Standards – Salient Features

This standard uses a rectangular symbol for all devices instead of a different symbol shape for each device. For instance, all logic gates (OR, AND, NAND, NOR) will be represented by a rectangular block.

A right triangle is used instead of a bubble to indicate inversion of a logic level. Also, the right triangle is used to indicate whether a given input or output is active LOW. The absence of a triangle indicates an active HIGH input or output. As far as logic gates are concerned, a special notation inside the rectangular block describes the logic relationship between output and inputs. A '1' inside the block indicates that the device has only one input. An AND operation is expressed by '&', and an OR operation is expressed by the symbol '≥ 1'.

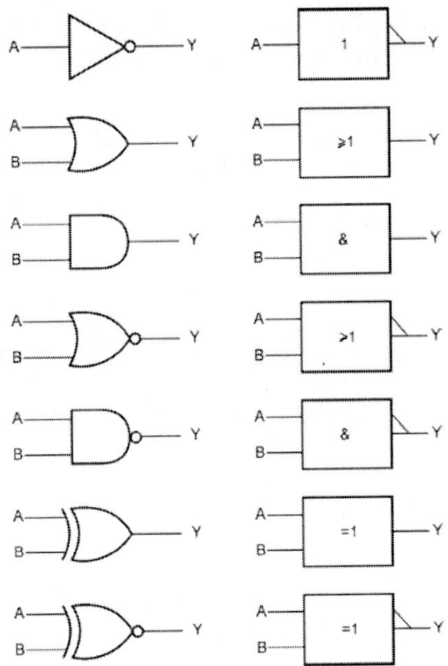

Fig. 2.39: IEEE / ANSI symbols.

Fig. 2.39 shows the ANSI counterparts of various logic gates. A '≥1' symbol indicates that the output is HIGH when one or more than one input is HIGH. An '&' symbol indicates that the output is HIGH only when all the inputs are HIGH. The two-input EX-OR is represented by the symbol '=1' which implies that the output is HIGH only when one of its inputs is HIGH.

A special dependency notation system is used to indicate how the outputs depend upon the input. This notation contains almost the entire functional information of the logic device in question. This will be clearer as we illustrate this new standard with the help of ANSI symbols for some of the actual devices belonging to the category of flip-flops, counters, etc., in the following chapters. All those control inputs that control the timing of change in output states as per logic status of inputs are designated by the letter 'C'. These are ENABLE inputs in latches or CLOCK inputs in flip-flops. Most of the digital ICs contain more than one similar function on one chip such as IC 7400 (quad two-input NAND), IC 7404 (hex inverter), IC 74112 (dual-edge triggered JK flip-flop), IC 7474 (dual D-type latch), IC 7475 (quad D-type latch) and so on. Those inputs to such ICs that are common to all the functional blocks or in other words similarly affect various individual but similar functions are represented by a separate notched rectangle on the top of the main rectangle.

2.9 Some Common Applications of Logic Gates

In this section, we will briefly look at some common applications of basic logic gates. The applications discussed here include those where these devices are used to provide a specific function in a larger digital circuit. These also include those where one or more logic gates, along with or without some external components, can be used to build some digital building blocks.

Applications of OR Gate

An OR gate can be used in all those situations where the occurrence of any one or more than one event needs to be detected or acted upon. One such example is an industrial plant where any one or more than one parameter exceeding a preset limiting value should lead to initiation of some kind of protective action.

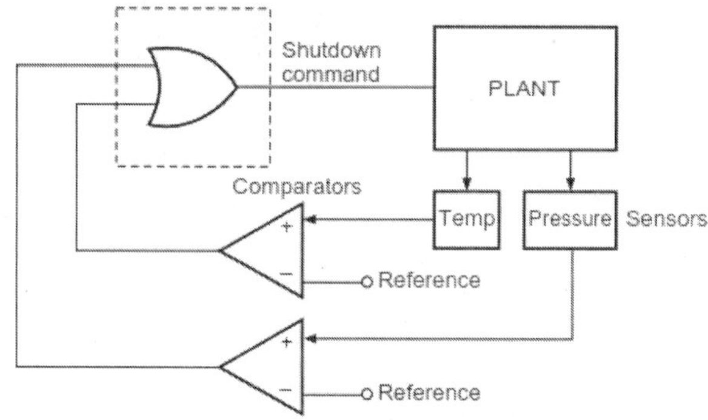

Fig. 2.40: Application of an OR gate.

Fig. 2.40 shows a typical schematic where the OR gate is used to detect either temperature or pressure exceeding a preset threshold value and produce the necessary command signal for the system.

Applications of AND Gate

An AND gate is commonly used as an ENABLE or INHIBIT gate to allow or disallow passage of data from one point in the circuit to another. One such application of enabling operation, for instance, is in the measurement of the frequency of a pulsed waveform or the width of a given pulse with the help of a counter. In the case of frequency measurement, a gating pulse of known width is used to enable the passage of the pulse waveform to the counter's clock input. In the case of pulse width measurement, the pulse is used to enable the passage of the clock input to the counter. Fig. 2.41 shows the arrangement.

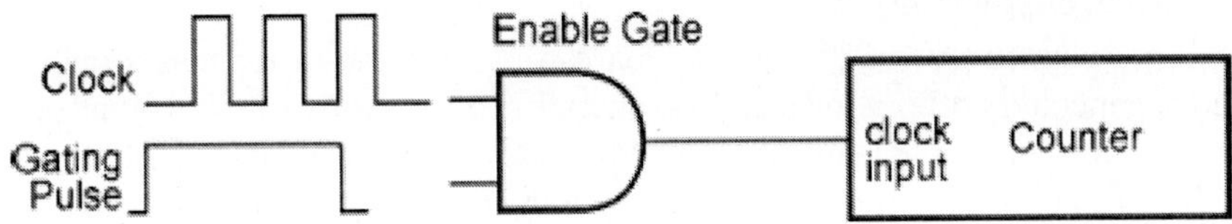

Fig. 2.41: Application of an AND gate.

Applications of EX-OR/EX-NOR Gate

EX-OR and EX-NOR logic gates are commonly used in parity generation and checking circuits. Figures Fig. 2.42 (a) and (b) respectively show even and odd parity generator circuits for four-bit data. The circuits are self-explanatory.

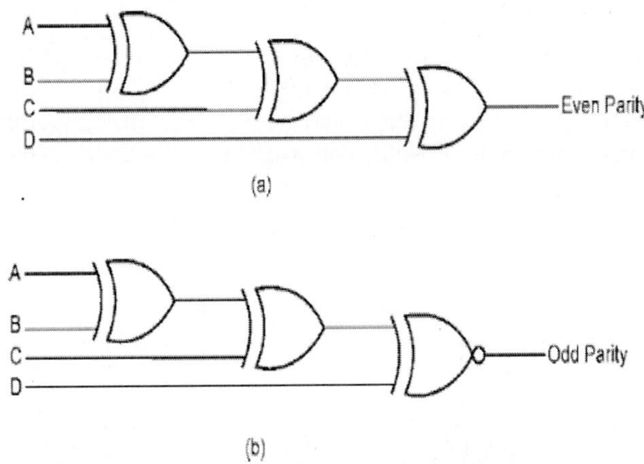

Fig.2.42: Parity generation using EX-OR/EX-NOR gates.

The parity check operation can also be performed by similar circuits. Fig. 2.43 (a) and (b) respectively show simple even and odd parity check circuits for a four-bit data stream. In the circuits shown in Fig. 2.43, a logic '0' at the output signifies correct parity and a logic '1' signifies one-bit error. Parity generator/checker circuits are available in IC form. 74180 in TTL and 40101 in CMOS are nine-bit odd/even parity generator/checker ICs. Parity generation and checking circuits are further discussed in coming section on arithmetic circuits.

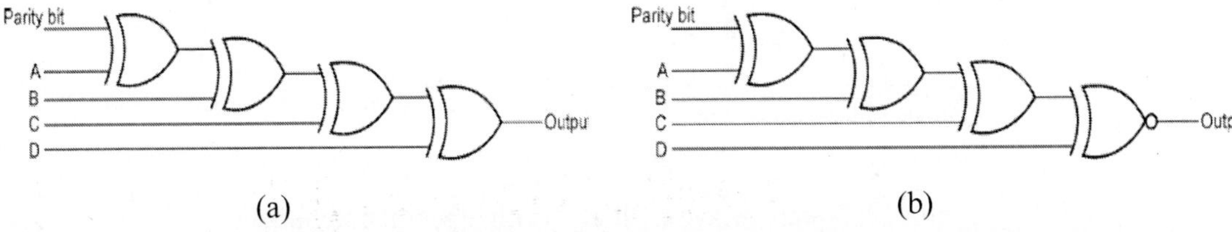

(a) (b)

Fig. 2.43: Parity check using (a)EX-OR and (b)EX-NOR gates.

Applications of NOT gate (Inverter)

CMOS inverters are commonly used to build square-wave oscillators for generating clock signals. These clock generators offer good stability, operation over a wide supply voltage range (3–15 V) and frequency range (1 Hz to in excess of 15 MHz), low power consumption and an easy interface to other logic families. The most fundamental circuit is the ring configuration of any odd number of inverters.

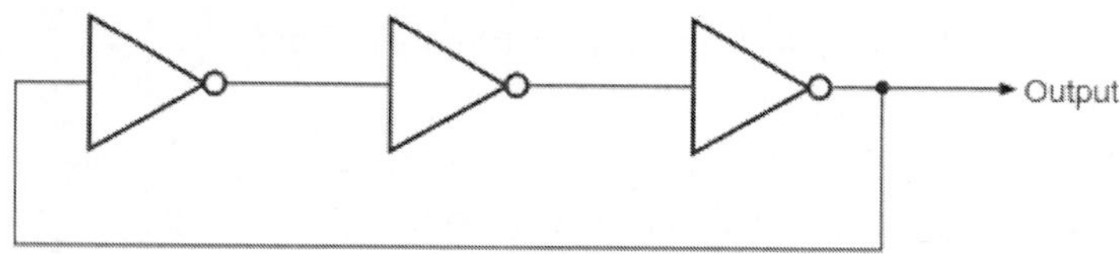

Fig. 2.44: Square- wave oscillator using a ring configuration.

Fig. 2.44 shows one such circuit using three inverters. Inverting gates such as NAND and NOR gates can also be used instead. This configuration does not make a practical oscillator circuit as its frequency of oscillation is highly susceptible to variation with temperature, supply voltage and external loading. The frequency of oscillation is given by the equation,

$$f = 1/2nt_p$$

$where, n\ is\ the\ number\ of\ inverters\ and\ t_p is\ the\ propogation\ delay\ per\ gate$

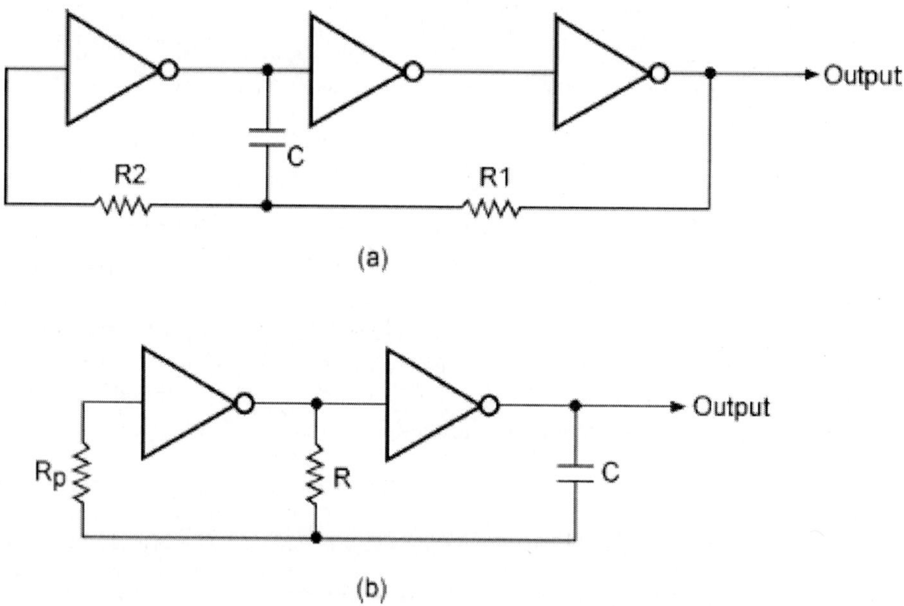

Fig. 2.45: Square- wave oscillator with external components.

Fig. 2.45 (a) shows a practical oscillator circuit. The frequency of oscillation in this case is given by Equation below (the duty cycle of the waveform is approximately 50 %):

$$f = 1/2C(0.405R_{eq} + 0.693R_1) \qquad where: R_{eq} = \frac{R_1.R_2}{R_1 + R_2}$$

Fig. 2.45 (b) shows another circuit using two inverters instead of three inverters. The frequency of oscillation of this circuit is given by the equation

$$f = 1/2.2RC$$

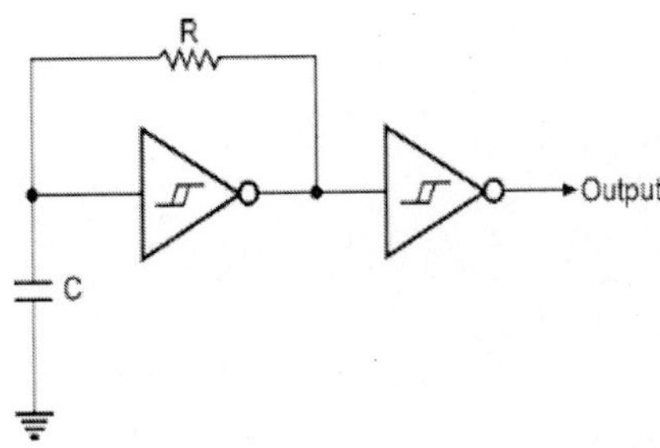

Fig. 2.46: Schmitt inverter-based oscillator.

The circuits shown in Fig. 2.45 are not as sensitive to supply voltage variations as the one shown in Fig. 2.44. Fig. 2.46 shows yet another circuit that is configured around a single Schmitt inverter. The capacitor charges (when the output is HIGH) and discharges (when the output is LOW) between the two threshold voltages. The frequency of oscillation, however, is sensitive to supply voltage variations. It is given by the equation

$$f = 1/RC$$

Fig. 2.47 shows a crystal oscillator configured around a single inverter as the active element. Any odd number of inverters can be used. A larger number of inverters limits the highest attainable frequency of oscillation to a lower value.

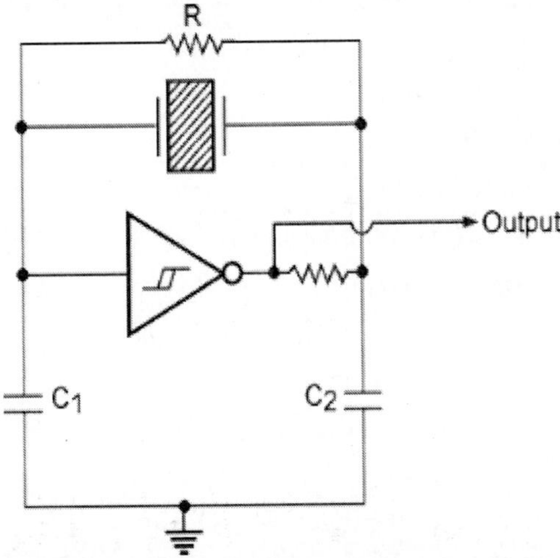

Fig. 2.47: Crystal oscillator configured around a single inverter

2.10 Application-Relevant Package Information

Table 2.2 lists the commonly used type nubers along with the functional description and the logic family. The pin connection diagrams and the functional tables of the more popular type numbers are given in the companion website.

Table 2.2: Functional index of logic gates.

Type (IC) No.	Function	Logic Family
7400	Quad two-input NAND gate	TTL
7401	Quad two-input NAND gate (open collector)	TTL
7402	Quad two-input NOR gate	TTL
7403	Quad two-input NAND gate (open collector)	TTL
7404	Hex inverter	TTL
7405	Hex inverter (open collector)	TTL
7408	Quad two-input AND gate	TTL
7409	Quad two-input AND gate (open collector)	TTL
7410	Triple three-input NAND gate	TTL
7411	Triple three-input AND gate	TTL
7412	Triple three-input NAND gate (open collector)	TTL
7413	Dual four-input Schmitt NAND gate	TTL
7414	Hex Schmitt trigger inverter	TTL
7418	Dual four-input Schmitt NAND gate	TTL
7419	Hex Schmitt trigger inverter	TTL
7420	Dual four-input NAND gate	TTL
7421	Dual four-input AND gate	TTL
7422	Dual four-input NAND gate (open collector)	TTL
7427	Triple three-input NOR gate	TTL
7430	Eight-input NAND gate	TTL
7432	Quad two-input OR gate	TTL
7451	Dual two-wide two-input three-input AND-OR-INVERT gate	TTL
7454	Four-wide two-input AND-OR-INVERT gate	TTL
7455	Two-wide four-input AND-OR-INVERT gate	TTL
7486	Quad two-input EX-OR gate	TTL
74125	Quad tristate noninverting buffer (LOW ENABLE)	TTL
74126	Quad tristate noninverting buffer (HIGH ENABLE)	TTL
74132	Quad two-input Schmitt trigger NAND gate	TTL
74133	13-input NAND gate	TTL
74136	Quad two-input EX-OR gate (open collector)	TTL
74240	Octal tristate inverting bus/line driver	TTL

Type (IC) No.	Function	Logic Family
74241	Octal tristate bus/line driver	TTL
74242	Quad tristate inverting bus transceiver	TTL
74243	Quad tristate noninverting bus transceiver	TTL
74244	Octal tristate noninverting driver	TTL
74245	Octal tristate noninverting bus transceiver	TTL
74266	Quad two-input EXCLUSIVE-NOR gate (open collector)	TTL
74365	Hex tristate noninverting buffer with common ENABLE	TTL
74366	Hex tristate inverting buffer with common ENABLE	TTL
74367	Hex tristate noninverting buffer, four-bit and two-bit	TTL
74368	Hex tristate inverting buffer, four-bit and two-bit	TTL
74386	Quad two-input EX-OR gate	TTL
74465	Octal tristate noninverting buffer Gated ENABLE inverted	TTL
74540	Octal tristate inverting buffer/line driver	TTL
74541	Octal tristate noninverting buffer/line driver	TTL
74640	Octal tristate inverting bus transceiver	TTL
74641	Octal tristate noninverting bus transceiver (open collector)	TTL
74645	Octal tristate noninverting bus transceiver	TTL
4001B	Quad two-input NOR gate	CMOS
4002B	Dual four-input NOR gate	CMOS
4011B	Quad two-input NAND gate	CMOS
4012B	Dual four-input NAND gate	CMOS
4023B	Triple three-input NAND gate	CMOS
4025B	Triple three-input NOR gate	CMOS
4030B	Quad two-input EX-OR gate	CMOS
4049B	Hex inverting buffer	CMOS
4050B	Hex noninverting buffer	CMOS
40097B	Tristate hex noninverting buffer	CMOS
40098B	Tristate inverting buffer	CMOS
4069UB	Hex inverter	CMOS
4070B	Quad two-input EX-OR gate	CMOS
4071B	Quad two-input OR gate	CMOS
4081B	Quad two-input AND gate	CMOS
4086B	Four-wide two-input AND-OR-INVERT gate	CMOS
4093B	Quad two-input Schmitt NAND	CMOS

Type (IC) No.	Function	Logic Family
10100	Quad two-input NOR gate with strobe	ECL
10101	Quad two-input OR/NOR gate	ECL
10102	Quad two-input NOR gate	ECL
10103	Quad two-input OR gate	ECL
10104	Quad two-input AND gate	ECL
10113	Quad two-input EX-OR gate	ECL
10114	Triple line receiver	ECL
10115	Quad line Receiver	ECL
10116	Triple Line receiver	ECL
10117	Dual two-wide two- to three-input OR-AND/OR-AND-INVERT gate	ECL
10118	Dual two-wide three-input OR-AND gate	ECL
10123	Triple 4-3-3 input bus driver	ECL
10128	Dual bus driver	ECL
10129	Quad bus driver	ECL
10188	Hex buffer with ENABLE	ECL
10192	Quad bus driver	ECL
10194	Dual simultaneous transceiver	ECL
10195	Hex buffer with invert/non-invert control	ECL

2.11 Boolean Algebra

Boolean algebra is a basic mathematics tool available to the logic designer and thus can be effectively used for simplification of complex logic expressions. Boolean algebra, quite interestingly, is simpler than ordinary algebra.

2.11.1 Introduction

Boolean algebra is composed of a set of symbols and a set of rules as like ordinary algebra to manipulate these symbols. However, this is the only similarity between the two. The differences are many. These include the following:

- In ordinary algebra, the letter symbols can take on any number of values including infinity. In Boolean algebra, they can take on either of two values, that is, 0 and 1.
- The values assigned to a variable have a numerical significance in ordinary algebra, whereas in its Boolean counterpart they have a logical significance.
- While '.' and '+' are respectively the signs of multiplication and addition in ordinary algebra, in Boolean algebra '.' means an AND operation and '+' means an OR operation. For instance, A+B in ordinary algebra is read as A plus B, while the same in Boolean algebra is read as A OR B.

- Basic logic operations such as AND, OR and NOT have already been discussed. More specifically, Boolean algebra captures the essential properties of both logic operations such as AND, OR and NOT and set operations such as intersection, union and complement.

Table 2.3: A brief description of the various Laws of Boolean Algebra with A and B are representing variable input.

Boolean Expression	**Description**	**Equivalent Switching Circuit**	**Boolean Algebra Law or Rule**
A + 1 = 1	A in parallel with closed = "CLOSED"		Annulment
A + 0 = A	A in parallel with open = "A"		Identity
A . 1 = A	A in series with closed = "A"	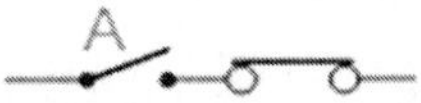	Identity
A . 0 = 0	A in series with open = "OPEN"		Annulment
A + A = A	A in parallel with A = "A"		Idempotent
A . A = A	A in series with A = "A"		Idempotent
NOT $\bar{A}$ = A	NOT NOT A (double negative) = "A"		Double Negation
A + $\bar{A}$ = 1	A in parallel with NOT A = "CLOSED"	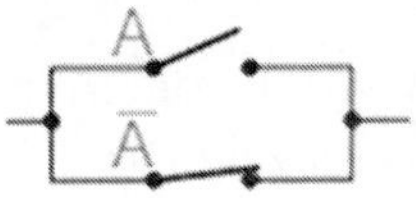	Complement
A . $\bar{A}$ = 0	A in series with NOT A = "OPEN"		Complement
A+B = B+A	A in parallel with B = B in parallel with A	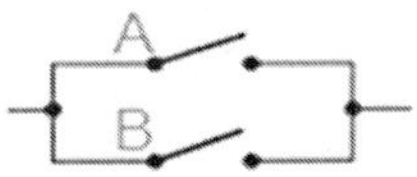	Commutative
A.B = B.A	A in series with B = B in series with A		Commutative
$\overline{A+B} = \bar{A}.\bar{B}$	invert and replace OR with AND		de Morgan's Theorem
$\bar{A}.\bar{B} = \overline{A+B}$	invert and replace AND with OR		de Morgan's Theorem

- As an illustration, the logical assertion that both a statement and its negation cannot be true has a counterpart in set theory, which says that the intersection of a subset and its complement is a null (or empty) set.
- Boolean algebra may also be defined to be a set A supplied with two binary operations of logical AND (.), logical OR (+), a unary operation of logical NOT (') and two elements, namely logical FALSE (0) and logical TRUE (1). This set is such that, for all elements of this set, the postulates or axioms relating to the associative, commutative, distributive, absorption and complementation properties of these elements hold good. These postulates are described in the following pages.

Table 2.4: A brief description of the various Boolean Postulates are a set of Mathematical Laws which can be used in the simplification of Boolean Expression.

Boolean Expression	Description
0.0 = 0	A 0 AND'ed with itself is always equal to 0.
1.1=1	A 1 AND'ed with itself is always equal to 1.
1.0=0	A 1 AND'ed with a 0 is always equal to 0.
0+0=0	A 0 OR'ed with itself is always equal to 0.
1+1=1	A 1 OR'ed with itself is always equal to 1.
1+0=1	A 1 OR'ed with a 0 is always equal to 1.
$\overline{1} = 0$	The inverse (complement) if a 1 is always equal to 0.
$\overline{0} = 1$	The inverse (complement) if a 0 is always equal to 1.

Using the information above, simple 2-input AND, OR and NOT Gates can be represented by 16 possible functions as shown in the following table.

Table 2.5: Boolean Algebra Functions for AND, OR and NOT Gates

S.No.	Function Description	Expression	S.No.	Function Description	Expression
1.	NULL	0	9.	NOT A AND B	$\bar{A}$. B
2.	IDENTITY	1	10.	NOT AND (NAND)	$\overline{A . B}$
3.	Input A	A	11.	A OR B (OR)	A + B
4.	Input B	B	12.	A OR NOT B	A + $\bar{B}$
5.	NOT A	$\bar{A}$	13.	NOT A OR B	$\bar{A}$ + B
6.	NOT B	$\bar{B}$	14.	NOT OR (NOR)	$\overline{A + B}$
7.	A AND B (AND)	A.B	15.	Exclusive-OR	A.$\bar{B}$ + $\bar{A}$.B
8.	A AND NOT B	A.$\bar{B}$	16.	Exclusive-NOR	A.B + $\bar{A}$.$\bar{B}$

2.11.2 *Variables, Literals and Terms in Boolean Expressions*

- ***Variables*** (different symbols) in a Boolean expression, may take on the value '0' or '1'. For instance, the below expression uses 3 variables (A, B, and C) as well as using four variables (P, Q, R, and S). The complement of a variable is not considered to be a separate variable.

$$F_1 = \bar{A} + A.B + A.\bar{C} + \bar{A}.B.C$$
$$F_2 = (P + Q).(R + \bar{S}).(P + \bar{Q} + R)$$

- Each occurrence of a variable or its complement is called a ***literal***. Expressions above for F_1 and F_2 have eight and seven literals respectively.
- ***A term*** is an expression formed by literals and operations at one level. Expression for F_1 above has five terms including four AND terms and one OR term that combines the first-level AND terms.

2.11.3 *Equivalent and Complement of Boolean Expressions*

Two given Boolean expressions are said to be equivalent if one of them equals '1' only when the other equals '1' and also one equals '0' only when the other equals '0'. They are said to be the complement of each other if one expression equals '1' only when the other equals '0', and vice versa.

The complement of a given Boolean expression is obtained by complementing each literal, changing all '.' to '+' and all '+' to '.', all 0s to 1s, and all 1s to 0s. The examples below give some Boolean expressions and their complements:

Given Boolean expression if

$$F = \bar{A}.B + A.\bar{B}$$

Then,

$$\bar{F} = \overline{(\bar{A}.B + A.\bar{B})} = (A + \bar{B}).(\bar{A} + B)$$

If

$$F = (A + B)(\bar{A} + \bar{B})$$

Then,

$$\bar{F} = \bar{A}.\bar{B} + AB$$

When ORed with its complement the Boolean expression yields a '1', and when ANDed with its complement it yields a '0'. The '.' sign is usually omitted in writing Boolean expressions and is implied merely by writing the literals in juxtaposition. For instance, A.B would normally be written as AB.

2.11.4 *Dual of a Boolean Expression*

The dual of a Boolean expression is obtained by replacing all '.' operations with '+' operations, all '+' operations with '.' operations, all 0s with 1s and all 1s with 0s and leaving all literals unchanged. The examples below give some Boolean expressions and the corresponding dual expressions:

If

$$F = \bar{A}.B + A.\bar{B}$$

Then its corresponding Dual is

$$Dual(F) = \bar{A}.B + A.\bar{B}$$

If Given Boolean expression

$$F = (A + B).(\bar{A} + \bar{B})$$

Then its corresponding Dual is

$$Dual(F) = A.B + \bar{A}.\bar{B}$$

Duals of Boolean expressions are mainly of interest in the study of Boolean postulates and theorems. Otherwise, there is no general relationship between the values of dual expressions. That is, both of them may equal '1' or '0'. One may even equal '1' while the other equals '0'. The fact that the dual of a given logic equation is also a valid logic equation leads to many more useful laws of Boolean algebra. The principle of duality has been put to ample use during the discussion on postulates and theorems of Boolean algebra. The postulates and theorems, to be discussed in the paragraphs to follow, have been presented in pairs, with one being the dual of the other.

Example 2.7: Find (a) Find the dual of $A.\bar{B} + B.\bar{C} + C.\bar{D}$

(b) Find the complement of $F = [(A.\bar{B} + \bar{C}).D + \bar{E}].F$

Solution:

(a) The dual of $A.\bar{B} + B.\bar{C} + C.\bar{D}$ is given by $(A + \bar{B}).(B + \bar{C}).(C + \bar{D})$

(b) The complement of $[(A.\bar{B} + \bar{C}).D + \bar{E}].F$ is given by $[\{(\bar{A} + B).C) + \bar{D}\}.\bar{E}] + \bar{F}$

2.11.5 Postulates of Boolean Algebra

The following are the important postulates of Boolean algebra:

1. $1.1 = 1;\quad and\ 0.0 = 0$
2. $1 + 1 = 1;\ and\ 0 + 0 = 0$
3. $1.0 = 0.1 = 0$
4. $0 + 1 = 1 + 0 = 1$
5. $\bar{1} = 0;\ and\ \bar{0} = 1$

Many theorems of Boolean algebra are based on these postulates, which can be used to simplify Boolean expressions. These theorems are discussed in the next section.

2.11.6 Theorems of Boolean Algebra

The theorems of Boolean algebra can be used to simplify many a complex Boolean expression and also to transform the given expression into a more useful and meaningful equivalent expression. The theorems are presented as pairs, with the two theorems in a given pair being the dual of each other.

Table 2.6: Postulates and Theorem for Boolean Algebra

Postulates 1	$\boldsymbol{x + 0 = 0}$	$\boldsymbol{x.1 = 1}$
Postulates 2	$x + x' = 1$	$x.x' = 0$
Postulates 3, Commutative	$x + y = y + x$	$xy = yx$
Postulates 4, Distributive	$x(y + z) = xy + xz$	$x(yz) = (xy)z$
Theorem 1, Identity	$x + x = x$	$x.x = x$
Theorem 2	$x + 1 = 1$	$x.0 = 0$
Theorem 3, involution	$(x')' = x$	
Theorem 4, Associative	$x + (y + z) = (x + y) + z$	$x(yz) = (xy)z$
Theorem 5, DeMorgans	$(x + y)' = x'.y'$	$(xy)' = x' + y'$
Theorem 6, Absorption	$x + xy = x$	$x(x + y) = x$

These theorems can be very easily verified by the method of 'perfect induction'. According to this method, the validity of the expression is tested for all possible combinations of values of the variables involved. Also, since the validity of the theorem is based on its being true for all possible combinations of values of variables, there is no reason why a variable cannot be replaced with its complement, or vice versa, without disturbing the validity.

Another important point is that, if a given expression is valid, its dual will also be valid. Therefore, in all the discussion to follow in this section, only one of the theorems in a given pair will be illustrated with a proof. Proof of the other being its dual is implied.

1. Theorem 1 (Operations with '0' and '1') i.e., (a) 0.X=0 and (b) 1+X=1

The theorem says $x.0 = 0, and\ x + 1 = 1$, where x is not necessarily a single variable, it could be a term or even a large expression.

Proof

x	$x = 0$	$x = 1$	Remark
$x.0$	0.0=0	1.0=0	Due to AND operation whatever the value of x the net outcome after AND operation is 0
$x + 1$	0+1=1	1+1=1	Due to OR operation whatever the value of x the net outcome after OR operation is 1

Theorem 1 can be proved by substituting all possible values of x, that is, 0 and 1, into the given Expression.

2. Theorem 2 (Operations with '0' and '1') i.e., $1.x = x; and\ 0 + x = x$

The theorem says $1.x = x, and\ 0 + x = x$, *Where x could be a variable, a term or even a large expression. According to this, ANDing a Boolean expression to '1' or ORing '0' to it makes no difference to the expression.*

Proof

x	$x = 0$	$x = 1$	Remark
$1.x$	1.0=0	1.1=1	Due to AND operation value of x will decide the net outcome
$0 + x$	0+0=0	0+1=1	Due to OR operation the value of x will decide the net outcome.

If x, denotes a Boolean expression then,

then, 1. (Boolean expression) = Boolean expression and 0 + (Boolean expression) = Boolean expression

3. *Theorem 3 (Idempotent or Identity Laws) i.e.,* $x.x = x; and\ x + x = x$

The theorem says $x.x = x; and\ x + x = x$, this theorem also known as idempotent laws, also known as identity laws. Theorem directly outcome of an AND/OR gate operation. When all the inputs of the gate have been tied together.

Proof

x	$x = 0$	$x = 1$	Remark
$x.x$	0.0=0	1.1=1	Due to AND operation value of x will decide the net outcome
$x + x$	0+0=0	1+1=1	Due to OR operation the value of x will decide the net outcome.

The scope of idempotent laws can be expanded further by considering x to be a term or an expression. For example, let us apply idempotent laws to simplify the following Boolean expression:

$$F = (A.\bar{B}.\bar{B} + C.C).(A.\bar{B}.\bar{B} + A.\bar{B} + C.C)$$

$$F = (A.\bar{B} + C).(A.\bar{B} + A.\bar{B} + C)$$

$$F = (A.\bar{B} + C).(A.\bar{B} + C)$$

$$F = (A.\bar{B} + C)$$

4. *Theorem 4 (Complementation Law) i.e.,* $x.\bar{x} = 0; and\ x + \bar{x} = 1$

The theorem says $x.\bar{x} = 0; and\ x + \bar{x} = 1$, according to this theorem, in general, any Boolean expression when ANDed to its complement yields a '0' and when ORed to its complement yields a '1', irrespective of the complexity of the expression:

Proof

x	$x = 0$	$x = 1$	Remark
$x.\bar{x}$	0.1=0	1.0=1	Due to AND operation the net outcome will always 0.
$x + \bar{x}$	0+1=0	1+0=1	Due to OR operation the net outcome will always 1.

5. *Theorem 5 (Commutative Laws)* i.e., $x + y = y + x;\ \ and\ \ x.y = y.x$

Theorem 5 implies that the order in which variables are added or ORed is immaterial. That is, the result of A OR B is the same as that of B OR A. Theorem also implies that the order in which variables are ANDed is also immaterial. The result of A AND B is same as that of B AND A.

Proof

Expressions	$x = 0$		$x = 1$	
	Y=0	*Y=1*	Y=0	Y=1
$x + y$	0+0=0	0+1=1	1+0=1	1+1=1
$y + x$	0+0=0	1+0=1	0+1=1	1+1=1

$x.y$	0.0=0	0.1=1	1.0=0	1.1=1
$y.x$	0.0=0	1.0=1	0.1=0	1.1=1

6. *Theorem 6 (Associative Laws)* i.e., $x + (y + z) = (x + y) + z$, and $x.(y.z) = (x.y).z$

Theorem 6 says that, when three variables are being ORed, it is immaterial whether we do this by ORing the result of the first and second variables with the third variable or by ORing the first variable with the result of ORing of the second and third variables or even by ORing the second variable with the result of ORing of the first and third variables.

The same concept applied to AND operation as well, when three variables are being ANDed, it is immaterial whether you do this by ANDing the result of ANDing of the first and second variables with the third variable or by ANDing the result of ANDing of the second and third variables with the first variable or even by ANDing the result of ANDing of the third and first variables with the second variable.

For example,

$$\bar{A}.B + (C.\bar{D} + \bar{E}.\bar{F}) = C.\bar{D} + (\bar{A}.B + \bar{E}.\bar{F}) = \bar{E}.\bar{F} + (\bar{A}.B + C.\bar{D})$$

Also

$$\bar{A}.B.(C.\bar{D}.\bar{E}.\bar{F}) = C.\bar{D}.(\bar{A}.B.\bar{E}.\bar{F}) = \bar{E}.\bar{F}.(\bar{A}.B.C.\bar{D})$$

Theorems 6 can be further illustrated by the logic diagrams in Figs 2.48(a) and (b).

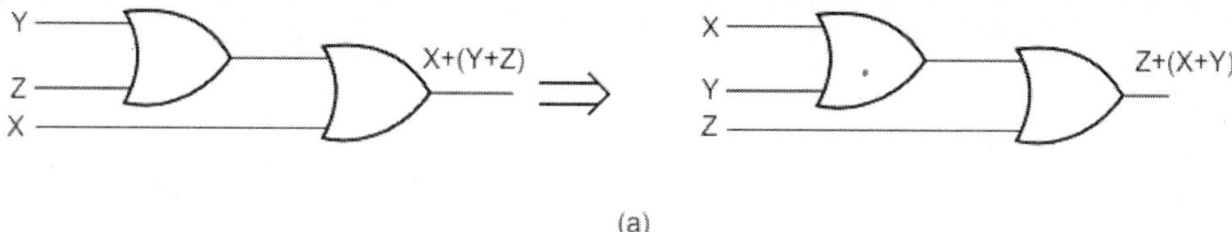

(a)

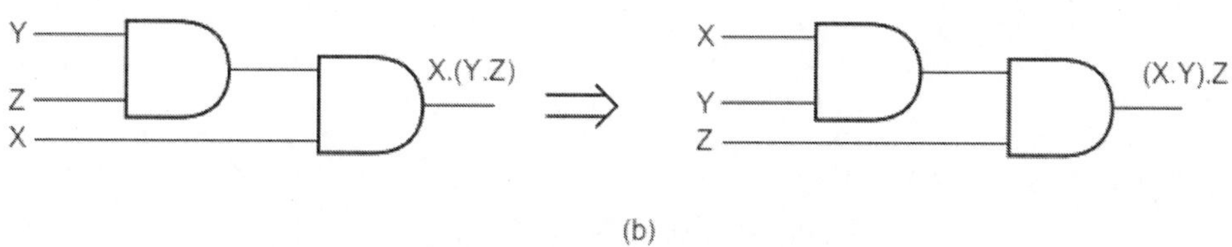

(b)

Fig. 2.48: Associative Laws.

7. Theorem 7 (Distributive Laws)

$$(a)\ x.(y+z) = x.y + x.z$$

$$(b) x + y.z = (x+y).(x+z)$$

Theorem 7(b) is the dual of theorem 7(a). The distribution law implies that a Boolean expression can always be expanded term by term. Also, in the case of the expression being the sum of two or more than two terms having a common variable, the common variable can be taken as common as in the case of ordinary algebra. Table 2.7 gives the proof of theorem 7(a) using the method of perfect induction.

Proof

Table 2.7: Proof of Distributive law

x	y	z	$y+z$	$x.y$	$x.z$	$x.(y+z)$	$xy+xz$
0	0	0	0	0	0	0	0
0	0	1	1	0	0	0	0
0	1	0	1	0	0	0	0
0	1	1	1	0	0	0	0
1	0	0	0	0	0	0	0
1	0	1	1	0	1	1	1
1	1	0	1	1	0	1	1
1	1	1	1	1	1	1	1

Theorem 7(b) is the dual of theorem 7(a) and therefore its proof is implied. Theorems 7(a) and (b) are further illustrated by the logic diagrams in Figs 2.49(a) and (b).

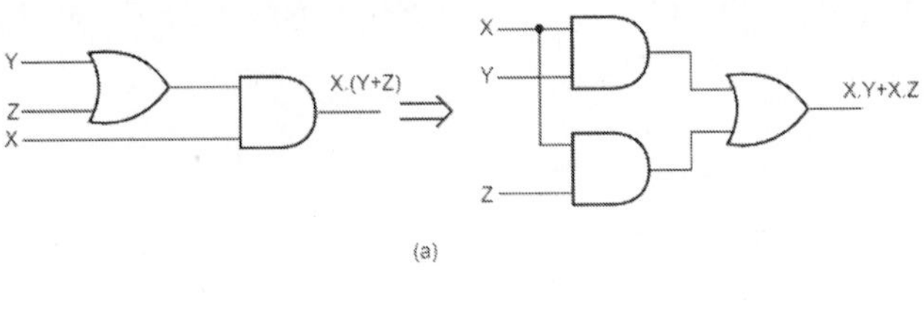

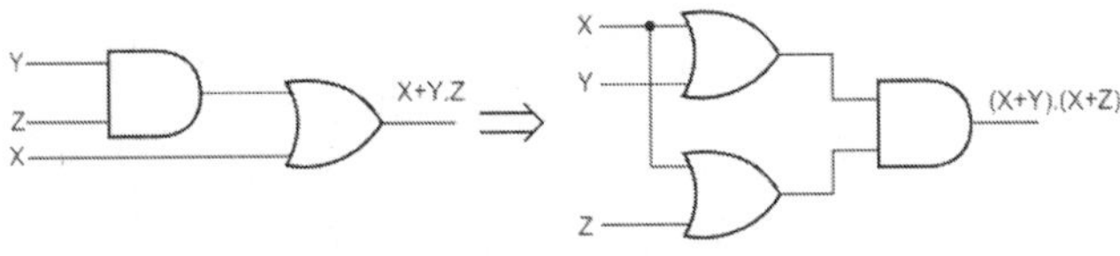

Fig. 2.49: Distributive Laws.

8. Theorem 8: Special case of theorem 7

$$(a)x.y + x.\overline{y} = x$$
$$(b)(x + y).(x + \overline{y}) = x$$

This special case has very interesting interpretation. Referring to above, there are two two-variable terms in the LHS expression. One of the variables, Y, is present in all possible combinations in this expression, while the other variable, X, is a common factor. The expression then reduces to this common factor.

9. Theorem 9

$$(a)(x + \overline{y}).y = x.y \quad and \quad (b)x.\overline{y} + y = x + y$$

Proof

$$(a) \quad (x + \overline{y}).y = x.y + \overline{y}.y = x.y + 0 = x.y$$

Theorem 9(b) is the dual of theorem 9(a) and hence stands proved.

10. Theorem 10 (Absorption Law or Redundancy Law)

$$(a) \ x + x.y = x \quad and \quad (b)x.(x + y) = x$$

Proof

$$(a) \ x + x.y = x(1 + y) = x.1 = x$$

Theorem 10(b) is the dual of theorem 10(a) and hence stands proved. The crux of this simplification theorem is that, if a smaller term appears in a larger term, then the larger term is redundant. The following examples further illustrate the underlying concept:

$$F = A + A.\bar{B} + A.\bar{B}.\bar{C} + A.B.C + A.B.\bar{C} = A.(1 + \cdots + \cdots) = A.1 = A$$

In the same way

$$F = (\bar{A} + B + \bar{C}).(\bar{A} + B).(\bar{A} + B + C) = \bar{A} + B$$

11. Theorem 11

$$(a) \ z.x + z.\overline{x}.y = z.x + z.y$$
$$(b)(z + x).(z + \overline{x} + y) = (z + x).(z + y)$$

Proof

Table 2.8: Proof of theorem 11

x	y	z	zx	zy	$z\bar{x}$	$z\bar{x}y$	$zx + z\bar{x}y$	$zx + zy$
0	0	0	0	0	0	0	0	0
0	0	1	0	0	1	0	0	0
0	1	0	0	0	0	0	0	0
0	1	1	0	1	1	1	1	1
1	0	0	0	0	0	0	0	0
1	0	1	1	0	0	0	1	1
1	1	0	0	0	0	0	0	0
1	1	1	1	1	0	0	1	1

Table 2.8 gives the proof of theorem 11(a) using the method of perfect induction. Theorem 11(b) is the dual of theorem 11(a) and hence stands proved. A useful interpretation of this theorem is that, when a smaller term appears in a larger term except for one of the variables appearing as a complement in the larger term, the complemented variable is redundant.

As an example,

$$F = (A + \bar{B}).(\bar{A} + \bar{B} + C).(\bar{A} + \bar{B} + D)$$

can be simplified as follows:

$$F = (A + \bar{B}).(\bar{A} + \bar{B} + C).(\bar{A} + \bar{B} + D) = (A + \bar{B}).(\bar{B} + C).(\bar{A} + \bar{B}.+ D)$$
$$= (A + \bar{B}).(\bar{B} + C).(\bar{B} + D)$$

12. Theorem 12 (Consensus Theorem)

$$\boldsymbol{(a)\ x.y + \bar{x}.z + y.z = x.y + \bar{x}.z}$$
$$\boldsymbol{(b)(x + y).(\bar{x} + z).(y + z) = (x + y).(\bar{x} + z)}$$

Proof

Table 2.9: Proof of theorem 12(a).

x	y	z	xy	$\bar{x}z$	yz	$x.y + \bar{x}z + y.z$	$x.y + \bar{x}.z$
0	0	0	0	0	0	0	0
0	0	1	0	1	0	1	1
0	1	0	0	0	0	0	0
0	1	1	0	1	1	1	1
1	0	0	0	0	0	0	0
1	0	1	0	0	0	0	0
1	1	0	1	0	0	1	1
1	1	1	1	0	1	1	1

Table 2.9 shows the proof of theorem 12(a) using the method of perfect induction. Theorem 12(b) is the dual of theorem 12(a) and hence stands proved. A useful interpretation of theorem 12 is as follows. If in a given Boolean expression, we can identify two terms with one having a variable and the other having its complement, then the term that is formed by the product of the remaining variables in the two terms in the case of a sum-of-products expression or by the sum of the remaining variables in the case of a product-of-sums expression will be redundant.

Example 2.8: Proof that $ABCD + AB\bar{C}\bar{D} + ABC\bar{D} + AB\bar{C}D + ABCDE + AB\bar{C}\bar{D}\bar{E} + AB\bar{C}DE = AB$

Solution

Let

$$F = ABCD + AB\bar{C}\bar{D} + ABC\bar{D} + AB\bar{C}D + ABCDE + AB\bar{C}\bar{D}\bar{E} + AB\bar{C}DE$$

Rearrange the terms

$$F = (ABCD + ABCDE) + (AB\bar{C}\bar{D} + AB\bar{C}\bar{D}\bar{E}) + ABC\bar{D} + (AB\bar{C}D + AB\bar{C}DE)$$
$$F = ABCD(1+E) + AB\bar{C}\bar{D}(1+\bar{E}) + ABC\bar{D} + AB\bar{C}D(1+E)$$
$$= ABCD + AB\bar{C}\bar{D} + ABC\bar{D} + AB\bar{C}D$$

Rearrange the terms

$$= ABCD + ABC\bar{D} + AB\bar{C}\bar{D} + +AB\bar{C}D$$
$$= ABC(D+\bar{D}) + AB\bar{C}(\bar{D}+D)$$
$$= ABC + AB\bar{C}$$
$$= AB(C+\bar{C})$$
$$= AB$$

13. Theorem 13 (DeMorgan's Theorem)

$$(\boldsymbol{a})\ \overline{\boldsymbol{x_1 + x_2 + x_3 + \cdots + \cdots + x_n}} = \overline{\boldsymbol{x_1}}.\overline{\boldsymbol{x_2}}.\overline{\boldsymbol{x_3}} \ldots\ldots.\overline{\boldsymbol{x_n}}$$

$$(\boldsymbol{b})\ \overline{\boldsymbol{x_1.x_2.x_3.\ldots\ldots.x_n}} = \overline{\boldsymbol{x_1}} + \overline{\boldsymbol{x_2}} + \overline{\boldsymbol{x_3}} + \cdots + \cdots + \overline{\boldsymbol{x_n}}$$

According to the first theorem the complement of a sum equals the product of complements, while according to the second theorem the complement of a product equals the sum of complements.

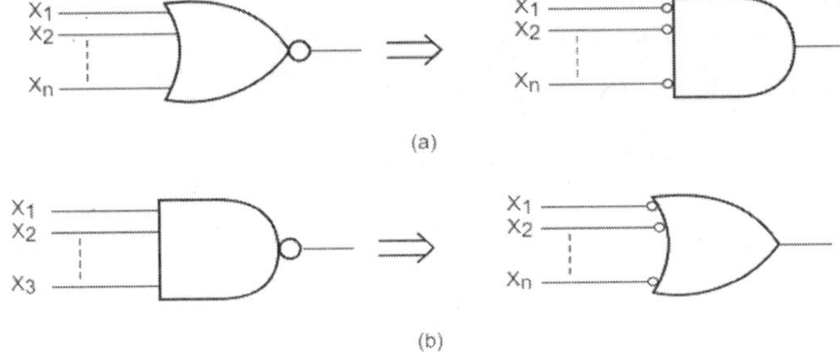

Fig.2.50: DeMorgan's theorem.

Fig. 2.50 (a) and (b) show logic diagram representations of De Morgan's theorems. While the first theorem can be interpreted to say that a multi-input NOR gate can be implemented as a multi-input bubbled AND gate, the second theorem, which is the dual of the first, can be interpreted to say that a multi-input NAND gate can be implemented as a multi-input bubbled OR gate.

DeMorgan's theorem can be proved as follows. Let us assume that all variables are in a logic '0' state. In that case

$$\text{LHS} = [\overline{X_1 + X_2 + X_3 + \cdots + X_n}] = [\overline{0+0+0+\cdots+0}] = \bar{0} = 1$$

$$\text{RHS} = \overline{X_1}.\overline{X_2}.\overline{X_3}.\ldots.\overline{X_n} = \bar{0}.\bar{0}.\bar{0}.\ldots.\bar{0} = 1.1.1.\ldots.1 = 1$$

Since theorem 13(b) is the dual of theorem 13(a), the same also stands proved. Theorem 13(b), though, can be proved on similar lines.

14. Theorem 14 (Transposition Theorem)

$$(a)\ x.y+\bar{x}.z=(x+z).(\bar{x}+y)$$
$$(b)\ (x+y).(\bar{x}+z)=x.z+\bar{x}.y$$

This theorem can be applied to any sum-of-products or product-of-sums expression having two terms, provided that a given variable in one term has its complement in the other.

Table 2.10: Proof of theorem 14(a).

x	y	z	xy	$\bar{x}z$	$x+z$	$\bar{x}+y$	$x.y+\bar{x}z$	$(x+z).(\bar{x}+y)$
0	0	0	0	0	0	1	0	0
0	0	1	0	1	1	1	1	1
0	1	0	0	0	0	1	0	0
0	1	1	0	1	1	1	1	1
1	0	0	0	0	1	0	0	0
1	0	1	0	0	1	0	0	0
1	1	0	1	0	1	1	1	1
1	1	1	1	0	1	1	1	1

Table 2.10 gives the proof of theorem 14(a) using the method of perfect induction. Theorem 14(b) is the dual of theorem 14(a) and hence stands proved.

15. Theorem 15 (Involution Law)

Involution law says that the complement of the complement of an expression leaves the expression unchanged. Also, the dual of the dual of an expression is the original expression. This theorem forms the basis of finding the equivalent product-of-sums expression for a given sum-of-products expression, and vice versa. It may be given as

$$\bar{\bar{X}}=X$$

Example 2.9: Starting with the Boolean expression for a two-input OR gate, apply Boolean laws and theorems to modify it in such a way as to facilitate the implementation of a two-input OR gate by using two-input NAND gates only.

Solution:

- A two input OR gate is represented by the Boolean Expression $Y=A+B$, here *A*, and *B* are inputs and *Y* is the output.
- Now, $(A+B)=\bar{\bar{A}}+\bar{\bar{B}}$ $\ldots\ldots\ldots\ldots\ldots\ldots$ *Involution Law*

  $=\overline{\bar{A}+\bar{B}}$ $\ldots\ldots\ldots\ldots\ldots\ldots$ *DeMorgans Law*

  $=\overline{\overline{(A.A)}+\overline{(B.B)}}$ $\ldots\ldots$ *Idempotent Law*
- Fig. 2.51 below shows the NAND gate implementation of a two input OR gate

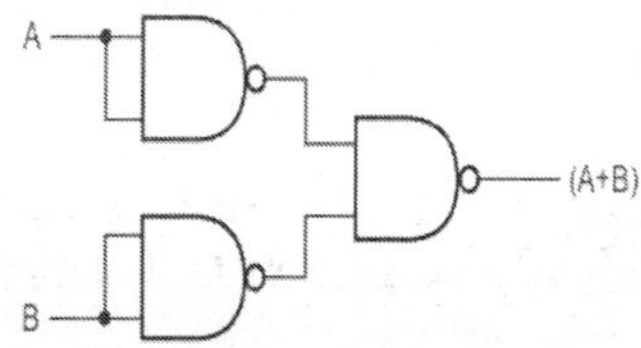

Fig. 2.51: Example 6.7

2.12 Simplification Techniques

In this section, we will discuss techniques other than the application of laws and theorems of Boolean algebra discussed in the preceding paragraphs of this chapter for simplifying or more precisely minimizing a given complex Boolean expression. The primary objective of all simplification procedures is to obtain an expression that has the minimum number of terms. Obtaining an expression with the minimum number of literals is usually the secondary objective. If there is more than one possible solution with the same number of terms, the one having the minimum number of literals is the choice.

The techniques to be discussed include:

a) The Quine–McCluskey tabular method;

b) The Karnaugh map method.

Before we move on to discuss these techniques in detail, it would be relevant briefly to describe sum-of-products and product-of-sums Boolean expressions. The given Boolean expression will be in either of the two forms, and the objective will be to find a minimized expression in the same or the other form.

2.12.1 Sum-of-Products Boolean Expressions

A sum-of-products expression contains the sum of different terms, with each term being either a single literal or a product of more than one literal. It can be obtained from the truth table directly by considering those input combinations that produce a logic '1' at the output. Each such input combination produces a term. Different terms are given by the product of the corresponding literals. The sum of all terms gives the expression. For example, for the given truth table

Table 2.11: Truth Table

A	*B*	*C*	*Y*
0	0	0	1
0	0	1	0
0	1	0	0
0	1	1	1
1	0	0	0
1	0	1	1
1	1	0	1
1	1	1	0

For the given truth table as shown in table 2.11, it can be represented by the Boolean expression

$$Y = \bar{A}.\bar{B}.\bar{C} + \bar{A}.B.C + A.B.\bar{C} + A.\bar{B}.C$$

Considering the first term, the output is '1', when $A = 0, B = 0, and\ C = 0$. This is possible only when $\bar{A}, \bar{B}, and\ \bar{C}\ are\ ANDed$. Also, for the second term, the output is '1' only when $B, C, and\ \bar{A}\ are\ ANDed$. Other terms can be explained similarly. A sum of products expression is also known as $minterms\ expression$.

2.12.2 Product-of-Sums Expressions

A product-of-sums expression contains the product of different terms, with each term being either a single literal or a sum of more than one literal. It can be obtained from the truth table by considering those input combinations that produce a logic '0' at the output. Each such input combination gives a term, and the product of all such terms gives the expression. Different terms are obtained by taking the sum of the corresponding literals. Here, '0' and '1' respectively mean the uncomplemented and complemented variables, unlike sum-of-products expressions where '0' and '1' respectively mean complemented and uncomplemented variables.

Table 2.12: Minterms and Maxterms for three binary variables

x	y	z	Minterms		Maxterms	
			Term	Designation	Term	Designation
0	0	0	$x'y'z'$	m_0	$x+y+z$	M_0
0	0	1	$x'y'z$	m_1	$x+y+z'$	M_1
0	1	0	$x'yz'$	m_2	$x+y'+z$	M_2
0	1	1	$x'yz$	m_3	$x+y'+z'$	M_3
1	0	0	$xy'z'$	m_4	$x'+y+z$	M_4
1	0	1	$xy'z$	m_5	$x'+y+z'$	M_5
1	1	0	xyz'	m_6	$x'+y'+z$	M_6
1	1	1	xyz	m_7	$x'+y'+z'$	M_7

To illustrate this further, consider once again the Table 2.12. Since each term in the case of the product-of-sums expression is going to be the sum of literals; this implies that it is going to be implemented using an OR operation. Now, an OR gate produces a logic '0' only when all its inputs are in the logic '0' state, which means that the first term corresponding to the second row of the truth table will be $A+B+\bar{C}$. The product-of-sums Boolean expression for this truth table is given by

$$F=(A+B+\bar{C}).(A+\bar{B}+C).(\bar{A}+B+C).(\bar{A}+\bar{B}+\bar{C})$$

Transforming the given product-of-sums expression into an equivalent sum-of-products expression is a straightforward process. Multiplying out the given expression and carrying out the obvious simplification provides the equivalent sum-of-products expression:

$$F=(A+B+\bar{C}).(A+\bar{B}+C).(\bar{A}+B+C).(\bar{A}+\bar{B}+\bar{C})$$
$$F=\bar{A}.\bar{B}.\bar{C}+\bar{A}.B.C+A.B.\bar{C}+A.\bar{B}.C$$

A given sum-of-products expression can be transformed into an equivalent product-of-sums expression by (a) taking the dual of the given expression, (b) multiplying out different terms to get the sum-of-products form, (c) removing redundancy and (d) taking a dual to get the equivalent product-of-sums expression. As an illustration, let us find the equivalent product-of-sums expression of the sum-of-products expression

$$F=A.B+\bar{A}.\bar{B}$$
$$Dual\ (F)=(A+B)(\bar{A}+\bar{B})$$
$$(A+B)(\bar{A}+\bar{B})=A.\bar{A}+A.\bar{B}+B.\bar{A}+B.\bar{B}=0+A.\bar{B}+B.\bar{A}+0$$

Therefore,

$$(A+B)(\bar{A}+\bar{B})=A.\bar{B}+\bar{A}.B$$

Example 2.10: Express the Boolean Expression $F=x.y+x'.z$ as a product of Maxterms.
Solution:

- First convert the function into the OR terms using Distributive Law.

$$F=x.y+x'.z=(x.y+x').(x.y+z)$$
$$F=(x+x')(y+x')(x+z)(y+z)=(x'+y)(x+z)(y+z)$$

- The function has three variables $x, y, and\ z$. Each term is missing with one variable. Hence,

$$x' + y = x' + y + zz' = (x' + y + z)(x' + y + z')$$
$$x + z = x + z + yy' = (x + z + y)(x + z + y')$$
$$y + z = y + z + xx' = (y + z + x)(y + z + x')$$

- Combining all the terms and while not repeating the terms which appeared more than once, hence finally,

$$F = (x + y + z)(x + y' + z)(x' + y + z)(x' + y + z')$$
$$F = M_0 M_2 M_4 M_5$$

2.12.3 Expanded Forms of Boolean Expressions (Canonical Form)

Expanded sum-of-products (SOP) and product-of-sums (POS) forms of Boolean expressions are useful not only in analyzing these boolean expressions but also in the application of minimization techniques such as the Quine–McCluskey tabular method and the Karnaugh mapping method for simplifying given Boolean expressions. The expanded form, sum-of-products or product-of-sums, is obtained by including all possible combinations of missing variables.

Example 2.11: Consider the following sum-of-products expression and convert it to canonical form.

$$F = A.\bar{B} + B.\bar{C} + A.B.\bar{C} + \bar{A}.C$$

Solution: It is a three-variable expression. Expanded (Canonical) versions of different Minterms can be find as follows

- $A.\bar{B} = A.\bar{B}.(C + \bar{C}) = A.\bar{B}.C + A.\bar{B}.\bar{C}$
- $B.\bar{C} = B.\bar{C}.(A + \bar{A}) = A.B.\bar{C} + \bar{A}.B.\bar{C}$
- $A.B.\bar{C}$ is a complete term and has no missing values.
- $\bar{A}.C = \bar{A}.C.(B + \bar{B}) = \bar{A}.B.C + \bar{A}.\bar{B}.C$

Expanded sum-of-product is therefore given by

$$F = A.\bar{B}.C + A.\bar{B}.\bar{C} + A.B.\bar{C} + \bar{A}.B.\bar{C} + \bar{A}.B.C + \bar{A}.\bar{B}.C$$

Example 2.12: Consider the following Product-of-Sums expression and convert it to canonical form.

$$F = (\bar{A} + B)(\bar{A} + B + \bar{C} + \bar{D})$$

Solution: It is a four-variable expression $(A, B, C, and\ D)$. Expanded (Canonical) versions of different Maxterms can be find as follows.

- $(\bar{A} + B) = (\bar{A} + B)(C + \bar{C}) = (\bar{A} + B + C)(\bar{A} + B + \bar{C})$
  - $(\bar{A} + B + C)(D + \bar{D}) = (\bar{A} + B + C + D)(\bar{A} + B + C + \bar{D})$
  - $(\bar{A} + B + \bar{C})(D + \bar{D}) = (\bar{A} + B + \bar{C} + D)((\bar{A} + B + \bar{C} + \bar{D})$
- $(\bar{A} + B + \bar{C} + \bar{D})$ is a complete term and has no missing values.

Expanded Product-of-Sums is therefore given by

$$F = (\bar{A} + B + C + D)(\bar{A} + B + C + \bar{D})(\bar{A} + B + \bar{C} + D)(\bar{A} + B + \bar{C} + \bar{D})$$

2.12.4 $\sum$ and $\prod$ Nomenclature

$\sum$ and $\prod$ notations are respectively used to represent sum-of-products and products-of-sums Boolean expressions. We will illustrate these notations with the help of examples. Let us consider the following Boolean function:

$$f(A, B, C, D) = A.\bar{B}.\bar{C} + A.B.C.D. + \bar{A}.B.\bar{C}.D + \bar{A}.\bar{B}.\bar{C}.D$$

We will represent this function using $\sum$ notation. The first step is to write the expanded sum-of-products given by

$$f(A, B, C, D) = A.\bar{B}.\bar{C}.(D + \bar{D}) + A.B.C.D. + \bar{A}.B.\bar{C}.D + \bar{A}.\bar{B}.\bar{C}.D$$
$$= A.\bar{B}.\bar{C}.D + A.\bar{B}.\bar{C}.\bar{D} + A.B.C.D + \bar{A}.B.\bar{C}.D + \bar{A}.\bar{B}.\bar{C}.D$$

Different terms are then arranged in ascending order of the binary numbers represented by various terms, with true variables representing a '1' and a complemented variable representing a '0'. The expression becomes

$$f(A,B,C,D) = A.\bar{B}.\bar{C}.D + A.\bar{B}.\bar{C}.\bar{D} + A.B.C.D. + \bar{A}.B.\bar{C}.D + \bar{A}.\bar{B}.\bar{C}.D$$

The different terms represent 0001, 0101, 1000 and 1111. The decimal equivalent of these terms enclosed in the $\sum$ then gives the $\sum$ notation for the given Boolean function. That is,

$$f(A,B,C,D) = \sum(1,5,8,9,15)$$

The complement of $f(A,B,C,D)$, that is, $f'(A,B,C,D)$, *can be directly determined from* $\sum$ notation by including the left -out entries from the list of all possible numbers for a four-variable function.

$$f'(A,B,C,D) = \sum(0,2,3,4,6,7,10,11,12,13,14)$$

Let us now take the case of a product-of-sums Boolean function and its representation in $\prod$ nomenclature. Let us consider the Boolean function

$$f(A,B,C,D) = (A + B + \bar{C} + \bar{D}).(\bar{A} + B + \bar{C} + \bar{D}).(\bar{A} + \bar{B} + C + D).(A + \bar{B} + \bar{C} + \bar{D})$$

The binary numbers represented by the different sum terms are 0011, 1011, 1100 and 0111(true and complemented variables here represent 0 and 1repectively). When arranged in ascending order, these numbers are 0011, 0111,1011 and 1100.Therefore,

$$f(A,B,C,D) = \prod(3,7,11,12)$$

$$f'(A,B,C,D) = \prod(0,1,2,4,5,6,8,9,10,13,14,15)$$

An interesting corollary of what we have discussed above is that, if a given Boolean function $f(A,B,C)$ is given by

$$f(A,B,C) = \sum(0,1,4,7) = \prod(2,3,5,6)$$

Then,

$$f(A,B,C) = \prod(2,3,5,6), then, f'(A,B,C) = \sum(2,3,5,6) = \prod(0,1,4,7)$$

Optional combinations can also be incorporated into $\sum$ and $\prod$ nomenclature using suitable identifiers; $'\emptyset'$ or 'd' are used as identifiers for don't care terms. For example, if

$$f(A,B,C) = \sum(0,4,5) + \sum_{\emptyset}(3,7) = \sum(0,4,5) + \sum_{d}(3,7)$$

$$f(A,B,C) = \prod 1,2,6 + \prod_{\emptyset} 3,7 = \prod 1,2,6 + \prod_{d} 3,7$$

Example 2.13: For a Boolean expression $f(A,B) = \sum 0,2$; Prove that

$$f(A,B) = \prod 1,3 \text{ and } f'(A,B) = \sum 1,3 = \prod 0,2$$

Solution:

- $f(A,B) = \sum 0,2 = \bar{A}\bar{B} + A\bar{B} = \bar{B}.(A + \bar{A}) = \bar{B}$

- $f(A,B) = \prod 1,3 = (A+\bar{B}).(\bar{A}+\bar{B}) = A.\bar{A} + A.\bar{B} + \bar{B}.\bar{A} + \bar{B}.\bar{B} = \bar{B}$

- $f(A,B) = \sum 1,3 = \bar{A}.B + A.B = B.(\bar{A}+A) = B$
  - $f(A,B) = \prod 0,2 = (A+B)(\bar{A}+B) = A.\bar{A} + A.B + \bar{A}.B + B.B = B$

Hence, $\sum 1,3 = \prod 0,2$

2.12.5 Quine–McCluskey Tabular Method

The Quine–McCluskey tabular method of simplification is based on the complementation theorem, which says that

$$X.Y + \bar{X}.Y = X$$

Where X represents either a variable or a term or an expression and Y is a variable. This theorem implies that, if a Boolean expression contains two terms that differ only in one variable, then they can be combined together and replaced with a term that is smaller by one literal. The same procedure is applied for the other pairs of terms wherever such a reduction is possible. All these terms reduced by one literal are further examined to see if they can be reduced further. The process continues until the terms become '*irreducible*'.

The irreducible terms are called prime implicants. An optimum set of prime implicants that can account for all the original terms then constitutes the minimized expression. The technique can be applied equally well for minimizing sum-of-products and product-of-sums expressions and is particularly useful for Boolean functions having more than six variables as it can be mechanized and run on a computer.

On the other hand, the Karnaugh mapping method, to be discussed later, is a graphical method and becomes very cumbersome when the number of variables exceeds six. The step-by-step procedure for application of the tabular method for minimizing Boolean expressions, both sum-of-products and product-of-sums, is outlined as follows:

1. The Boolean expression to be simplified is expanded if it is not in expanded form.
2. Different terms in the expression are divided into groups depending upon the number of 1s they have.
3. True and complemented variables in a sum-of-products expression mean '1' and '0' respectively.
4. The reverse is true in the case of a product-of-sums expression.
5. The groups are then arranged, beginning with the group having the least number of 1s in its included terms.
6. Terms within the same group are arranged in ascending order of the decimal numbers represented by these terms.
   - As an illustration, consider the expression given below. The grouping of different terms and the arrangement of different terms within the group are shown below:

$$Y = A.B.C + \bar{A}.B.C + A.\bar{B}.\bar{C} + A.\bar{B}.C + \bar{A}.\bar{B}.\bar{C}$$

$\bar{A}.\bar{B}.\bar{C}$	000	**First Group**	**Total No. of 1's=0**
$A.\bar{B}.\bar{C}$	100	Second Group	Total No. of 1's=1
$\bar{A}.B.C$	011	Third Group	Total No. of 1's=2
$A.\bar{B}.C$	101		
$A.B.C$	111	Fourth Group	Total No. of 1's=3

- As further another illustration, consider a product-of-sums expression given below. The formation of groups and the arrangement of terms within different groups for the product-of-sums expression are as follows:

$$Y = (\bar{A} + \bar{B} + \bar{C} + \bar{D})(\bar{A} + \bar{B} + \bar{C} + D)(\bar{A} + B + \bar{C} + D)(A + B + \bar{C} + \bar{D})(A + B + C + D)$$
$$(A + \bar{B} + \bar{C} + D)((A + \bar{B} + \bar{C} + \bar{D})$$

$\boldsymbol{A.B.C.D}$	**0000**	**First Group**	**Total No. of 1's=0**
$\boldsymbol{A.B.\bar{C}.\bar{D}}$	0011		
$\boldsymbol{A.\bar{B}.C.\bar{D}}$	0101	Second Group	Total No. of 1's=2
$\boldsymbol{\bar{A}.B.\bar{C}.D}$	1010		
$\boldsymbol{A.\bar{B}.\bar{C}.\bar{D}}$	0111	Third Group	Total No. of 1's=3
$\boldsymbol{\bar{A}.\bar{B}.\bar{C}.D}$	1110		
$\boldsymbol{\bar{A}.\bar{B}.\bar{C}.\bar{D}}$	1111	Fourth Group	Total No. of 1's=4

- It may be mentioned here that the Boolean expressions that we have considered above did not contain any optional terms. If there are any, they are also considered while forming groups. This completes the first table.

7. The terms of the first group are successively matched with those in the next adjacent higher-order group to look for any possible matching and consequent reduction. The terms are considered matched when all literals except for one match. The pairs of matched terms are replaced with a single term where the position of the unmatched literals is replaced with a dash (—). These new terms formed as a result of the matching process find a place in the second table. The terms in the first table that do not find a match are called the prime implicants and are marked with an asterisk (∗). The matched terms are ticked (√).
8. Terms in the second group are compared with those in the third group to look for a possible match. Again, terms in the second group that do not find a match become the prime implicants.
9. The process continues until we reach the last group. This completes the first round of matching. The terms resulting from the matching in the first round are recorded in the second table.
10. The next step is to perform matching operations in the second table. While comparing the terms for a match, it is important that a dash (—) is also treated like any other literal, that is, the dash signs also need to match. The process continues on to the third table, the fourth table and so on until the terms become irreducible any further.
11. An optimum selection of prime implicants to account for all the original terms constitutes the terms for the minimized expression. Although optional (also called 'don't care') terms are considered for matching, they do not have to be accounted for once prime implicants have been identified.

Example 2.14: Consider the following sum-of-products expression, and express it in minimum number of literals.

$$Y = \bar{A}.B.\bar{C} + \bar{A}.\bar{B}.D + A.\bar{C}.D + B.\bar{C}.\bar{D} + \bar{A}.B.\bar{C}.D$$

Solution:

1. First step, we write the expanded version of the given expression. It can be written as follows:

$$Y = \bar{A}.\bar{B}.\bar{C}.D + \bar{A}.\bar{B}.C.D + \bar{A}.B.\bar{C}.\bar{D} + \bar{A}.B.\bar{C}.D + \bar{A}.B.C.\bar{D} + \bar{A}.B.C.D + A.\bar{B}.\bar{C}.D + A.B.\bar{C}.\bar{D} + A.B.\bar{C}.D$$

2. The formation of groups, the placement of terms in different groups and the first-round matching are shown as follows:

Minterm and its value

A	B	C	D	Minterm
0	0	0	1	$\bar{A}.\bar{B}.\bar{C}.D$
0	0	1	1	$\bar{A}.\bar{B}.C.D$
0	1	0	0	$\bar{A}.B.\bar{C}.\bar{D}$
0	1	0	1	$\bar{A}.B.\bar{C}.D$
0	1	1	0	$\bar{A}.B.C.\bar{D}$
0	1	1	1	$\bar{A}.B.C.D$
1	0	0	1	$A.\bar{B}.\bar{C}.D$
1	1	0	0	$A.B.\bar{C}.\bar{D}$
1	1	0	1	$A.B.\bar{C}.D$

Forming Groups based on no. of 1's

A	B	C	D	
0	0	0	1	No. of 1's=1
0	1	0	0	
0	0	1	1	No. of 1's=2
0	1	0	1	
0	1	1	0	
1	0	0	1	
1	1	0	0	
0	1	1	1	No. of 1's=3
1	1	0	1	

Compare the groups for change in bits i.e., 0 to 1 and 1 to 0 (First Round)

A	B	C	D	
0	0	_	1	Compare G1 vs G2
0	_	0	1	
_	0	0	1	
0	1	0	_	
0	1	_	0	
_	1	0	0	
0	_	1	1	Compare G2 vs G3
0	1	_	1	
_	1	0	1	
0	1	1	_	
1	_	0	1	
1	1	0	_	

3. The second round of matching begins from the comparison output obtained from the previous table. Each term in the first group is compared with every term in the second group.

Comparison (Second Round)

A	B	C	D
0	_	_	1
_	_	0	1
0	1	_	_
_	1	0	_

- For instance, the first term in the first group 0 0 _ 1 match with the second term in the second group 0 1 _ 1 (As we are seeking for one-bit change) to yield 0 _ _ 1which is recorded in the table shown below. The process continues until all terms have been compared for a possible match.

4. Since this new table has only one group, the terms contained therein are all prime implicants. In the present example, the terms in the first and second tables have all found a match. But that is not always the case.
5. The next table is what is known as the prime implicant table. The prime implicant table contains all the original terms in different columns and all the prime implicants recorded in different rows as shown below:

$A.B.C.D$	0001	0011	0100	0101	0110	0111	1001	1100	1101	Prime Implicant
0 _ _ 1	√	√		√		√				$\bar{A}.D$
_ _ 0 1	√			√			√		√	$\bar{C}.D$
0 1 _ _			√	√	√	√				$\bar{A}.B$
_ 1 0 _			√	√				√	√	$B.\bar{C}$

6. Each prime implicant is then examined one by one and the terms it can account for are ticked as shown. The next step is to write a product-of-sums expression using the prime implicants to account for all the terms. In the present illustration, the minimized expression is

$$F = \bar{A}.D + \bar{C}.D + \bar{A}.B + B.\bar{C}$$

NOTE: What has been described above is the formal method of determining the optimum set of prime implicants. In most of the cases where the prime implicant table is not too complex, the exercise can be done even intuitively. The exercise begins with identification of those terms that can be accounted for by only a single prime implicant. In the present example, 0011, 0110, 1001 and 1100 are such terms. As a result, P, Q, R and S become the essential prime implicants. The next step is to find out if any terms have not been covered by the essential prime implicants. In the present case, all terms have been covered by essential prime implicants. In fact, all prime implicants are essential prime implicants in the present example.

Example 2.15: Express the given product-of-sums expression in minimum number of literals

$$F = (\bar{A} + \bar{B} + \bar{C} + \bar{D}).(\bar{A} + \bar{B} + \bar{C} + D).(\bar{A} + \bar{B} + C + \bar{D}).(A + \bar{B} + \bar{C} + \bar{D}).(A + \bar{B} + C + \bar{D})$$

Solution:

1. The procedure is similar to that described for the case of simplification of sum-of-products expressions. The resulting tables leading to identification of prime implicants are as follows:

Maxterm and its value					**Forming Groups based on no. of 1's**					**Comparison (First Round)**			
Maxterm	A	B	C	D	A	B	C	D	1's	A	B	C	D
$\boldsymbol{A + \bar{B} + C + \bar{D}}$	0	1	0	1	0	1	0	1	1	0	1	_	1
$\boldsymbol{A + \bar{B} + \bar{C} + \bar{D}}$	0	1	1	1	0	1	1	1		_	1	0	1
$\boldsymbol{\bar{A} + \bar{B} + C + \bar{D}}$	1	1	0	1	1	1	0	1	3	_	1	1	1
$\boldsymbol{\bar{A} + \bar{B} + \bar{C} + D}$	1	1	1	0	1	1	1	0		1	1	_	1
$\boldsymbol{\bar{A} + \bar{B} + \bar{C} + \bar{D}}$	0	1	1	1	1	1	1	1	4	1	1	1	_

2. The prime implicant table is constructed after all prime implicants have been identified to look for the optimum set of prime implicants needed to account for all the original terms. The prime implicant table shows that both the prime implicants are the essential ones:

$A + B + C + D$	0101	0111	1101	1110	1111	Prime Implicants
1 1 1 _				√	√	$\bar{A} + \bar{B} + \bar{C}$
_ 1 _ 1	√	√	√		√	$\bar{B} + \bar{D}$

Hence, the minimized expression is

$$F = (\bar{A} + \bar{B} + \bar{C}).(\bar{B} + \bar{D})$$

2.12.6 Tabular Method for Multi-Output Functions

A network that has more than one output is called a multi-output logic network. In such a case sharing of some logic blocks between different functions are highly probable. Hence, for an optimum logic implementation of the multi-output function, different functions cannot be and should not be minimized in isolation because a possible common term that could have been shared may not turn out to be a prime implicant if the functions are worked out individually. The method of applying the tabular approach to multioutput functions is to get a minimized set of expressions that would lead to an optimum overall system. The idea is best explained by the example below.

Example 2.16: Consider a logic system with two outputs that is described by the following Boolean expressions:

$$Y_1 = \bar{A}.B.\bar{D} + \bar{A}.C.D + \bar{A}.\bar{C}.\bar{D}$$

$$Y_2 = \bar{A}.B.C + A.C.D + A.\bar{B}.C.\bar{D} + \bar{A}.\bar{B}.C.\bar{D}$$

Express the output with minimum number of literals.

Solution:

1. Find out the expanded form first. Hence, the expanded form for the above two functions can be given as

$$Y_1 = \bar{A}.B.C.D + \bar{A}.B.\bar{C}.D + \bar{A}.\bar{B}.C.D + \bar{A}.B.\bar{C}.\bar{D} + \bar{A}.\bar{B}.\bar{C}.\bar{D}$$

$$Y_2 = \bar{A}.B.C.D + \bar{A}.B.C.\bar{D} + A.B.C.D + A.\bar{B}.C.D + A.\bar{B}.C.\bar{D} + \bar{A}.\bar{B}.C.\bar{D}$$

2. The rows representing different terms are arranged in the usual manner, with all the terms contained in the two functions finding a place without repetition, as shown in the table below:

Availability of the terms			
$A.B.C.D$	Y_1	Y_2	
0000	√		√
0010		√	√
0100	√		√
0011	√		√
0101	√		√
0110		√	√
1010		√	√
0111	√	√	√
1011		√	√
1111		√	√

3. Each term is checked under the column or columns depending upon the functions in which it is contained. For instance, if a certain term is contained in the logic expressions for both output 1 and output 2, it will be checked in both output columns. The matching process begins in the same way as described earlier for the case of single-output functions, with some modifications outlined as follows:
    a. Only those terms can be combined that have at least one checkmark in the output column in common. For instance, 0000 cannot combine with 0010 but can combine with 0100.

b. In the resulting row, only the common outputs are checked. For instance, when 0101 is matched with 0111, then, in the resulting term 01–1, only output 1 will be checked.

c. A combining term can be checked off only if the resulting term accounts for all the outputs in which the term is contained.

4. The table below shows the results of the first round of matching:

Availability of the terms

$A.B.C.D$	Y_1	Y_2	
0_00	√		*
0_10		√	*
_010		√	*
010_	√		*
0_11	√		*
01_1	√		*
011_		√	*
101_		√	*
_111		√	*
1_11		√	*

5. No further matching is possible. The prime implicant table is shown below:

		Y_1					Y_2				$A.B.C.D$
0000	0011	0100	0101	0111	0010	0110	0111	1010	1011	1111	
√		√									0_00
					√	√					0_10
					√			√			_010
		√	√								010_
	√			√							0_11
			√	√							01_1
						√	√				011_
								√	√		101_
							√			√	_111
									√	√	1_11

6. For each prime implicant, check marks are placed only in columns that pertain to the outputs checked off for this prime implicant. For instance, 0-00 has only output 1 checked off. Therefore, the relevant terms under output 1 will be checked off. The completed table is treated as a whole while marking the required prime implicants to be considered for writing the minimized expressions. The minimized expressions are as follows:

$$Y_1 = \bar{A}.B.\bar{C} + \bar{A}.C.D + \bar{A}.\bar{C}.\bar{D}$$
$$Y_2 = B.C.D + A.\bar{B}.C + \bar{A}.C.\bar{D}$$

Example 2.17: using the Quine-McCluskey tabular method, find the minimum literals to express the given sum-of-products expression.

$$f(A,B,C,D) = \sum(1,2,3,9,12,13,14) + \sum_{\emptyset}(0,7,10,15)$$

Solution: The different steps to finding the solution to the given problem are tabulated below.

1. As we can see, eight prime implicants have been identified. These prime implicants along with the inputs constitute the prime implicant table. Remember that optional inputs are not considered while constructing the prime implicant table:

A	*B*	*C*	*D*		*A*	*B*	*C*	*D*		*A*	*B*	*C*	*D*	
0	0	0	0	√	0	0	0	_	√	0	0	_	_	*
					0	0	_	0	√	1	1	_	_	*
0	0	0	1	√										
0	0	1	0	√	0	0	_	1	√					
					_	0	0	1	*					
0	0	1	1	√	0	0	1	_	√					
1	0	0	1	√	_	0	1	0						
1	0	1	0	√										
1	1	0	0	√	0	_	1	1	*					
					1	_	0	1	*					
0	1	1	1	√	1	_	1	0	*					
1	1	0	1	√	1	1	0	_	√					
1	1	1	0	√	1	1	_	0	√					
1	1	1	1	√	_	1	1	1	*					
					1	1	_	1	√					
					1	1	1	_	√					

2. The product-of-sums expression that tells about the combination of prime implicants can be represent by the table

0001	**0010**	**0011**	**1001**	**1100**	**1101**	**1110**	**Prime Implicant**	**Named as**
√			√				_001	L
	√						_010	M
		√					0_11	N
			√		√		1_01	P
						√	1_10	Q
							_111	R
√	√	√					00__	S
				√	√	√	11__	T

3. After obvious simplification, this reduces to the expression

$$(L+S).(M+S).(N+S).(L+P).(T).(P+T).(Q+T)$$

Can be simplified as

$(L+S).(M+S).(N+S).(L+P).(T) = T.(L+S).(M+S).(N+S).(L+P)$

$$= T.(LM + LS + MS + S).(LN + PN + \ LS + PS)$$
$$= T.(LM + S).(LN + PN + \ LS + PS)$$
$$= T.(LM + LS + MS + S).(LN + PN + \ LS + PS)$$
$$= T(LMN + LMPN + LMS + LMPS + LNS + PNS + LS + PS)$$
$$= T.(LMN + LMPN + LS + PS)$$
$$= TLMN + TLMPN + TLS + TPS$$

4. The sum-of-products Boolean expression above states that all the input combinations can be accounted by the prime implicants (T , L, M, N) or (T , L, M, P, N) or (T , L, S) or (T , P, S). The most optimum expression would result from either TLS or TPS. Therefore, the minimized Boolean function is given by

$$f(A,B,C,D) = A.B + \bar{B}.\bar{C}.D + \bar{A}.\bar{B}$$

OR

$$f(A,B,C,D) = A.B + \bar{A}.\bar{B} + A.\bar{C}.D$$

Example 2.18: A logic system has three inputs $A, B, and\ C\ and\ two\ outputs\ Y_1\ and\ Y_2$. The output functions $Y_1\ and\ Y_2$ are expressed as

$$Y_1 = \bar{A}.B.C + B.\bar{C} + \bar{A}.\bar{C} + A.\bar{B}.C + A.B.C$$
$$Y_2 = \bar{A}.B + A.\bar{C} + A.B.C$$

Solution:

1. The expanded form of $Y_1\ and\ Y_2$ can be written as follows

$$Y_1 = \bar{A}.B.C + A.B.\bar{C} + \bar{A}.B.\bar{C} + \bar{A}.\bar{B}.\bar{C} + A.\bar{B}.C + A.B.C$$
$$Y_2 = \bar{A}.B.C + \bar{A}.B.\bar{C} + A.B.\bar{C} + A.\bar{B}.\bar{C} + A.B.C$$

2. Further, construction of the prime implicant table has been created as shown in tabular form below:

A	***B***	***C***	**1**	**2**		***A***	***B***	***C***	**1**	**2**		***A***	***B***	***C***	**1**	**2**	
0	0	0	√		√	0	_	0	√		*	_	1	_	√	√	*
0	1	0	√	√	√	0	1	_	√	√	√						
1	0	0		√	√	_	1	0	√	√	√						
						1	_	0		√	*						
0	1	1	√	√	√												
1	0	1	√		√	_	1	1	√	√	√						
1	1	0	√	√	√	1	_	1	√		*						
						1	1	_	√	√	√						
1	1	1	√	√	√												

3. From the prime implicant table, the minimized output Boolean functions can be written as follows:

		Y_1						Y_2			
000	010	011	101	110	111	010	011	100	110	111	$A.B.C$
√	√										0_0
								√	√		1_0

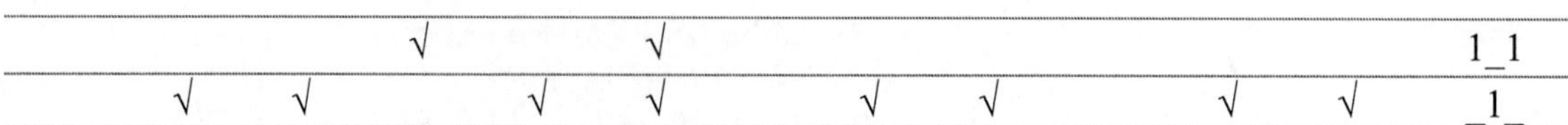

Hence, the expression can be given as

$$Y_1 = B + \bar{A}.\bar{C} + A.C$$
$$Y_1 = B + A.\bar{C}$$

Unit summary

Logic gates are electronics circuits that can be used to implement the most elementary logic expression also known as Boolean expressions. The logic gate is the most basic building block of combinational logic. There are three basic logic gates, namely the OR gate, the AND gate and the NOT gate. Other logic gates that are derived from these basic gates are the NAND gate, the NOR gate, the EXCLUSIVE-OR gate and the EXCLUSIVE-NOR gate. This chapter deals with logic gates and some related devices such as buffers, drivers, etc., as regards their basic functions. This chapter contain type of logic gates their functional description and Boolean expressions.

Solved Examples

Example 2.19: Simplify the following Boolean expressions to a minimum number of literals.

1) $xy + xy'$
2) $(x + y)(x + y')$
3) $xyz + x'y + xyz'$
4) $(A + B)'(A' + B')$
5) $(a + b + c')(a'b' + c)$
6) $a'bc + abc' + abc + a'bc'$

Solution:

1) $xy + xy' = x(y + y') = x.1 = x$
2) $(x + y)(x + y') = xx + xy' + xy + yy' = x$
3) $xyz + x'y + xyz' = xy(z + z') + x'y = y(x + x') = y$
4) $(A + B)'(A' + B')' = A'B'(AB) = 0$
5) $(a + b + c')(a'b' + c) = aa'b' + ac + ba'b' + bc + a'b'c' + cc' = ac + bc + a'b'c'$
6) $a'bc + abc' + abc + a'bc' = a'b(c + c') + ab(c + c') = a'b + ab = b(a + a') = b$

Example 2.20: Simplify the following Boolean expressions to a minimum number of literals.

1) $ABC + A'B + ABC'$
2) $x'yz + xz$
3) $(x + y)'(x' + y')$
4) $xy + x(wz + wz')$
5) $(BC' + A'D)(AB' + CD')$
6) $(a' + c')(a + b' + c')$

Solution:

1) $ABC + A'B + ABC' = AB + A'B = B(A + A') = B$
2) $x'yz + xz = z(x'y + x) = z(x' + x)(x + y) = z(x + y)$
3) $(x + y)'(x' + y') = x'y'(x' + y') = x'y'$
4) $xy + x(wz + wz') = xy + x\big(w(z + z')\big) = x(w + y)$
5) $(BC' + A'D)(AB' + CD') = BC'AB' + BC'CD' + A'DAB' + A'DCD' = 0$
6) $(a' + c')(a + b' + c') = a'a + a'b' + a'c' + c'a + c'b' + c'c' = c' + a'b'$

Example 2.21: Fins the complement of the given expression for 'F', also proof that

$$FF' = 0, and\ F + F' = 1$$

$$F = wx + yz$$

Solution:

$$F' = (wx + yz)' = (w.x)'.(y.z)' = (w' + x').(y' + z')$$

a) $F.F' = [(wx + yz)].[(w' + x').(y' + z')]$

$$= wx.(w' + x').(y' + z') + yz.(w' + x').(y' + z') = 0$$

b) $F + F' = [(wx + yz)] + [(w' + x').(y' + z')] = A + A' = 1$

Example 2.22: Draw the logic diagram to implement the following Boolean expression

1) $y = [(u + x')(y' + z)]$
2) $y = (u \oplus y)' + x$
3) $y = (u' + x')(y + z')$
4) $y = u(x \oplus z) + y'$
5) $y = u + yz + uxy$
6) $y = u + x + x'(u + y')$

Solution:

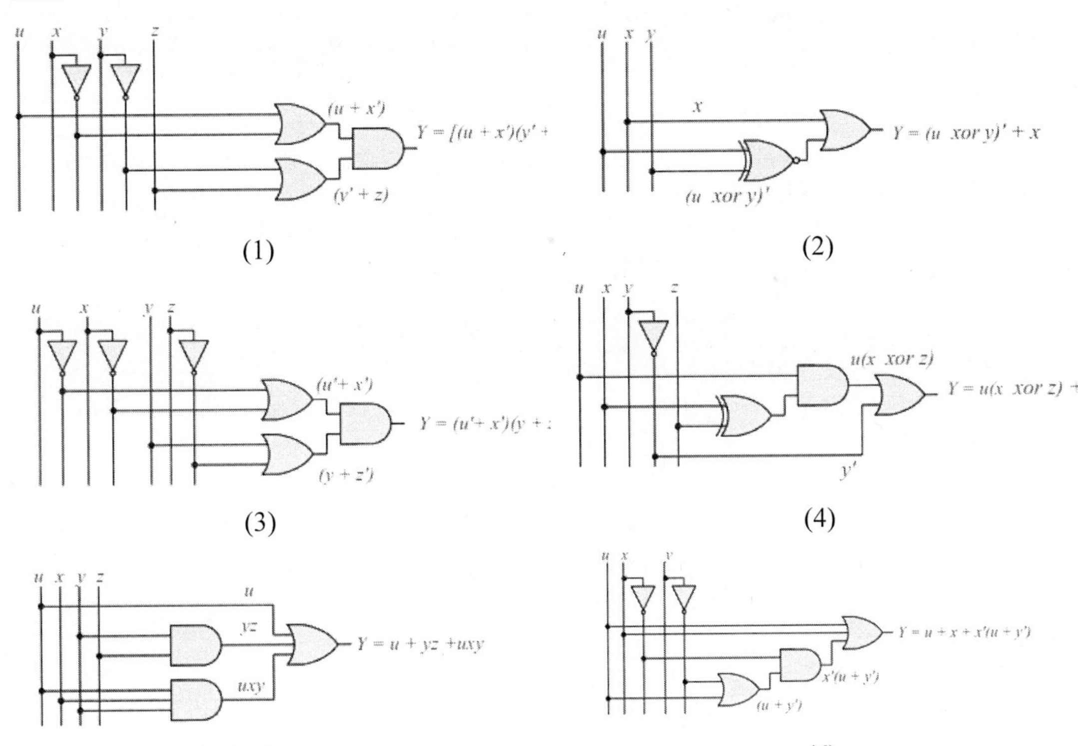

Example 2.23: Simplify the Boolean expression T_1 and T_2 represent by the table below to a minimum number of literals.

A	B	C	T_1	T_2
0	0	0	1	0
0	0	1	1	0
0	1	0	1	0
0	1	1	0	1
1	0	0	0	1
1	0	1	0	1
1	1	0	0	1
1	1	1	0	1

Solution:

$$T_1 = A'B'C' + A'B'C + A'BC' = A'B' + A'C'$$

$$T_2 = ABC' + ABC + AB'C' + AB'C + A'BC = AB(C + C') + AB'(C + C') + BC(A + A')$$
$$= AB + AB' + BC = A + BC$$

Example 2.24: Express the following functions in sum-of-minterms and product-of-maxterms.

1) $F = (b + cd)(c + bd)$
2) $F = (cd + b'c + bd')(b + d)$
3) $F = (c' + d)(b + c')$
4) $F = bd' + acd' + ab'c + a'c'$

Solution: To solve such problems all the brackets will be open, further convert it to canonical form and compare it with equivalent minterm or maxterm.

1) $F = (b + cd)(c + bd) = bc + bd + cd + bcd =$

$$F = \sum(3, 5, 6, 7, 11, 14, 15) = \prod(0,1,2,4, 8,9,10,12,13)$$

$$F' = \sum(0,1,2,4, 8,9,10,12,13)$$

2) $F = (cd + b'c + bd')(b + d) = cd + bd' =$

$$F = \sum 3,4,7,11,12,14,15 = \prod 0,1,2,5,6,8,9,10,13$$

$$F' = \sum 0,1,2,5,6,8,9,10,13$$

3) $F = (c' + d)(b + c') = c' + bd =$

$$F = \sum 0,1,4,5,7,8,12,13,15 = \prod 2,3,6,9,10,11,14$$

$$F' = \sum 2,3,6,9,10,11,14$$

4) $F = bd' + acd' + ab'c + a'c' =$

$$F = \sum 0,1,4,5,10,11,14 = \prod 2,3,6,7,8,9,12,13,15$$

$$F' = \sum 2,3,6,7,8,9,12,13,15$$

Example 2.25: for the given Boolean function

$$F = xy'z + x'y'z + w'xy + wx'y + wxy$$

1) Obtain the truth table using F
2) Draw the logic diagram using F

3) Use the Boolean algebra to reduce the function to minimum no. of literals
4) Obtain the truth table of the function from the simplified expression and show that it is the same as the one obtained from (1)
5) Draw the logic diagram from the simplified version and compare the total number of gates used with (2)

Solution:

1) Obtain the truth table (Input/output relationship): the output $F = 1$ for the minterms present in the expression. Hence,

$$F = xy'z + x'y'z + w'xy + wx'y + wxy$$

$$F = xy'z(w + w') + x'y'z(w + w') + w'xy(z + z') + wx'y(z + z') + wxy(z + z')$$

$$F = wxy'z + w'xy'z + wx'y'z + w'x'y'z + w'xyz + w'xyz' + wx'yz + wx'yz' + wxyz + wxyz'$$

w	x	y	z	F	w	x	y	z	F
0	0	0	0	0	1	0	0	0	0
0	0	0	1	1	1	0	0	1	1
0	0	1	0	0	1	0	1	0	1
0	0	1	1	0	1	0	1	1	1
0	1	0	0	0	1	1	0	0	0
0	1	0	1	1	1	1	0	1	1
0	1	1	0	1	1	1	1	0	1
0	1	1	1	1	1	1	1	1	1

$$F = \sum(1,5,6,7,9,10,11,13,14,15)$$

2) Logic diagram using F

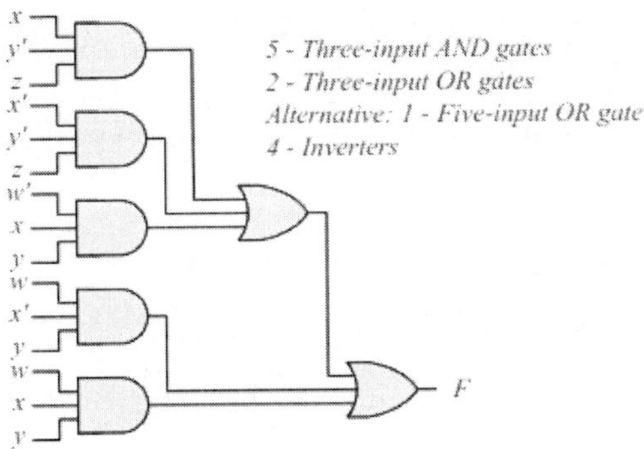

3) Boolean algebra to reduce the function to minimum no. of literals

$$F = xy'z + x'y'z + w'xy + wx'y + wxy =$$

$$F = xy(w + w') + y'z(x + x') + wy\,(x + x') = xy + y'z + wy = y'z + y(w + x)$$

4) Obtain the truth table of the function from the simplified expression

$$F = y'z + y(w + x)$$

$$F = \sum(1,5,6,7,9,10,11,13,14,15)$$

5) Logic diagram from the simplified version

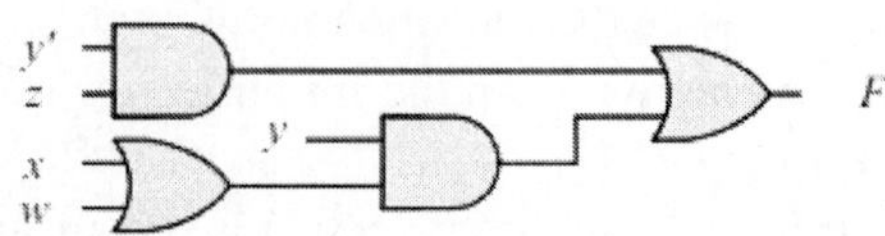

Comparison from Canonical form and Simplified form

Gate type	Canonical Form	Simplified form
AND Gates	5 (3 Input)	2 (2 Input)
OR Gates	2 (3 Input)	2 (2 Input)
Inverters	4	1

Example 2.26: Express the following function as a sum of minterms and as a product of maxterms

$$F(A,B,C,D) = B'D + A'D + BD$$

Solution:

$ABCD$ $-B'-D$	Equivalent Value	$ABCD$ $A'--D$	Equivalent Value	$ABCD$ $-B-D$	Equivalent Value
0001	1	0001	1	0101	5
0011	3	0011	3	0111	7
1001	9	0101	5	1101	13
1011	11	0111	7	1111	15

Hence,

$$F(A,B,C,D) = \sum(1,3,5,7,9,11,13,15) = \prod(0,2,4,6,8,10,12,14)$$

Example 2.27: Express the complement of the following functions:

$$F(A,B,C,D) = \sum(2,4,7,10,12,14)$$
$$F(x,y,z) = \prod(3,5,7)$$

Solution:

1)

$$F'(A,B,C,D) = \sum(0,1,3,5,6,8,9,11,13,15)$$

2)

$$F(x,y,z) = \prod(3,5,7) = \sum(0,1,2,4,6)$$
$$F'(x,y,z) = \sum(3,5,7)$$

Example 2.28: Convert each of the following to the other canonical form

$$F(x,y,z) = \sum(1,3,5)$$
$$F(A,B,C,D) = \prod(3,5,8,11)$$

Solution:

$$F(x,y,z) = \sum(1,3,5) = \prod(0,2,4,6,7)$$
$$F(A,B,C,D) = \prod(3,5,8,11) = \sum(0,1,2,4,6,7,9,10,12,13,14,15)$$

Example 2.29: Convert each of the following expressions into sum of products and product of sums

1) $F = (u + xw)(x + u'v)$

2) $F = x' + x(x + y')(y + z')$

Solution:

1) $F = (u + xw)(x + u'v) = ux + uu'v + xw + xwu'v = ux + xw(1 + u'v) = ux + xw$

$$F = ux + xw \;\ldots\ldots\ldots \text{SOP Form}$$
$$F = x(u + w) \;\ldots\ldots\ldots \textit{POS Form}$$

2) $F = x' + x(x + y')(y + z') = x' + x(xy + xz' + yy' + y'z') = x' + xy + xz'$

$$F = x' + xy + xz' \;\ldots\ldots\ldots \text{SOP Form}$$
$$F = (x' + y + z') \;\ldots\ldots\ldots \textit{POS Form}$$

Example 2.30: Show that a positive logic NAND gate is a negative logic NOR gate and vice versa.

Solution:

Gate		NAND (Positive logic)		NOR (Negative logic)	
x y	z	x y	z	x y	z
L L	H	0 0	1	1 1	0
L H	H	0 1	1	1 0	0
H L	H	1 0	1	0 1	0
H H	L	1 1	0	0 0	1

Gate		NOR (Positive logic)		NAND (Negative logic)	
x y	z	x y	z	x y	z
L L	H	0 0	1	1 1	0
L H	L	0 1	0	1 0	1
H L	L	1 0	0	0 1	1
H H	L	1 1	0	0 0	1

Example 2.31: Write Boolean expressions and construct the truth tables describing the outputs of the circuits described by the logic diagrams shown below

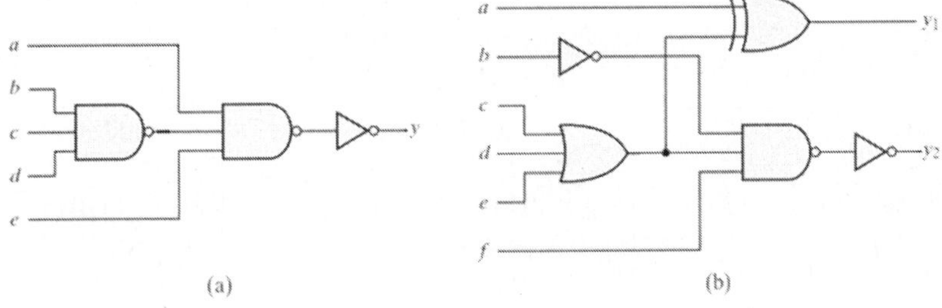

(a) (b)

Solution:

(a)

$$y = a(bcd)'e = ae(b' + c' + d') = aeb' + aec' + aed'$$

Truth Table can be constructed as

$ab' - -e$	Equivalent Value	$a - c' - e$	Equivalent Value	$a - -d'e$	Equivalent Value
10001	17	10001	17	10001	17
10011	19	10011	19	10101	21
10101	21	11001	25	11001	25
10111	23	11011	27	11101	29

$$y = \sum(17, 19, 21, 23, 25, 27, 29)$$

(b)

$$y_1 = a \oplus (c + d + e) = a'(c + d + e) + a(c'd'e')$$

$$= a'c + a'd + a'e + ac'd'e'$$

$a' - c - -$	Value	$a' - -d -$	Value	$a' - - - e$	Value	$a' - - - e$	Value
001000	8	**001000**	8	000010	2	100000	32
001001	9	**001001**	9	000011	3	100001	33
001010	10	**001010**	10	000110	6	110000	34
001011	11	**001011**	11	000111	7	110001	35
001100	12	**001100**	12	001010	10		
001101	13	**001101**	13	001011	11		
001110	14	**001110**	14	001110	14		
001111	15	**001111**	15	001111	15		
011000	24	**010100**	20	010010	18		
011001	25	**010101**	21	010011	19		
011010	26	**010110**	22	010110	22		
011011	27	**010111**	23	010111	23		
011100	28	**011100**	28	011010	26		
011101	29	**011101**	29	011001	27		
011110	30	**011110**	30	011110	30		
011111	31	**011111**	31	011111	31		

Hence,

$$y_1 = \sum(2,3,6,7,8,9,10,11,12,13,14,15,18,19,22,23,24,25,26,27,28,29,30,31,32,33,34,35)$$

$$y_2 = b'(c + d + e)f = b'cf + b'df + b'ef$$

$-b'c - -f$	Value	$-b' - d - f$	Value	$-b' - -ef$	Value
001001	9	001001	9	000011	3
001011	11	001011	11	000111	7
001101	13	001101	13	001011	11
001111	15	001111	15	001111	15
101001	41	101001	41	100011	35
101011	43	101011	43	100111	39
101101	45	101101	45	101011	51
101111	47	101111	47	101111	55

Hence,

$$y_1 = \sum(3,7,9,13,1535,39,41,43,45,47,51,55)$$

<u>Review Questions:</u>

1. How do you distinguish between positive and negative logic systems? Prove that an OR gate in a positive logic system is an AND gate in a negative logic system.
2. Give brief statements that would help one remember the truth table of AND, NAND, OR, NOR, EX-OR and EX-NOR logic gate functions, irrespective of the number of inputs used.
3. Why are NAND and NOR gates called universal gates? Justify your answer with example.

4. What are Schmitt gates? How does a Schmitt gate overcome the problem of occurrence of an erratic output for slow varying input transitions?
5. What are logic gates with open collector or open drain outputs? What are the major advantages and disadvantages of such devices?
6. Draw the circuit symbol and the associated truth table for the following:
   a. a tristate noninverting buffer with an active HIGH ENABLE input;
   b. a tristate inverting buffer with an active LOW ENABLE input;
   c. a three-input NAND with an open collector output;
   d. a four-input INHIBIT gate.
7. What is the main significance of IEEE/ANSI symbols when compared with the conventional ones?
8. Draw the ANSI symbols for four-input OR, two-input AND, two-input EX-OR and two-input NAND gates.
9. Draw the truth table of the logic circuit shown in Fig. 2.52.

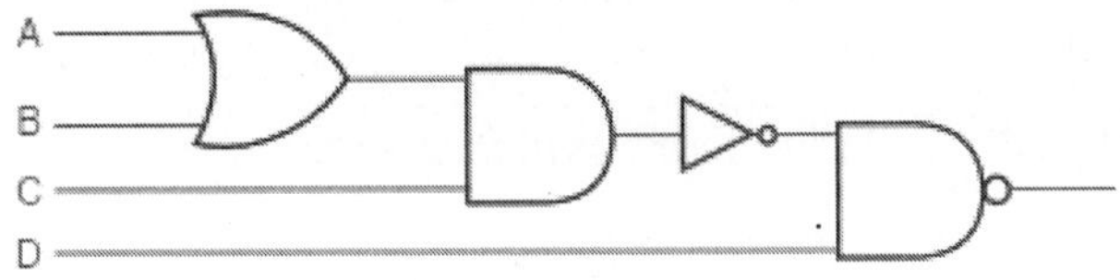

Fig. 2.52: Problem 8

10. Redraw the logic circuit of Fig. 2.53 using IEEE/ANSI symbols.

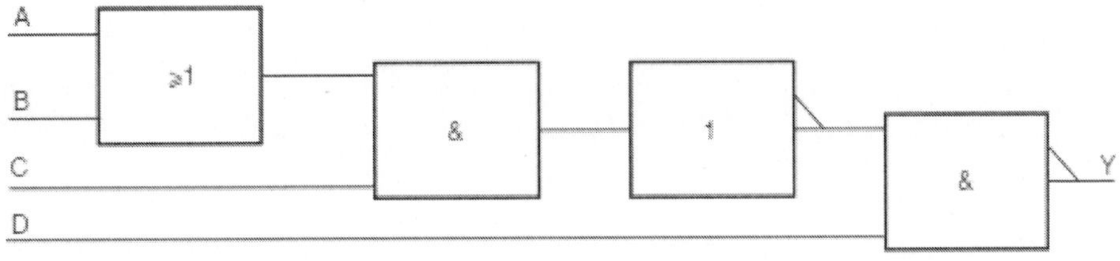

Fig. 2.53: Problem 9

11. Draw the logic diagram to implement an eight-input EX-NOR function using the minimum number of two-input logic gates.
12. Simplify the following Boolean functions to a minimum number of literals.
    a. $f = xy + xy'$
    b. $f = (x + y)(x + y')$
    c. $f = xyz + x'y + xyz'$
    d. $f = (A + B)'(A' + B')'$
13. Show that $(A + B)\overline{AB}$ is equivalent to $A \oplus B$
14. Show that $AB + (\overline{A + B})$ is equivalent to $A \odot B$

15. Find the Logical equivalent of the following expressions

(a) $A \oplus 0$ (b) $A \oplus 1$ (c) $A \odot 0$

(d) $A \odot 1$ (e) $1 \oplus \bar{A}$ (f) $0 \oplus \bar{A}$

16. Determine which of the following Expression are equivalent to $A \oplus B$ and which to $A \odot B$

(a) $\bar{A} \oplus B$ (b) $\bar{A} \odot B$ (c) $\bar{A} \oplus \bar{B}$

(d) $\bar{A} \odot \bar{B}$ (e) $A \oplus \bar{B}$ (f) $A \odot \bar{B}$

Multiple Choice Questions (MCQs)

1. Boolean algebra is also called
   a. switching algebra
   b. arithmetic algebra
   c. linear algebra
   d. algebra
2. To perform product of max terms Boolean function must be brought into
   a. AND terms
   b. OR terms
   c. NOT terms
   d. NAND terms
3. According to the Boolean algebra absorption law, which of the following is correct?
   a. x+xy=x
   b. (x+y)=xy
   c. xy+y=x
   d. x+y=y
4. The expression for Absorption law is given by
   a. A + AB = A
   b. A + AB = B
   c. AB + AA' = A
   d. A + B = B + A
5. X*y = y*x is the
   a. commutative law
   b. inverse property
   c. associative law
   d. identity element
6. Minterms is also called
   a. standard sum
   b. standard product
   c. standard division
   d. standard subtraction
7. Maxterms is also called
   a. standard sum
   b. standard product
   c. standard division
   d. standard subtraction

8. In Boolean algebra Multiplicative inverse is
   a. 0
   b. 1
   c. 1/a
   d. a
9. MCQ: (a+b+c)' is equal to
   a. a'b'c'
   b. a'+b'+c'
   c. abc
   d. a+b+c
10. The complement of function (A+B+C)' using theorem and laws is
    a. (A')+B+C
    b. (A+B)'+C
    c. A+B+C
    d. A'B'C'
11. A Boolean function can be converted from algebraic expressions to a product of max terms by using
    a. graphical representation
    b. Truth table
    c. Canonical Conversion Method
    d. Both b and c
12. DE Morgan's theorem states that
    a. (AB)' = A' + B'
    b. (A + B)' = A' * B
    c. A' + B' = A'B'
    d. A' + B' = A'B'
13. The Boolean function A + BC is a reduced form of
    a. A'B + AB'C
    b. AB + BC
    c. (A + B)(A + C)
    d. (A + C)B
14. According to the Boolean algebra theorems x.x is equal to
    a. x
    b. 1
    c. 0
    d. x'
15. Boolean algebra is an algebraic structure with two arithmetic operations
    a. addition and subtraction
    b. subtraction and multiplication
    c. addition and multiplication
    d. addition and division

16. The logical sum of two or more logical product terms is called
    a. Sum of Product (SOP)
    b. Product of Sum (POS)
    c. OR operation
    d. NAND operation
17. X+X.Y is equal to
    a. y
    b. 1
    c. 0
    d. x
18. X+0=0+X =X is an example of
    a. commutative property
    b. inverse property
    c. associative property
    d. Identity element
19. (x*y)*z=x*(y*z) is the
    a. commutative property
    b. inverse property
    c. identity element
    d. associative property
20. In Boolean algebra, the OR operation is performed by which of the following properties?
    a. associative properties
    b. commutative properties
    c. distributive properties
    d. identity element
21. Exclusive-OR (XOR) logic gates can be constructed from ....logic gates.
    a. OR gates only
    b. AND gates and NOT gates
    c. AND gates, OR gates, and NOT gates
    d. OR gates and NOT gates
22. What is the one's complement for the binary number 011001
    a. 000111
    b. 100110
    c. 111001
    d. 110001
23. What is the Truth Table?
    a. It is a table representing what output will one get for a given input.
    b. It is a table representing what input and output will be given by a given boolean operator.
    c. Both of these
    d. None of these

24. What will be the output of the given logic gate?

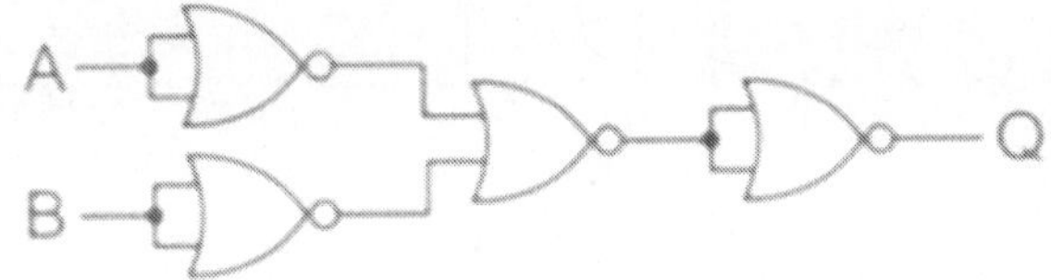

   a. NOR
   b. NAND
   c. AND
   d. OR

25. The output of a logic gate is I when all its inputs are at logic 0, the gate is either
   a. a NAND or an EX-OR
   b. an OR or an EX-NOR
   c. an AND or an EX-OR
   d. a NOR or an EX-NOR

26. How many two - input AND and OR gates are required to realize Y = CD + EF + G?
   a. 2,2.
   b. 2,3.
   c. 3.3.
   d. none of these

27. DeMorgan's first theorem shows the equivalence of
   a. OR gate and Exclusive OR gate.
   b. NOR gate and Bubbled AND gate.
   c. NOR gate and NAND gate.
   d. NAND gate and NOT gate

References and Suggested Readings

- *A. Anand Kumar, Fundamentals of digital circuits Second Edition, PHI Learning Private Limited, ISBN (978-81-203-3679-7)*
- *Floyd, T. L. (2005) Digital Fundamentals, Prentice-Hall Inc., USA.*
- *Morris Mano, M. and Kime, C. R. (2003) Logic and Computer Design Fundamentals, Prentice-Hall Inc., USA.*
- *Malvino, A. P. and Leach, D. P. (1994) Digital Principles and Applications, McGraw-Hill Book Company. USA.*
- *Rafiquzzaman, M. (2005) Fundamentals of Digital Logic and Microcomputer Design, Wiley-Interscience, New York, USA.*
- *Tocci, R. J. and N. S. Widmer. 2004. Digital Systems Principles and Applications, 9th ed. Upper Saddle River, NJ: Prentice Hall.*
- *Tokheim, R. L. (1994) Schaum's Outline Series of Digital Principles, McGraw-Hill Companies Inc., USA.*
- *Yarbrough, John M. Digital logic: Applications and design. Eagan: West Publishing Company, 1997.*

3 Digital Logic Families

UNIT SPECIFICS

Through this unit we have discussed the following aspects:

- *To understand significance of logic family and various parameters associated.*
- *To Describe the characteristics of major logic families, such as TTL, CMOS, ECL, BiCMOS.*
- *To analyse and design logic circuits using TTL.*
- *To analyse and design logic circuits using CMOS.*
- *To understand the concept of fan-out and its impact on circuit performance.*

RATIONALE

Logic families are groups of logic circuits that are based on particular types of elements (resistors, transistors, and so forth). Families are identified by the manner in which the elements are connected, and, in some cases, by the types of elements used.

Unipolar Logic Family: In unipolar logic families, unipolar devices are the key element. MOSFET (Metal Oxide Semiconductor Field Effect Transistor) is a unipolar device, in which the current flows because of only one type of charge carriers (that is, either electrons or holes). The examples of unipolar families include PMOS, NMOS, and CMOS

Bipolar Logic Family: Transistors and diodes are bipolar devices, in which the current flows because of both the charge carriers (electrons and holes). In bipolar logic families, transistors and diodes are used as key element.

For each family, variations of a basic gate circuit are used to design a wide range of logic circuits with compatible input and output logical levels. In the design of a complete logic system, it is generally necessary to use logic circuits of one family only.

Along with the various logic families, it is appropriate to discuss the general terminology used to describe logic families. These characteristics include logic assignments, logic voltage levels, supply voltage, noise margins, operating speeds, fan-in and fan-out, operating temperature, and power dissipation.

PRE-REQUISITES

Logic Gates, Boolean Expressions, Basic Binary Arithmetic.

UNIT OUTCOMES

List of outcomes of this unit is as follows:

U3-O1: Compare their advantages and disadvantages of Logic Families

U3-O2: Understand speed, power consumption, noise immunity, etc. for Logic Families.

U3-O3: Understand how to interface between different logic families like TTL & CMOS.

U3-O4: Implement and realization of TTL, CMOS, ECL etc.

U3-O5: Identify various application relevant information for Logic Families.

Unit-3 Outcomes	EXPECTED MAPPING WITH COURSE OUTCOMES (1- Weak Correlation; 2- Medium correlation; 3- Strong Correlation)					
	CO-1	CO-2	CO-3	CO-4	CO-5	CO-6
U3-O1	3	3	3	-	3	1
U3-O2	1	1	2	2	1	-
U3-O3	2	1	3	1	2	1
U3-O4	-	-	3	1	2	2
U3-O5	3	3	3	-	3	1

➢ Scan the below QR codes for more information on the topic written below the QR code.

Description of Logic Families

Analysis of logic families

Quantitative Discussion on TTL circuits

Unit outcomes

In this section discussed based on digital devices, various digital circuits, referred to as logic families. we will briefly describe the parameters used to characterize different logic families and also used to compare different logic families. In this section, significance and types of logic families will be discussed. Further, the positive and negative logic and its significance are also discussed. In addition to this, different characteristics which are the key parameters in deciding the logic family for any circuit design are discussed in detail. The module is concluded with explanation of the brief history of the logic family in terms of discrete logic circuits. Several logic-circuit groups or families have been introduced. They differ primarily in the methods for carrying out the logic and the coupling to the inverter stages. For example, the transistor-transistor logic (TTL) uses a multi-emitter transistor instead of the diodes found in DTL circuits. In emitter-coupled logic (ECL), the circuits are coupled by a common-emitter resistor, and complementary transistor logic (CTL) uses a combination of PNP and NPN transistors.

3.1 Significance and Types of Logic Families

There are a variety of circuit configurations or more appropriately various approaches used to produce different types of digital integrated circuit. Each such fundamental approach is called a logic family. The idea is that different logic functions, when fabricated in the form of an IC with the same approach, or in other words belonging to the same logic family, will have identical electrical characteristics. These characteristics include supply voltage range, speed of response, power dissipation, input and output logic levels, current sourcing and sinking capability, fan-out, noise margin, etc. In other words, the set of digital ICs belonging to the same logic family are electrically compatible with each other.

A digital system in general comprises digital ICs performing different logic functions, and choosing these ICs from the same logic family guarantees that different ICs are compatible with respect to each other and that the system as a whole performs the intended logic function. In the case where the output of an IC belonging to a certain family feeds the inputs of another IC belonging to a different family, we must use established interface techniques to ensure compatibility. Understanding the features and capabilities of different logic families is very important for a logic designer who is out to make an optimum choice for his new digital design from the available logic family alternatives. A not so well thought out choice can easily underkill or overkill the design with either inadequate or excessive capabilities.

3.1.1 Types of Logic Family

The entire range of digital ICs is fabricated using either bipolar devices or MOS devices or a combination of the two. The classification is shown in Fig..

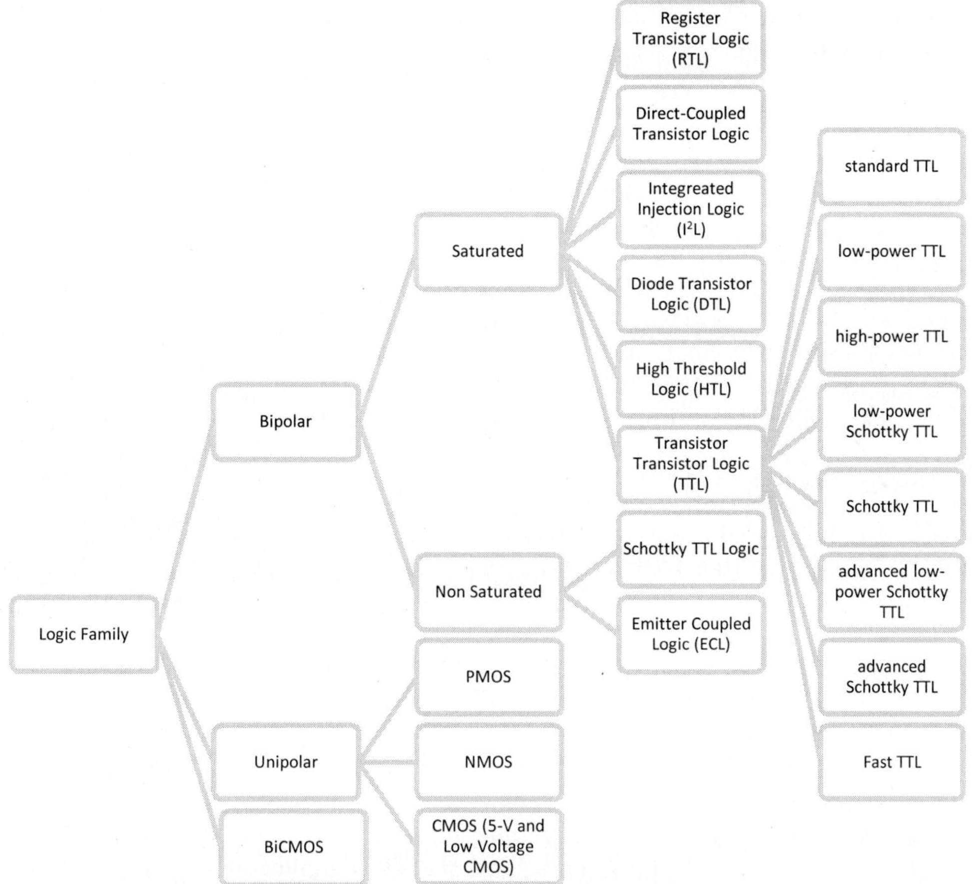

Fig. 3.1: Classification of Logic Families

- Bipolar Logic Families have used resistors, Diodes (Which are also Capacitors), and Transistors. Basically, they are of two types saturated and non-Saturated. In saturated logic, the transistors in the IC are driven to saturation, whereas in the case of non-saturated logic, the transistors are not driven into saturation.
- MOS devices are unipolar devices and only MOSFETs are employed in the MOS logic circuits. While in PMOS only p-channel MOSFETs are used and in NMOS only n-channel MOSFETs are used, in complementary MOS (CMOS), both p-channel and n-channel MOSFETs are fabricated on the same silicon chip.
- BiCMOS logic circuits use CMOS devices for input and logic operations whereas bipolar devices are used for output.

Furthermore, all the logic families listed above, the first three, that is, diode logic (DL), resistor transistor logic (RTL), and diode transistor logic (DTL) are of historical importance only. Both RTL and DTL suffered from large propagation delay owing to the need for the transistor base charge to leak out if the transistor were to switch from conducting to a nonconducting state.

Fig. 3.2 below shows the simplified schematics of a two-input AND gate using DL [Fig. 3.2 (a)], a two-input NOR gate using RTL [Fig. 3.2 (b)] and a two-input NAND gate using DTL [Fig. 3.2 (c)]. The DL, RTL, and DTL families, however, were rendered obsolete very shortly after their introduction in the early 1960s owing to the arrival on the scene of Transistor-Transistor Logic (TTL).

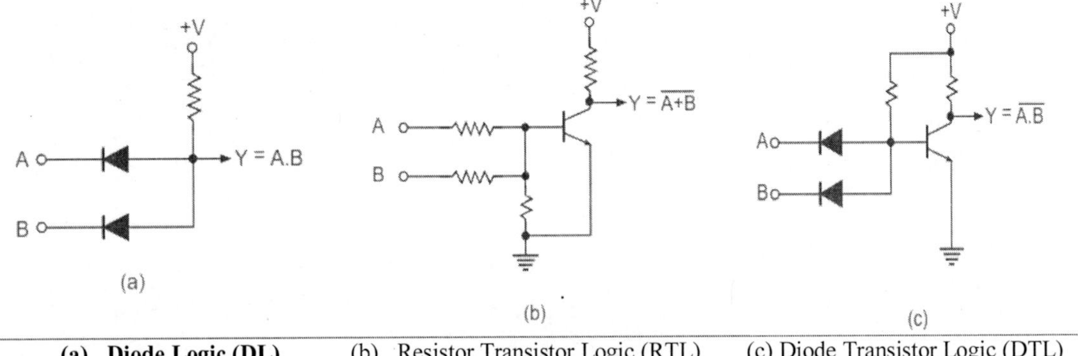

(a) Diode Logic (DL) (b) Resistor Transistor Logic (RTL) (c) Diode Transistor Logic (DTL)

Fig.3.2: Simplifies schematic for Logic Families DL, RTL and DTL

Logic families that are still in widespread use include TTL, CMOS, ECL, NMOS and Bi-CMOS. The PMOS and I2L logic families, which were mainly intended for use in custom large-scale integrated (LSI) circuit devices, have also been rendered more or less obsolete, with the NMOS logic family replacing them for LSI and VLSI applications.

3.1.2 Characteristic Parameters:

There are various logic families and the selection of a family for a particular application is based on its characteristics. Following are the parameters used to compare the performance of digital Ics:

- **High-level input current, I_{IH}.** This is the current flowing into (taken as positive) or out of (taken as negative) an input when a High-level input voltage is equal to the minimum High-level output voltage specified for the family is applied. In the case of bipolar logic families such as TTL, the circuit design is such that this current flows into the input pin and is therefore specified as positive. In the case of CMOS logic families, it could be either positive or negative, and only an absolute value is specified in this case.
- **Low-level input current, I_{IL}.** The Low-level input current is the maximum current flowing into (taken as positive) or out of (taken as negative) the input of a logic function when the voltage applied at the input equals the maximum Low-level output voltage specified for the family. In the case of bipolar logic families such as TTL, the circuit design is such that this current flow out of the input pin and is therefore specified as negative. In the case of CMOS logic families, it could be either positive or negative. In this case, only an absolute value is specified.

High-level and Low-level input current or loading are also sometimes defined in terms of unit load (UL). For devices of the TTL family, 1 UL (HIGH) = 40 μA and 1 UL (LOW) = 1.6 mA.

- **High-level output current, I_{OH}** : This is the maximum current flowing out of an output when the input conditions are such that the output is in the logic HIGH state. It is normally shown as a negative number. It tells about the current sourcing capability of the output. The magnitude of I_{OH} determines the number of inputs the logic function can drive when its output is in the logic HIGH state. For example, for the standard TTL family, the minimum guaranteed I_{OH} is -400 μA, which can drive 10 standard TTL inputs with each requiring 40 A in the HIGH state, as shown in Fig. 3.3.

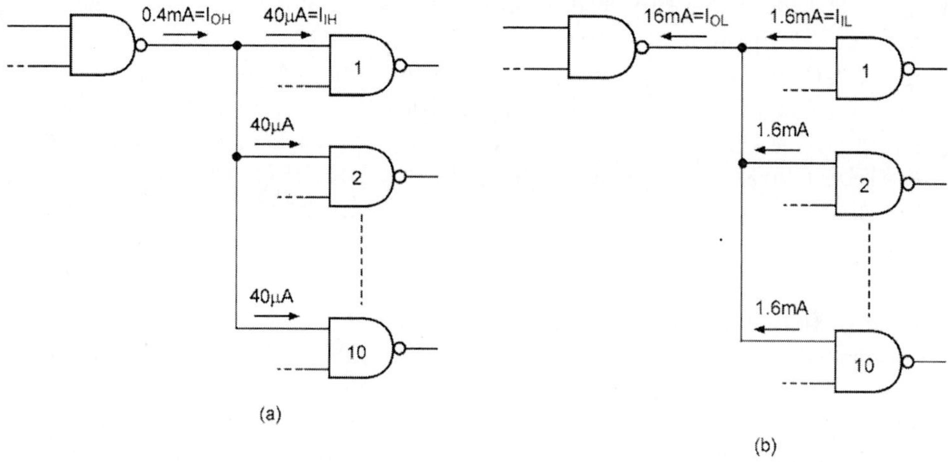

Fig.3.3: Input and output current (I_{OH}, I_{IH} , I_{OL} , **and** I_{IL})specifications.

- **Low-level output current, I_{OL}** : This is the maximum current flowing into the output pin of a logic function when the input conditions are such that the output is in the logic LOW state. It tells about the current-sinking capability of the output. The magnitude of I_{OL} determines the number of inputs the logic function can drive when its output is in the logic LOW state. For example, for the standard TTL family, the minimum guaranteed is 16 mA, which can drive 10 standard TTL inputs with each requiring 1.6 mA in the LOW state, as shown in fig. 3.3 (b).
- **High-level off-state (high-impedance state) output current, I_{OZH}**: This is the current flowing into an output of a tristate logic function with the ENABLE input chosen so as to establish a high-impedance state and a logic HIGH voltage level applied at the output. The input conditions are chosen so as to produce logic LOW if the device is enabled.
- **LOW-level off-state (high-impedance state) output current, I_{OZL}:** This is the current flowing into an output of a tristate logic function with the ENABLE input chosen so as to establish a high-impedance state and a logic LOW voltage level applied at the output. The input conditions are chosen so as to produce logic HIGH if the device is enabled.

- **HIGH-level input voltage, V_{IH}.** This is the minimum voltage level that needs to be applied at the input to be recognized as a legal HIGH level for the specified family. For the standard TTL family, a 2 V input voltage is a legal HIGH logic state.
- **LOW-level input voltage, V_{IL}.** This is the maximum voltage level applied at the input that is recognized as a legal LOW level for the specified family. For the standard TTL family, an input voltage of 0.8 V is a legal LOW logic state.
- **HIGH-level output voltage, V_{OH}.** This is the minimum voltage on the output pin of a logic function when the input conditions establish logic HIGH at the output for the specified family. In the case of the standard TTL family of devices, the HIGH-level output voltage can be as low as 2.4 V and still be treated as a legal HIGH logic state. It may be mentioned here that, for a given logic family, the VOH specification is always greater than the VIH specification to ensure output-to-input compatibility when the output of one device feeds the input of another.
- **LOW-level output voltage, V_{OL}.** This is the maximum voltage on the output pin of a logic function when the input conditions establish logic LOW at the output for the specified family. In the case of the standard TTL family of devices, the LOW-level output voltage can be as high as 0.4 V and still be treated as a legal LOW logic state. It may be mentioned here that, for a given logic family, the V_{OL} specification is always smaller than the V_{IL} specification to ensure output-to-input compatibility when the output of one device feeds the input of another.

The different input/output current and voltage parameters are shown in fig. below, with HIGH-level current and voltage parameters in 3.4(a) and LOW-level current and voltage parameters in 3.4 (b). It may be mentioned here that the direction of the LOW-level input and output currents shown in 3.4(b) is applicable to logic families with current-sinking action such as TTL.

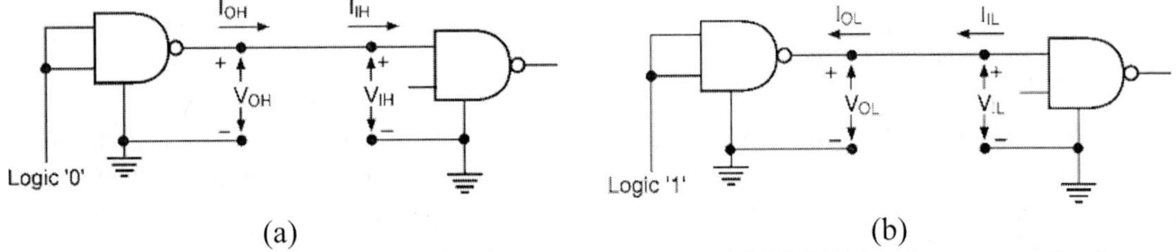

Fig. 3.4: (a) HIGH-level current and voltage parameters and (b) LOW-level current and voltage parameters.

- **Supply current, I_{CC}.** The supply current when the output is HIGH, LOW and in the high-impedance state is respectively designated as I_{CCH}, I_{CCL} and I_{CCZ}.
- **Rise time, t_r:** This is the time that elapses between 10 and 90 % of the final signal level when the signal is making a transition from logic LOW to logic HIGH.
- **Fall time, t_f :** This is the time that elapses between 90 and 10 % of the signal level when it is making HIGH to LOW transition.

- **Propagation delay t_p:** The propagation delay is the time delay between the occurrence of change in the logical level at the input and before it is reflected at the output. It is the time delay between the specified voltage points on the input and output waveforms. Propagation delays are separately defined for LOW-to-HIGH and HIGH-to-LOW transitions at the output.
    - Propagation delay t_{pLH}: This is the time delay between specified voltage points on the input and output waveforms with the output changing from LOW to HIGH.
    - Propagation delay t_{pHL}: This is the time delay between specified voltage points on the input and output waveforms with the output changing from HIGH to LOW. Fig 3.5 shows the two types of propagation delay parameter.

In addition, we also define enable and disable time delays that occur during transition between the high-impedance state and defined logic LOW or HIGH states.

- **Disable time from the HIGH state, t_{pHZ}.:** Defined for a tristate device, this is the time delay between specified voltage points on the input and output waveforms with the tristate output changing from the logic HIGH level to the high-impedance state.
- **Disable time from the LOW state, t_{pLZ}:** Defined for a tristate device, this is the time delay between specified voltage points on the input and output waveforms with the tristate output changing from the logic LOW level to the high-impedance state.

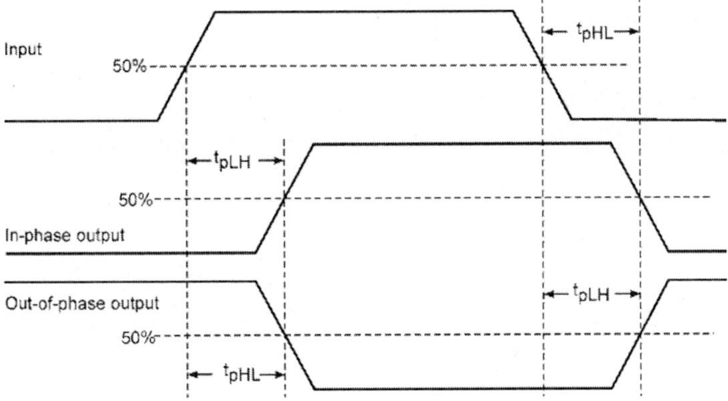

Fig. 3.5: Propagation delay parameters.

- **Enable time from the HIGH state, t_{pZH}:** Defined for a tristate device, this is the time delay between specified voltage points on the input and output waveforms with the tristate output changing from the high-impedance state to the logic HIGH level.
- **Enable time from the LOW state, t_{pZL}:** Defined for a tristate device, this is the time delay between specified voltage points on the input and output waveforms

with the tristate output changing from the high-impedance state to the logic LOW level.

- **Maximum clock frequency, f_{max}** : This is the maximum frequency at which the clock input of a flip-flop can be driven through its required sequence while maintaining stable transitions of logic level at the output in accordance with the input conditions and the product specification. It is also referred to as the maximum toggle rate for a flip-flop or counter device.
- **Power dissipation.** The power dissipation parameter for a logic family is specified in terms of power consumption per gate and is the product of supply voltage VCC and supply current I_{CC}. The supply current is taken as the average of the HIGH-level supply current I_{CCH} and the LOW-level supply current I_{CCL}.
- **Speed–power product.** The speed of a logic circuit can be increased, that is, the propagation delay can be reduced, at the expense of power dissipation. We will recall that, when a bipolar transistor switches between cut-off and saturation, it dissipates the least power but has a large associated switching time delay. On the other hand, when the transistor is operated in the active region, power dissipation goes up while the switching time decreases drastically. It is always desirable to have in a logic family low values for both propagation delay and power dissipation parameters. A useful figure-of-merit used to evaluate different logic families is the speed–power product, expressed in picojoules, which is the product of the propagation delay (measured in nanoseconds) and the power dissipation per gate (measured in milliwatts).
- **Fan-in:** Fan in or gate is the number of inputs that can practically be supported without degrading practically input voltage level. For example, a four-input gate will have a fan-in is equal to 4, as shown in Fig. 3.6(a).

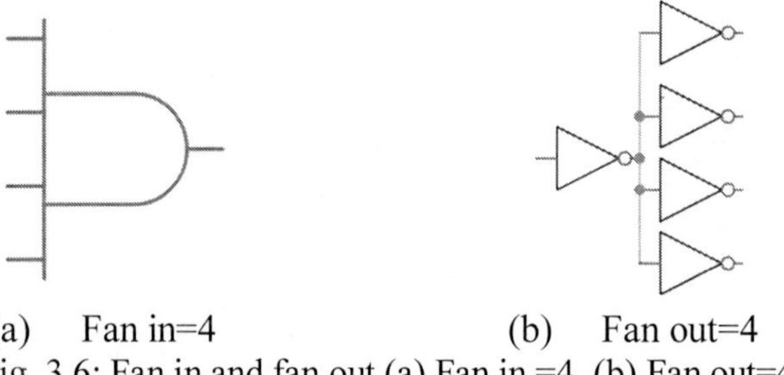

(a) Fan in=4 (b) Fan out=4

Fig. 3.6: Fan in and fan out (a) Fan in =4, (b) Fan out=4

- **Fan-out.** The fan-out is the number of inputs of a logic function that can be driven from a single output without causing any false output, as shown in Fig 3.6(b). It is a characteristic of the logic family to which the device belongs. It can be computed from I_{OH}/I_{IH} in the logic HIGH state and from I_{OL}/I_{IL} in the logic LOW state. If, in a certain case, the two values I_{OH}/I_{IH} and I_{OL}/I_{IL} are different, the fan-out is taken as the smaller of the two. This description of the fan-out is true for bipolar logic families like TTL and ECL. When determining the fan-out of CMOS logic devices, we should also take into consideration how much input load capacitance can be driven from the output without exceeding the acceptable value of propagation delay.

- **Noise margin.** This is a quantitative measure of noise immunity offered by the logic family. When the output of a logic device feeds the input of another device of the same family, a legal HIGH logic state at the output of the feeding device should be treated as a legal HIGH logic state by the input of the device being fed. Similarly, a legal LOW logic state of the feeding device should be treated as a legal LOW logic state by the device being fed. We have seen in earlier paragraphs while defining important characteristic parameters that legal HIGH and LOW voltage levels for a given logic family are different for outputs and inputs.

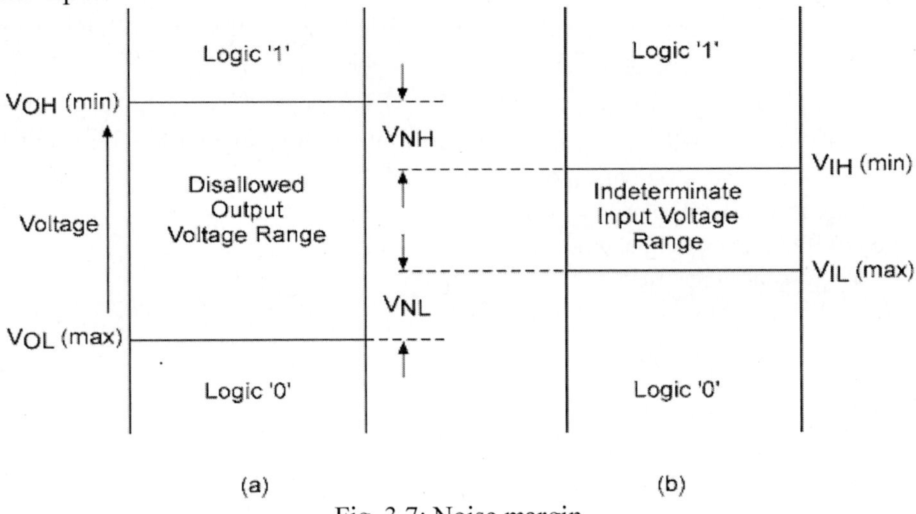

Fig. 3.7: Noise margin.

Fig. 3.7 shows the generalized case of legal HIGH and LOW voltage levels for output [Fig. 3.7 (a)] and input [Fig. 3.7(b)]. As we can see from the two diagrams, there is a disallowed range of output voltage levels from $V_{OL(max)}$ to $V_{OH(min)}$ and an indeterminate range of input voltage levels from $V_{IL(max)}$ to $V_{IH(min)}$. Since $V_{IL(max)}$ is greater than $V_{OL(max)}$the LOW output state can therefore tolerate a positive voltage spike equal to $V_{IL(max)}$ $V_{OL(max)}$ still be a legal LOW input. Similarly,$V_{OH(min)}$ is greater than $V_{IH(min)}$, and the HIGH output state can tolerate a negative voltage spike equal to $V_{OH(min)}$ $V_{IH(min)}$ still be a legal HIGH input. Here, $V_{IL(max)}$, $V_{OL(max)}$ and $V_{OH(min)}$, and $V_{IH(min)}$are respectively known as the LOW-level and HIGH-level noise margin.

Let us illustrate it further with the help of data for the standard TTL family. The minimum legal HIGH output voltage level in the case of the standard TTL is 2.4 V. Also, the minimum legal HIGH input voltage level for this family is 2 V. This implies that, when the output of one device feeds the input of another, there is an available margin of 0.4 V. That is, any negative voltage spikes of amplitude less than or equal to 0.4 V on the signal line do not cause any spurious transitions.

Similarly, when the output is in the logic LOW state, the maximum legal LOW output voltage level in the case of the standard TTL is 0.4 V. Also, the maximum legal LOW input voltage level for this family is 0.8 V. This implies that, when the output of one device feeds the input of another, there is again an available margin of 0.4 V. That is, any positive voltage spikes of amplitude less than or equal to 0.4 V on the signal line do not cause any spurious transitions. This leads to the standard TTL family offering a noise margin of 0.4 V. To generalize, the noise margin offered by a logic family, as outlined earlier, can be computed from the HIGH-state noise margin, $V_{NH} = V_{OH(min.)}V_{IH(min.)}$ and the LOW-state noise margin, $V_{NL} = V_{IL(max.)}V_{OL(max.)}$. If the two values are different, the noise margin is taken as the lower of the two.

Example 3.1: The data sheet of a quad two-input NAND gate specifies the following parameters: $I_{OH\ (max)} = 0.4\ mA,\ V_{OH\ (min)} = 2.7\ V,\ V_{IH\ (min)} = 2\ V,\ V_{IL\ (max)} = 0.8\ V$,, $V_{OL\ (max)} = 0.4\ V,\ I_{OL\ (max)} = 8\ mA,\ I_{IL\ (max)} = 0.4\ mA,\ I_{IH\ (max)} = 20\ \mu A,\ I_{CCH(max)} = 1.6\ mA,\ I_{CCL(max)} = 4.4\ mA,\ t_{pLH} = t_{pHL} = 15ns$ and a supply voltage range of 5V. Determine

a) The average power dissipation of a single NAND gate,
b) The maximum average propagation delay of a single gate,
c) The HIGH-state noise margin and
d) The LOW-state noise margin

Solution:

a) The average supply current=$I_{CCH} + \frac{I_{CCL}}{2} = \frac{1.6+4.4}{2} = 3mA$
The Supply Voltage=$V_{CC} = 5V$
Therefore, the power dissipation for all gates in the IC= $5\ X\ 3 = 15mW$
The Average power dissipation per gate is=$\frac{15}{4} = 3.75mW$
b) The propagation delay=$15ns$
c) The HIGH-state noise margin=$V_{OH\ (min)} - V_{IH\ (min)} = 2.7 - 2.0 = 0.7V$
d) The LOW-state noise margin=$V_{IL\ (max)\cdot} - V_{OL\ (max)} = 0.8 - 0.4 = 0.4V$

Example 3.2: Refer to example 3.1 how many NAND gates input can be driven from the output of a NAND gate of this type?

Solution:

This can be calculated while considering the worst-case scenario maximum number is to calculate.

The HIGH-state fan out=$\frac{I_{OH\ (max)}}{I_{IH\ (max)}} = \frac{400}{20} = 20$

The LOW-state fan out=$\frac{I_{OL\ (max)}}{I_{IL\ (max)}} = \frac{8}{0.4} = 20$

Therefore, the number of inputs that can be driven from a single output=20

Example 3.3: Determine the fan-out of IC 74LS04 given the following data: input loading factor (HIGH state) =0.5 UL, input loading factor (LOW state) =0.25 UL, output loading factor (HIGH state) = 10 UL, output loading factor (LOW state) =5 UL. *where UL is the unit load.*

Solution

- The HIGH-state fan-out can be computed from: fan-out =

$$\frac{\text{output loading factor (HIGH)}}{\text{input loading factor (HIGH)}} \frac{10\,UL}{0.5\,UL} = 20 =$$

- The LOW-state fan-out can be computed from: fan-out

$$\frac{\text{output loading factor (LOW)}}{\text{input loading factor (LOW)}} = \frac{5\,UL}{0.25\,UL} = 20$$

Since the fan-out in the two cases turns out to be the same, it follows that the fan-out = 20.

Example 3.4: A certain TTL gate has $I_{IH} = 20\mu A, I_{IL} = 0.1\ mA, I_{OH} = 0.4\ mA, I_{OL} = 4\ mA$. Determine the input and output loading in the HIGH and LOW states in terms of UL.

Solution:

1 UL (LOW state) = 1.6 mA and 1 UL (HIGH state) $=40\mu A$.

The input loading factor (HIGH state) = $20\mu A = \frac{20}{40} = 0.5$UL.

The input loading factor (LOW state) =0.1 mA=0.1/1.6 = 1/16 UL

The output loading factor (HIGH state) =0.4mA = 0.4/0.04= 10 UL.

The output loading factor (LOW state) =4mA=4/1.6 = 2.5 UL.

Transistor as a switch:

A circuit that can turn on/off current in electrical circuit is referred to a switching circuit and transistor can be employed as an electronic switch.

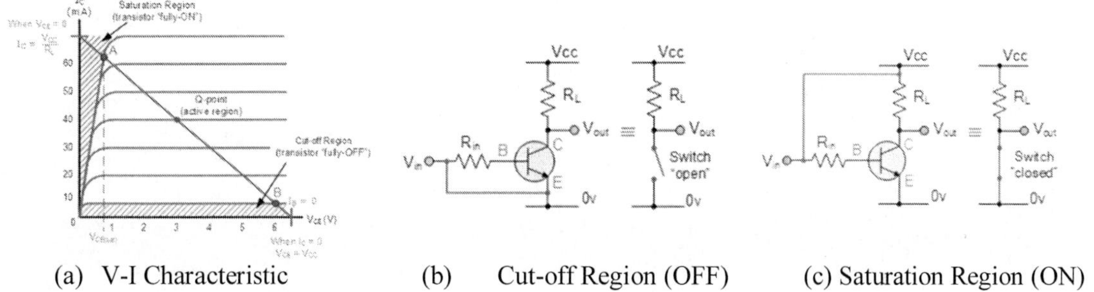

(a) V-I Characteristic (b) Cut-off Region (OFF) (c) Saturation Region (ON)

Fig. 3.8: Transistor used as switch (Logic 0 and Logic 1)

The cut off region for a transistor can be termed as OFF state where both the junctions are reverse biased. Here the operating conditions of the transistor are zero input base current (I_B), zero output collector current (I_C) and maximum collector voltage (V_{CE}) which results in a large depletion layer and no current flowing through the device. Therefore, the transistor is switched "Fully-OFF".

Hence, $I_C = 0$ and $V_{BE} < 0.7$ v

The saturation region can be termed as ON State, Here the transistor will be biased so that the maximum amount of base current is applied, resulting in maximum collector current resulting in the minimum collector emitter voltage drop which results in the depletion layer being as small as possible and maximum current flowing through the transistor. Therefore, the transistor is switched "Fully-ON".

Hence, I_C = maximum and V_{BE} >0.7v

3.2 Description of Logic Families

Logic circuits are normally connected in cascade; that is, the output from one gate is connected to the input of the following gate, and so on. Thus, the switching behavior of one circuit depends not only on its own output characteristic, but may also depend on the input characteristic of the next gate. Two possible logic assignments, positive logic or negative logic, are used to implement a Boolean function. For positive logic, a logical "one" is represented by a "high" voltage level, and a logical "zero" is represented by a "low" voltage. The reverse is used to represent negative logic.

3.2.1 Direct-Coupled Transistor Logic (DCTL)

Transistorized digital circuits basically fulfill the three logical functions of AND (or NAND) gating, OR (or NOR) gating, and signal inversion (NOT gate). An additional function usually performed, though not logical in nature but nevertheless a practical necessity, is signal amplification. Other logical blocks, such as NOR, NAND, and flip-flops, are easily obtained using these three fundamental functional blocks. Several different circuit configurations can be used for these functional blocks.

The figure below shows three DCTL inverters in cascade. In this circuit the collector resistors R_1, R_2, and R_3 serve as constant current sources. They supply current to their respective transistors' collectors, when they are on or to the base of the next transistor when they are off.

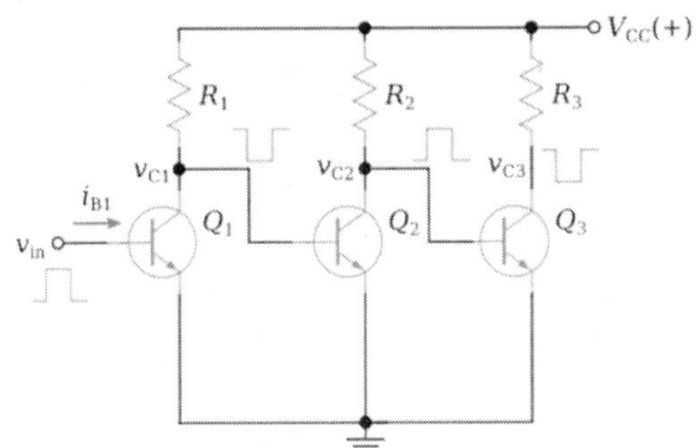

Fig. 3.9: DCTL inverters in cascade.

When the input voltage v_{in} to the base of Q_1 is near ground—that is at $V_{CE(SAT)}$ of the previous stage on transistor—voltage v_{C1} tends to approach the supply voltage V_{CC}. Current is supplied to the base of Q_2 through R_1, and this turns Q_2 on. The clamping action of the base-emitter diode of Q_2 holds v_{C1} at the value determined by V_{BE2}. Since Q_2 is on, v_{C2} is now determined by the $V_{CE(SAT)}$ of Q_2. If Q_2 has a sufficiently low saturation drop, then v_{C2} will not be positive enough to turn Q_3 on. The reverse situation holds true when a sufficiently positive voltage, v_{in}, turns Q_1 on. In this case v_{C1} will maintain Q_2 off, which in turn will turn Q_3 on. From this brief description several significant points are apparent. A low $V_{CE(SAT)}$ is a desirable feature of transistors used for DCTL. If the $V_{CE(SAT)}$ is high, then there is always the possibility that the next stage transistor may be erroneously turned on. Furthermore, in order to assure that all the fan-out transistors are held off, the $V_{CE(SAT)}$ must be smaller than the smallest $V_{BE(ON)}$ of the succeeding transistors. It is readily seen that the supply voltage can be relatively small because the output voltage swing varies between the $V_{CE(SAT)}$ of the on transistor and the $V_{BE(ON)}$ of the following stage transistors.

Disadvantages of DCTL

Current Hogging

One of the most undesirable features of DCTL is what is commonly known as current hogging, and this phenomenon arises because of the spread in $V_{BE(ON)}$ of the various driven (fan-out) transistors. No two transistors will ever have identical input characteristics, and it is always desirable to use transistors with as small a production spread as possible on $V_{BE(ON)}$.

The Crosstalk or the Noise Problem

In any high-speed system, pulses with fast rise times are likely to produce potential differences in the ground system mainly because of ground inductances. These voltages are likely to interfere with stable operation of the system. DCTL systems are very susceptible to these noise voltages because the operating and signal voltages are naturally low in these systems. If several transistors on one end of a ground system are turned on, the resulting pulse generated in the ground system can supply a positive or negative pulse (depending on the polarity) which can cause faulty turn-on or turn-off of other transistors further down the ground system.

One solution is to mount the transistors very close together, thereby minimizing ground inductances. Besides ground noise, DCTL is also vulnerable to noise on power supplies and stray noise picked up by connecting leads. Of the various logic schemes, DCTL has one of the lowest noise margins, typically 0.1 V at 125°C to about 0.2 V at room temperature, depending on the fan-out and, whether the transistor is on or off.

3.2.2 Resistor-Transistor Logic (RTL)

Resistor-transistor logic (RTL) is constructed from resistors and transistors. It is shown in fig. below

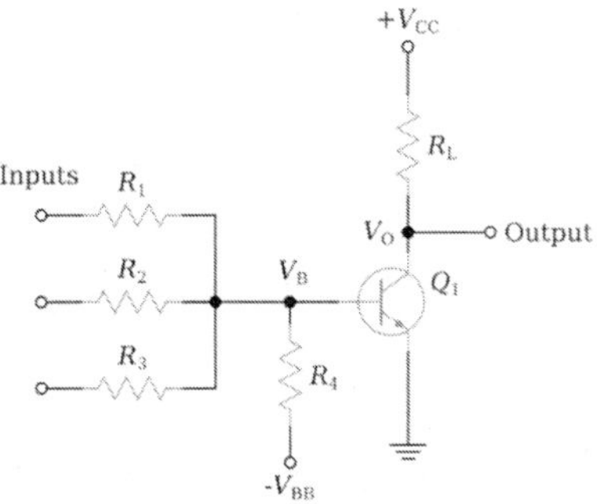

Fig. 3.10: Basic one-transistor RTL NOR circuit.

The figure below shows a simple 3-input RTL circuit which is a NOR gate. The circuit has each input combined through a coupling resistor (R_1, R_2, or R_3) and fed to the base of Q_1. The coupling resistors provide isolation that is less effective than that provided by transistors or diodes. This isolation makes the RTL sensitive to the value of the coupling resistors and the number of inputs (fan-in).

The operation of the circuit is quite simple. When all three inputs are zero (i.e., at ground level), the voltage V_B at the base of the transistor Q_1 is slightly more negative than ground since the base is tied to a negative voltage, $-V_{BB}$ through resistor R_4. Q_1 is cut off and the output voltage V_0 tends to fall to the supply voltage $+V_{CC}$. If any of the three inputs are positive, the current flow through its corresponding resistor pulls the voltage V_B to some positive value. This drives Q_1 into saturation. Here, the output voltage V_0 is determined by the saturation voltage $V_{CE(SAT)}$ of Q_1.

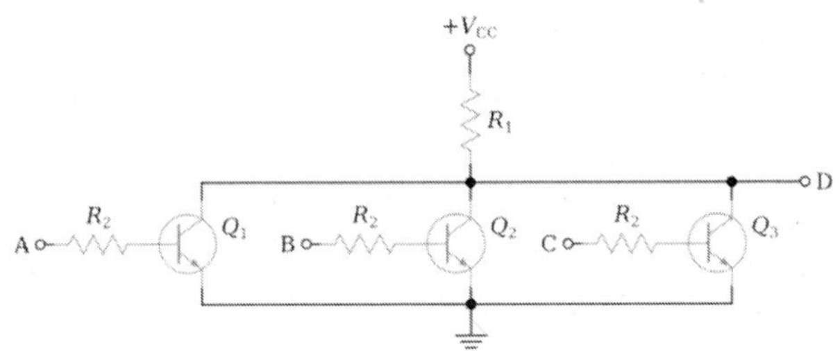

Fig. 3.11: Multi-transistor RTL NOR circuit.

The multi-transistor configuration is shown in the figure below. The configuration is the same as that of the corresponding DCTL NOR circuit, except for the addition of an input resistor, R_2. The addition of R_2 forces some of the input voltage to appear across R_2, and the base currents are more evenly divided between parallel inputs. By increasing R_2, high fan-out is achieved with no current-hogging; however, as R_2 is increased, the switching speed of the circuit is reduced. A design compromise must be made between high fan-out and high switching speed when using an RTL circuit.

The major con is that it has been technologically obsolete for over 50 years. It is slow, has a low fan-out, is not power efficient, and was only produced in low levels of integration.

3.2.3 Diode-Transistor Logic (DTL)

Diode-transistor logic (DTL) was one of the most popular manufactured circuits. The basic DTL circuit is a diode AND with a transistor inverter. The figure below shows the basic DTL configuration which is a NAND circuit.

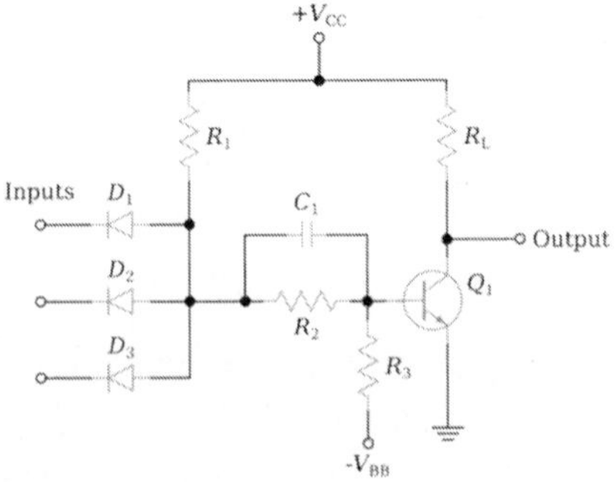

Fig. 3.12: Basic diode-transistor logic circuit (NAND).

The divider network, consisting of R1, R2, and R3, is designed such that if all the input voltages are HIGH (logic 1), the base of Q1 is relatively positive and Q1 is on. The output voltage at the collector of Q1 is almost zero. If any of the inputs now swing to 0 volt, that particular diode conducts. This applies a relatively negative voltage to the base of the transistor which is now cutoff. The collector tends to rise to the supply voltage +VCC. Capacitor C1 is used to provide an overdrive current during switching time. This reduces the switching time to some extent.

Demerits of DTL

- Limited operating speed.
- Poor noise immunity.
- High temperature sensitive.

3.2.4 Transistor-Transistor Logic (TTL)

Transistor-transistor logic (TTL) integrated circuits were introduced in the late 1960s. TTL is best represented by the SN 54/74 series of digital integrated circuits. Texas Instruments Corporation originally developed this series of logic ICs but a number of manufacturers make them and use the same numbering system. The SN 54 (SN stands for semiconductor network) series is more expensive and designed for greater temperature range than the SN 74 series. The SN 54 devices will operate over a range of temperatures from -55 to +125 °C, whereas the SN 74 devices operate over a range from 0 to 70 °C. TTL ICs operate on a power supply voltage of 5 volts and require a well-regulated 5-volt power source.

There are two variations in the output of this circuit namely open collector and tri-state. These are discussed in detail in the successive sections. The Transistor Transistor Logic (TTL) family has heavily dependence on transistors to provide basic operation. Figure 1 shows internal schematic of basic two inputs TTL NAND gate. This gate performs the positive NAND function. Other TTL functions are implemented with this basic circuit or a modified version of it.

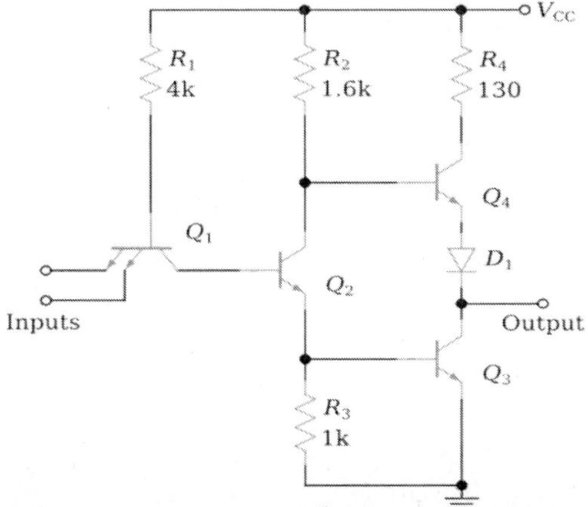

Fig. 3.13: Transistor Transistor logic (TTL)

The TTL circuit, shown in the figure above, is analogous to the DTL (diode-transistor logic) circuit in certain respects. When TTL was developed, it made use of the economic feasibility of using transistors instead of diodes, and also solved the capacitive loading problem.

The input transistor (Q_1) offers a significant advantage in switching time over the diode input of DTL. The addition of the phase-splitter driver (Q_2) and, transistor Q_4 sits above Q_3, forms the totem pole arrangement this push-pull buffered-output circuit provides good noise rejection and excellent drive characteristics into capacitive loads. Referring to the figure above, the TTL configuration employs a circuit which is a very low output impedance for high-capacity drive capability, high fan-out and good noise immunity. When the inputs of the circuit are high, Q_2 and Q_3 are driven "on" as common-emitter amplifiers. Transistor Q_4 and its emitter diode are both lightly forward-biased but conduct only a negligible amount of current. When one of the inputs is low, Q_2 and Q_3 are both cut off, and Q_4 conducts as an emitter-follower. The output is, therefore, pushed and pulled up and down by a transistor turning "on".

3.2.5 Classification of TTL Logic Family

TTL logic family is further classified based on the performance characteristics namely, speed and propagation delay. Accordingly, it is classified as Low Power Schottky TTL, Schottky TTL etc. The subfamilies are then compared based on these key parameters. TTL is the abbreviation of Transistor- Transistor Logic. TTL logic families are classified based on the types of output configurations. There are three Types of TTL family

1. Open collector output
2. Totem pole or standard output
3. Three state (or tri state) output

Open Collector output:

In the open collector TTL gate, the output stage does not have the active pull-up transistor i.e. Q4 and the output is taken at the collector of transistor Q3. In order to generate required HIGH and LOW states, an external pull-up resistor is connected to +Vcc from the collector of Q3. The Figure 2 shows the internal schematic of a NAND gate with an open collector output.

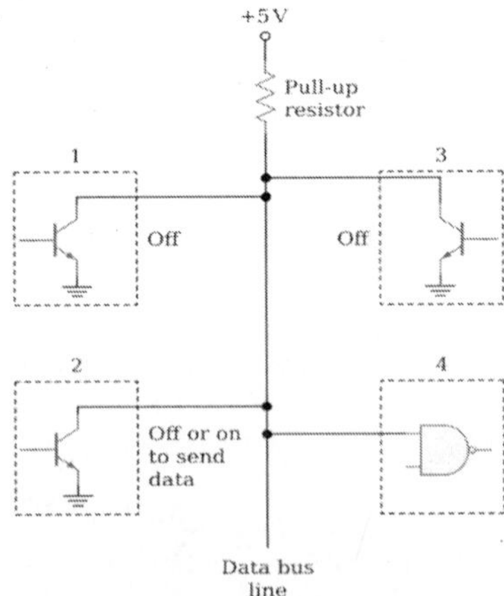

Fig. 3.14: Open collector output configuration of TTL NAND gate

When transistor Q3 is OFF, the output is pulled to +Vcc through the external resistor typically of value 10 kΩ. When Q3 is ON, the transistor is in saturation and the output is nearly at ground potential.

While open collector TTL circuits can be used in this way, many of the benefits of TTL circuits are lost by eliminating the active pull-up transistor. The high speed and high noise immunity associated with standard TTL circuits are lost when open collector circuits are used. To gain back these advantages and maintain the high speed and superior noise immunity of TTL circuits, the three-state TTL gate was developed.

Tristate Gate Output

In addition to totem-pole configuration and wire ANDing operation, there is third configuration i.e. Tristate gate, which is also popular. Tristate logic gates have three possible output states, i.e. the logic '1' state, the logic '0' state and a high-impedance state. The high-impedance state is controlled by an external ENABLE input. The ENABLE input decides whether the gate is active or in the high-impedance state.

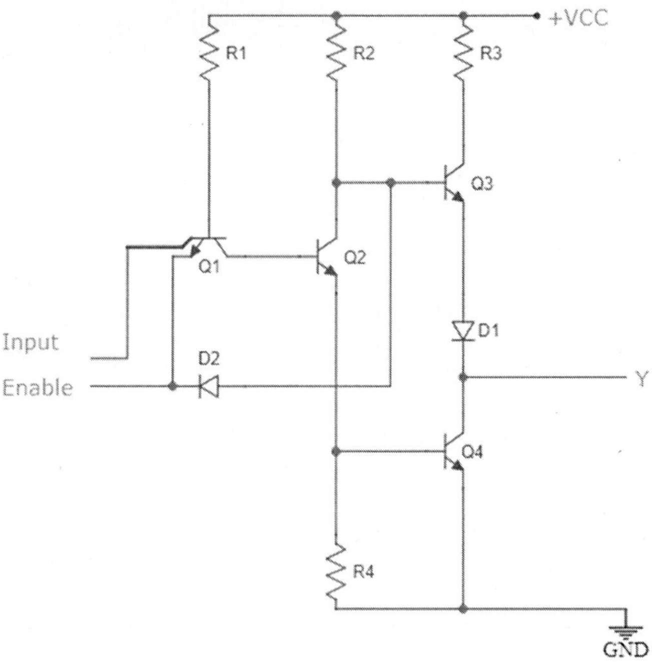

Fig. 3.15: TTL Inverter with tristate output

Figure 3 shows the circuit configuration for Tristate TTL inverter. The standard TTL NAND circuit is modified so as to act as an inverter with tristate logic. When enable input is active, the output changes the state and can be '0' or '1' depending upon input conditions. One of the main advantages of these gates is that, their inputs and outputs can be connected in parallel to a common bus line.

- When the enable input is HIGH, the circuit operates as normal inverter because transistor Q1 and Q2 have no effect and diode D1 is reverse-biased. In this enabled operation, the circuit behaves like an inverter.
- When the enable input is LOW, diode D1 becomes forward biased. A LOW enables input forces transistor Q2 to cut-off and which in turn forces Q4 to turn-off. Also, when enable is low a forward-biased D1 forces Q3 to cut-off. When both output transistors are in cut-off, the output essentially is an open circuit. Thus, the output presents high output impedance state.

Schottky TTL (SN 54S/74S)

A significant improvement in speed of the basic TTL circuit was made by implementing it with Schottky transistors. A Schottky transistor is essentially a bipolar transistor which has a Schottky barrier or hot carrier diode connected between its base and collector as shown in view A of the figure below. This diode prevents the transistors in the TTL circuit from saturating and thereby, improves their ability to switch at higher speeds. View B of the figure below shows thc symbol that is used to represent a Schottky transistor.

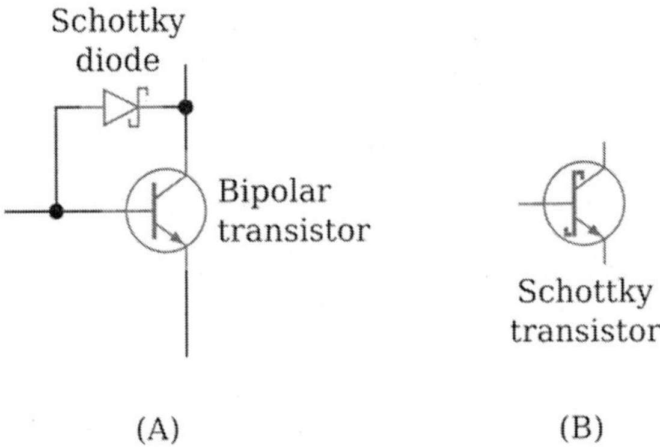

Fig. 3.16: (A) A bipolar transistor with a Schottky diode between collector and base becomes a Schottky transistor; (B) the symbol used to represent a Schottky transistor.

The Schottky TTL removes the storage time of transistors by preventing them from going into saturation. In this way, the speed of operation increases without excessive increase in power dissipation from Schottky TTL. Basic TTL family has a speed limitation due to turn off delays of transistors which are caused by transition from saturation to cutoff. This limitation can be removed by replacing the transistors of TTL gate by Schottky transistors. These transistors are prevented from entering the saturation and hence there is saving in turn-off time. The propagation delays of Schottky TTL is of the order of 2 ns. In comparison, the propagation delay for standard TTL is 10 ns. The Schottky TTL therefore is a non- saturating bipolar logic.

The Schottky diode between the base and the collector of the bipolar transistor prevents saturation and, therefore, eliminates this problem. The Schottky diode switches very fast and requires only 0.1 to 0.3 volts of forward bias depending on temperature. When this diode is added, speed is increased because the time required to get a transistor out of saturation is not a factor. The diode will conduct before the transistor is allowed to saturate.

By using Schottky transistors and decreasing the values of the resistances in the standard TTL circuit, the speed of the basic TTL circuit can be greatly improved. A typical Schottky TTL gate circuit has a propagation delay of about 3 nanoseconds, which is less than one-third that of the standard TTL gate. This increased speed is obtained with an increase in power dissipation to about 20 milliwatts and, of course, the higher cost of the Schottky technique.

Low-Power Schottky TTL (SN 54LS/74LS)

A low-power version of the Schottky TTL gate is also available. This device uses much larger circuit resistances, thereby considerably reducing the power consumption. At the same time, Schottky transistors are used to improve the switching speed. A typical low-power Schottky TTL gate has a propagation delay of about 10 nanoseconds, with a power dissipation of 2 milliwatts. A low-power Schottky gate has the same propagation delay as a standard TTL gate, but its power dissipation is one-fifth of the standard gate. These improved speed-power characteristics are obtained at only a slight increase in cost over the standard circuit. At present, the low-power Schottky TTL circuit is perhaps the optimum TTL circuit. This accounts for its high popularity with digital designers.

The low power Schottky TTL sacrifices speed for a reduced power dissipation. It has the same propagation delay as standard TTL, but the power dissipation is reduced to a 1/5 that of standard TTL. It is the most popular version of in new digital circuit system design. The TTL versions are available in SSI, MSI and LSI packages.

The listing below indicates the different versions of TTL gates available, the power requirements per gate, propagation delay, and maximum operating speed (Refer Table). As mentioned earlier, the TTL versions do not vary functionally; they vary in terms of the values of resistors used and the type of transistors that the basic gate uses. These subfamilies are compared on the basis of figure of merit as shown in the Table 1.

Table 3.1: Comparison of TTL subfamilies

Version	Abbreviation	IC Nos.	Delay (ns)	Power/ gate (mW)	Speed (MHz)	Figure of Merit (PJ)
Standard TTL	STTL	(SN 54/74)	10	10	35	100
High speed TTL	HTTL	(SN 54H/74H)	6	22	50	132
Low power TTL	LTTL	(SN 54L/74L)	33	1	3	33
Schottky TTL	STTL	(SN 54S/74S)	3	20	125	57
Low-power Schottky TTL	LSTTL	(SN 54LS/74LS)	10	2	40	19

It has been observed from the above table that analyzing the power and speed characteristics shows that if you want low power in a TTL IC, you will have to sacrifice speed and vice-versa. Moreover, for all subfamilies input and output voltage and current level requirements are almost same.

However, one can distinguish them on the basis of total propagation delay. In Low power, TTL it is about 30ns whereas for Schottky TTL is very small of the order of 2.5ns. In low power Schottky TTL it is about 15 ns. It is a trade- off between the LTTL and STTL. A comparison is given in Table 2

Table 3.2: Electrical Characteristics of TTL subfamilies

Parameter		5400/ 7400	54H00/ 74H00	54L00/ 74L00	54S00/ 74S00	54LS00/ 74LS00	UNITS
V_{IH}		2	2	2	2	2	*volts*
V_{IL}	54	0.8	0.8	0.7	0.8	0.7	*volts*
	74	0.8	0.8	0.8	0.8	0.8	*volts*
V_{OH}	54	*2.4*	*2.4*	*2.4*	*2.5*	*2.5*	*volts*
	74	*2.4*	*2.4*	*2.4*	*2.7*	*2.7*	*volts*
V_{OL}	54	0.4	0.4	0.3	0.5	0.4	*volts*
	74	0.4	0.4	0.4	0.5	0.5	*volts*
I_{IH}		40	50	10	50	20	μA
I_{IL}		-1.6	-2.0	-0.18	-2.0	-0.36	mA
I_{OH}		16	20	2	20	4	μA
I_{IL}	54	16	20	2	20	4	mA
	74	16	20	3.6	20	8	mA
$I_{cc}(1)$		8	16.8	0.8	16	1.6	mA
$I_{cc}(0)$		22	40	2.04	36	4.4	mA
t_{pHL}		15	10	60	5	15	ns
t_{pLH}		22	10	60	4.5	15	ns

Fan-out capabilities of different subfamilies, are dependent on the output source current capabilities. It is observed from the table that, if the source device is S TTL and output device is L TTL maximum number of gates can be connected (Refer Table 3.3).

Table 3.3: Summary of TTL fan out capabilities

Source TTL Device	5400/ 7400	54H00/ 74H00	54L00/ 74L00	54S00/ 74S00	54LS00/ 74LS00
5400/7400	10	8	40	8	20
54H00/74H00	12	10	50	10	25
54L00/74L00	2	1	20	1	10
54S00/74S00	12	10	100	10	50
54LS00/74LS00	5	4	40	4	20

Handling unused TTL inputs

The floating input of TTL family devices behaves as if a logic HIGH has been applied to the input. Such a behaviour is explained from the input circuit of a TTL device as shown in Figure 5 When the input is HIGH, the input emitter-base junction is reverse biased, and the current that flows into the input is the reverse-biased diode leakage current. The input diode will be reverse biased, even when the input terminal is left unconnected or floating. This implies that a floating input behaves as if there were a logic HIGH applied to it.

As an initial thought, we may tend to believe that it should not make any difference if we leave the unused inputs of NAND and AND gates as floating, as a logic HIGH- like behaviour of the floating input makes no difference to the logical behaviour of the gate, as shown in Fig. 3.17 (a) and (b).

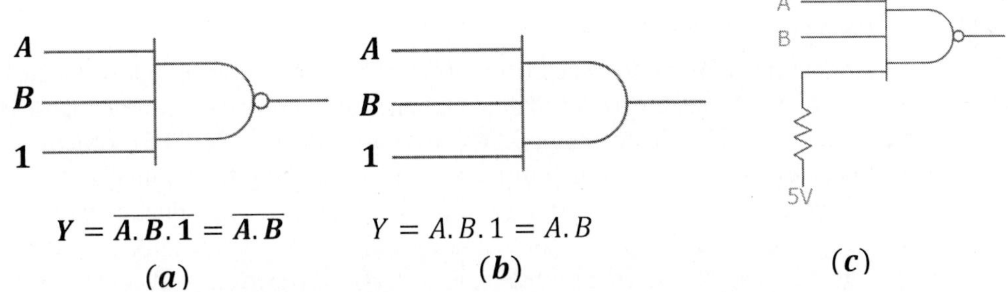

$\boldsymbol{Y = \overline{A.B.1} = \overline{A.B}}$ $\qquad Y = A.B.1 = A.B$

$(\boldsymbol{a})$ $\qquad (\boldsymbol{b})$ $\qquad (\boldsymbol{c})$

Fig. 3.17: Handling unused pins of AND as well as NAND gates

In spite of this, it is strongly recommended that the unused inputs of AND and NAND gates be connected to a logic HIGH input [Fig. 3.18 (c)] because, floating input behaves as an antenna and may pick up stray noise and interference signals, thus causing the gate to function improperly. 1kΩ resistance is connected to protect the input from any current spikes caused by any spikes on the power supply line. More than one unused input (up to 50) can share the same 1 kΩ resistance, if needed.

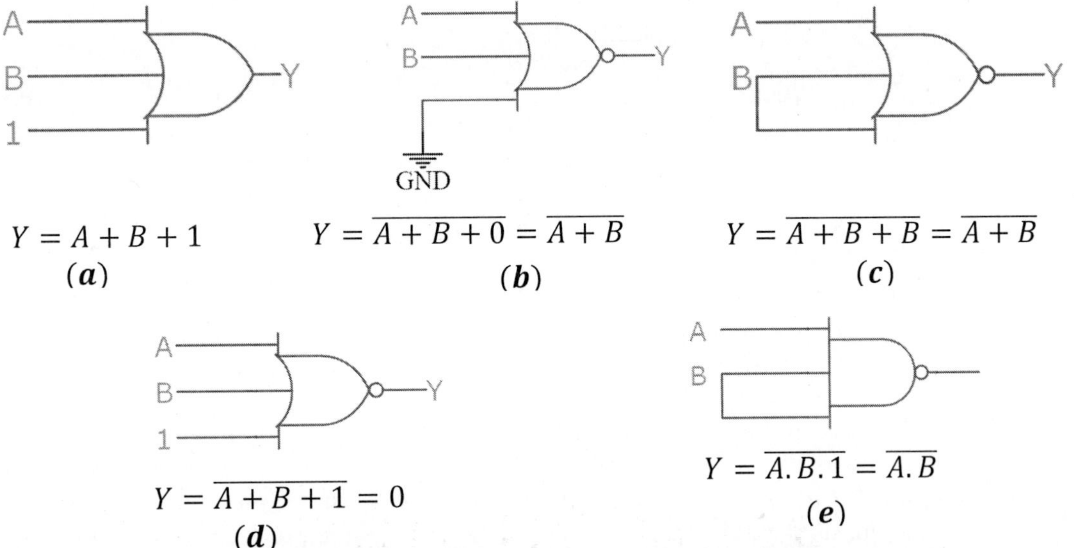

$Y = A + B + 1$ $\qquad Y = \overline{A + B + 0} = \overline{A + B}$ $\qquad Y = \overline{A + B + B} = \overline{A + B}$

$(\boldsymbol{a})$ $\qquad (\boldsymbol{b})$ $\qquad (\boldsymbol{c})$

$Y = \overline{A + B + 1} = 0$ $\qquad Y = \overline{A.B.1} = \overline{A.B}$

$(\boldsymbol{d})$ $\qquad (\boldsymbol{e})$

Fig. 3.18: Unused inputs of NAND and NOR gates

In the case of OR shown in Fig. 3.18(a) and (c) for NOR gates, unused inputs are connected to ground (logic LOW), as shown in Fig. 3.18 (b) for obvious reasons. A floating input or an input tied to logic HIGH in this case produces a permanent logic HIGH (for an OR gate) and LOW (for a NOR gate) at the output as shown in Fig. 3.18 (a) and (b) respectively. An alternative solution is shown in Fig. 3.18 (d) and (e), where the unused input is tied to one of the used inputs. This solution works well for all gates, but one has to be conscious of the fact that the fanout capability of the output driving the tied inputs is not exceeded.

Current Transients and Power Supply Decoupling

TTL family devices are prone to occurrence of narrow-width current spikes on the power supply line. Current transients are produced, when the totem-pole output stage of the device undergoes a transition from a logic LOW to a logic HIGH state. The problem becomes severe when in a digital circuit, a large number of gates are likely to switch states at the same time. These current spikes produce voltage spikes due to any stray inductance present on the line. On account of the large rate of change in current in the current spike, even a small value of stray inductance produces voltage spikes large enough to adversely affect the circuit performance.

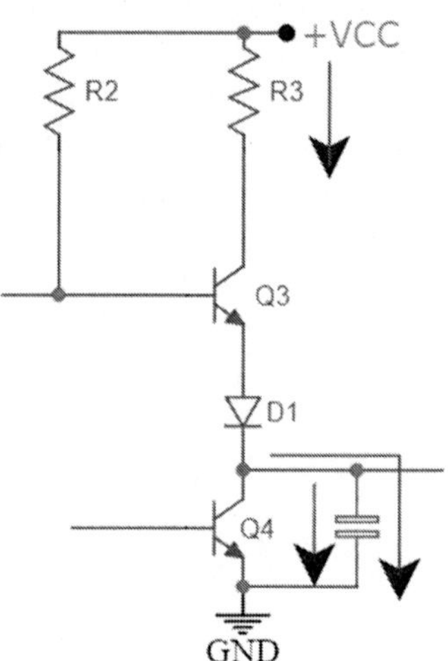

Fig. 3.19: Current transient and power supply decoupling

Above Fig. 3.19 illustrates the phenomenon. When the output changes from LOW to HIGH, there is a small-time interval during which both the transistors are conducting; because the pull-up transistor Q3 has switched on and the pull-down transistor Q4 has not yet come out of saturation. During this time interval, there is an increase in current drawn from the supply; ICCL experiences a positive spike before it settles down to a usually lower ICCH. The presence of any stray capacitance C across the output owing to any stray wiring capacitance or capacitance loading of the circuit being fed also adds to the problem. The problem of voltage spikes on the power supply line is usually overcome by connecting small-value, low-inductance, high frequency capacitors between VCC terminal and ground. It is standard practice to use a 0.01 or 0.1 μF ceramic capacitor from VCC to ground. This capacitor is also known by the name of power supply decoupling capacitor, and it is recommended to use a separate capacitor for each IC.

A decoupling capacitor is connected as close to the VCC terminal as possible, and its leads are kept to a bare minimum to minimize lead inductance. In addition, a single relatively large-value capacitor in the range of 1–22 μF is also connected between VCC and ground on each circuit card to take care of any low-frequency voltage fluctuations in the power supply line.

3.2.6 Emitter Coupled Logic (ECL):

Emitter-coupled logic (ECL) is a BJT-based logic family which is generally considered as the fastest logic available. ECL achieves its high-speed operation by employing a relatively small voltage swing and preventing the transistors from entering the saturation region. The ECL (emitter coupled logic) was first invented at IBM in August 1956 by Hannon S. Yourke. This logic is also known as current mode logic, used in the computers of IBM 7090 & 7094. ECL family is very fast as compared to digital logic families.

An implementation of ECL utilizes a positive supply voltage which is known as PECL or positive-referenced ECL. In early ECL gates, a negative voltage supply is used due to the noise immunity. After that, positive-referenced ECL became very famous due to its more compatible logic levels as compared to TTL logic families.

Emitter Coupled Logic Features

The features of ECL will make them used in many high-performance-based applications.

- ECL provides two outputs which are complements of each other always because, the operation of the circuit is based on a differential amplifier.
- This logic family is mainly suitable for monolithic fabrication methods because logic levels are a function of resistor ratios.
- The devices of the ECL family generate the right & complementary output of the proposed function without using any outside inverters. Consequently, it decreases the package count, and requirements of power & also decreases problems occurring from time delays.
- ECL devices in differential amplifier design offer broad performance flexibility, so ECL circuits allow being used both as digital and linear circuits.
- The design of the ECL gate has normally high & low input impedance, which is extremely conducive to attaining large fan-out as well as drive capability.
- ECL devices generate a constant current drain on the power supply to simplify the design of the power supply.
- The devices of ECL including open emitter outputs simply allow them to include transmission line drive capacity.

Inverter circuit of emitter-coupled logic

The circuit shown below represents the emitter-coupled logic circuit of an inverter. It has two NPN transistors connected in differential single-ended input mode. Both the emitters are connected together with common resistance R_E. It is a current limiting resistance, used to prevent the transistor from entering into saturation.

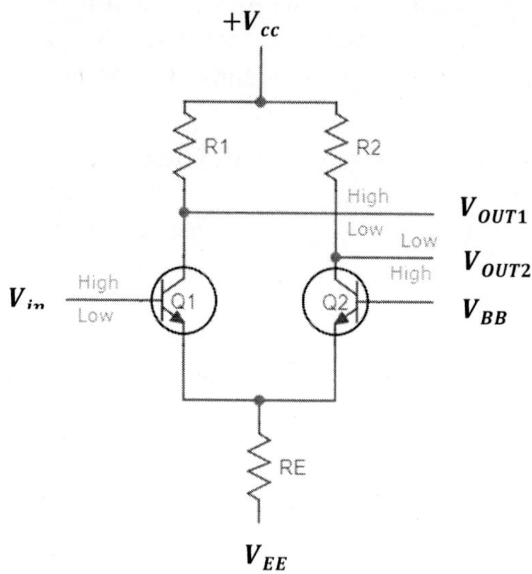

Fig. 3.20: ECL based inverter circuit

It has two outputs: inverting output (V_{OUT1}) and non-inverting output (V_{OUT2}). V_{IN} is the input terminal, where LOW or HIGH input is given. When the input is HIGH, it will turn ON the transistor Q_1 but not saturated and the transistor Q_2 is turned OFF. This will pull the output V_{OUT2} to HIGH but due to the drop in resistant R_1, the output at terminal V_{OUT1} will be at LOW value.

On the other side, when the input V_{IN} is given LOW value, it will turn OFF the transistor Q_1 and the transistor is turned ON. The transistor Q_2 will not enter into saturation. It will make the output at terminal V_{OUT1} to be pulled HIGH value. Due to the drop-in resistance R_2, the output at terminal V_{OUT2} will have LOW value.

Two input ECL OR/NOR gate

The following circuit is the Emitter-coupled logic circuit of the 2-input OR/NOR gate.

- It is the slight modification of the inverter circuit given above.
- In this, an additional transistor is used at the input side. The operation is simple as explained above.
- If the input at both the transistors Q_1 and Q_2 are LOW, it will make V_{OUT1} to HIGH value. It corresponds to the NOR gate output.
- At the same time, transistor Q_3 is turned ON, which will make the V_{OUT2} to be LOW. It corresponds to the OR gate output.
- Similarly, if both the input of transistors Q_1 and Q_2 are HIGH, it will turn on both the transistors. It will drive the output at terminal V_{OUT1} to be LOW.
- The transistor Q_3 is turned OFF during this operation. It will drive the output at terminal V_{OUT2} to be HIGH.

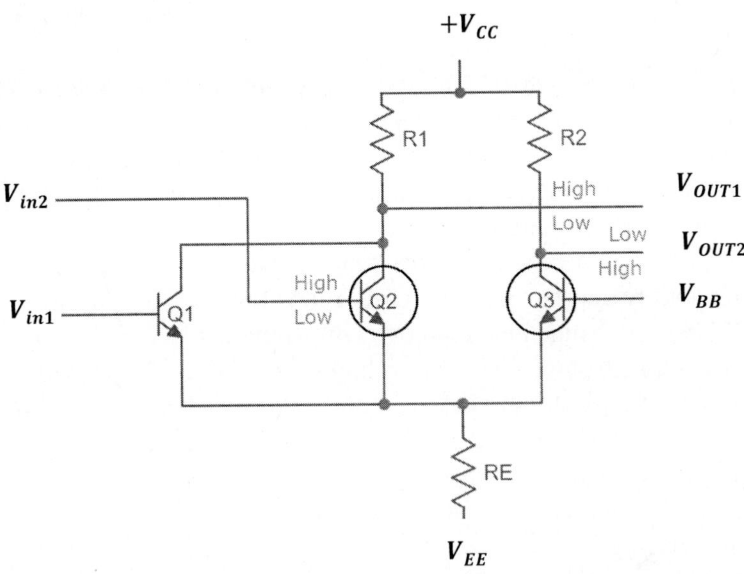

Fig. 3.21: ECL based OR/NOR Gate

ECL logic family implements the gates in differential amplifier configuration in which transistors are never driven in the saturation region thereby improving the speed of circuit to a great extent. The functional table, Truth table and Logic symbol is shown below

Functional and Truth table for ECL based OR/NOR Gate

Functional Table										**Truth Table**			
X	Y	V_x	V_y	Q_1	Q_2	Q_3	V_E	V_{OUT1}	V_{OUT2}	X	Y	$OUT1$	$OUT2$
L	L	3.6	3.6	OFF	OFF	ON	3.4	5.0	4.2	L	L	H	L
L	H	3.6	4.4	OFF	ON	OFF	3.8	4.2	5.0	L	H	L	H
H	L	4.4	3.6	ON	OFF	OFF	3.8	4.2	5.0	H	L	L	H
H	H	4.4	4.4	ON	ON	OFF	3.8	4.2	5.0	H	H	L	H

The input impedance is high and the output impedance is low. As a result, the transistors change states quickly, gate delays are low, and the fan out capability is high.

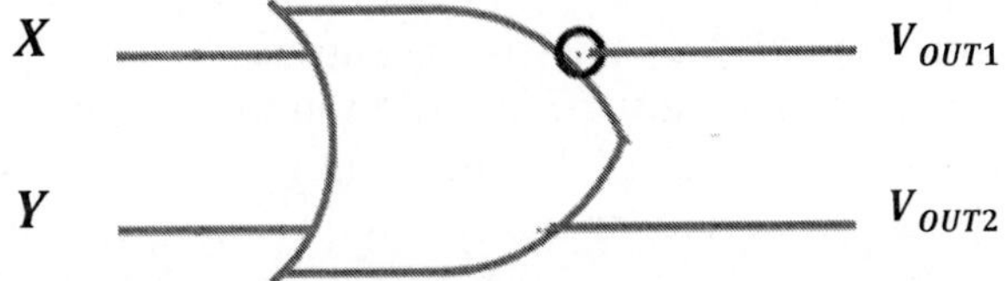

Fig. 3.22: Logic Symbol for ECL based OR/NOR Gate

Advantages, Disadvantages and Applications of ECL Logic family: The advantages of emitter-coupled logic are discussed below.

- The fanout of ECL is 25 which is better as compared to TTL & it is low as compared to CMOS.
- The average propagation delay time of ECL is 1 to 4 ns which is better as compared to both CMOS &TTL. Thus it is called as fastest logic family.
- When the BJTs in emitter coupled logic gates work in the active region, then they have the maximum speed as compared to all logic families.
- ECL gates generate complementary outputs.
- Current switching spikes are not there in the power supply leads.
- Outputs can be coupled jointly to provide the wired-OR function.
- The parameters of ECL do not change much through temperature.
- The no. of functions accessible from an only chip is high.

The disadvantages of emitter-coupled logic are discussed below.

- It has an extremely less noise margin i.e, ±200 mV.
- Power dissipation is high as compared to other logic gates.
- To interface with other logic families, level shifters are necessary.
- Fanout limits capacitive loading.
- As compared to TTL, ECL gates are expensive.
- As compared to CMOS & TTL, ECL noise immunity is worst.

Applications

- The applications of emitter-coupled logic include the following.
- Emitter-coupled logic is used as a logic & interface technology within extremely high-speed communications devices like fiber-optic transceiver interfaces, Ethernet & ATM (Asynchronous Transfer Mode) networks.
- ECL is a logic family based on BJT where its high-speed operation can be achieved by using a relatively small voltage swing & avoiding the transistors from moving into the saturation region.
- ECL is used in making the ASLT circuits within the IBM 360/91.
- ECL avoids the utilization of stacked transistors by using a single-ended bias i/p & positive feedback between primary & secondary transistors to attain an inverter function.
- ECL is used in extremely high-speed electronics.

3.2.7 CMOS Logic Family:

Complementary metal oxide semiconductor (CMOS) technology employs both P-channel and N-channel MOS transistors on the same silicon substrate. Both types are enhancement-mode devices; that is, gate voltage must be increased in the direction that inverts the surface in order for the device to conduct. An enhancement-mode device is normally turned off. The N-channel and P-channel devices are connected in series. Only one device is turned on at a time, resulting in extremely low power dissipation. Dissipation is primarily from the switching of devices through the active region and the charging and discharging capacitances.

A typical CMOS inverter circuit is shown in the figure below. With the input at zero volts or ground, the upper P-channel MOSFET is properly biased and, therefore, conducts.

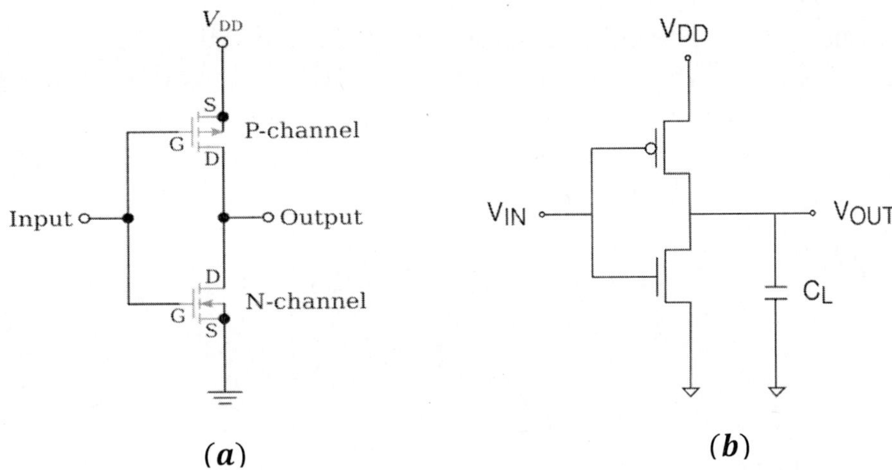

Fig. 3.23: A CMOS inverter (a) Basic Diagram (b) Circuit Schematic.

Basic Operation: The lower N-channel device is cut off at this time. The output is about the same as the supply voltage VDD. When the input voltage is a positive value approaching the supply voltage, the upper P-channel device cuts off. The lower N-channel device conducts and the output approaches zero volts. This basic circuit can be modified by adding additional P-channel and N-channel devices to form a variety of logic gates and storage elements.

Hence

$$If, V_{IN} = 0; than\ V_{OUT} = V_{DD}$$

i.e.,

$V_{IN} = 0 \Rightarrow V_{OUT} = V_{DD}$

- $V_{GSn} = 0\ (< V_{Tn}) \Rightarrow$ NMOS OFF
- $V_{SGp} = V_{DD}\ (> -V_{Tp}) \Rightarrow$ PMOS ON

$V_{IN} = V_{DD} \Rightarrow V_{OUT} = 0$

- $V_{GSn} = V_{DD}\ (> V_{Tn}) \Rightarrow$ NMOS ON
- $V_{SGp} = 0\ (< -V_{Tp}) \Rightarrow$ PMOS OFF

CMOS NAND gate

The circuit shown below shows the circuit of the 2-input CMOS NAND gate. It has two p-channel MOSFETs (Q_1, Q_2) and two n-channel MOSFETs (Q_3 and Q_4). A and B are two inputs. The input A is given to the gate terminal of Q_1 and Q_3. The input B is given to the gate terminal of Q_2 and Q_4. The output is obtained from the terminal V_O. When both the inputs are given LOW input, it will turn ON Q_1, Q_2 and turn OFF the MOSFETs Q_3 and Q_4. The output terminal is connected to the supply voltage V_{DD} and the output will be HIGH. It is shown in the below diagram(a). When either one of the inputs is high, for e.g., let us consider A is given HIGH input and B is given LOW input. In this case, MOSFETs Q_1 and Q_4 get turned

ON, whereas Q_2 and Q_3 are turned OFF as shown in the diagram(b) below. This will make a path for the supply voltage to be connected to the load, making the output to be HIGH.

If both the inputs A and B are HIGH inputs, which make the MOSFETs Q_3, Q_4 to be turned ON and Q_1, Q_2 to be turned OFF. Thus, the output is connected to the ground alone as shown below(c). Thereby, the output will be at LOW value.

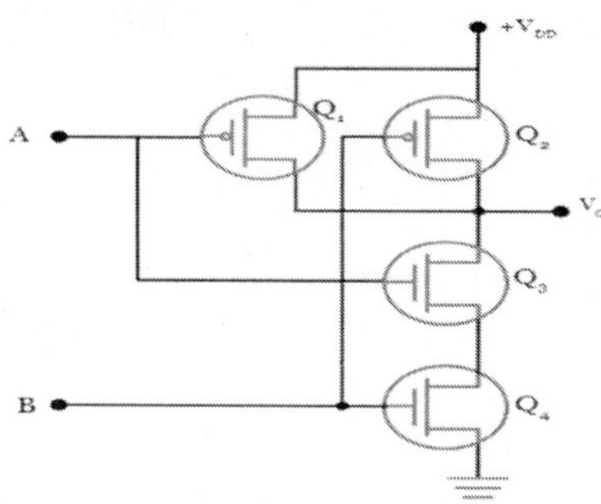

Fig. 3.24: CMOS NAND Gate

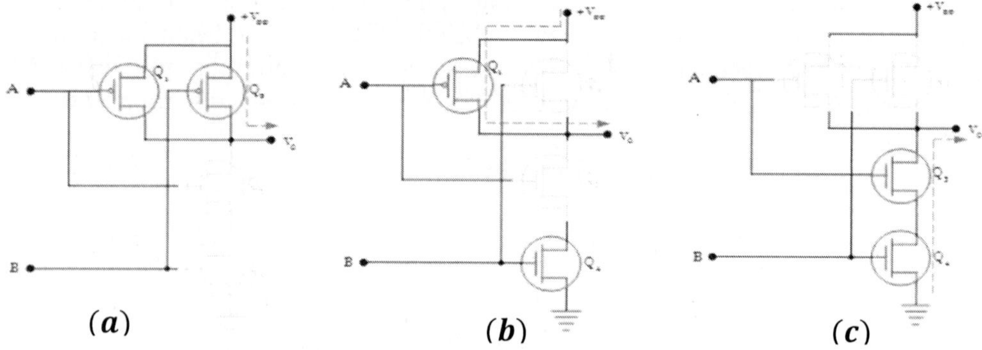

Fig. 3.25: CMOS NAND gate operations

The table below represents the operation for NAND gate using CMOS (Refer Table 3.5)

Table 3.5: Operation for NAND gate using CMOS

Inputs		**Status of MOSFETs**				**Output**	
A	*B*	Q_1	Q_2	Q_3	Q_4	V_o	Statuts
0	0	ON	On	OFF	OFF	1	High
0	1	ON	OFF	OFF	ON	1	High
1	0	OFF	ON	ON	OFF	1	High
1	1	OFF	OFF	OFF	OFF	0	Low

Merits and Demerits of CMOS Logic Family

Merits:

- Lower power supply requirement.

- Excellent noise immunity.
- High packaging density.
- Simpler and cheaper in fabrication.
- Consume very little power.

Demerits:

- Low operating speed in comparison to TTL ICs.
- No Bipolar.

Hence, Complimentary MOS (CMOS), have lower energy consumption than TTL and ECL, which has made portable electronics possible. CMOS is the most widely used family for large-scale devices as it combines high speed with low power consumption. CMOS usually operates from a single supply of 5 – 15 V, with excellent noise immunity of about 30% of supply voltage. Moreover, CMOS can be connected to a large number of gates (about 50).

3.3 Interfacing between CMOS and TTL

To achieve optimum performance in a digital system, devices from more than one logic family can be used, taking advantages of the superior characteristics of each family for different parts of the system. For example, CMOS logic ICs can be used in those parts of the system where low power dissipation is required, whereas TTL can be used for those portions of the system which require high speed of operation. Also, some function may be easily available in TTL and others may be available in CMOS. Therefore, it is necessary to examine the interface between CMOS and TTL devices.

CMOS and TTL are the two most widely used logic families. Although ICs belonging to the same logic family have no special interface requirements, that is, the output of one can directly feed the input of the other, the same is not true if we have to interconnect digital ICs belonging to different logic families. Incompatibility of ICs belonging to different families mainly arises from different voltage levels and current requirements associated with LOW and HIGH logic states at the inputs and outputs.

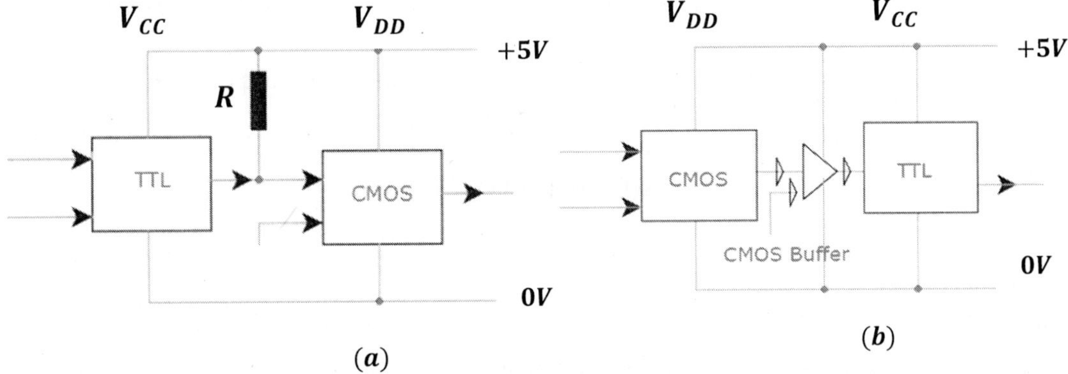

Fig. 3.26: Interfacing (a) TTL to CMOS and (b) CMOS to TTL

Figure-1 depicts TTL to CMOS interfacing and CMOS to TTL interfacing circuits. When 5V supply is given to TTL and CMOS ICs, logic levels of TTL and CMOS are different. One TTL IC can drive any number of CMOS ICs. However, TTL output in 'high state' yields 2.4 Volt which is lower than the minimum voltage required by CMOS IC (which is 3.5V) . For TTL to CMOS interfacing, standard pull up resistor is connected which solves the interfacing problem as mentioned. This is shown in figure-1(a). A CMOS IC can easily drive any low power Schottky TTL IC directly. But to interface standard TTL IC, buffer is provided in between CMOS and TTL ICs. This is shown in figure-1(b).

CMOS driving TTL

The first possible type of CMOS-to-TTL interface is the one where both ICs are operated from a common supply. We have read in earlier sections that the TTL family has a recommended supply voltage of 5 V, whereas the CMOS family devices can operate over a wide supply voltage range of 3V-18 V. In the present case, both ICs would operate from 5 V. As far as the voltage levels in the two logic states are concerned, the two have become compatible. Figure 4.40 shows a CMOS gate driving N TTL gates

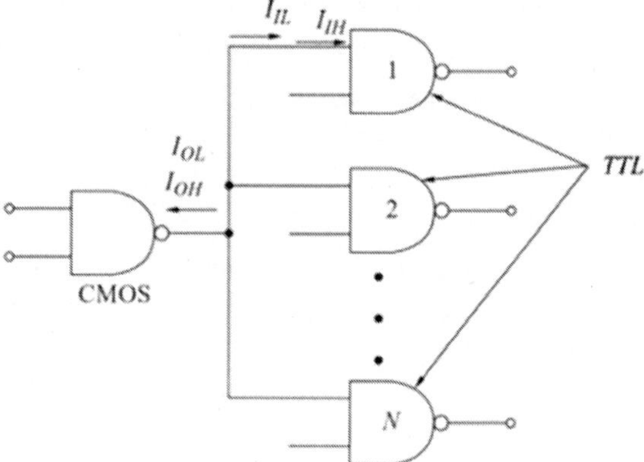

Fig. 3.27: A CMOS gate driving N TTL Gate

. For such an arrangement to operate properly the following conditions are required to be satisfied,

$$V_{OH}(CMOS) \geq V_{IH}(TTL)$$

$$V_{OL}(CMOS) \leq V_{IL}(TTL)$$

$$-I_{OH}(CMOS) \geq NI_{IH}(TTL)$$

$$I_{OL}(CMOS) \geq -NI_{IL}(TTL)$$

The CMOS output has a $V_{OH\,(min.)}$ of 4.95V (for V_{CC}=5 V) and a $V_{OL\,(max.)}$ of 0.05 V, which is compatible with $V_{IH\,(min.)}$ and $V_{IL\,(max.)}$ requirements of approximately 2 and 0.8V respectively for TTL family devices. In fact, in a CMOS-to- TTL interface, with the two devices operating on the same VCC, voltage level compatibility is always there. It is the current level compatibility that needs attention. That is, in the LOW state, the output current-sinking capability of the CMOS IC in question must at least equal the input current-sinking requirement of the TTL IC being driven. Similarly, in the HIGH state, the HIGH output current drive capability of the CMOS IC must equal or exceed the HIGH-level input current requirement of TTL IC. For a proper interface, both the above conditions must be met. As a rule of thumb, a CMOS IC belonging to the 4000B family (the most widely used CMOS family) can feed one LS TTL or two low-power TTL unit loads. When a CMOS IC needs to drive a standard TTL or a Schottky TTL device, a CMOS buffer (4049B or 4050B) is used. 4049B and 4050B are hex buffers of inverting and noninverting types respectively, with each buffer capable of driving two standard TTL loads.

TTL driving CMOS

Figure 4.41 shows a TTL gate driving N CMOS gates.

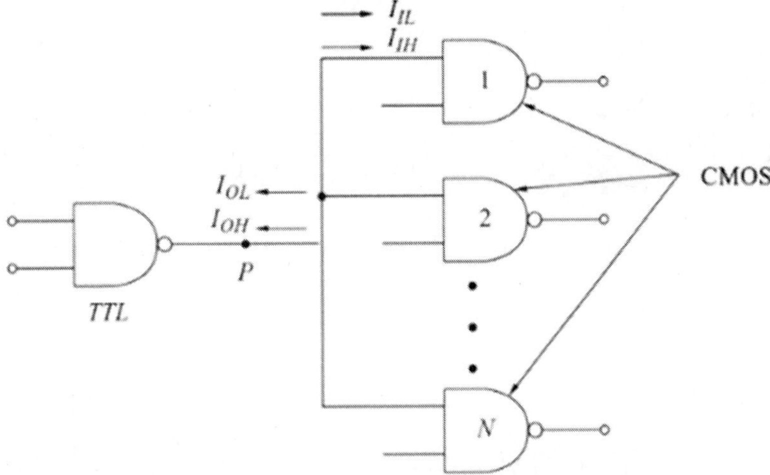

Fig. 3.28: A TTL gate driving N CMOS Gate

For such an arrangement to operate properly, the following conditions are required to be satisfied:

$$V_{OH}(TTL) \geq V_{IH}(CMOS)$$

$$V_{OL}(TTL) \leq V_{IL}(CMOS)$$

$$-I_{OH}(TTL) \geq NI_{IH}(CMOS)$$

$$I_{OL}(TTL) \geq -NI_{IL}(CMOS)$$

A digital designer selects a "default" logic family to use in a system, based on general requirements of speed, power, cost, and so on. However, the designer may select devices from other families in some cases because of availability or other special requirements. However, some factors to be consider like Noise Margin, Fanout capability, and Capacitive Loading.

3.4 Comparison of Different Logic Families:

Digital IC gates are classified not only by their logic operation, but also by the specific logic circuit family to which it belongs. Each logic family has its own basic electronic circuit upon which more complex digital circuits and functions are developed.

Table 3.5: Some statistical characteristics for different logic families

S.No.	Parameter	Logic family							
		DTL	DCTL	TTL	RTL	ECL	MOS	CMOS	IIL
1	Basic Gate (Positive Logic)	NAND	NAND	NAND	NOR	OR-NOR	NAND	NOR/ NAND	NOR
2.	Fan out (Minimum)	8	10	10	5	24	20	>50	Depends#
3	Power dissipation per gate (mW)	8-12	55	12-22	12	40-55	0.2-10	0.01	.07-6
4.	Noise Immunity	Good	Excellent	Very Good	Medium	Good	Medium	Very Good	Poor
5.	Propagation delay (ns)	30	90	12-6	12	4-1	300	70	25-250
6.	Clock rate* (MHZ)	12-30	4	15-60	8	60-400	2	5	
7.	Number of Functions	Fairly High	Medium	Very High	High	High	Low	Low	
8.	Speed Power product (PJ)	300		100	144	100	60	d.c.-0.7	<1

*(Minimum frequency at which Flip-flop operate)
Depends on Injector Current

In a summarized way we may say

1. RTL and DTL families have become obsolete because of their low speed, high power dissipation, and low fan-out.
2. TTL is the most popular general-purpose logic family. It is available in seven different series with a wide range of operating speed, power dissipation and fan-out. There are a large number of functions in SSI and MSI available in TTL.
3. TTL ICs are available with totem-pole output (which reduces speed-power product), open-collector output (which makes possible wired-AND connection and bus operation), and tri-state (TSL) outputs (which are ideally suited for bus operation).
4. HTL/CMOS are best suited for an industrial environment where electrical noise level is high.
5. ECL is the fastest logic family. Its main drawbacks are: low noise-margins and high-power dissipation. For interfacing other logic family's level-shifting networks are required.

6. I^2L is the only saturated bipolar logic family suitable for LSI because of small silicon chip area requirement and low power consumption. Because of low voltage require-ment, it is highly suitable for battery operated systems.
7. MOS devices occupy a very small fraction of silicon chip area as compared to bipolar devices and need very small power. So, MOS logic is most popular logic for LSI. The main disadvantage of MOS logic family is low speed, which is being improved upon by improvements in the technology of MOS fabrication.
8. CMOS has the lowest speed-power product and need very small power. CMOS is increasingly becoming popular in the area of MSI and LSI.

Unit summary

This unit has discussed various logic families with the interfacing problems between ICs of same logic families and between those of different logic families. This module mainly discusses about the TTL i.e. Transistor-Transistor family which is the next step of DTL family. It uses transistors, diodes and diffusive resistors as the basic components. These are classified on the basis of improvement in the performance parameter which is figure merit of the family. Low power TTL uses resistors of low value while Schottky TTL uses Schottky transistors as the basic components. Other alternatives are high power TTL and Low power Schottky TTL. These are referred as subfamilies of TTL. TTL is further classified on the basis of output configuration as a standard Totem pole arrangement: Open collector and tri-state. In order to eliminate the shortcomings of totem pole arrangement, where wired AND is not possible, open collector configuration is most suitable. However, tri-state configuration has the additional input known as Enable which disconnects output for logic circuit itself. Different characteristics of the TTL family such as voltage and current requirements, noise margin, propagation delay, power dissipation, fan out etc are compared for different subfamilies

Solved examples

1. What is the difference between a bipolar logic family and unipolar logic family?
   Logic family using BJTs is known as bipolar logic family whereas the logic family using unipolar transistors (MOSFETs) is known as unipolar logic family.
2. Differentiate between saturated and non-saturated logic.
   The logic circuits, in which transistors are driven into saturation, are known as saturated logic circuits or simply the saturated logic and the logic circuits avoiding saturation of their transistors are known as non-saturated logic.
3. What is meant by breadth of a logic family?
   The number of various functions available in a logic family is known as the breadth of the logic family.
4. What is tristate logic?
   Instate logic is a logic that has three states of output viz. high impedance state in addition to 0 and 1 states of output.
5. Which is the fastest of all logic families?
   ECL is the fastest of all logic families.
6. Which type of TTL gates can drive CMOS gates?

TTL with open-collector can drive CMOS.

7. On what factors the switching speed of a BJT and MOSFET depends?
   The switching speed of a BJT depends upon the time required to charge base-emitter capacitance and input capacitances whereas the switching speed of MOSFET depends upon internal capacitance of the device and the internal impedance of the gate drive circuit.
8. What determines the fan-out limitations of MOS logic circuits?
   The input capacitance restricts the number of MOS inputs that one MOS output can drive and thus determines the fan-out limitations of MOS logic circuits.
   The large fan-out enhances the MOS capacitance at the output of the driving gate which reduces the speed of MOS gates.
9. Define fan-out. Which factor is responsible for the limit of fan out in TTL circuits?
   The fan-out (also sometimes called the loading factor) is defined as the maximum number of standard logic inputs that an output can drive reliably. For example, a logic gate that is specified to have a fan-out of 8 can drive 8 standard logic inputs. If this number exceeds, the output logic-level voltages cannot be guaranteed.
   The fan out in TTL circuits is limited by the amount of current that the output transistor can sink in low state or can source in high state.
10. What do you understand by fan-out of a gate?
    The fan-out (also sometimes called the loading factor) is defined as the maximum number of standard logic inputs that an output can drive reliably. For example, a logic gate that is specified to have a fan-out of 8 can drive 8 standard logic inputs. If this number exceeds, the output logic-level voltages cannot be guaranteed.
11. What is the disadvantage of using back-to-back diodes in place of multi-emitter transistor in TTL totem-pole output circuit?
    During turn-off multi-emitter transistor reduces storage time considerably. Thus, the drawback of using back-to-back diodes in place of multi-emitter transistor is the increase in storage time.
12. What is the function of protecting diodes connected to multi-emitter transistor in a TTL totem-pole circuit?
    The function of protecting diodes in multi-emitter transistor is to protect the circuit from the negative spikes appearing at the input terminal.
13. Why the totem-pole output cannot be wire-ANDed?
    The totem-pole output cannot be wire-ANDed, because, in case, if one of the output is HIGH and another is LOW, then a large current flows from supply to ground through Q3 of the HIGH state and Q4 of the LOW state. This large current can damage any of the transistors.
14. Why TTL totem-pole circuit is faster than open-collector logic circuit?
    TTL totem-pole circuit is faster than open-collector logic circuit because the totem-pole logic circuit makes use of active pull-up circuit while the open-collector circuit uses passive pull-up resistor, and therefore, propagation delay in totem-pole logic circuit is reduced.
15. What determines the fan-out limitations of the TTL totem-pole circuit?
    The fan-out in TTL totem-pole circuit is limited by the amount of current that the output transistor can sink in LOW state or can source in HIGH state.

16. "Loading an output with more than its rated fan-out has several effects" — Write at least five effects.

    Loading an output with more than its rated fan-out affect the (i) rise and fall times of the gates (ii) propagation delay time of each gate (iii) power dissipation in each gate (iv) speed of operation (v) working of transistors.

Review Questions:

1. Describe the advantages and disadvantages of totem – pole output of the TTL logic IC.
2. At which state the TTL acts as current sinking and current source?
3. Explain the term: Current sinking and Current sourcing transistor
4. Explain the term: Unit load
5. Which type of output does the TTL consists of?
6. Explain the term: Passive pull up
7. Why the totem – pole TTL is faster than the open collector TTL logic?
8. Why the open collector TTL logic is not suitable for fast switching speed application?
9. What is wired AND operation of the TTL? What is its purpose?
10. Why the wired AND operation cannot be done in the totem – pole TTL ?
11. In which type of output, the wired AND operation is possible in the TTL?
12. Describe the disadvantage of wired AND operation of TTL?
13. Explain the term: Tri state TTL
14. Describe the advantage of the tri – state configuration of the TTL.
15. Explain the term: Buffer / Driver
16. What are the characteristics of Resistor Transistor logic (RTL)?
17. What are the characteristics of complementary metal oxide semiconductor (CMOS)?
18. What do you mean by current hogging and which logic family has this problem?
19. What is the use of Schottky TTL?
20. In the RTL NOR gate of Figure below, calculate the average power supplied by V_{cc} to the driver gate when it is driving 5 gates. Assume $V_{BE,sat} \approx 0.8V$, $V_{CE,sat} \approx 0.2V$, $h_{FE} = 10$. Neglect leakage currents.

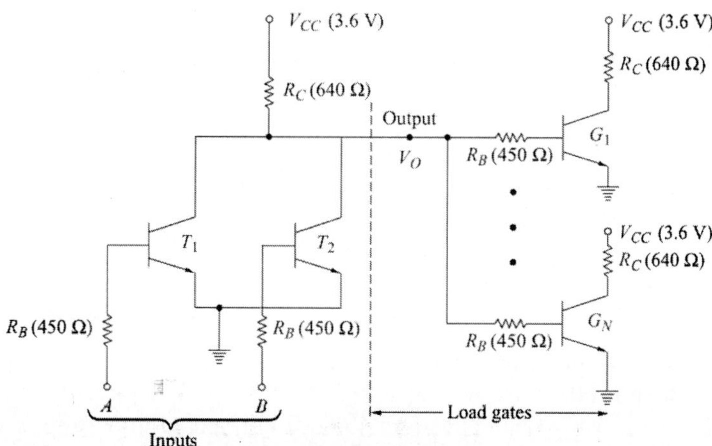

Fig. 3.29: Figure for Q. 20

21. In the circuit of Figure shown above, calculate
    (a) Output Voltage V_o and the Noise Margin Δ for $N = 5,6,7,8,9,10$. Assume $h_{FE} = 10$
    (b) Repeat (a) for $h_{FE} = 20$
    (c) Comment on the effect of h_{FE} on the fan-out and the noise margin of the circuit.
    (d) Comment on the effect of N on the noise margin for a given h_{FE}.
22. In the circuit given Below, the fan-out of RTL NOR gates P and Q is 5 each.

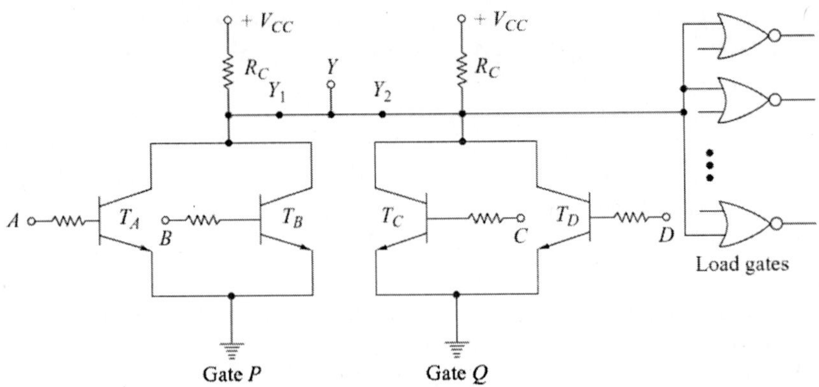

    (a) Calculate the fan-out of the combined gate.
    (b) Evaluate the propagation delay time constant and power dissipation, and comment on the effect of wired-logic on these.
23. A buffer is used to increase the output drive capability of a logic circuit. An RTL buffer inverter is shown in Figure below.

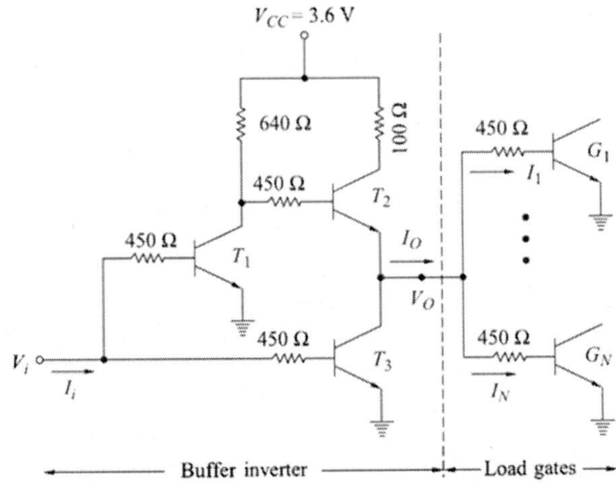

    (a) Explain the operation of this circuit.
    (b) Calculate the fan-out. Assume $h_{FE} = 30$.
    (c) Consider outputs of two such buffers 4 and B connected in parallel. Let the input to buffer A be logic 1 and the input to buffer B B be logic 0. Calculate the current flowing in T_3 of buffer A.

4 What will happen in the DTL circuit of figure below if

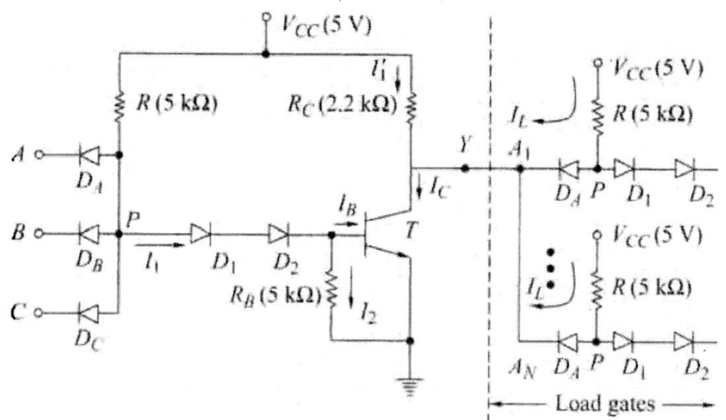

(a) one of the diodes D_1,or D_2, is removed,

(b) one more diode D_3, is inserted in series with D_1,or D_2.

For a open collector TTL gate the specifications are $V_{OH} = 2.4V, V_{OL} = 0.4V, I_{OH} = 250\mu A$, $I_{OL} = 16mA$, $I_{IH} = 40\mu A$, $I_{IL} = -1.6mA$. Calculate the value of R_C required for the open collector gate. Assume $V_{CC} = 5V$ and a fan-out of 8.

5 Design a circuit for interfacing an ECL 2-input NOR gate with a TTL inverter to obtain NOR function of the combined circuit.

6 What happens if output accidently gets shorted to ground in

(a) NMOS?

(b) CMOS?

27. Is it possible to use TTL-to-ECL translator for CMOS-to-ECL interfacing? Justify your answer
28. Is it possible to use ECL-to-TTL translator for ECL-to-CMOS interfacing? Justify your answer.
29. Give Reason for the Following

    (a) The temperature coefficient of resistance of a semiconductor is negative while that of a metal is positive.

    (b) A semiconductor behaves as an insulator at 0 K while it has some conductivity at room temperature.
30. (a) In a silicon p-n junction, calculate the increase in voltage across the diode if the forward current is doubled. Assume V_T, = 26 mV.

    (b) If the diode voltage in part (a) for the lower current is 700 mV, find the percent change in diode voltage.
31. The voltage across a silicon diode at room temperature (300 K) is 0.7 V when 2 mA current flows through it. If the voltage increases to 0.75 V; calculate

    (a) the diode current.

    (b) per cent change in diode current.
32. If the current flowing through a p-n junction diode increases ten times, what is the increase in diode voltage? Assume forward-biased silicon diode operating at room temperature.

Multiple Choice Questions (MCQs)

1. The propagation time delay of the silicon gate CMOS is _____
   a. 2ns
   b. 7ns
   c. 10ns
   d. 8ns
2. Which IC is quad two-input AND gate?
   a. 74LS08
   b. 74LS00
   c. 74LS01
   d. All of the above
3. The maximum input current of the metal gate CMOS is _________
   a. 0.0001mA
   b. -0.0001mA
   c. ±0.0001mA
   d. None of the above
4. The maximum clock of the metal gate CMOS is _________
   a. 20MHz
   b. 40 MHz
   c. 12 MHz
   d. 25 MHz
5. What is the standard form of IIL?
   a. Injection Integrated Logic
   b. Integrated Injection Logic
   c. Integrated-Integrated Logic
   d. None of the above
6. The noise immunity is excellent in _____ logic family
   a. CMOS
   b. TTL
   c. ECL
   d. None of the above
7. The maximum input current of the Schottky TTL gates is _________
   a. -2.0mA
   b. 2.0mA
   c. -2.5mA
   d. -1.0mA
8. The maximum clock of low power Schottky TTL is __________
   a. 20MHz
   b. 40 MHz
   c. 12 MHz
   d. 25 MHz
9. The propagation delay of the low power Schottky TTL is __________
   a. 2ns
   b. 7ns

   c. 10ns
   d. 105ns

10. The fanout of the CMOS logic family is ______
    a. >10
    b. >20
    c. >30
    d. >50
11. What are the components used in the RTL logic family?
    a. Resistors
    b. Transistors
    c. Both a and b
    d. None of the above
12. The TTL family is classified into __________
    a. Two
    b. Four
    c. Six
    d. Seven
13. Which IC is a quad two-input NOR gate?
    a. 74LS02
    b. 74LS01
    c. 74LS00
    d. None of the above
14. The maximum input current of the low power Schottky TTL is ____________
    a. 0.2mA
    b. 0.3mA
    c. -0.4mA
    d. 0.1mA
15. The maximum clock of Schottky TTL gate is __________
    a. 20MHz
    b. 40 MHz
    c. 12 MHz
    d. 125 MHz
16. The propagation delay of Schottky TTL gate is ________
    a. 3ns
    b. 7ns
    c. 10ns
    d. 105ns
17. The fan-out of TTL logic family is __________
    a. 2
    b. 5
    c. 8
    d. 10
18. What are the components used in the DCTL logic family?

a. Resistors
b. transistors
c. Diodes
d. Both a and b

19. Which IC is a triple three-input AND gate?
a. 74LS11
b. 74LS10
c. 74LS12
d. None of the above

20. The maximum input current of the advanced low power Schottky TTL gate is ________
a. 0.2mA
b. 0.3mA
c. -0.4mA
d. -0.1mA

21. The maximum clock of advanced low power Schottky TTL is __________
a. 70MHz
b. 40 MHz
c. 12 MHz
d. 25 MHz

22. The propagation delay of advanced low power Schottky TTL gate is ________
a. 3ns
b. 4ns
c. 10ns
d. 105ns

23. What are the components used in the TTL logic family?
a. Resistors, Transistors
b. Diodes
c. Both a and b
d. None of the above

24. The RTL logic family has ______
a. Poor noise margin
b. Poor fan-out
c. Higher speed and power dissipation
d. All of the above

25. Which IC is a triple three-input NAND gate?
a. 74LS11
b. 74LS10
c. 74LS27
d. None of the above

26. The maximum input current of the advanced Schottky TTL gate is _______
a. 0.2mA
b. 0.3mA
c. -0.5mA
d. -0.1mA

27. The maximum clock of advanced low power Schottky TTL is __________
    a. 200MHz
    b. 40 MHz
    c. 12 MHz
    d. 25 MHz
28. The propagation delay of advanced Schottky TTL gate is ________
    a. 3ns
    b. 1.5ns
    c. 10ns
    d. 105ns
29. The fan-out of the ECL logic family is ________
    a. 10
    b. 20
    c. 25
    d. None of the above
30. What are the components used in the DTL logic family?
    a. Resistors
    b. Diodes
    c. Transistors
    d. All of the above
31. Which IC is a triple three-input NOR gate?
    a. 74LS10
    b. 74LS11
    c. 74LS27
    d. All of the above
32. What are the components used in the ECL logic family?
    a. Resistors
    b. Transistors
    c. Both a and b
    d. None of the above
33. The power dissipation in RTL logic family is ______
    a. 30mv
    b. 40mv
    c. 50mv
    d. 60mv
34. In which logic family the power dissipation is very low?
    a. RTL
    b. DCTL
    c. CMOS
    d. TTL
35. Which IC is a dual four-input NAND gate?
    a. 74LS20
    b. 74LS21

c. 74LS27

d. None of the above

36. What are the components used in CMOS logic family?

a. PMOS

b. NMOS

c. Both a and b

d. None of the above

37. The power dissipation in the DTL logic family is ________

a. 1-12mv

b. 4-12mv

c. 6-12mv

d. 8-12mv

38. Which one is a saturated bipolar logic family?

a. RTL

b. Schottky TTL

c. ECL

d. None of the above

39. The power dissipation in the ECL logic family is ______

a. 30mv

b. 40-55mv

c. 50-80mv

d. 60-90mv

40. The basic gate of the ECL logic family is _____

a. NAND

b. NOT

c. OR/ NOR

d. None of the above

41. The basic gate of the CMOS logic family is _____

a. NAND

b. NOT

c. NAND/ NOR

d. None of the above

42. The noise immunity of TTL logic family is _________

a. Strong

b. Good

c. Very strong

d. Poor

43. Which IC is a hex NOT gate?

a. 74LS04

b. 74LS19

c. 74LS23

d. None of the above

44. The power dissipation of TTL logic family is ____________

a. 30mv

b. 40mv
c. 10mv
d. 60mv

45. The noise margin in TTL logic family is _______
    a. Moderate
    b. Low
    c. High
    d. None of the above
46. What are the disadvantages of the ECL logic family?
    a. Noise immunity is worst
    b. Power consumption is high
    c. Power consumption is low
    d. Both a and b

References and Suggested Readings

- *A. Anand Kumar, Fundamentals of digital circuits Second Edition, PHI Learning Private Limited, ISBN (978-81-203-3679-7)*
- *Floyd, T. L. (2005) Digital Fundamentals, Prentice-Hall Inc., USA.*
- *Morris Mano, M. and Kime, C. R. (2003) Logic and Computer Design Fundamentals, Prentice-Hall Inc., USA.*
- *Malvino, A. P. and Leach, D. P. (1994) Digital Principles and Applications, McGraw-Hill Book Company. USA.*
- *The TTL Data Book, Volume 1. 1985. Dallas: Texas Instruments.*
- *Tokheim, R. L. (1994) Schaum's Outline Series of Digital Principles, McGraw-Hill Companies Inc., USA.*
- *Yarbrough, John M. Digital logic: Applications and design. Eagan: West Publishing Company, 1997.*

4 Combinational Logic Circuits

UNIT SPECIFICS

Through this unit we have discussed the following aspects:

- *To understand K-map representation for 2, 3, 4, and more variables.*
- *To apply K-map to minimize the logic function while considering don't care.*
- *To understand various combinational logic circuits like Adder & Subtractor.*
- *To implement and realization of Decoder Encoder, Multiplexer, Demultiplexer etc.*
- *To identify various application relevant information for combinational circuits.*

RATIONALE

The K-map is a systematic way of simplifying Boolean expressions. With the help of the K-map method, we can find the simplest POS and SOP expression, which is known as the minimum expression. The K-map method is used for expressions containing 2, 3, 4, and 5 variables. For a higher number of variables, there is another method used for simplification called the Quine-McClusky method. In K-map, the number of cells is similar to the total number of variable input combinations.

A combinational circuit is the digital logic circuit in which the output depends on the combination of inputs at that point of time with total disregard to the past state of the inputs. Combinational Logic Circuits are memoryless digital logic circuits whose output at any instant in time depends only on the combination of its inputs, the result is that combinational logic circuits have no feedback, and any changes to the signals being applied to their inputs will immediately have an effect at the output.

Combinational logic circuits can be very simple or very complicated and any combinational circuit can be implemented with only NAND and NOR. Initially the logic has been defined using Boolean Expression, further a truth table has been developed, and finally the logic diagram will be created.

PRE-REQUISITES

Logic Gates, Boolean Algebra, Binary Opeartions.

UNIT OUTCOMES

List of outcomes of this unit is as follows:

U4-O1: Describe K-map for different number of variables.

U4-O2: Apply K-map to minimize the logic function while considering don't care as well.

U4-O3: Implement various combinational logic circuits like Adder & Subtractor.

U4-O4: implement and realization of Decoder Encoder, Multiplexer, Demultiplexer.

U4-O5: Identify various application relevant information for combinational circuits.

Unit-4 Outcomes	***EXPECTED MAPPING WITH COURSE OUTCOMES*** *(1- Weak Correlation; 2- Medium correlation; 3- Strong Correlation)*					
	CO-1	***CO-2***	***CO-3***	***CO-4***	***CO-5***	***CO-6***
U4-O1	*3*	*3*	*3*	-	*3*	*1*
U4-O2	*1*	*1*	*2*	*2*	*1*	-
U4-O3	*2*	*1*	*3*	*1*	*2*	*1*
U4-O4	-	-	*3*	*1*	*2*	*2*
U4-O5	*3*	*3*	*3*	-	*3*	*1*

➢ Scan the below QR codes for more information on the topic written below the QR code.

K-map representation

Combinational Circuits

Priority Encoder

Application of combinational Logic

4. Unit outcomes

A combinational logic circuit is one where the output or outputs depend upon the present state of a combination of the logic inputs. The logic gates discussed in chapter 2 constitute the most fundamental building block of a combinational circuit. More complex combinational circuits such as adders and subtractors, multiplexers and demultiplexers, magnitude comparators, etc., can be implemented using a combination of logic gates. Combinational circuit may be implemented using a combination of various logic function available in IC form. This unit deals with the discussion devices used to perform various arithmetic and other related operations. This unit includes explaining about designing aspects of combinational circuits like adders, subtractors, magnitude comparators and look- ahead carry generators etc.

4.1 K-map representation

In the previous section two standard forms of logic functions and their realizations using gates have been discussed. Further, the need for the simplification of the Boolean expression have also been established by applying algebraic method of simplification using Boolean algebraic theorems. Sometimes it is difficult to be sure that a logic expression can be simplified. There is another technique, which is graphical, known as the **Karnaugh map** or **K-map** technique which provides a systematic method for simplifying and manipulating Boolean expressions.

In this technique, the information contained in a truth table or available in POS or SOP form is represented on Karnaugh map(K-map). This is the perhaps the most extensively used tool for simplification of Boolean functions. Although the technique may be used for any number of variables, it is generally used up to six variables beyond which it becomes very cumbersome. In the upcoming sections designing of K-map for 2-, 3- and 4- variables have been introduced.

In an n-variable K-map there are 2^n cells. Each cell corresponds to one of the Combinations of n variables, since there are 2^n combinations of n variables. Therefore, for each row of the truth table i.e., for each minterm as well as for each maxterm there is one specific cell in the K-map. The variables have been designed as A, B, C, and D, and the binary numbers formed by them are taken as AB, ABC, and ABCD for 2-, 3-, and 4-variables respectively.

In each map the variables and all possible values of the variables are indicated (the first bit corresponds to the first variable and the second bit corresponds to the second variable) to identify the cells. *Gray code has been used for the identification of cells.* The reason for using Gray code will become clear when we discuss the application of K-map.

4.1.1 Introduction

Logic operation and Boolean algebra have already been discussed in chapter 2. Boolean algebra theorems are used for the manipulations of logic expressions. It has also been demonstrated that a logic expression can be realized using logic gates. In general, a logic expression has to get reduced considerably. If the expression can be simplified, accordingly the number of gates and the number of input terminals also gets minimized for the logic expression realization. Therefore, the simplification of logic expression is very important as it saves the hardware required to design a specific system. A large number of functions are available in

IC form and therefore, we should be able to make optimum use of these ICs in the design of digital systems. Our aim should be to minimize the number of IC packages. Some of logic gates available in ICs have been discussed previous chapters. We will be discussing in upcoming section, some of the MSI digital circuits available in IC form and design of digital system using these ICs.

Generally, digital logic circuits are divided into two broad categories:

- Combinational circuits, and
- Sequential circuits.

In *combinational circuits*, the outputs at any instant of time depend upon the inputs present at that instant of time. This means there is no memory in these circuits. In another category of logic circuits in which the output at any instant of time depends upon the present inputs as well as past inputs/outputs. This means that there is element used to store past information. These elements are known as memory element. Such circuits are known as *sequential circuits*. A sequential logic system may have combinational logic sub-system. The design of combinational circuits will be discussed here. Sequential circuits design will be discussed later.

The design requirement of combinational circuits may be specified in one of the following ways:

- A set of statements,
- Boolean expression,
- Truth table
- Circuit realization

The aim is to design a circuit using the gates or some other circuits which are derived from the basic gates. As is usual in any engineering design, the number of components used should be minimum to ensure low cost, saving in space, power requirements etc. There can be two different approaches to the design of combinational circuits. One of these is the Boolean expression or the truth table is simplified by using standard methods and the simplified expression is realized using the gates. The other method normally does not require any simplification of the logic expression or truth table, instead the complex logic functions available in Medium Scale Integrated (MSI) circuits can be directly used.

The following methods can be used to simplify the Boolean functions:

- Algebraic method,
- Karnaugh-map technique,
- Quine-McCluskey method, and
- Variable Entered Mapping (VEM) technique.

The algebraic method, the Karnaugh-map (K-map) technique, and the Quine-McCluskey method have been discussed here. the K-map is the simplest and most commonly used method. It is a manual method and depends to a great extent upon human intuition. This method can be used conveniently up to six variables beyond which it is very cumbersome. The Quine-McCluskey method is suitable for computer mechanization and is seldom used by logic designers manually.

4.1.2 Standard representation for logic functions

Logic functions are expressed in terms of logical variables. The values assumed by the logic functions as well as the logic variables are in the binary form. Any arbitrary logic function can be expressed in the following forms:

- Sum-of-products (SOP), and
- Product-of-sums (POS).

These two forms are conveniently suited in arriving at the standard methods for designing the circuits which will become clear from the following discussions.

$$Y = AB + A\bar{C} + BC \ldots\ldots\ldots SOP\ format$$
$$Y = (A + B)(A + C)(B + \bar{C}) \ldots\ldots\ldots POS\ format$$

In above expression of SOP and POS form each individual term in canonical SOP form is called as *minterm* and in canonical POS form as *maxterm*. SOP form can be converted to canonical SOP by ANDing the terms in the expression with terms formed by ORing the variable and its complement which are not present in that term.

For example, for a three- variable expression with variables $A, B\ and\ C$. If, there are two variables B and C are missing, then we AND two terms $(B + \bar{B})$ and $(C + \bar{C})$ with A. Therefore, we get

$$Y = A.(B + \bar{B}).(C + \bar{C}) = ABC + A\bar{B}C + AB\bar{C} + A\bar{B}\bar{C}$$

Example 4.1: Convert $Y = AB + A\bar{C} + BC$ into canonical SOP form.

Solution:

$$Y = AB + A\bar{C} + BC = AB(C + \bar{C}) + A\bar{C}(B + \bar{B}) + BC(B + \bar{B})$$
$$= ABC + AB\bar{C} + AB\bar{C} + A\bar{B}\bar{C} + ABC + \bar{A}BC$$
$$= ABC + AB\bar{C} + A\bar{B}\bar{C} + \bar{A}BC$$

Example 4.2: Convert $Y = (A + B)(A + C)(B + \bar{C})$ into canonical POS form.

Solution:

$$Y = (A + B)(A + C)(B + \bar{C}) = (A + B + C\bar{C})(A + C + B\bar{B})(B + \bar{C} + A\bar{A})$$
$$= (A + B + C)(A + B + \bar{C})(A + B + C)(A + \bar{B} + C)(A + B + \bar{C})(\bar{A} + B + \bar{C})$$
$$Y = (A + B + C)(A + \bar{B} + C)(A + B + \bar{C})(\bar{A} + B + \bar{C})$$

The concept of *minterm* and *maxterm* introduced above allows us to introduce a very convenient shorthand notation to express logical functions. Tab gives the *minterms* and *maxterms* for a four variable logical function where the number of *minterms* as well as *maxterms* is $2^4 = 16$.

For an n variable logical function there are 2^n minterms and an equal number of maxterms. While writing a particular minterm or maxterm, we shall always write it with the variables in an orderly way as can be seen from Tab.

Each minterm is represented by m_i and maxterm is represented by M_i, where subscript i is the decimal equivalent of the natural binary number. In case of minterm with normal (uncomplemented) variables taken as the 1's and the complemented variables taken as 0's. Further, in case of maxterms normal (uncomplemented) variables taken as 0's and complemented variables taken as 1's. Moreover, A variable in uncomplemented or complemented form is known as a literal.

Table 4.1: List of minterms and maxterms for n = 4 variable

Variables (4-Bit)				**Minterm**	**Maxterm**
A	B	C	D	m_i	M_i
0	0	0	0	$\bar{A}\bar{B}\bar{C}\bar{D} = m_0$	$A + B + C + D = M_0$
0	0	0	1	$\bar{A}\bar{B}\bar{C}D = m_1$	$A + B + C + \bar{D} = M_1$
0	0	1	0	$\bar{A}\bar{B}C\bar{D} = m_2$	$A + B + \bar{C} + D = M_2$
0	0	1	1	$\bar{A}\bar{B}CD = m_3$	$A + B + \bar{C} + \bar{D} = M_3$
0	1	0	0	$\bar{A}B\bar{C}\bar{D} = m_4$	$A + \bar{B} + C + D = M_4$
0	1	0	1	$\bar{A}B\bar{C}D = m_5$	$A + B + C + \bar{D} = M_5$
0	1	1	0	$\bar{A}BC\bar{D} = m_6$	$A + \bar{B} + \bar{C} + D = M_6$
0	1	1	1	$\bar{A}BCD = m_7$	$A + \bar{B} + \bar{C} + \bar{D} = M_7$
1	0	0	0	$A\bar{B}\bar{C}\bar{D} = m_8$	$\bar{A} + B + C + D = M_8$
1	0	0	1	$A\bar{B}\bar{C}D = m_9$	$\bar{A} + B + C + \bar{D} = M_9$
1	0	1	0	$A\bar{B}C\bar{D} = m_{10}$	$\bar{A} + B + \bar{C} + D = M_{10}$
1	0	1	1	$A\bar{B}CD = m_{11}$	$\bar{A} + B + \bar{C} + \bar{D} = M_{11}$
1	1	0	0	$AB\bar{C}\bar{D} = m_{12}$	$\bar{A} + \bar{B} + C + D = M_{12}$
1	1	0	1	$AB\bar{C}D = m_{13}$	$\bar{A} + B + C + \bar{D} = M_{13}$
1	1	1	0	$ABC\bar{D} = m_{14}$	$\bar{A} + \bar{B} + \bar{C} + D = M_{14}$
1	1	1	1	$ABCD = m_{15}$	$\bar{A} + \bar{B} + \bar{C} + \bar{D} = M_{15}$

The canonical SOP can be written as

$$Y = \bar{A}BC + A\bar{B}\bar{C} + AB\bar{C} + ABC$$
$$= m_3 + m_4 + m_6 + m_7 = \sum m\,(3, 4, 6, 7)$$

Similarly , the canonical POS can be written as

$$Y = (A + B + C)(A + B + \bar{C})(A + \bar{B} + C)(\bar{A} + B + \bar{C})$$
$$= M_{0,}\, M_{1,}\, M_{2,}\, M_5 = \prod M(0, 1, 2, 5)$$

We notice that there is a complementary type of relationship between a function expressed in terms of *minterms* and in terms of *maxterms*. Here, we are dealing with a three-variable function where the number of minterms and maxterms are $2^3 = 8$ and the corresponding decimal numbers are 0 through 7. Out of this the terms corresponding decimal numbers 3, 4, 6, and 7 are minterms, and the terms corresponding decimal numbers 0, 1, 2, and 5 are maxterms. Hence if a logic function is specified in terms of minterm/maxterm, its maxterm/minterm representation can be determined by using this *complementary property*.

For example. for a four-variable case if

$$Y = \sum m(0, 3, 6, 7, 10, 12, 15)$$

$$Y = \prod M\ (1, 2, 4, 5, 8, 9,11,13, 4)$$

Example 4.3: For the given the logic equation Design a circuit using gates to realize this function.

$$Y = (A + BC)(B + A\bar{C})$$

Solution:

There are three input logic variables $A, B\ and\ C$ as input variable whereas Y is the output variable. The variable C appears as C in one term and as $\bar{C}$ in the other term. A variable in uncomplemented or complemented form is known as a literal.

- A circuit using gates can simply be designed by looking at the expression and finding out the basic gates which can be used to realize the various terms and them connect these gates appropriately.
- The first term A has only one literal A and the second term BC has two literals B and C. This term is recognized as an ***AND*** operation and can be realized by using 2 input ***AND*** gate. The combination of two ***OR*** terms will be realized by using 2 input OR gate. As shown in Fig. 4.1 (a).
- The third term B is again a single literal term and the fourth term $A\bar{C}$ has two literals. The combination of these terms is similar to the combination of the first two terms and is realized in a similar way. This realization is shown in Fig. 4.1 (b).
- Now, the compete realization is obtained by using a 2-input AND gate with $Y_1$and Y_2 as the inputs and the output of this gate will be required output Y. This realization is given as in Fig. 4.1 (C). this design requires three, 2- input AND gates and two 2-input OR gates.

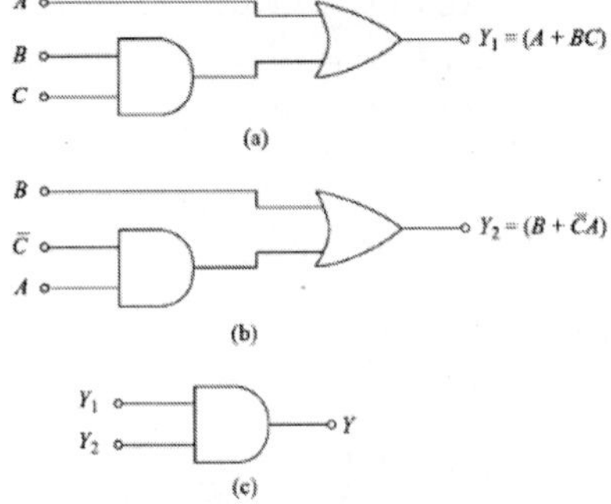

Fig. 4.1: Logic Circuits for the Realization of Ex. 4.3.

4.1.3 Simplification of logic functions using K-map

Simplification of logic functions with K-map is based on the principle of combining terms in adjacent cells. *Two cells are said to be adjacent if they differ in only one variable.*

For example, in the two-variable K-maps of Fig. 4.2 (a) and (b), the top two cells are adjacent and the bottom two cells are adjacent. Also, the left two cells and the right two cells are adjacent. It can be verified that in adjacent cells one of the literals is same, whereas the other literal appears in uncomplemented form in one and in the complemented form in the other cell.

Similarly, we observe adjacent cells in the 3-variable and 4-variable K-maps as shown in Fig. 4.3 and Fig. 4.4 respectively, they show the adjacent cells in 3- variable, and 4-variable K-maps. From this it becomes clear that if the gray code is used for the identification of cells in K-map, physically adjacent (horizontal and vertical but not diagonal) cells differ in only one variable. Also, the left- most cells are adjacent to their corresponding bottom cells. The simplification of logical function is achieved by grouping adjacent 1's or 0's in groups of 2^i, where i = 1, 2, ..., n and n is the number of variables.

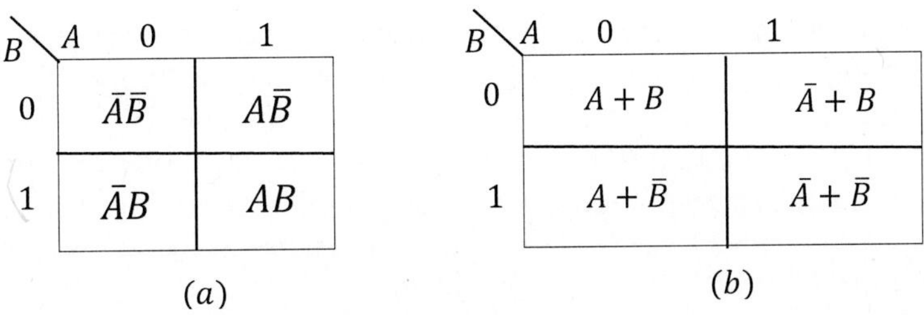

Fig. 4.2: 2-Variable K-map represented in form of (a) Minterms, (b)Maxterms.

C \ AB	00	01	11	10
0	$\bar{A}\bar{B}\bar{C}$	$\bar{A}B\bar{C}$	$AB\bar{C}$	$A\bar{B}\bar{C}$
1	$\bar{A}\bar{B}C$	$\bar{A}BC$	ABC	$A\bar{B}C$

(a)

AB \ C	0	1
00	$\bar{A}\bar{B}\bar{C}$	$\bar{A}\bar{B}C$
01	$\bar{A}B\bar{C}$	$\bar{A}BC$
11	$AB\bar{C}$	ABC
10	$A\bar{B}\bar{C}$	$A\bar{B}C$

(b)

Fig. 4.3: 3-Variable K-map representation

CD \ AB	00	01	11	10
00	$\bar{A}\bar{B}\bar{C}\bar{D}$	$\bar{A}B\bar{C}\bar{D}$	$AB\bar{C}\bar{D}$	$A\bar{B}\bar{C}\bar{D}$
01	$\bar{A}\bar{B}\bar{C}D$	$\bar{A}B\bar{C}D$	$AB\bar{C}D$	$A\bar{B}\bar{C}D$
11	$\bar{A}\bar{B}CD$	$\bar{A}BCD$	$ABCD$	$A\bar{B}CD$
10	$\bar{A}\bar{B}C\bar{D}$	$\bar{A}BC\bar{D}$	$ABC\bar{D}$	$A\bar{B}C\bar{D}$

(a)

CD \ AB	00	01	11	10
00	0000	0100	1100	1000
01	0001	0101	1101	1001
11	0011	0111	1111	1011
10	0010	0110	1110	1010

(b)

Fig. 4.4: 4-Variable K-map representation

The process of simplification involves grouping of minterms and identifying *Prime-Implicants (PI)* and *Essential Prime-Implicants (EPI)*. A prime-implicant is a group of minterms that cannot be combined with any other minterm or groups. An essential prime-implicant is a prime-implicant in which one or more minterms are unique; i.e. it contains at least one minterm which is not contained in any other prime-implicant.

4.1.4 Grouping of bits in K-Map

Grouping two adjacent ones:

If there are two adjacent ones on the map, these can be grouped together, this is called ***Pair***. This results in an output minterm/maxterm, which will delete one variable from the output term. It means that if it's a four variable k map than output equation will have only three variables, and so on. The only condition to be fulfilled is to be only one bit change at a time in a sequence. It can be observed in Fig. 4.5 for such kind of groupings using two ones.

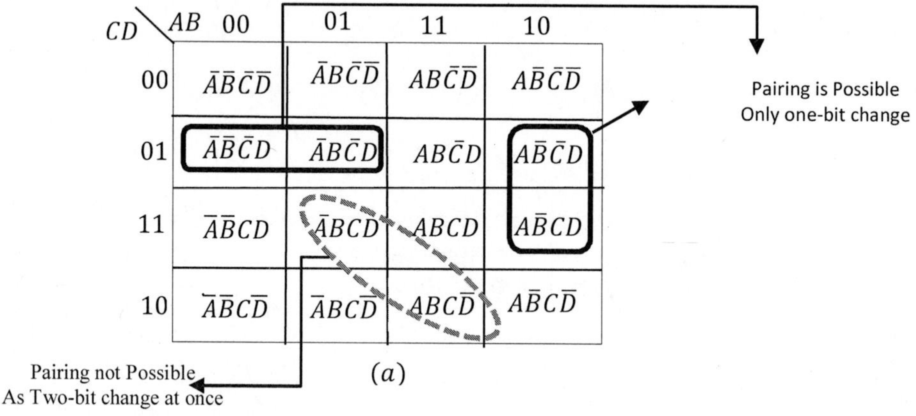

Fig. 4.5: Grouping of two adjacent ones (Pairing)

Grouping Four adjacent ones:

Four cells from a group of four adjacent ones, called ***Quad***.

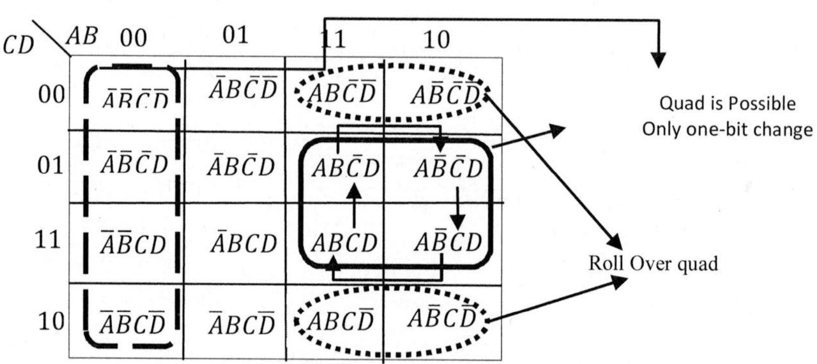

Fig. 4.6: A three variable k map

This results in an output minterm/maxterm, which will delete two variables from the output term. It means that if it's a four variable k map than output equation will have only two variables, and so on. The only condition to be fulfilled is to be only one bit change at a time in a sequence.

Note: In case of 2-variable map, there is only one possibility corresponding to entry 1 in all the four cells, and the simplified expression will be Y = 1. That is, Y always equals 1 (independent of the variables).

On the basis of grouping of 4 adjacent ones given in Fig. 4.6, shows the possible combination for making a Quad in the case of four variable K-Map. However, same can be applied to two and three variable K-Map. we can find the grouping in K-maps of four or more variables.

<u>*Grouping eight adjacent ones:*</u>

A group of eight adjacent ones in Eight cells forms ***Octave***.

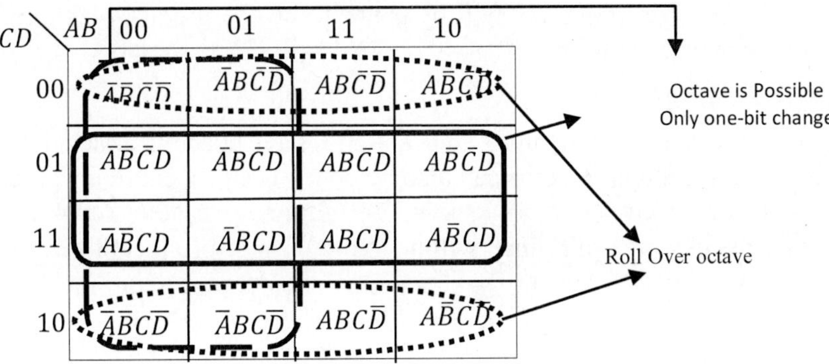

Fig. 4.7: A four -variable K-map Illustrating the Grouping of Eight Adjacent Ones.

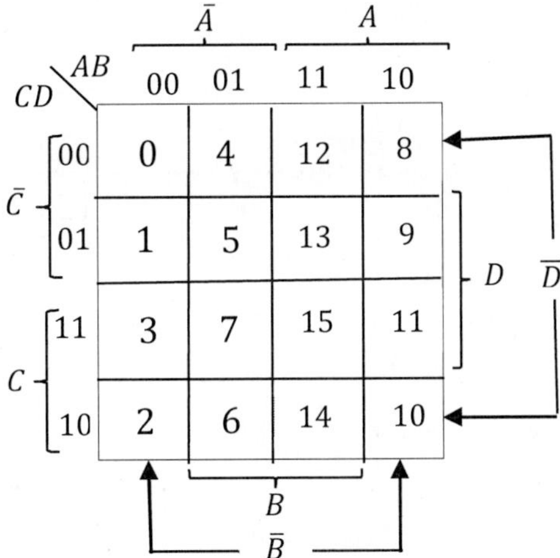

Fig. 4.8: Simplified expression for each of the groupings of eight ones for a 4 variable K-map

This results in an output minterm/maxterm, which will delete three variables from the output term. It means that if it's a four variable k map than output equation will have only one variable, and so on. The only condition to be fulfilled is to be only one bit change at a time in a sequence. Fig. 4.7 shows the possible combination for making a Quad in the case of four variable K-Map.

Note: In case of 3-variable K-map, there is only one possibility of eight ones appearing in the K-map and this corresponds to output equal to 1, irrespective of the values of the input variables.

Fig. 4.8 shows the simplified expression for each of the groupings of eight ones for a 4 variable K-map. From an understanding of this, we can easily find out such combinations for 5- and 6-variables K-maps. When eight adjacent ones are combined, the resulting equation will have only one term with the number of literals there less than the number of literals in the original minterms. Similar to the groupings of adjacent two and four ones, the literals which are common in all the eight minterms will be present and the literals which are not same get eliminated in the resulting term.

<u>Grouping 2, 4, and 8 adjacent zeros:</u>

In the previous section groups of 2, 4, and 8 adjacent ones have been discussed. Instead of making the groups of ones, groups of zeros can also be done. The procedure for grouping of zeroes is similar to the one used above. Fig. shows 2-variable, 3-variable, and 4-variable K-maps. In such a case the outcome of K-map will be in the form of product of sums.

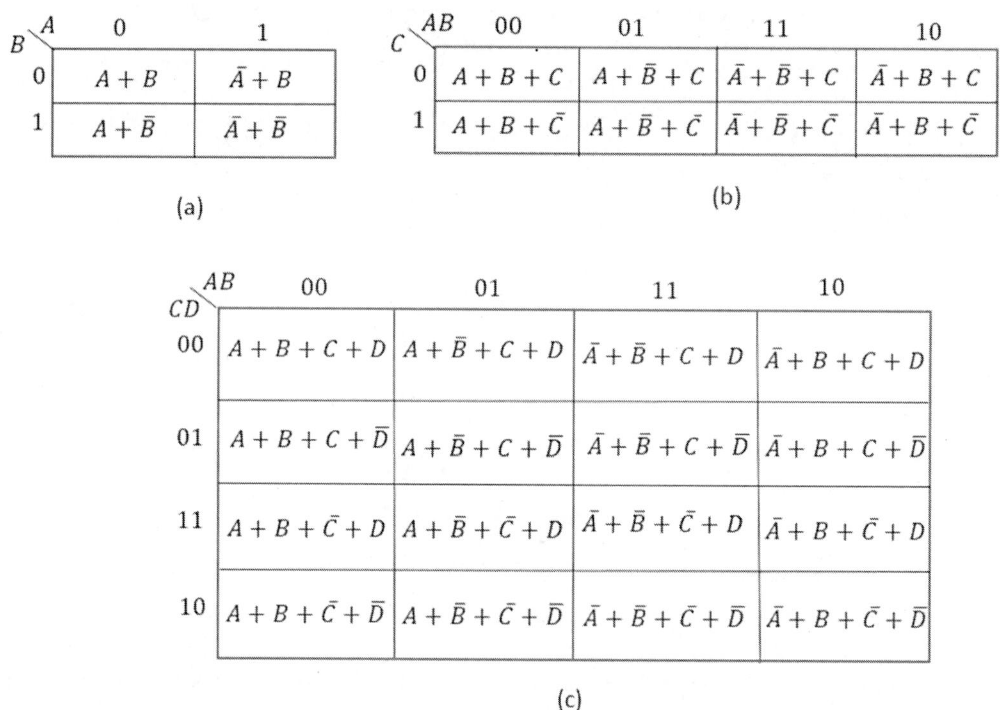

B \ A	0	1
0	$A + B$	$\bar{A} + B$
1	$A + \bar{B}$	$\bar{A} + \bar{B}$

(a)

C \ AB	00	01	11	10
0	$A + B + C$	$A + \bar{B} + C$	$\bar{A} + \bar{B} + C$	$\bar{A} + B + C$
1	$A + B + \bar{C}$	$A + \bar{B} + \bar{C}$	$\bar{A} + \bar{B} + \bar{C}$	$\bar{A} + B + \bar{C}$

(b)

CD \ AB	00	01	11	10
00	$A + B + C + D$	$A + \bar{B} + C + D$	$\bar{A} + \bar{B} + C + D$	$\bar{A} + B + C + D$
01	$A + B + C + \bar{D}$	$A + \bar{B} + C + \bar{D}$	$\bar{A} + \bar{B} + C + \bar{D}$	$\bar{A} + B + C + \bar{D}$
11	$A + B + \bar{C} + D$	$A + \bar{B} + \bar{C} + D$	$\bar{A} + \bar{B} + \bar{C} + D$	$\bar{A} + B + \bar{C} + D$
10	$A + B + \bar{C} + \bar{D}$	$A + \bar{B} + \bar{C} + \bar{D}$	$\bar{A} + \bar{B} + \bar{C} + \bar{D}$	$\bar{A} + B + \bar{C} + \bar{D}$

(c)

Fig. 4.9: K-map using maxterm (a) Two-variable (b) Three-variable (c) Four-variable.

4.1.5 Minimization of logical functions

If we consider the combination of inputs for which the Boolean function is '1', then we will get the Boolean function, which is in **standard sum of products** form after simplifying the K-map. Similarly, if we consider the combination of inputs for which the Boolean function is

'0', then we will get the Boolean function, which is in **standard product of sums** form after simplifying the K-map. Follow these **rules for simplifying K-maps** in order to get standard sum of products form.

- Select the respective K-map based on the number of variables present in the Boolean function.
- If the Boolean function is given as sum of min terms form, then place the ones at respective min term cells in the K-map. If the Boolean function is given as sum of products form, then place the ones in all possible cells of K-map for which the given product terms are valid.
- Check for the possibilities of grouping maximum number of adjacent ones. It should be powers of two. Start from highest power of two and up to least power of two. Highest power is equal to the number of variables considered in K-map and least power is zero.
- Each grouping will give either a literal or one product term. It is known as **prime implicant**. The prime implicant is said to be **essential prime implicant**, if at least single '1' is not covered with any other groupings but only that grouping covers.
- Note down all the prime implicants and essential prime implicants. The simplified Boolean function contains all essential prime implicants and only the required prime implicants.

Note 1 – *If outputs are not defined for some combination of inputs, then those output values will be represented with* ***don't care symbol 'x'****. That means, we can consider them as either '0' or '1'.*
Note 2 – *If don't care terms also present, then place don't care 'x' in the respective cells of K-map. Consider only the don't cares 'x' that are helpful for grouping maximum number of adjacent ones. In those cases, treat the don't care value as '1'.*
Note 3 *- If don't care terms also present, then place don't care 'x' in the respective cells of K-map. Consider only the don't cares 'x' that are helpful for grouping maximum number of adjacent zeroes. In those cases, treat the don't care value as '0'.*

Example 4.4: Simplify the following Boolean function using K-map.

$$f(w,x,y,z) = wx'y' + wy + w'yz'$$

Solution: The given Boolean function is in sum of products form. It is having 4 variables W, X, Y & Z. So, we require 4 variable K-map.

Table 4.2: Truth table for example 4.4

Function	variables	minterms	Possible minterms	Decimal equivalent
$wx'y'$	$wx'y'z$	$wx'y'z$	1001	9
	$wx'y'z'$	$wx'y'z'$	1000	8
wy	wxy	$wxyz$	1111	15
		$wxyz'$	1110	14
	$wx'y$	$wx'yz$	1011	11
		$wx'yz'$	1010	10
$w'yz'$	$w'xyz'$	$w'xyz'$	0110	6
	$w'x'yz'$	$w'x'yz'$	0010	2

The 4 variable truth table for the above expression is given by Table 4.2. Its K-map, its decimal values representation, and different groupings of ones corresponding to the given product terms is shown in the Fig. 4.10.

WX \ YZ	00	01	11	10
00	0	1	3	2
01	4	5	7	6
11	12	13	15	14
10	8	9	10	11

(a)

WX \ YZ	00	01	11	10
00	0	0	0	1
01	0	0	0	1
11	0	0	1	1
10	1	1	1	1

(b)

WX \ YZ	00	01	11	10
00	0	0	0	1
01	0	0	0	1
11	0	0	1	1
10	1	1	1	1

(c)

Fig.4.10: Example 4.4

Here, 1s are placed in the following cells of K-map.

- The cells, which are common to the intersection of Row 4 and columns 1 & 2 are corresponding to the product term, **WX'Y'**.
- The cells, which are common to the intersection of Rows 3 & 4 and columns 3 & 4 are corresponding to the product term, **WY**.
- The cells, which are common to the intersection of Rows 1 & 2 and column 4 are corresponding to the product term, **W'YZ'**.

There are no possibilities of grouping either 16 adjacent ones or 8 adjacent ones. There are three possibilities of grouping 4 adjacent ones. After these three groupings, there is no single one left as ungrouped. So, we no need to check for grouping of 2 adjacent ones. The **4 variable K-map** with these three **groupings** is shown in Fig. 4.10(c).

Here, we got three prime implicants WX', WY & YZ'. All these prime implicants are **essential** because of following reasons.

- Two ones **(m_8 & m_9)** of fourth row grouping are not covered by any other groupings. Only fourth row grouping covers those two ones.
- Single one **(m_{15})** of square shape grouping is not covered by any other groupings. Only the square shape grouping covers that one.
- Two ones **(m_2 & m_6)** of fourth column grouping are not covered by any other groupings. Only fourth column grouping covers those two ones.
- Therefore, the simplified Boolean function is

$$f(w,x,y,z) = wx' + wy + yz'$$

Example 4.5: Write the logic equation in the canonical SOP form for the K-map of Fig. 4.11.

AB \ CD	$\bar{A}\bar{B}$ 00	$\bar{A}B$ 01	AB 11	$A\bar{B}$ 10
$\bar{C}\bar{D}$ 00	1	0	1	0
$\bar{C}D$ 01	0	1	0	1
CD 11	0	1	1	0
$C\bar{D}$ 10	0	0	1	0

Fig. 4.11: K-map for example 4.8

Solution:

$$Y = \bar{A}\,\bar{B}\,\bar{C}\,\bar{D} + AB\bar{C}\,\bar{D} + \bar{A}\,B\bar{C}\,D + A\bar{B}\,\bar{C}\,D + \bar{A}\,BCD + ABCD + ABC\bar{D}$$

Hence,

$$Y = \sum m(0,5,7,9,12,14,15)$$

Follow these **rules for simplifying K-maps** in order to get standard product of sums form.

- Select the respective K-map based on the number of variables present in the Boolean function.
- If the Boolean function is given as product of Max terms form, then place the zeroes at respective Max term cells in the K-map. If the Boolean function is given as product of sums form, then place the zeroes in all possible cells of K-map for which the given sum terms are valid.
- Check for the possibilities of grouping maximum number of adjacent zeroes. It should be powers of two. Start from highest power of two and up to least power of two. Highest power is equal to the number of variables considered in K-map and least power is zero.
- Each grouping will give either a literal or one sum term. It is known as **prime implicant**. The prime implicant is said to be **essential prime implicant**, if at least single '0' is not covered with any other groupings but only that grouping covers.
- Note down all the prime implicants and essential prime implicants. The simplified Boolean function contains all essential prime implicants and only the required prime implicants.

Note *– If don't care terms also present, then place don't care 'x' in the respective cells of K-map. Consider only the don't cares 'x' that are helpful for grouping maximum number of adjacent zeroes. In those cases, treat the don't care value as '0'.*

4.1.6 Don't care Conditions

Don't care denote the set of inputs that never occurs for given digital circuits. Therefore, to simplify the Boolean output expressions, the 'don't care' are used. In Boolean expression, when minterms/maxterms along with don't care has been used, the switching of the state is reduced. This reduced the required memory space resulting in low power consumption. Hence, to reduce the no. of gates that are used to implement the given expression. The use of don't care terms makes the logic design more economical.

To distinguish the don't -care condition from 1's and 0's, an X will be used. While selecting adjacent square to simplify the function in the map, the X's may be assumed to be either 0 or 1, whichever gives the simplest expression. An X need not be used at all if it does not contribute to covering a larger area. in each case, the choice deepens only on the simplification that can be achieved.

With don't care terms the designer has a flexibility and it is left to him whether to assume a 0 or a 1 as output for each of these combinations. This condition is known as don't -care condition and can be represented on the K-map as a X mark in the corresponding cell. The X mark in a cell may be assumed to be a 1 or a 0 depending upon which one leads to a simpler expression. The function can be specified in one of the following ways:

- Don't care in a Karnaugh map, or truth table, may be either 1s o 0s, as long as we don't care what the output is for an input condition we never expect to see.
- When forming groups of cells, treat the don't care cell as either a 1 or 0, or ignore the don't cares. This is helpful if it allows us to form a larger group than would otherwise be possible without the don't cares. There is no requirement to group all or any of the don't cares.
- The system designer, do not care what output the logic circuit produces for these don't cares. Only use them in a group if it simplifies the logic.

Example 4.6: Simplify the following Boolean function using K-map.

$$f(A,B,C) = \prod M(0,1,2,4)$$

Solution:

The given Boolean function is in product of maxterms form. It is having 3 variables X, Y & Z. So, we require 3 variable K-map. The given maxterms are M_0, M_1, M_2 & M_4. The 3 **variable K-map** with zeroes corresponding to the given maxterms and possible grouping is shown in Fig. 4.12.

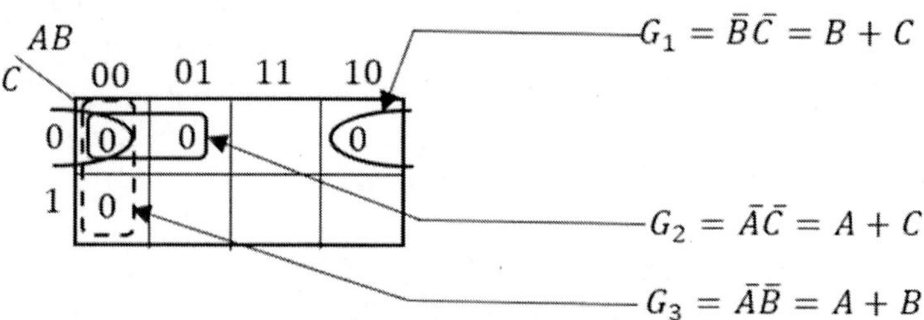

Fig. 4.12: K-map for Example 4.6

From Fig. 4.12, it is observed that there are no possibilities of grouping either 8 adjacent zeroes or 4 adjacent zeroes. There are three possibilities of grouping 2 adjacent zeroes. After these three groupings, there is no single zero left as ungrouped.

Here, we got three prime implicants $(B + C), (A + C), and\ (A + B)$. All these prime implicants are essential because one zero in each grouping is not covered by any other groupings except with their individual groupings. Therefore, the simplified Boolean function in product of sum form

$$f(A,B,C) = Y = (B + C).(A + C).(A + B)$$

Note: However, the same equation can be written in sum of product form, but it will be a complemented output,

$$f(A,B,C) = \bar{Y} = \bar{B}\bar{C} + \bar{A}\bar{C} + \bar{A}\bar{B}$$

Example 4.7 Minimize the Boolean function using the mapping method in both minimized Sum-of-products and Product-of-sums forms.

$$f(A,B,C) = \sum 0,1,3,\ 5 + \sum_{\emptyset} 2,7$$

Solution:

$$f(A,B,C) = \sum 0,1,3,5 + \sum_{\emptyset} 2,7 = \prod 4,6 + \prod_{\emptyset} 2,7$$

From given Boolean function in $\sum$ and $\prod$ notation, we can write Sum-of-products and Product-of-sums Boolean expression as follows:

$$f(A,B,C) = \bar{A}.\bar{B}.\bar{C} + \bar{A}.\bar{B}.C + \bar{A}.B.C + A.\bar{B}.C$$

$$f(A,B,C) = (\bar{A} + B + C).(\bar{A} + \bar{B} + C)$$

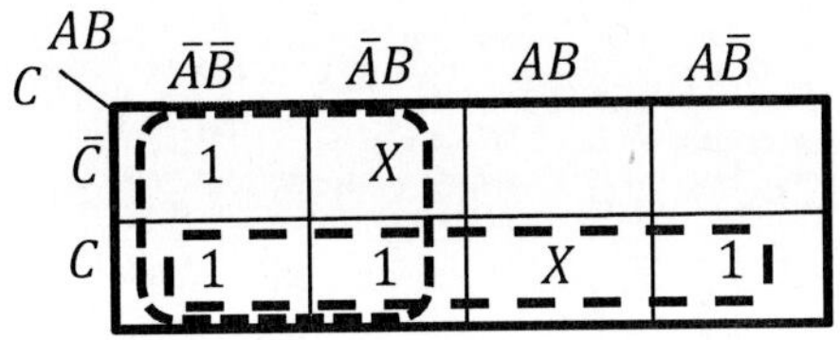

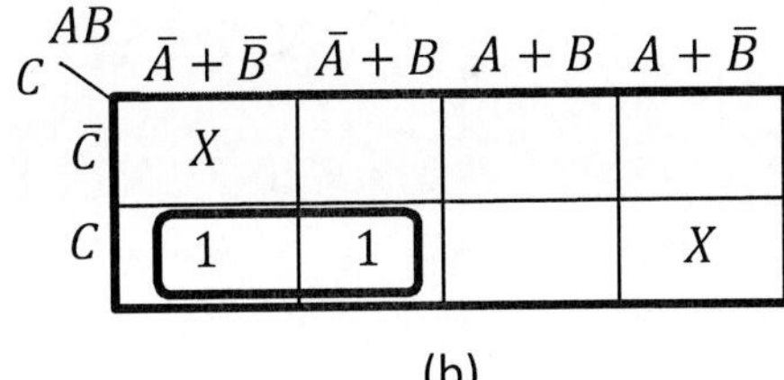

(b)

Fig. 4.13: K-map for Example 4.7

- The 'don't care' input combinaton for the sum-oproducts Boolean expression are $\bar{A}.B.\bar{C}$ *and* $A.B.C$.
- The 'don't care' input combinaton for the product-of-sums expression are $(A + \bar{B} + C).(\bar{A} + \bar{B} + \bar{C})$.

The Karnaugh maps for the two cases are shown in Fig. 4.13 (a) and (b). The minimized sum-of-products and product-of-sums Boolean functions are respectively given by the equations

$$f(A,B,C) = C + \bar{A}$$

$$f(A,B,C) = \bar{A} + C$$

Example 4.8 Write the simplified Boolean expression given by the Karnaugh map shown in Fig. 4.14.

Solution:

The k-map is shown in Fig. 4.15. Consider the group of four 1s at the top left of the map. It yields a term $\bar{A}.\bar{C}$. Now consider the another group of four 1s, two on the exterme left and two on th exterme right. This group yields a term $\bar{A}.\bar{D}$. The third group of two 1s is in the third row of the map. The Third row corresponds to the intersection of A and B , as is clear from the map. Therefore , this group yields a term ABC.

$CD \backslash AB$	$\bar{A}\bar{B}$	$\bar{A}B$	AB	$A\bar{B}$
$\bar{C}\bar{D}$	1	1	0	1
$\bar{C}D$	1	1	0	1
CD	0	0	1	1
$C\bar{D}$	0	0	0	0

Fig. 4.14: K-map for Example 4.8

Solution :

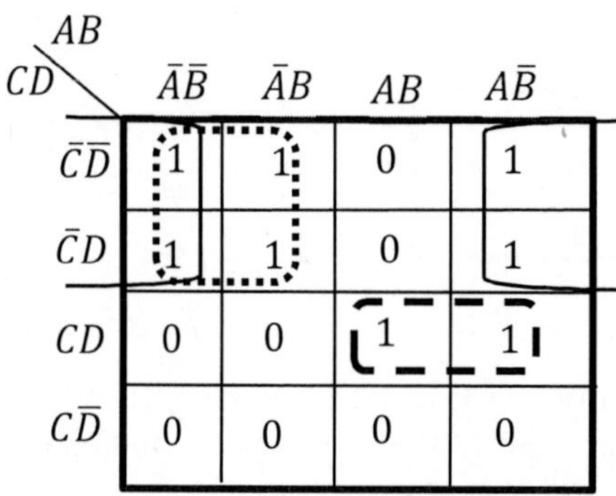

Fig. 4.15: Grouping of 1's for K-map

The simplified Boolean expression is given by

$$Y = \bar{A}.\bar{C} + \bar{A}.\bar{D} + A.B.C.$$

Example 4.9 Minimizing a given Boolean expression using the Quine-McCluskey tabular method yields the following prime implicants *(_0_0, _1_1, 1_10, 0_00)*. Draw the corresponding Karnaugh map.

Solution: As is clear from the prime implicants, the expression has four variables. If the variables are assumed to be A, B, C and D, then the given prime implicants correspond to the folowing terms:

Table 4.3: Example 4.9

Prime Implicant	Given minterms	Possible options	Minterms	Decimal Equivalent
_0_0	$\bar{B}.\bar{D}.$	$A.\bar{B}.C.\bar{D}.$	1010	10
		$A.\bar{B}.\bar{C}.\bar{D}$	1000	8
		$\bar{A}.\bar{B}.C.\bar{D}$	0010	2
		$\bar{A}.\bar{B}.\bar{C}.\bar{D}.$	0000	0
_1_1	$B.D.$	$A.B.C.D$	1111	15
		$A.B.\bar{C}.D$	1101	13

Prime Implicant	Given minterms	Possible options	Minterms	Decimal Equivalent
		$\bar{A}.B.C.D$	0111	7
		$\bar{A}.\bar{B}.\bar{C}.D$	0001	1
1_10	$A.C.\bar{D}$	$A.B.C.\bar{D}.$	1110	14
		$A.\bar{B}.C.\bar{D}.$	1010	10
0_00	$\bar{A}.\bar{C}.\bar{D}.$	$\bar{A}.B.\bar{C}.\bar{D}.$	0100	4
		$\bar{A}.\bar{B}.\bar{C}.\bar{D}.$	0000	0

$$Y = \sum m(0,1,2,4,7,8,10,13,14,15)$$

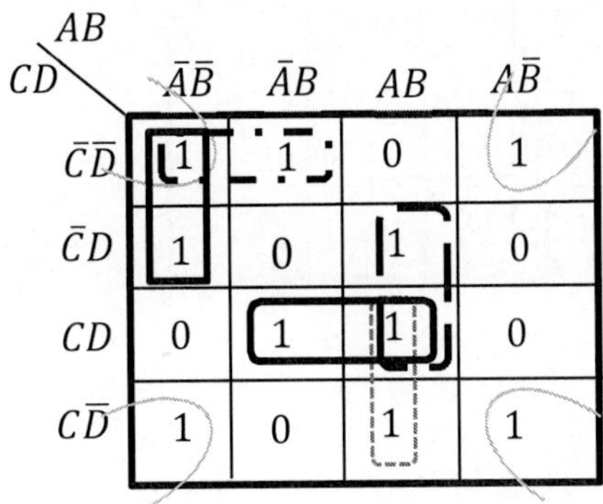

Fig. 4.16: Solution to example 4.9.

Example 4.10 $\bar{A}.\bar{B} + C.D$ is a simplified Boolean expression of the expression $A.B.C.D + \bar{A}.\bar{B}.C.D + \bar{A}.B$. Determine if there are any 'don't care' entries.

Solution:

The expression given by the equation has the following possible minterms

Table 4.4: Example 4.10

S.No.	Given minterms	Possible options	Minterms	Decimal Equivalent
1	$A.B.C.D$	$A.B.C.D$	1111	15
2	$\bar{A}.\bar{B}.C.D$	$\bar{A}.\bar{B}.C.D$	0011	3
3	$\bar{A}.B$	$\bar{A}.B.C.D$	0111	7
		$\bar{A}.B.\bar{C}.D$	0101	5
		$\bar{A}.B.C.\bar{D}$	0110	6
		$\bar{A}.B.\bar{C}.\bar{D}.$	0100	4

CD \ AB	$\bar{A}\bar{B}$	$\bar{A}B$	AB	$A\bar{B}$
$\bar{C}\bar{D}$	0	1	0	0
$\bar{C}D$	0	1	0	0
CD	1	1	1	X
$C\bar{D}$	0	1	0	0

Fig. 4.17 :Solution to example 4.10

Example 4.11 Minimize the following Boolean function-

$$F(A,B,C,D) = \sum m(0,1,2,5,7,8,9,10,13,15)$$

Solution:

Since the given Boolean expression has 4 variables, so a 4 x 4 K-Map has to be drawn as shown in *Fig.*. Further, cells of K Map will be filled in accordance with the given Boolean function. Then, groups will be formed

Then, we have-

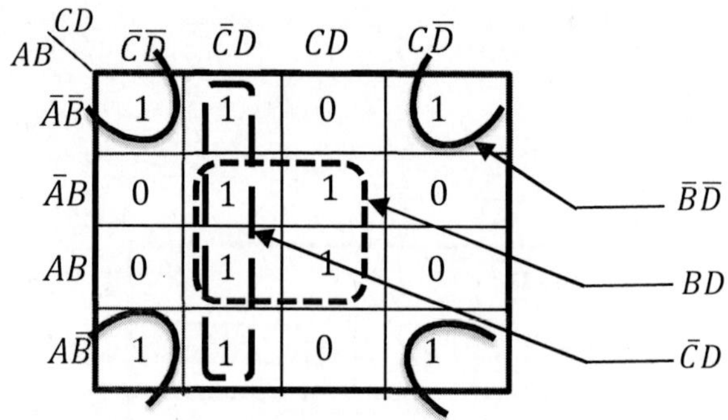

Fig. 4.18: Solution to example 4.11

From above K-Map, minimized Boolean expression can be written as

$$F(A,B,C,D) = BD + \bar{B}\bar{D} + \bar{C}D$$

Example 4.12 Minimize the following Boolean function

$$F(A,B,C,D) = \sum m(0,1,3,5,7,8,9,11,13,15)$$

Solution:

Since the given Boolean expression has 4 variables, so a 4 x 4 K-Map has to be drawn as shown in *Fig.*. Further, cells of K Map will be filled in accordance with the given Boolean function. Then, groups will be formed.

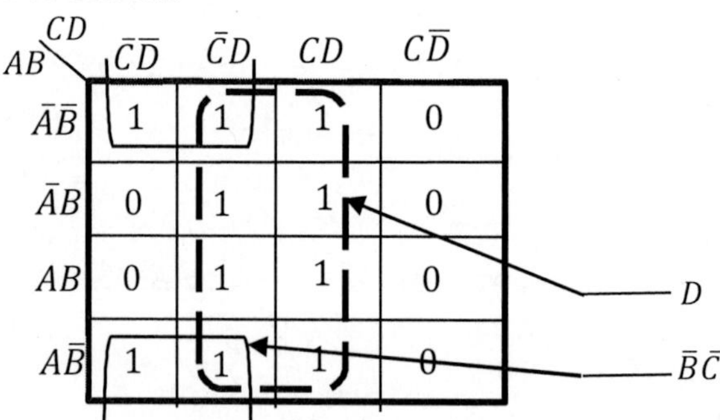

Fig. 4.19: Solution to example 4.12

From above K-Map, minimized Boolean expression can be written as

$$F(A,B,C,D) = \bar{B}\bar{C} + D$$

Example 4.13 Minimize the following Boolean function

$$F(A,B,C,D) = \sum m(1,3,4,6,8,9,11,13,15) + \sum d(0,2,14)$$

Solution:

Since the given Boolean expression has 4 variables, so a 4 x 4 K-Map has to be drawn as shown in Fig.. Further, cells of K Map will be filled in accordance with the given Boolean function. Then, groups will be formed.

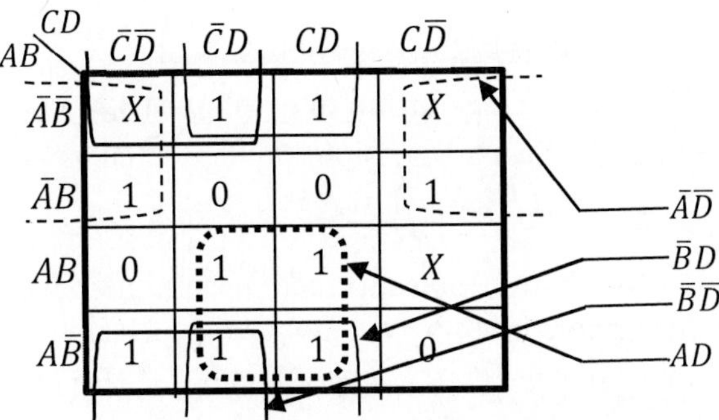

Fig. 4.20: Solution to example 4.13

From above K-Map, minimized Boolean expression can be written as

$$F(A,B,C,D) = \bar{A}\bar{D} + \bar{B}D + \bar{B}\bar{C} + AD$$

Example 4.14 Minimize the following Boolean function

$$F(A,B,C) = \sum m(0,1,6,7) + \sum d(3,5)$$

Solution-

Since the given Boolean expression has 3 variables, so a 4 x 2 K-Map has to be drawn. Further, cells of K Map will be filled in accordance with the given Boolean function. Then, groups will be formed.

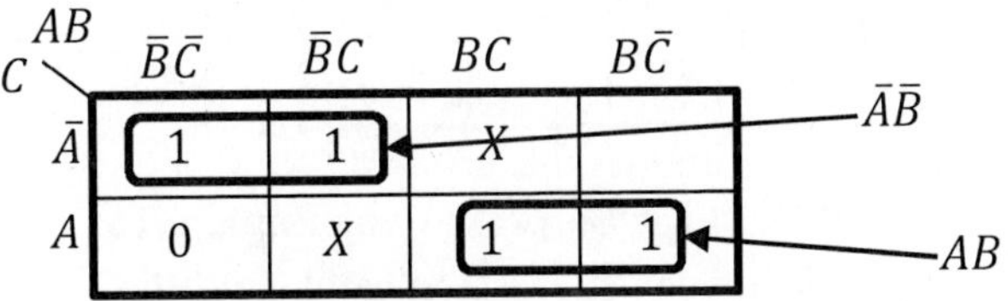

Fig. 4.21: Solution to example 4.14

From above K-Map, minimized Boolean expression can be written as

$$F(A,B,C) = \bar{A}\bar{B} + AB$$

Example 4.15 Minimize the following Boolean function

$$F(A,B,C) = \sum m(1,2,5,7) + \sum d(0,4,6)$$

Solution-

Since the given Boolean expression has 3 variables, so a 4 x 2 K-Map has to be drawn. Further, cells of K Map will be filled in accordance with the given Boolean function. Then, groups will be formed.

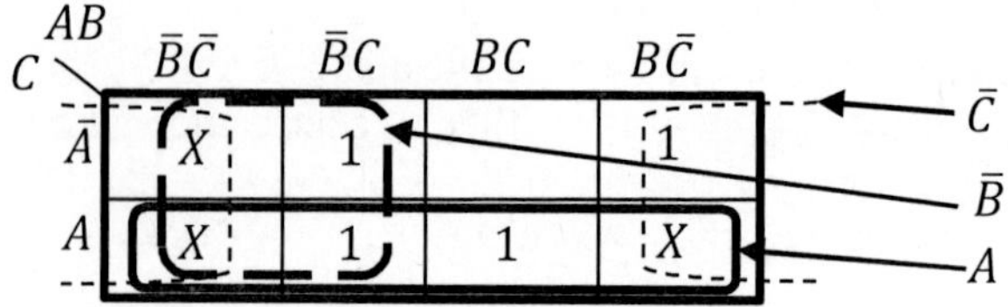

Fig. 4.22: Solution to example 4.14

From above K-Map, minimized Boolean expression can be written as

$$F(A,B,C) = A + \bar{B} + \bar{C}$$

Example 4.16 Consider the following Boolean function, minimize the expression and find the dependent, and independent variable for the given Boolean expression.

$$F(A,B,C,D) = \sum m(1,3,4,6,9,11,12,14)$$

Solution-

Since the given Boolean expression has 4 variables, so a 4 x 4 K-Map has to be drawn. Further, cells of K Map will be filled in accordance with the given Boolean function. Then, groups will be formed.

AB \ CD	$\bar{C}\bar{D}$	$\bar{C}D$	CD	$C\bar{D}$
$\bar{A}\bar{B}$	0	1	1	0
$\bar{A}B$	1	0	0	1
AB	1	0	0	1
$A\bar{B}$	0	1	1	0

$B\bar{D}$

$\bar{B}D$

Fig. 4.23: Solution to example 4.16

From above K-Map, minimized Boolean expression can be written as

$$F(A,B,C) = B\bar{D} + \bar{B}D = B \oplus D$$

Thus, minimized Boolean expression clearly, shows that the given Boolean function depends on only two variables B and D. Hence, it is independent of other two variables A and C.

Example 4.17 Minimise the following function in SOP minimal form using K-Maps:

$$F = \sum m(1,5,6,11,12,13,14) + \sum d(4)$$

solution**:** The SOP K-map for the given expression is:

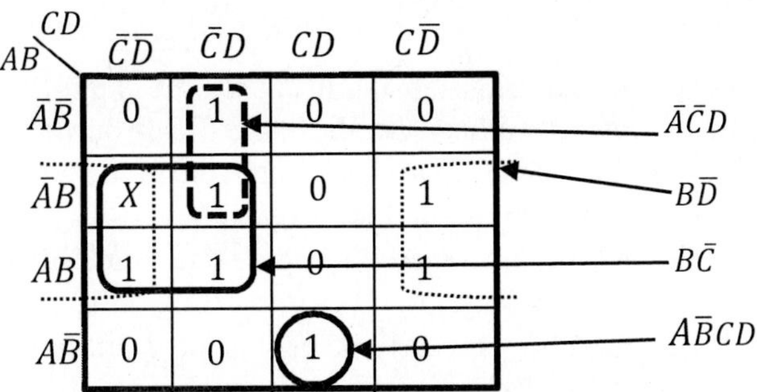

Fig. 4.24: Solution to example 4.17

The minterms of F are the variable combinations that make the function equal to 1.the minterms of d are the don't -care combinations known never to occur.This minimization is shown as in Fig. 4.24. The minterms of F are marked by 1's , those of d are marked by X's, and the remaining squares are filled with 0's. The 1's and X's are combined in any convenient manner so as to enclose the maximum number of adjacent squares. It is not necessary to include all or any of the X's , but only those useful for simplifying a term.

From above K-Map, minimized Boolean expression in SOP form can be written as

$$F(A,B,C,D) = B\bar{D} + B\bar{C} + \bar{A}\bar{C}D + A\bar{B}CD$$

Complementing again, we obtain a simplified POS function as:

$$F(A,B,C,D) = (\bar{B} + D)(\bar{B} + C)(A + C + \bar{D})(\bar{A} + B + \bar{C} + \bar{D})$$

If an X is used as a 1 when combining the 1's and again as a 0 when combning the 0's, the two resulting fuctions will not yields algebraically equal anewers. The selection of the don't-care condition as a 1 in the first case and as a 0 in the second results in different minterm expressions and thus different functions.

4.1.7 Hazards in combinational circuits

Generally, a hazard in a digital circuit is a temporary disturbance in ideal operation of the circuit which some time gets resolved itself. A hazard, if exists, in a digital circuit causes a temporary fluctuation in output of the circuit. These disturbances or fluctuations occur when different paths from the input to output have different delays and due to this fact, changes in input variables do not change the output instantly but do appear at output after a small delay caused by the circuit building elements, i.e., logic gates.

There are three different kinds of hazards found in digital circuits

- Static hazard
- Dynamic hazard
- Functional hazard

Out of above-mentioned hazards static hazard is the most common. Hence, static hazard only discussed in the upcoming sections.

Static hazard

A static hazard takes place when change in an input causes the output to change momentarily before stabilizing to its correct value. A static hazard can be detected by observing the products used for the function on a K-map. Based on what is the correct value, there are two types of static hazards, as shown below in the image:

- Static-1 Hazard: If the output is currently at logic state 1 and after the input changes its state, the output momentarily changes to 0 before settling on 1, then it is a Static-1 hazard.
- Static-0 Hazard: If the output is currently at logic state 0 and after the input changes its state, the output momentarily changes to 1 before settling on 0, then it is a Static-0 hazard.

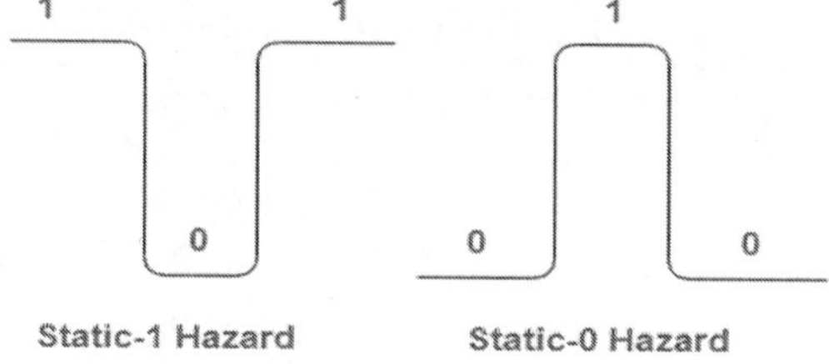

Fig. 4.25: Types of Hazard

Detection of Static hazards using K-map:

Let's consider static-1 hazard first. To detect a static-1 hazard for a digital circuit following steps are used:

Step-1: Write down the output of the digital circuit, say Y.

Step-2: Draw the K-map for this function Y and note all adjacent 1's.

Step-3: If there exists any pair of cells with 1's which do not occur to be in the same group (i.e., prime implicant), it indicates the presence of a static-1 hazard. Each such pair is a static-1 hazard.

Example 4.18: Consider the circuit shown below. Find the Hazard.

Solution:

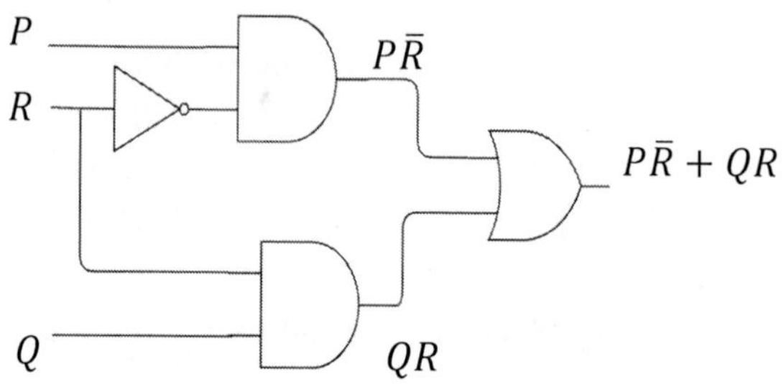

Fig. 4.26: Example 4.18

We have output, say F as

$$F(P, Q, R) = P\bar{R} + QR = \sum m(3, 4, 6, 7)$$

Let's draw the K-map for this Boolean function as follows:

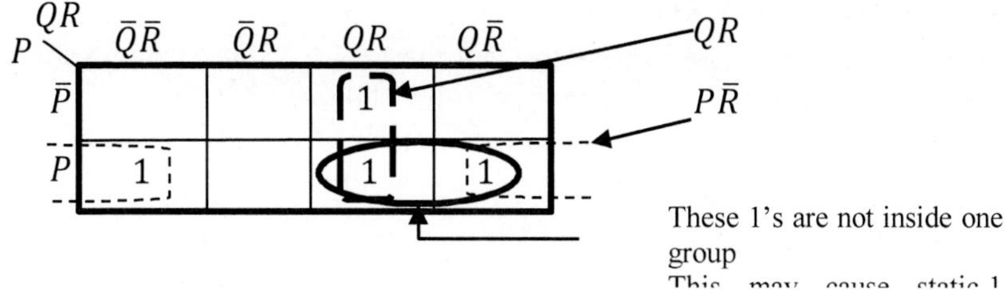

Fig. 4.27: Solution for example 4.18

The pair of 1's encircled as green are not part of the grouping/pairing provided by the output of this Boolean function. This will cause a static-1 hazard in this circuit.

Removal of static-1 hazard:

Once detected, a static-1 hazard can be easily removed by introducing some more terms (logic gates) to the function (circuit). The most common idea is to add the missing group in the existing Boolean function, as adding this term would not affect the function by any mean but it will remove the hazard. Since in above example the pair of 1's encircled with blue colour causes the static-1 hazard, we just add this as a prime implicant to the existing function as follows:

$$F(P,Q,R) = P\bar{R} + QR + PQ = \sum m(3,4,6,7)$$

Note that there is no difference in number of minterms of this function. The reason is that the static-1 hazards are based on how we group 1's (or 0's for static-0 hazard) for a given set of 1's in K-map. Thus, it does not make any difference in number of 1's in K-map. The circuit would look like as shown below in Gig. 4.28 with the change made for removal of static-1 hazard.

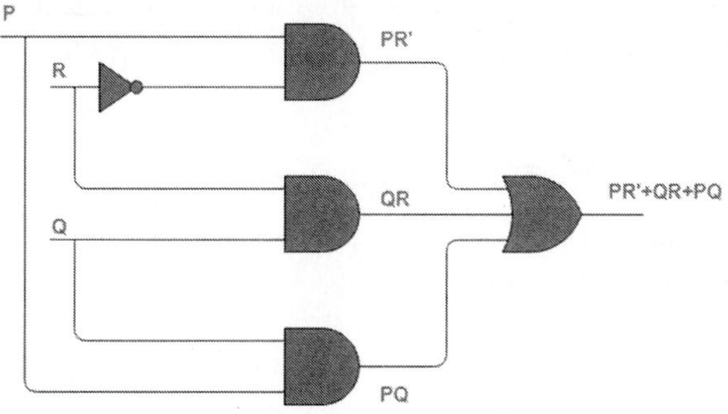

Fig. 4.28: Removal of static-1 hazard

Similarly, for **Static-0 Hazards** we need to consider 0's instead of 1's and if any adjacent 0's in K-map is not grouped into same group that may cause a static-0 hazard. The method to detect and resolve the static-0 hazard is completely same as the one we followed for static-1 hazard except that instead of SOP, POS will be used as we are dealing with 0's in this case.

4.2 COMBINATIONAL CIRCUITS

A combinational circuits is one where the output at any time depends only on the present combination of inputs at any instant of time. The combinational circuit do not use any memory. The previous state of input does not have any effect on the present state of the circuit.

A combinational circuit can have an $'n'$ number of inputs and $'m'$ number of outputs shown in Fig. 4.29.

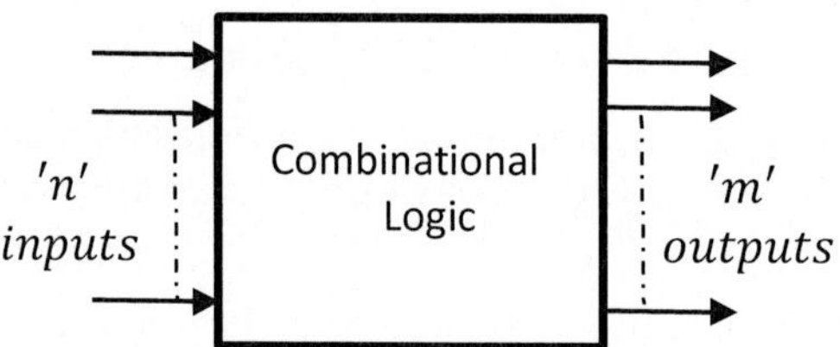

Fig. 4.29: Generalized Combinational Circuits

The logic gate is the most basic building block of combinational logic.the logical function performed by a combinational circuit is fully defined by a set of Boolean expressions. Fig. 4.29 shows the block schematic representation of a generalized combinational circuit having n input variable and m output variables or outputs. Since the number of input variables is n, there are 2^n possible combinations of bits at the input. In combinational circuits , input variable come from an external source and output variable feed an external destination.

4.2.1 Implementing Combinational Logic

The different steps involved in the design of a combinational logic circuit are as follows:

Step-1: Statement of the problem.

Step-2: Identification of input and output variables.

Step-3: Expressing the relationship between the input and output variables.

Step-4: Construction of a truth table to meet input-output requirements.

Step-5: Writing Boolean expression for various output variables in terms of input variables.

Step-6: Minimization of Boolean expressions.

Step-7: Implementation of minimized Boolean expressions.

There are various simplification technique available for minimizing Boolean expressions, which have been discussed in the previous section. And here are the following guidelines should be followed while choosing the preferred form for hardware implementation:

- The implementation should have the minimum number of gates, with the gates used having the minimum number of inputs.
- There should be a minimum number of interconnections, and the propagation time should be the shortest.
- Limitation on the driving capability of the gates should not be ignored.

4.2.2 ADDERS - Subtractor

<u>*Half adder*</u>

Half adder is a combinational logic circuit with two inputs and two outputs. Fig. 4.30 shows the half-adder is an airthmetic circuit block that can be used to add two bits. Such a circuit thus has two inputs that rpresent the two bits to be added and two outputs. With one producing the SUM output and the other producing the CARRY.figure 4.30 shows the truth table of a half-adder, showing all possible input combinations and the corresponding outputs.

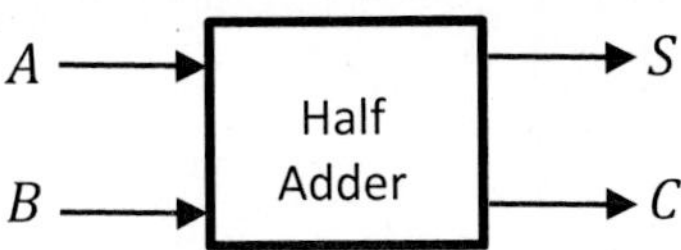

Fig. 4.30: Block diagram of Half adder.

Table 4.5: Truth table of a half-adder.

Input		**Output**	
A	B	S	C
0	0	0	0
0	1	1	0
1	0	1	0
1	1	0	1

From the truth table (Refer Table 4.5), we obtain the Boolean or logical expression for the SUM (S) and CARRY (C) outputs are given by the equations

$$Sum\ (S) = A.\bar{B} + \bar{A}.B$$
$$Carry\ (C) = A.B$$

While the equation representing the SUM output is that of an EX-OR gate,the equation representing the CARRY output is that of an AND gate (Refer Fig. 4.31).

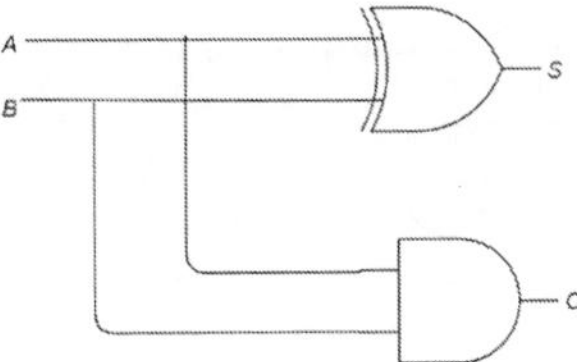

Fig. 4.31: Logic implementation of a half-adder.

Equations for *sum* and *carry* can be representated in different forms using various laws and theorem of Boolean algerbra with the flexibility for the designer has in hardware-implimenting as simple a combinational function as a half -adder.we have discussed in previous section on bolean algebra how varoious logic gates can be implemented in the form of either only NAND gaes or NOR gates.the simplest way to harsdware-implement a half-adder would be to use a tewo-input EX-OR gate for the SUM output and a two -input AND gate for the CARRY output, as shown in Fig. 4.31. It could also be implementation of a half-adder with NAND gates. Fig. 4.23 shows the implementation of a half-adder with NAND gates only.

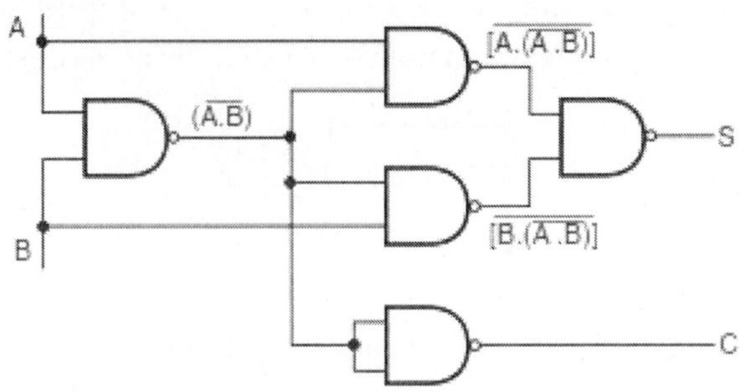

Fig. 4.32: Half-adder implementation using NAND gates.

In Logic diagram of Fig. 4.32 one part of the circuit implements a two-input EX-OR gate with two-input NAND gates. The AND gate required to generate CARRY output is implemented by complementing an already avilable NAND output of the input variables.

Full adder :

A full adder circuit is an arithmetic circuit block that can be used to add three bits to produce a SUM and a CARRY output. Such a buiding block becomes a necessity when it comes to adding binary numbers with a large number of bits.

A half -adder has only two inputs and there is no provision to add a carry coming from the lower order bits when multibit addition is performed. For this purpose , a third input terminal is added and this circuit is used to add A_n, B_n and C_{n-1}, where A_n and B_n are the nth order bits of the numbers A and B respectively and C_{n-1} is the carry generated from addition of $(n-1)^{th}$ order bits. This circuit is referred to as full-adder and its truth table is given in Fig. 4.33 andTable 4.6.

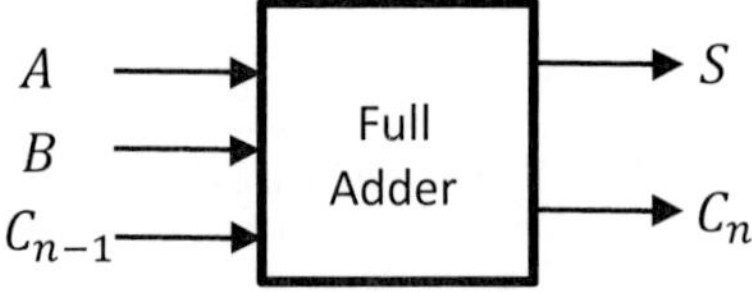

Fig. 4.33: circuit block for Full- adder

Table 4.6: Truth table for a Full- adder

Input			Output	
A_n	B_n	C_{n-1}	S_n	C_n
0	0	0	0	0
0	0	1	1	0
0	1	0	1	0
0	1	1	0	1
1	0	0	1	0
1	0	1	0	1
1	1	0	0	1
1	1	1	1	1

The figure below shows tha K-map for the Full Adder

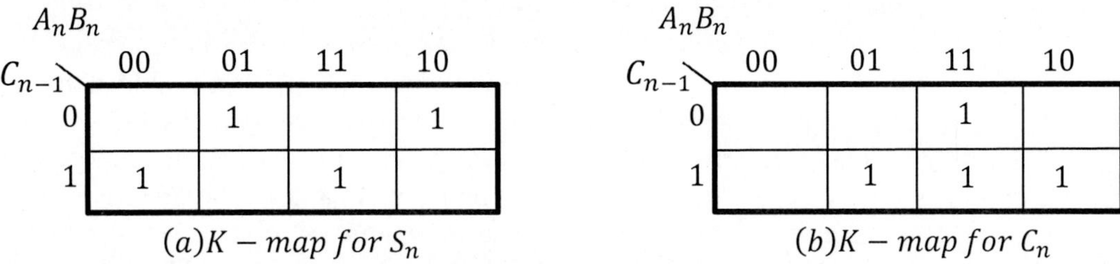

Fig. 4.34: K-maps for (a) Sum and (b) Carry

The K-maps for the outputs S_n and C_n are given in Fig. 4.34 and the minimsed expression are given by equations below.

$$S_n = \bar{A}_nB_n\bar{C}_{n-1} + \bar{A}_n\bar{B}_nC_{n-1} + A_n\bar{B}_n\bar{C}_{n-1} + A_nB_nC_{n-1}$$
$$C_n = A_nB_n + B_nC_{n-1} + A_nC_{n-1}$$

The NAND-NAND realisation if the full adder circuit is shown in Fig. 4.35.

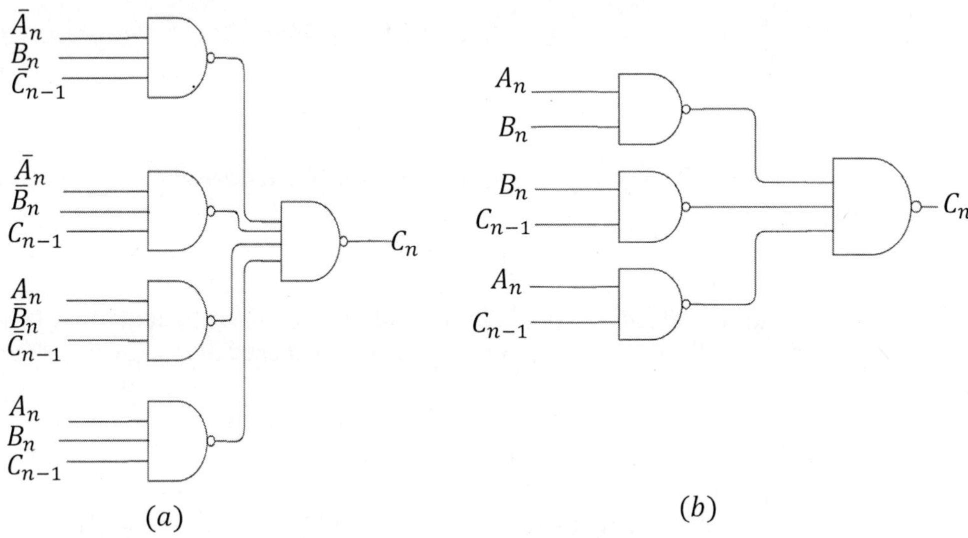

Fig. 4.35: NAND-NAND Realisation of (a) S_n (b) C_n.

Fig. 4.35 shows the logic diagram of the full-adder. A full adder can also be seen to comprise two half-adders and OR gate.the expression for SUM and CARRY Outputs can be rewritten as follows:

$$S_n = \bar{C}_{n-1}.(\bar{A}_n.B_n + A_n.\bar{B}_n) + C_{n-1}.(A_n.B_n + \bar{A}_n.\bar{B}_n)$$
$$S_n = \bar{C}_{n-1}.(\bar{A}_n.B_n + A_n.\bar{B}_n) + C_{n-1}.\left(\overline{\bar{A}_n.B_n + A_n \cdot \bar{B}_n}\right)$$

Similarly, the expression for CARRY output can be rewritten as follows:

$$C_n = B_n.C_{n-1}.(A_n + \bar{A}_n) + A_n.B_n + A_n.C_{n-1}.(B_n + \bar{B}_n)$$
$$C_n = A_n.B_n + A_n.B_n.C_{n-1} + \bar{A}_n.B_n.C_{n-1} + A_n.B_n.C_{n-1} + A_n.\bar{B}_n.C_{n-1}$$

$$C_n = A_n.B_n + A_n.B_n.C_{n-1} + \bar{A}_n.B_n.C_{n-1} + A_n.\bar{B}_n.C_{n-1}$$
$$C_n = A_n.B_n(1 + C_{n-1}) + C_{n-1}\,(\bar{A}_n.B_n. + A_n.\bar{B}_n$$
$$C_n = A_n.B_n + C_{n-1.}\,(\bar{A}_n.B_n. + A_n.\bar{B}_n$$

Boolean expressooin for sum can be implemented with a two-input EX-OR gate provided that one of the inputs is C_{n-1} and the other input is the output of another two-input EX-OR gate with A_n and B_n as its inputs.similarly, Boolean expression for carry can be implemented by ORing two minterms. One of them is the AND output of A_n and B_n.the other is also the output of an AND gate whose inputs are $C_{n-1}$and the output of an EX-OR operation on A_n and B_n. the whole idea of writing the Boolean repression in this modified form was to demonstrate the use of a half-adder circuit in building a full adder shown in Fig. 4.36.

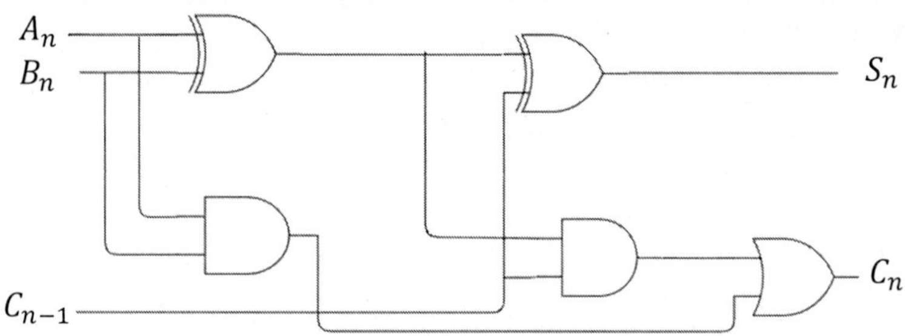

Fig. 4.36: implementation of a full-adder with two half-adders and an OR gate.

Four-bit binary adder :

- **Serial Adder:** A serial adder is used to add two binary numbers in serial form. The two binary numbers to be added serially are stored in two shift registers. The circuit adds one pair at a time with the help of one full adder. The carry output from the full adder is applied to a D flip-flop, the output of which is then used as a carry input for the next pair of significant bits. However, the sum bit S from the output of the full adder can be transferred into a third shift register will discuss in detail in next section.
- **Parallel_Adder:** A parallel adder is a combinational digital circuit that adds two binary numbers in parallel form. It consists of full adders connected in cascade, with the output carry from each full adder connected to the input carry of the next full adder.

Table 4.7: Difference between Serial Adder and Parallel Adder:

S. No.	Parameters	Serial Adder	Parallel Adder
1.	Addition manner	It is used to add two binary numbers in serial form.	It is used to add two binary numbers in parallel form.
2.	Type of Registers	A serial adder uses shift registers.	A parallel adder uses registers with parallel loads.
3.	Requirement	It requires a single full adder.	It requires multiple full adders.
4.	Usage of	A carry flip-flop is used in the serial adder.	Ripple carry adder is used in the parallel adder.
5.	Circuit Type	A serial adder is a sequential circuit.	A parallel adder is a combinational circuit.

S. No.	Parameters	Serial Adder	Parallel Adder
6.	Propagation Delay	In serial adder, propagation delay is less.	In parallel adder, propagation delay is present from input carry to output carry.
7.	Speed	The serial adder has a slow speed as compared to the parallel adder.	The parallel adder has fast speed as compared to the serial adder.
8.	Addition process	The addition process is carried out bit by bit. Therefore, addition time relies on bit count.	The addition process is carried out simultaneously. That implies all bits sum up simultaneously. Therefore, time does not rely on bit count.
9.	Requirement of Components	It necessitates fewer components.	It necessitates more components because of design complexity.
10.	Number of Full Adders	The number of required full adders is fixed i.e. one.	The number of required full adders is equal to the number of bits in the binary number.

- **Binary parallel adder-**The full adder as discussed in the above sections forms the sum of two bits and a previous carry. Two binary numbers of n bits each can be added by means of this circuit. The input carry C_i in the least significant position must be 0. The value of C_{i+1} in a given significant position is the output carry of the full-adder. This value is transferred into the input carry of the full-adder that adds the bits one higher significant position to the left.
  The sum bits are thus generated starting from the rightmost position and are available as soon as the corresponding previous carry bit is generated. The sum of two n-bit binary number ,A and B, can be generated in two ways, either in a serial or in parallel. The serial addition method uses only one full-adder circuit and a storage device to hold the generated output carry.the pair of bits in A and B are transferred serially, one at a time, through the single full-adder to produec a string of output bits for the sum.the stored output carry from one pair of bits is used as an input carry for the next pair of bits.

The parallel method uses n full-adder circuits, and all bits of A and B are applied simultaneously. The output carry from one full-adder is connected to the input carry of the full-adder one position to its left. As soon as the carries are generated, the correct sum bits emerge from the sum outputs of all full-adders.

A binary parallel adder is a combinational circuit that produces the airthmetic sum of two binary numbers in parallel. It consists of full-adders connected in cascade, with the output carry from one full-adder connected to the carry of the next full-adder.

Fig. 4.37 shows Four-bit binary adder such as arrangement ($A_3A_2A_1A_0$) and ($B_3B_2B_1B_0$) are the two binary numbers to be added, with A_0 and B_0 representing LSBs and A_3 and B_3 representing MSBs of the two numbers.

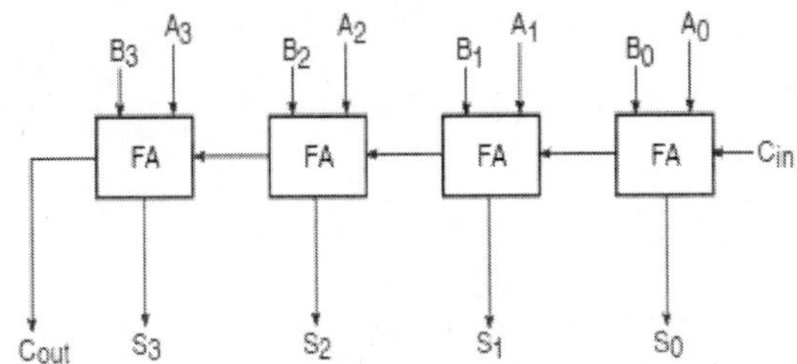

Fig. 4.37: Four-bit binary adder

The augend bits of A and the addend bits of B are designated by the subcript numbers from right to left, with subscrpt 0 denoting the low-order bit. The carries are connected in a chain through the full-adders. The input carry to the adder is C_{in} and the output carry is C_{out}. The S outputs generate the required sum bits. When the 4-bit full-adder circuit is enclosed within an IC package, it has four terminals for the augend bits, four terminals for the addend bits, four terminals for the sum bits. And two terminals for the input and output carries.

An *n-bit* parallel adder requires *n* full-adders. It can be constructed from 4-bit, 2-bit, and 1-bit full-adders ICs by cascading several packages. The output carry from one package must be connected to the input carry of the one with the next higher-order bits. The 4-bit full-adders is a typical example of an MSI function. It can be used in many applications involving arithmetic operations. The design of this circuit would require a truth trable with $2^9 = 512$ entries,since there are nine inputs. A decimal parallel adder that adds n decimal digits needs n BCD adder stages. The output carry from one stage must be connected to the input carry of the next higher order stages.

Half-subtractor

A half- subtractor is a combinational circuit that subtracts two bits and produces their difference, and borrow. In a half subtractor if minuend bit is designnate the by A and the subtrahend bit by B, than difference (D) and Borrow (B_o)can be calculated as

$$D = A - B$$

The truth table for the input-output relationships of a half-subtractor can now be derived as follows in Fig. 4.38 and its Truth Table is given by Table 4.8. Here A and B are the two inputs and D (difference) and B_o (borrow)are the two outputs.

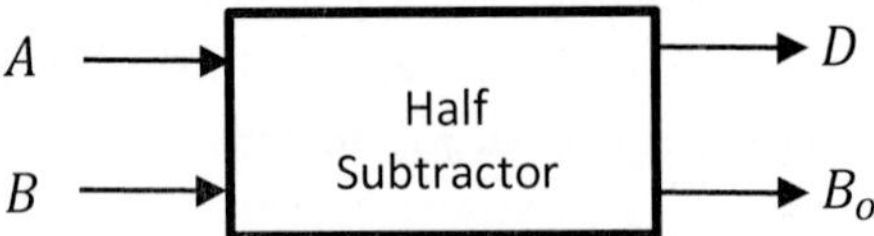

Fig. 4.38:Functional Block Diagram of a Half-subtractor

Table 4.8: Truth Table of a Half-subtractor

Input		Output	
A	*B*	*D*	B_o
0	0	0	0
0	1	1	1

1	0	1	0
1	1	0	0

The Boolean functions for the two outputs of the half-subtractor are derived from the truth table:

$$D = \bar{A}B + A\bar{B}$$
$$B_o = \bar{A}B$$

Using the above equation for difference and borrow the logic circuit can be designed and shown in Fig. 4.39. It is intresting to note that the logic for *D* is exactly the same as the logic for output *S* in the half-adder.

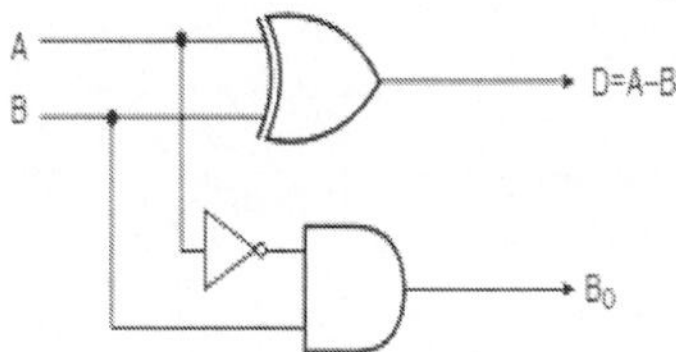

Fig. 4.39: Logic diagram of full-subtractor.

<u>*Full- Subtractor*</u>

A Full- subtractor is a combunational circuit that performs a subtraction between two bits, taking into account that a 1 may have been borrowed by a lower significant stage. This circuit has three inputs and two otputs. The three inpumts A, B, and B_{in}, demote the minuend, subtrahend, and previous borrow, respectively. The two outputs , D and B_o, represent the difference and output borrow, respectively. The block diagram and truth table for the full subtractor is given by *Fig.* and *Table* respectively.

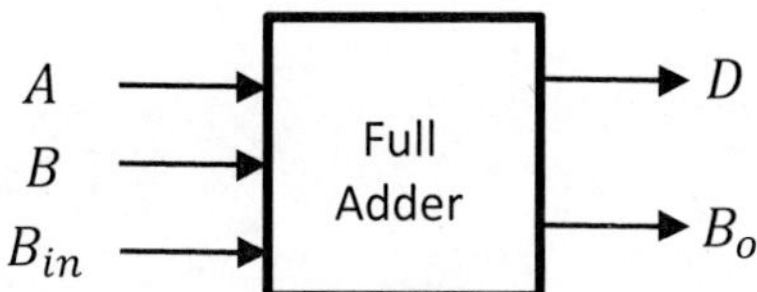

Fig.4.40: Functional Block Diagram of Full Subtractor

Table 4.9: Truth Table of Full Subtractor

Input			Output	
A	B	B_{in}	B_o	D
0	0	0	0	0
0	0	1	1	1
0	1	0	1	1
0	1	1	1	0
1	0	0	0	1
1	0	1	0	0
1	1	0	0	0
1	1	1	1	1

The Boolean expression for the two output variables are given by the equations

$$D = \bar{A}.\bar{B}.B_{in} + \bar{A}.B.\bar{B}_{in} + A.\bar{B}.\bar{B}_{in} + A.B.B_{in}$$

$$= \bar{A}.(\bar{B}.B_{in} + B.\bar{B}_{in}) + A.(\bar{B}.\bar{B}_{in} + B.B_{in}) = \bar{A}.(B_{in} \oplus B) + A.(B_{in} \odot B)$$

$$= \bar{A}.(B_{in} \oplus B) + A.\overline{(B_{in} \oplus B)} = A \oplus B \oplus B_{in}$$

$$B_0 = \bar{A}.\bar{B}.B_{in} + \bar{A}.B.\bar{B}_{in} + \bar{A}.B.B_{in} + A.B.B_{in} = \bar{A} \cdot B + \bar{A} \cdot B_{in} + B \cdot B_{in}$$

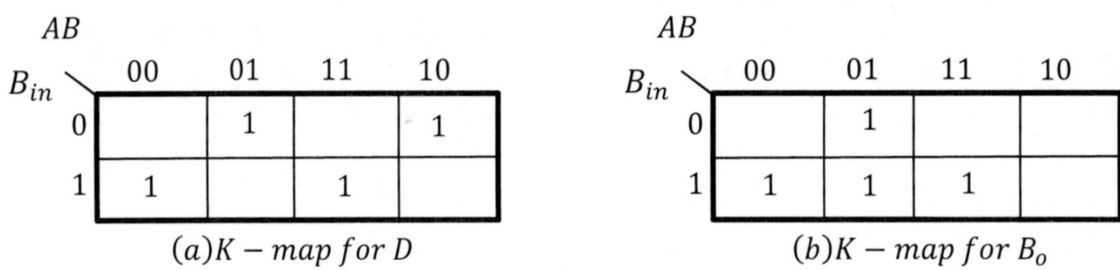

B_{in} \ AB	00	01	11	10
0		1		1
1	1		1	

(a) K − map for D

B_{in} \ AB	00	01	11	10
0		1		
1	1	1	1	

(b) K − map for B_o

Fig.4.41: Karnaugh maps for Difference (D) and Borrow (B_o) outputs.

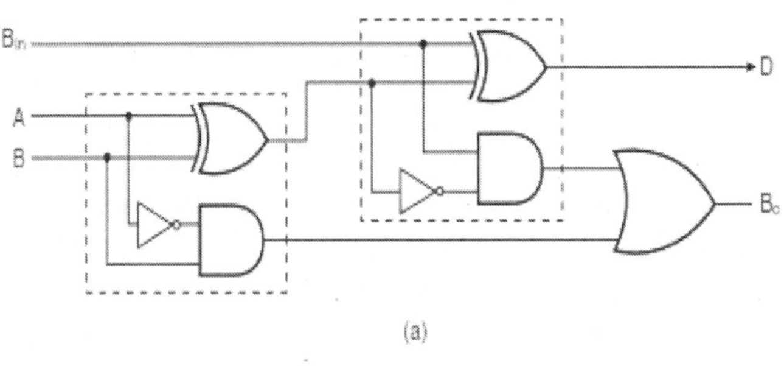

(a)

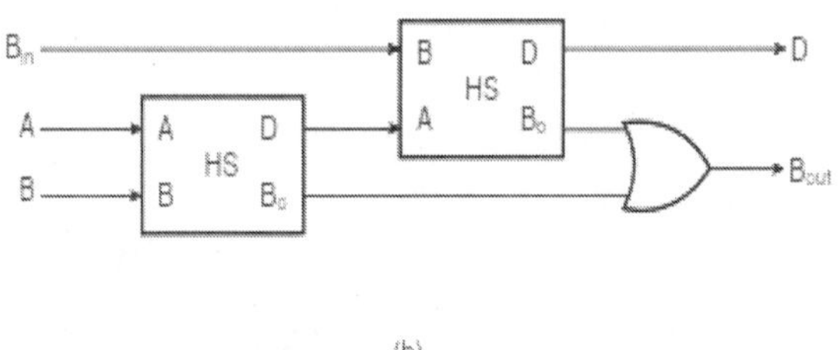

(b)

Fig. 4.42: Logic implementation of a full-subtractor with half-subtractors.

If we compare these expression in case of a full adder, then we find that the expression for difference (D) is the same as that for the sum (S) output. Also, the expression for BORROW output B_o is similar to the expression for CARRY-OUT C_n . In the case of half-subtractor, the *A* input is complemented. By a similar analysis it can be shown that a full subtractor can be implemented with half-subtractors in the same way as a full adder was constructed using half-adders.so more than one full subractor can be connected in casecade to perform subtraction on two larger binary numbers such as four-bit subtraction as shown inFig. 4.43.

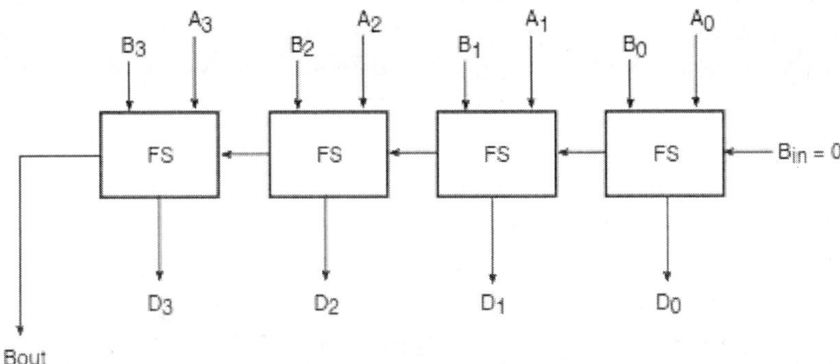

Fig. 4.43: Four-bit subtractor.

Adder-Subtractor:

In Digital Circuits, A **Binary Adder-Subtractor** is one which is capable of both addition and subtraction of binary numbers in one circuit. The operation being performed depends upon the binary value the control signal holds. It is one of the components of the ALU (Arithmetic Logic Unit). This Circuit shown in Figure below (Refer Fig. 4.44).

Let's consider two 4-bit binary numbers A and B as inputs to the Digital Circuit for the operation with digits $A_0A_1A_2A_3$ for A and $B_0B_1B_2B_3$ for B.

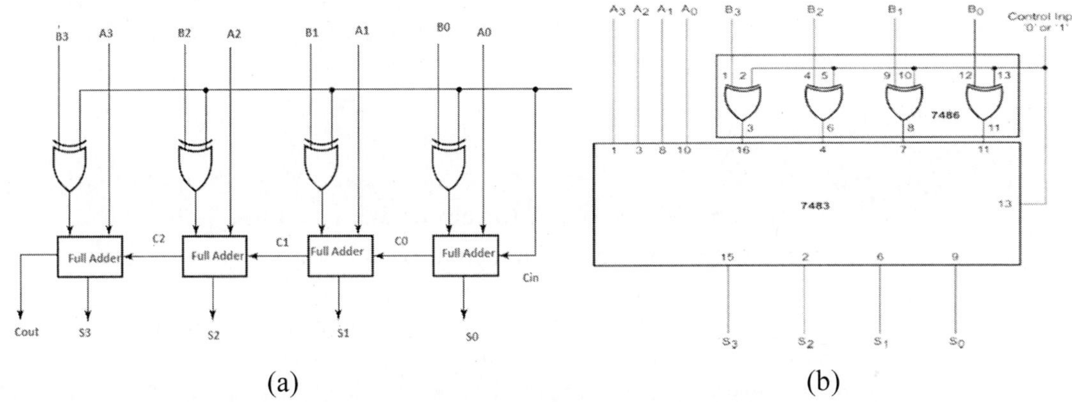

(a) (b)

Fig.4.44: Four-bit Binary Adder-Subtractor (a) Functional Block diagram (b) Using ICs.

As shown in the Fig. 4.44, the first full adder has control line given by input K.

- When the mode input (K) is at a low logic, i.e. '0', the circuit act as an adder and when the mode input is at a high logic, i.e. '1', the circuit act as a subtractor. The exclusive-OR gate connected in series receives input K and one of the inputs B, When, K is at a low logic, we have $B \oplus 0 = B$. The full-adders receive the value of B, the input carry is 0, and the circuit performs A plus B.
- When K is at a high logic, we have $B \oplus 1 = B'$ and $C0 = 1$. The B inputs are complemented, and a 1 is added through the input carry. The circuit performs the operation A plus the 2's complement of B.

- A commonly used four-bit adder in the TTL family is 7483. Also, type number 7486 is a quad two-input EX-OR gate in the TTL family. Fig. 4.44 shows a four-bit binary adder-subtractor circuit implemented with 7483 and 7486. Two each of 7483 and 7486 can be used to construct an eight-bit adder-subtractor circuit.

Concept of Overflow

- When two numbers with n digits each are added and the sum is a number occupying n + 1 digits, we say that an overflow occurred.
- Overflow is a problem in digital computers because the number of bits that hold the number is finite and a result that contains n + 1 bits cannot be accommodated by an n - bit word. For this reason, many computers detect the occurrence of an overflow, and when it occurs, a corresponding flip-flop (cell) is set that can then be checked by the user.
- The detection of an overflow after the addition of two binary numbers depends on whether the numbers are considered to be signed or unsigned. When two unsigned numbers are added, an overflow is detected from the end carry out of the most significant position.

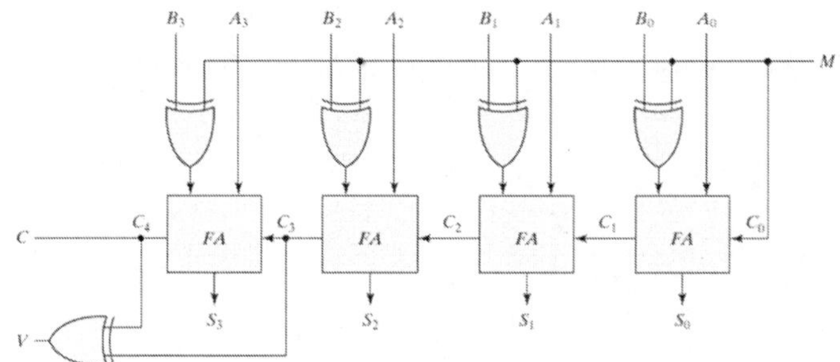

Fig. 4.45: Binary adder–subtractor circuit with overflow detection

- The binary adder–subtractor circuit with outputs C and V is shown in figure above (Refer Fig. 4.45).
    - If the two binary numbers are considered to be unsigned, then the C bit detects a carry after addition or a borrow after subtraction.
    - If the numbers are considered to be signed, then the V bit detects an overflow.

Example 4.19: Let's take two 3-bit numbers A=010 and B=011 and input them in the full adder with both values of control lines (Refer Fig. 4.45). Find the output for sum as well as Difference

Solution:

- For K=0:
    - $B_0 \oplus K = B_0$ *and* $C_0 = K = 0$
    - Thus, from first full adder
        - $A_0 + B_0 = 0 + 1 = 1,$ *hence* $S_0 = 1, C_1 = 1$
    - Similarly
        - $S_1 = 0, C_2 = 1$
        - $S_2 = 1, C_3 = 0$

Thus

$$A = 010 = 2$$
$$B = 011 = 3$$
$$Sum = 0101 = 5$$

- For K=1
  - $B_0 \oplus K = \overline{B_0}$ and $C_0 = K = 1$
    - Thus, from first full adder
      - $A_0 + \overline{B_0} + C_0 = 0 + 1 = 1,$ $hence$ $C_1 = 0$
  - Similarly
      - $S_1 = 1, C_2 = 0$
      - $S_2 = 1, C_3 = 0$

Thus

$$A = 010 = 2$$
$$B = 011 = 3$$
$$Difference = 1101 = -1$$

4.2.3 BCD adder and BCD arithmetic

BCD stands for binary coded decimal. A BCD adder is used to perform the addition of BCD numbers. A **BCD** digit can have any of the ten possible four-bit binary representations, represented by the Table 4.10.

Table 4.10: BCD representation

Binary	BCD	Binary	BCD
0000	0	0101	5
0001	1	0110	6
0010	2	0111	7
0011	3	1000	8
0100	4	1001	9

The only condition is when any number goes greater than nine (9) i.e., 1001, six (6), i.e., 0110 has been added to the number. Consider below example. However, a complete BCD adder is shown in Fig. 4.46. Fig. 4.46 (a) shows single digit BCD adder whereas Fig. 4.46 (b) represents three-digit BCD adder. Each 4-bit binary sequence (0000 to 1001) represents a BCD digit.

Decimal	Binary		
$\mathbf{7_{10}}$	0111		
$\mathbf{8_{10}}$	1000		
$\mathbf{15_{10}}$	1111		
	0110	$(+6)_{10}$	(Add 6 to convert to BCD)
	0001 0101		

***Note:** If the sum of two numbers is less than or equal to 9, then the value of BCD sum and binary sum will be same otherwise they will differ by 6 (0110 in binary). Now, let's move to the table and find out the logic when we are going to add "0110".*

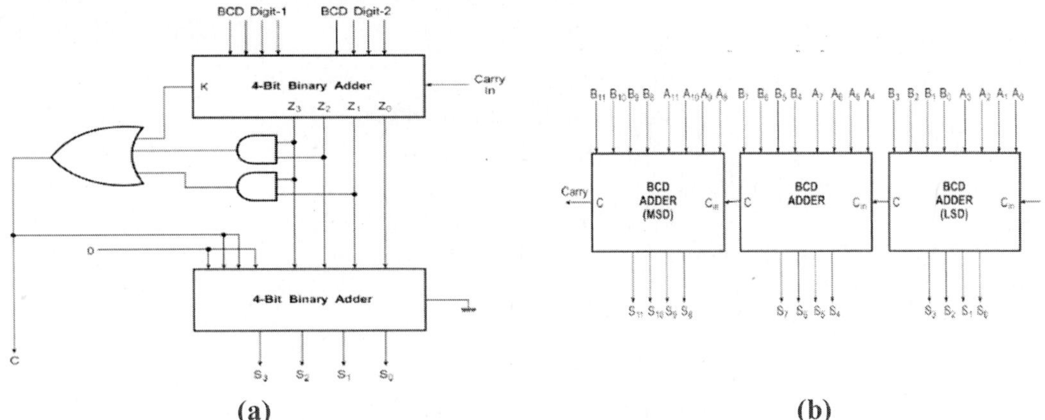

(a) (b)

Fig. 4.46: BCD adder (a) Single digit BCD adder (b) Three digit BCD adder.

Table 4.11: Results in binary and the expected results in BCD using a four-bit binary adder to perform addition of two BCD digits

Decimal	Binary sum					BCD sum				
sum	K	Z_3	Z_2	Z_1	Z_0	C	S_3	S_2	S_1	S_0
0	0	0	0	0	0	0	0	0	0	0
1	0	0	0	0	1	0	0	0	0	1
2	0	0	0	1	0	0	0	0	1	0
3	0	0	0	1	1	0	0	0	1	1
4	0	0	1	0	0	0	0	1	0	0
5	0	0	1	0	1	0	0	1	0	1
6	0	0	1	1	0	0	0	1	1	0
7	0	0	1	1	1	0	0	1	1	1
8	0	1	0	0	0	0	1	0	0	0
9	0	1	0	0	1	0	1	0	0	1
10	0	1	0	1	0	1	0	0	0	0
11	0	1	0	1	1	1	0	0	0	1
12	0	1	1	0	0	1	0	0	1	0
13	0	1	1	0	1	1	0	0	1	1
14	0	1	1	1	0	1	0	1	0	0
15	0	1	1	1	1	1	0	1	0	1
16	1	0	0	0	0	1	0	1	1	0
17	1	0	0	0	1	1	0	1	1	1
18	1	0	0	1	0	1	1	0	0	0
19	1	0	0	1	1	1	1	0	0	1

We can get the correct BCD sum and the desired carry output too. The B oolean expression with necessary correction is written as

$$C = K + Z_3.Z_2 + Z_3.Z_1$$

Equation represent the correction needs to be applied whenever $K = 1$. This takes care of the last four entries, Also, a correction needs to be applied whenever both Z_3 *and* Z_2 are '1'. This takes care of the next four entries from the bottom, corresponding to a decimal sum equal to 12,13,14, and 15.

For the remaining two entries corresponding to a decimal sum equal to 10 and 11, a correction is applied for both Z_3 and Z_1, being '1'. While hardware-implementing, 0110 can be added to the binary sum output with the help of a second four-bit binary adder.The correction logic should ensure that 0110 gets added only when the above expression is satisfied. Otherwise, the sum output of the first binary adder should be passed on as such to the final output, which can be accomplished by adding 0000 in the second adder.

Fig. 4.46 (a) shows the logic arrangement of a BCD adder capable of adding two BCD digits with the help of two four-bit binary adders and some additional combinational logic.

Example 4.20 For a half-adder circuit of Fig. 4.47 (a), the input applied at A and B are as shown in Fig. 4.47 (b), now plot the corresponding SUM (S) and CARRY (C) outputs on the same scale.
Solution

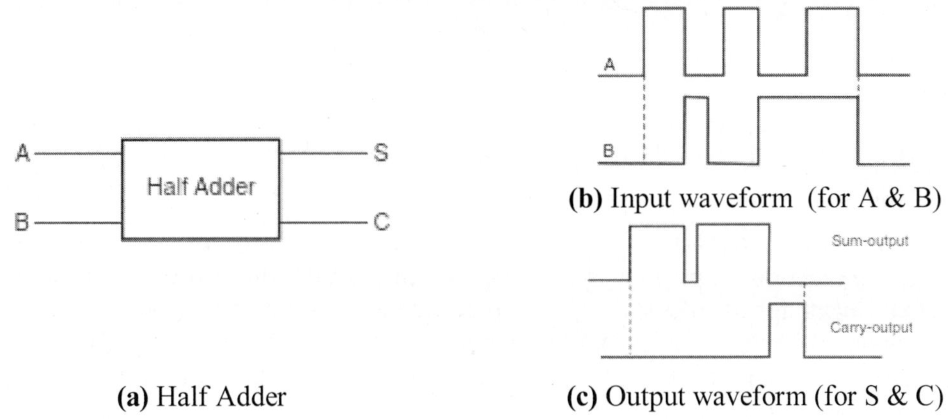

(b) Input waveform (for A & B)

(a) Half Adder **(c)** Output waveform (for S & C)

Fig. 4.47: Example 4.21

The SUM and CARRY waveforms can be plotted from our knowledge of the truth table of the half-adder. All that we need to remember to solve this problem is that 0 + 0 yields a '0' as the SUM output and a '0' as the CARRY. 0 + 1 or 1 + 0 yields '1' as the SUM output and '0' as the CARRY. 1 +1 produces a '0' as the SUM output and a '1' as the CARRY. The output waveforms are as shown in *Fig.* (c).

Example 4.21: Design a BCD adder circuit capable of adding BCD equivalents of two-digit decimal numbers. Indicate the IC type numbers used if the design has to be TTL logic family compatible.

Solution:

The desired BCD adder is a cascaded arrangement of two stages of the type of BCD adder discussed in the previous pages.

- A cascades arrangement of two such stages, where the output C of (CARRY-OUT) is fed to the CARRY-IN of the second stages, is shown in Fig. 4.48.
- In terms of IC type numbers, IC 7483 can be used four-bit-binary adders, IC 7408 can be used for implementing the required four two-input AND gates (IC 7408 is a quad two-input AND) and IC 7432 can be used to implement the required two three-input OR gate.
- IC 7432 is a quad two-input OR. Two two-input OR gates can be connected in cascade to get a three-input OR gate.

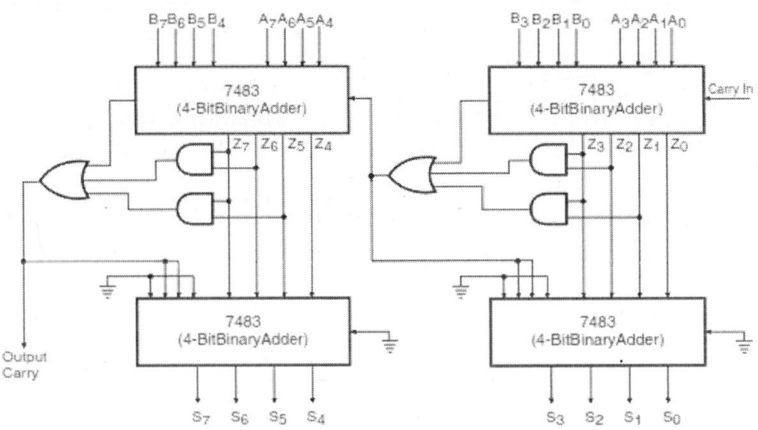

Fig. 4.48: Example 4.21

4.2.4 Carry look ahead adder

Carry Look-ahead Adder is the faster adder circuit. It reduces the propagation delay, which occurs during addition. In parallel adders, carry output of each full adder is given as a carry input to the next higher-order state. Hence, these adders it is not possible to produce carry and sum outputs of any state unless a carry input is available for that state. So, for computation to occur, the circuit has to wait until the carry bit propagated to all states. This induces carry propagation delay in the circuit.

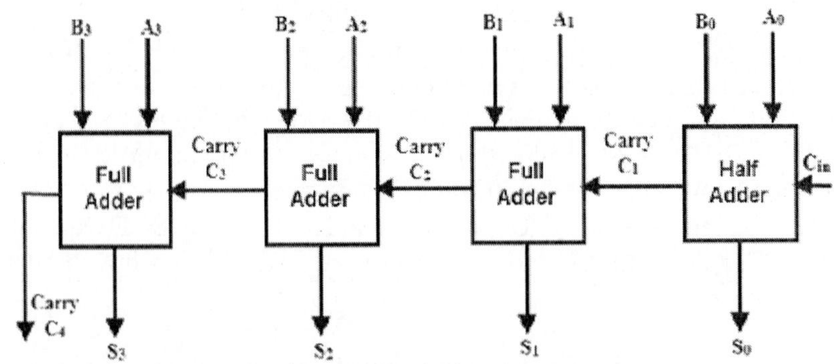

Fig. 4.49: Four-bit-ripple-carry adder.

Consider the 4-bit ripple carry adder circuit as shown in Fig. 4.49. Here the sum S_3 can be produced as soon as the inputs A_3 and B_3 are given. But carry C_3 cannot be computed until the carry bit C_2 is applied whereas C_2 depends on C_1. Therefore, to produce final steady-state results, carry must propagate through all the states. This increases the carry propagation delay of the circuit. The propagation delay of the adder is calculated as "the propagation delay of each gate times the number of stages in the circuit". For the computation of a large number of bits, more stages have to be added, which makes the delay much worse. Hence, to solve this situation, Carry Look-ahead Adder was introduced. This approach employs a *carry look-ahead* which solves this problem by calculating the carry signals in advance, based on the input signals.

This type of adder circuit is called a carry look-ahead adder. A carry look-ahead adder reduces the propagation delay by introducing more complex hardware. In this design, the ripple carry design is suitably transformed such that the carry logic over fixed groups of bits of the adder is reduced to two-level logic. Let us discuss the design in detail.

Table 4.12: Truth table for Full Adder

A	*B*	C_i	S_i	C_{i+1}	**Remark**
0	0	0	0	0	
0	0	1	1	0	No Carry Generate
0	1	0	1	0	
0	1	1	0	1	
1	0	0	1	0	No Carry Propogate
1	0	1	0	1	
1	1	0	0	1	Carry Generate
1	1	1	1	1	

Consider the full adder circuit shown in Fig. 4.50 below with corresponding truth table shown as Table 4.12.

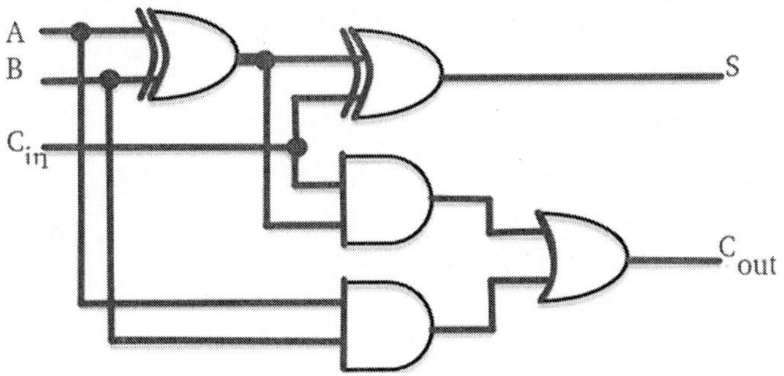

Fig. 4.50: A 2-Bit Full Adder

We define two variables as 'carry generate' $\boldsymbol{G_i}$ and 'carry propagate' $\boldsymbol{P_i}$ then,

$$P_i = A_i \oplus B_i$$
$$G_i = A_i . B_i$$

The sum output S_i, and carry output C_{i+1}, can be expressed in terms of carry generate $\boldsymbol{G_i}$ and 'carry propagate' $\boldsymbol{P_i}$

$$S_i = P_i \oplus C_i$$
$$C_{i+1} = G_i + P_i . C_i$$

From above expression of Sum and Carry a basic structure of a 4-bit look ahead carry adder can be design and developed as shown in Fig. 4.51.

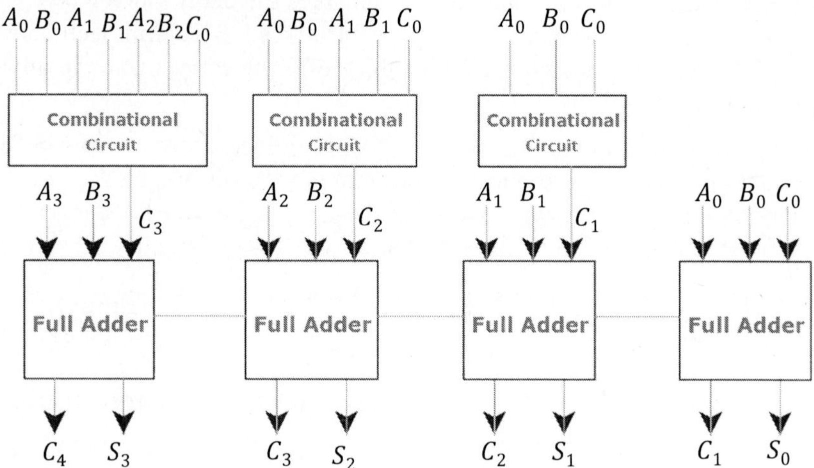

Fig. 4.51: A 4-bit Look ahead carry adder

In this adder, the carry input at any stage of the adder is independent of the carry bits generated at the independent stages. Here the output of any stage is dependent only on the bits which are added in the previous stages and the carry input provided at the beginning stage. Hence, the circuit at any stage does not have to wait for the generation of carry-bit from the previous stage and carry bit can be evaluated at any instant of time.

Using the various terms and their equations for as 'carry generate' $\boldsymbol{G_i}$, 'carry propagate' $\boldsymbol{P_i}$, the sum output S_i, and carry output C_{i+1} are given as below –

Therefore, the carry bits C_1, C_2, C_3, and C_4 can be calculated as

$$C_1 = C_0P_0 + G_0$$
$$C_2 = C_1P_1 + G_1 = (C_0P_0 + G_0)P_1 + G_1 = C_0P_0P_1 + G_0P_1 + G_1$$
$$C_3 = C_2P_2 + G_2 = (C_1P_1 + G_1)P_2 + G_2$$
$$C_4 = C_3P_3 + G_3 = (C_2P_2 + G_2)P_3 + G_3 = C_0P_0P_1P_2P_3 + G_0P_1P_2P_3 + G_1P_2P_3 + G_2P_3 + G_3$$

It can be observed from the equations that carry C_{i+1}only depends on the carry C_i, not on the intermediate carry bits. Hence, the above equations are implemented using two-level combinational circuits along with AND, OR gates, where gates are assumed to have multiple inputs (Refer Fig. 4.52).

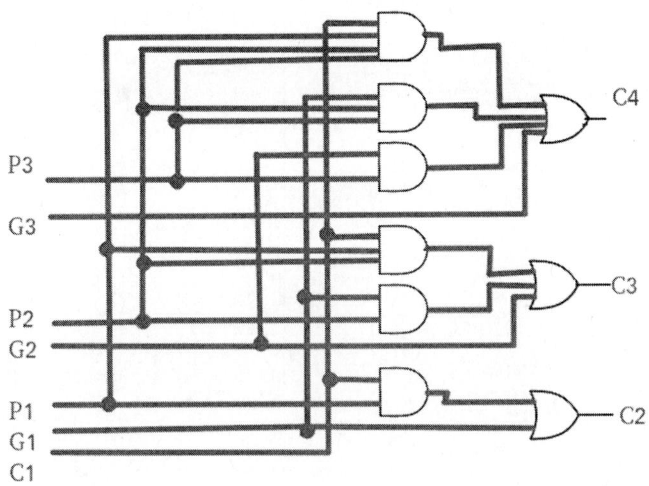

Fig. 4.52: Carry Output Generation Circuit of Carry Look-ahead Adder

The Carry Look-ahead Adder circuit for 4-bit is figure below (Refer Fig. 4.53).

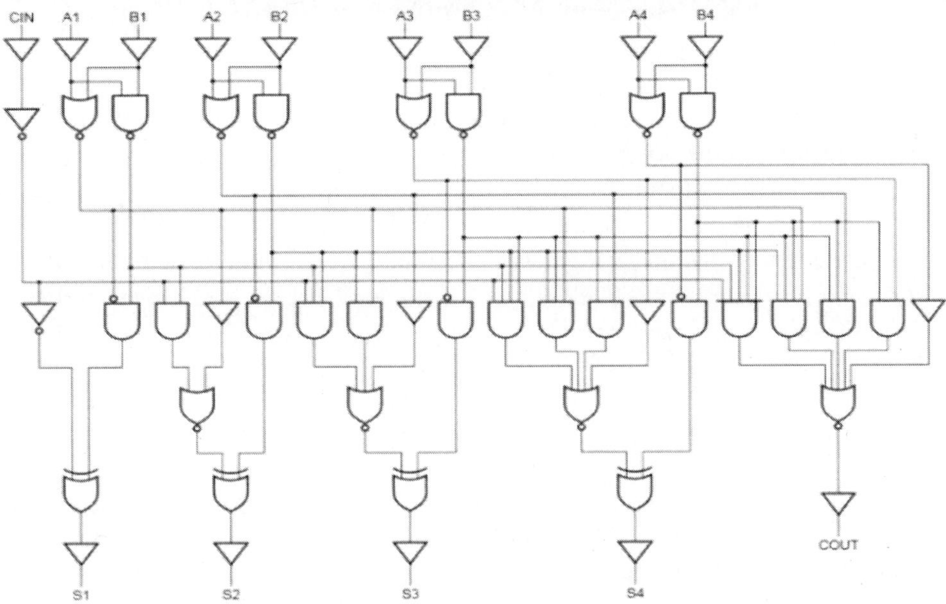

Fig. 4.53: 4-bit-Carry-Look-ahead-Adder-Circuit-Diagram

8-bit and 16-bit Carry Look-ahead Adder circuits can be designed by cascading the 4-bit adder circuit with carry logic. A 4-bit adder with look-ahead carry is shown in Fig. 4.54. Thus, we see that by generating all the individual carry terms needed by each full-adder in a 2-level circuit, the propagation delay time through the adder is considerably reduced and it becomes independent of the number of full-adders in the circuit.

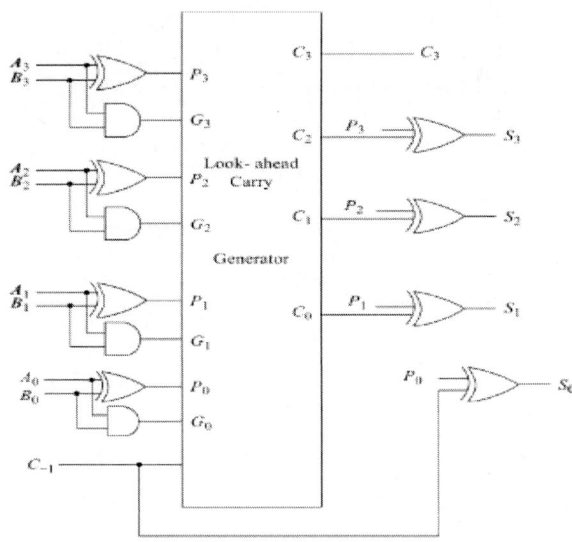

Fig. 4.54: A 4-bit Adder with Look-Ahead Carry

Advantages and Disadvantages of Carry Look-Ahead Adder:

Advantages:

- The propagation delay is reduced, as the carry output at any stage is dependent only on the initial carry bit of the beginning stage.
- It provides the fastest addition logic used for computation.
- Using this adder, it is possible to calculate the intermediate results.
- The increase in the number of gates is also moderate when used for higher bits.
- By combining two or more address calculations of higher bit Boolean functions can be done easily.

Disadvantages:

- The Carry Look-ahead adder circuit gets complicated as the number of variables increase.
- The circuit is costlier as it involves a greater number of hardware.

Applications

- For this Adder there is a trade-off between area and speed. When used for higher bit calculations, it provides high speed but the complexity of the circuit is also increased thereby increasing the area occupied by the circuit. This adder is usually implemented as 4-bit modules which are cascaded together when used for higher calculations. This adder is costlier compared to other adders.
- For Boolean computation in computers, adders are being used regularly. Charles Babbage implemented a mechanism for anticipating the carry bit in computers, to reduce the delay caused by the ripple carry adders. While designing a system, the speed of computation is the highest deciding factor for a designer.

Note: *In 1957, Gerald B. Rosenberger patented the modern Binary Carry Look-ahead Adder. Based on the analysis of gate delay and simulation, experiments are being conducted to modify the circuit of this adder to make it even faster.*

4.3 Code-converters

Sometimes it becomes necessary to use the output of one system as the input to another. Hence, a conversion circuit must be inserted between the two systems if each uses different codes for the same information. Thus, a converter circuit is used that makes the two systems compatible even though each uses a different binary code. These converters, which convert one code to other code are called as **code converters**. These code converters basically consist of Logic gates. To convert from binary code A to binary code B, the input lines must supply the bit combination of elements as specified by code A and the output lines must generate the corresponding bit combination of code B. To perform such operation combinational circuit are utilized. The design procedure of code converters will be illustrated by means of a specific example of conversion. Like, Binary to Gray code converter, BCD to excess-3 code etc.

Example 4.22: Design a binary to Gray code convertor.

Solution:

Let us implement a converter, which converts a 4-bit binary code $WXYZ$ into its equivalent Gray code $ABCD$.

Table 4.13: Truth table of a 4-bit binary code to Gray code converter.

Input Binary code				**Output Gray code**			
W	X	Y	Z	A	B	C	D
0	0	0	0	0	0	0	0
0	0	0	1	0	0	0	1
0	0	1	0	0	0	1	1
0	0	1	1	0	0	1	0
0	1	0	0	0	1	1	0
0	1	0	1	0	1	1	1
0	1	1	0	0	1	0	1
0	1	1	1	0	1	0	0
1	0	0	0	1	1	0	0
1	0	0	1	1	1	0	1
1	0	1	0	1	1	1	1
1	0	1	1	1	1	1	0
1	1	0	0	1	0	1	0
1	1	0	1	1	0	1	1
1	1	1	0	1	0	0	1
1	1	1	1	1	0	0	0

From Truth table (Refer Table 4.13), we can write the **Boolean functions** for each output bit of Gray code as below (see the position of 1's). write the equation of output $(ABCD)$ in terms of input $(WXYZ)$. Hence,

$$A = \sum m(8,9,10,11,12,13,14,15)$$
$$B = \sum m(4,5,6,7,8,9,10,11)$$
$$C = \sum m(2,3,4,5,10,11,12,13)$$
$$D = \sum m(1,2,5,6,9,10,13,14)$$

Let us simplify the above functions using 4 variable K-Maps. Fig. 4.55 shows the **4 variable K-Map** for simplifying **Boolean function** $A, B, C, and\ D$.

$AB \backslash CD$	$\bar{C}\bar{D}$	$\bar{C}D$	CD	$C\bar{D}$
$\bar{A}\bar{B}$	0	0	0	0
$\bar{A}B$	0	0	0	0
AB	1	1	1	1
$A\bar{B}$	1	1	1	1

(a) K-map for A

$AB \backslash CD$	$\bar{C}\bar{D}$	$\bar{C}D$	CD	$C\bar{D}$
$\bar{A}\bar{B}$	0	0	0	0
$\bar{A}B$	1	1	1	1
AB	0	0	0	0
$A\bar{B}$	1	1	1	1

(b) K-map for B

$AB \backslash CD$	$\bar{C}\bar{D}$	$\bar{C}D$	CD	$C\bar{D}$
$\bar{A}\bar{B}$	0	0	1	1
$\bar{A}B$	1	1	0	0
AB	1	1	0	0
$A\bar{B}$	0	0	1	1

(c) K-map for C

$AB \backslash CD$	$\bar{C}\bar{D}$	$\bar{C}D$	CD	$C\bar{D}$
$\bar{A}\bar{B}$	0	1	0	1
$\bar{A}B$	0	1	0	1
AB	0	1	0	1
$A\bar{B}$	0	1	0	1

(d) K-map for D

Fig. 4.55: K-map of a 4-bit binary code to Gray code converter

By grouping the adjacent ones, for the above K-maps the output can be calculated as

$$A = W$$
$$B = W'X + WX' = W \oplus X$$
$$C = X'Y + XY' = X \oplus Y$$
$$D = Y'Z + YZ' = Y \oplus Z$$

The Fig. 4.56 shows the **circuit diagram** of 4-bit binary code to Gray code converter.

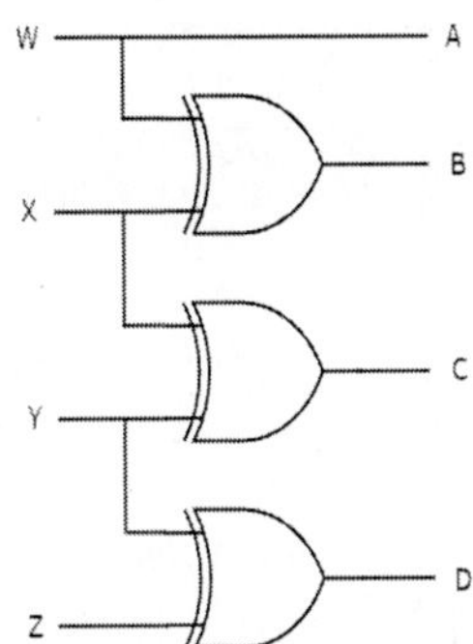

Fig. 4.56: 4-bit Binary code to Gray code converter

Since the outputs depend only on the present inputs, this 4-bit Binary code to Gray code converter is a combinational circuit. Similarly, you can implement other code converters.

Example 4.23: Design a BCD to excess-3 code converter.

Solution:

The bit combinations for the BCD to excess-3 codes are listed in table below. Since each code uses four bits to represent a decimal digit, there must be four input variables and four output variables. Let us consider the four input variables $W, X, Y, and\ Z$ and four output variables $A, B, C, and\ D$. The truth table relating the input and output variables is given below in Table 4.14.

We note that four binary variables may have 16-bit combinations, only 10 of which are listed in the truth table. The six-bit combinations not listed for the input variables are don't-care combinations. Since they will never occur, we are at liberty to assign to the output variables either a 1 or a 0, whichever gives a simpler circuit.

Table 4.14: Truth table for code-conversion example

Input BCD				**Output Excess-3 code**			
W	*X*	*Y*	*Z*	*A*	*B*	*C*	*D*
0	0	0	0	0	0	1	1
0	0	0	1	0	1	0	0
0	0	1	0	0	1	0	1
0	0	1	1	0	1	1	0
0	1	0	0	0	1	1	1
0	1	0	1	1	0	0	0
0	1	1	0	1	0	0	1
0	1	1	1	1	0	1	0
1	0	0	0	1	0	1	1
1	0	0	1	1	1	0	0

The maps in Fig. 4.57 are drawn to obtain a simplified Boolean function for each output.

$$A = \sum m(5,6,7,8,9) + \sum d(10,11,12,13,14,15)$$

$$B = \sum m(1,2,3,4,9) + \sum d(10,11,12,13,14,15)$$
$$C = \sum m(0,3,4,7,8) + \sum d(10,11,12,13,14,15)$$
$$D = \sum m(0,2,4,6,8) + \sum d(10,11,12,13,14,15)$$

Each of the four maps represents one of the four outputs of this circuits as a function of the four input variables. The six don't care combinations are marked by X's. one possible way to simplify the functions in sum of products is listed under the map of each variable.

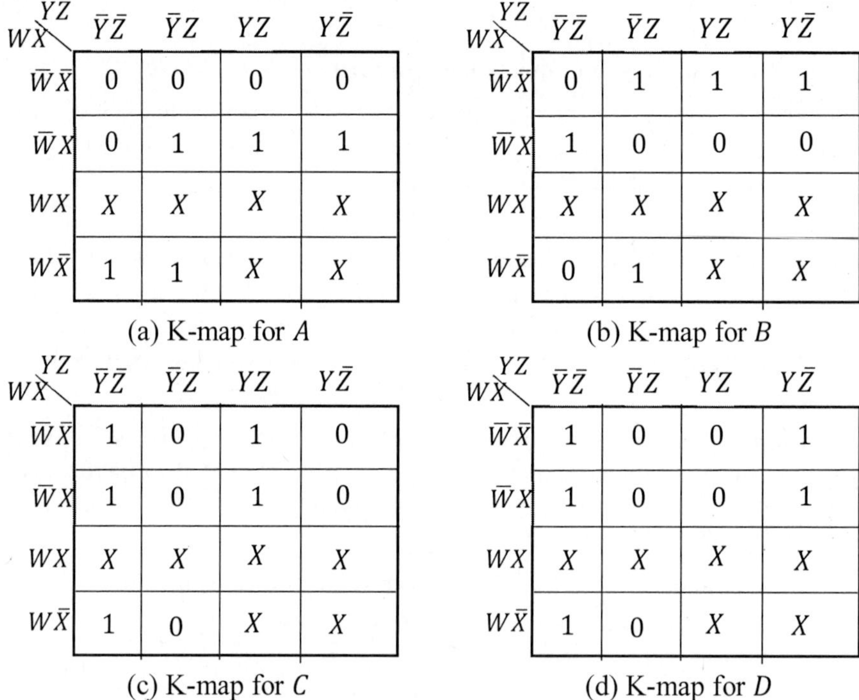

(a) K-map for A (b) K-map for B

(c) K-map for C (d) K-map for D

Fig. 4.57: K-maps for BCD-to-Excess-3 code converter.

A two-level logic diagram may be obtained directly from the Boolean expression derived by the maps. There are various other possibilities for a logic diagram that implements this circuit. The expressions obtained may be manipulated algebraically for the purpose of using common gates for two or more outputs. This manipulation, shown below, illustrates the flexibility obtained with multiple-output systems when implemented with three or more levels of gates.

$$A = W + XY + YZ = W + Y(X + Z)$$
$$B = X'Y + X'Z + XY'Z' = X'(Y + Z) + X(Y + Z)'$$
$$C = YZ + Y'Z'$$
$$D = Z'$$

The logic diagram that implements the above expression is shown in Fig. 4.58.

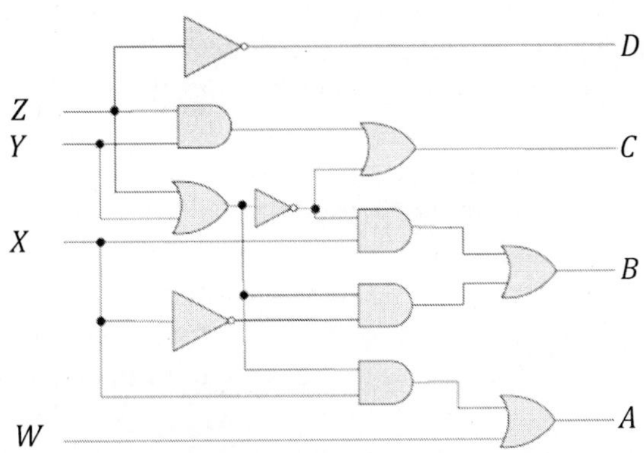

Fig. 4.58: Logic diagram for BCD-to-excess-3 code converter.

4.4 Arithmetic Logic Unit (ALU)

The arithmetic logic unit (ALU) is a digital building block capable of performing both arithmetic as well as logic operations. Arithmetic and Logic Units (or ALUs) are found at the core of microprocessors, where they implement the arithmetic and logic functions offered by the processor (e.g., addition, subtraction, AND'ing two values, etc.). An ALU is a combinational circuit that combines many common logic circuits in one block. Typically, ALU inputs are comprised of two N-bit busses, a carry-in, and M select lines that select between the 2^M ALU operations as shown in Fig. 4.59. ALU outputs include an N-bit bus for function output and a carry out.

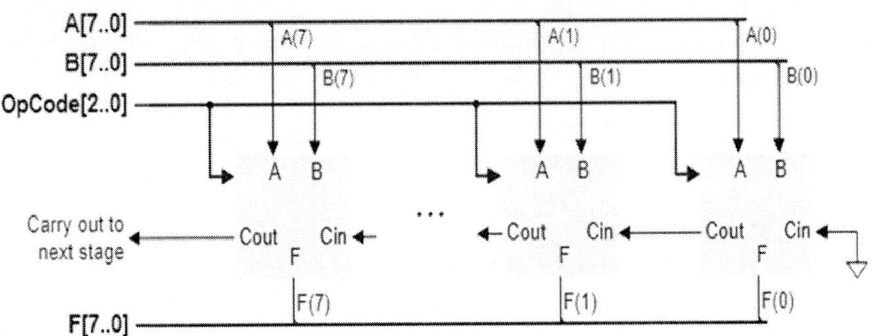

Fig. 4.59: Block diagram of ALU composed of bit-sliced blocks.

ALUs can be designed to perform a variety of different arithmetic and logic functions. Possible arithmetic functions include addition, subtraction, multiplication, comparison, increment, decrement, shift, and rotate; possible logic functions include AND, OR, XOR, XNOR, INV, CLR (for clear), and PASS (for passing a value unchanged). All of these functions find use in computing systems, although a complete description of their use is beyond the scope of this document. An ALU could be designed to include all of these functions, or a subset could be chosen to meet the specific needs of a given application. Either way, the design process is similar (but simpler for an ALU with fewer functions).

- An ALU is a combinational circuit that combines many common logic circuits in one block. Typically, ALU inputs are comprised of two N-bit busses, a carry-in, and M select lines that select between the $2M$ ALU operations.
- The three control bits used to select the ALU operation are called the "operation code" (or op code), because if this ALU were used in an actual microprocessor, these bits would come from the "opcodes" (or machine codes) that form the actual low-level computer programming code.
- ALU design should follow the same process as other bit-slice designs: first, define and understand all inputs and outputs of a bit slice (i.e., prepare a detailed block diagram of the bit slice); second, capture the required logical relationships in some formal method (e.g., a truth table); third, find minimal circuits (by using K-maps or espresso) or write VHDL code; and fourth, proceed with circuit design and verification.

The function to be performed is selectable from *function select* pins. Some of the popular type numbers of ALU include 74181, 74381, 74382, 74582 (all from the TTL logic family) and 40181 (from the CMOS logic family). More than one such IC can always be connected in cascade to perform arithmetic and logic operations on larger bit numbers.

4.5 Digital comparator

A magnitude digital Comparator is a combinational circuit that **compares two digital or binary numbers** in order to find out whether one binary number is equal, less than, or greater than the other binary number. We logically design a circuit for which we will have two inputs one for A and the other for B and have three output terminals, one for A > B condition, one for A = B condition, and one for A < B condition as shown in Fig. 4.60.

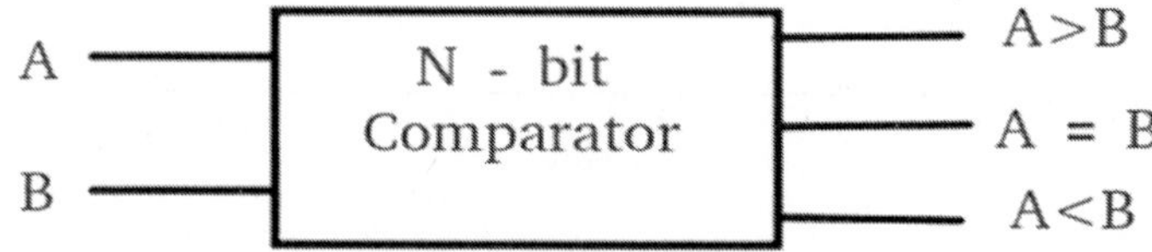

Fig. 4.60: Block diagram of N-bit comparator.

<u>*1-Bit Magnitude Comparator*</u>

A comparator used to compare two bits is called a single-bit comparator. It consists of two inputs each for two single-bit numbers and three outputs to generate less than, equal to, and greater than between two binary numbers. Truth table for a 1-bit comparator is given by Table 4.15.

Table 4.15: Truth table for a 1-bit comparator is given below:

A	B	$A < B$	$A = B$	$A > B$
0	0	0	1	0
0	1	1	0	0
1	0	0	0	1
1	1	0	1	0

From the above truth table logical expressions for each output can be expressed as follows:

$$A > B: AB'$$
$$A < B: A'B$$
$$A = B: A'B' + AB$$
$$A = B: [(A > B) + (A < B)]'$$

By using these Boolean expressions, we can implement a logic circuit for this comparator as shown in Fig. 4.61.

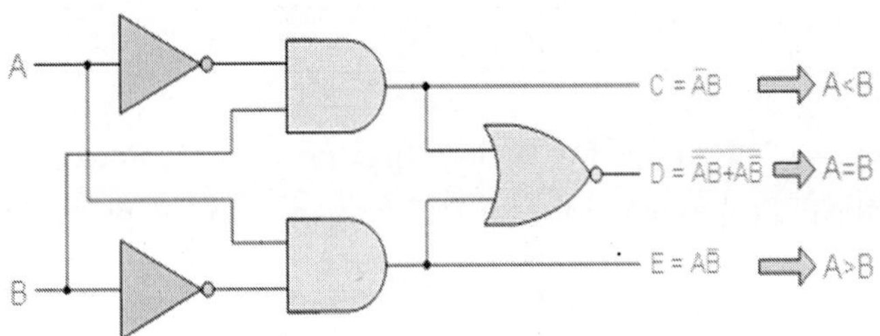

Fig. 4.61: Logic circuit for 1-bit magnitude comparator.

2-Bit Magnitude Comparator:

A comparator used to compare two binary numbers each of two bits is called a 2-bit Magnitude comparator. It consists of four inputs and three outputs to generate less than, equal to, and greater than between two binary numbers. The truth table for a 2-bit comparator is given by Table 4.16.

Table 4.16: Truth table for a 2-bit comparator

Input				Output		
A_1	A_0	B_1	B_0	$A < B$	$A = B$	$A > B$
0	0	0	0	0	1	0
0	0	0	1	1	0	0
0	0	1	0	1	0	0
0	0	1	1	1	0	0
0	1	0	0	0	0	1
0	1	0	1	0	1	0
0	1	1	0	1	0	0
0	1	1	1	1	0	0
1	0	0	0	0	0	1
1	0	0	1	0	0	1
1	0	1	0	0	1	0
1	0	1	1	1	0	0

Input				Output		
A_1	A_0	B_1	B_0	$A < B$	$A = B$	$A > B$
1	1	0	0	0	0	1
1	1	0	1	0	0	1
1	1	1	0	0	0	1
1	1	1	1	0	1	0

From the above truth table K-map for each output can be drawn as shown in Fig. 4.62.

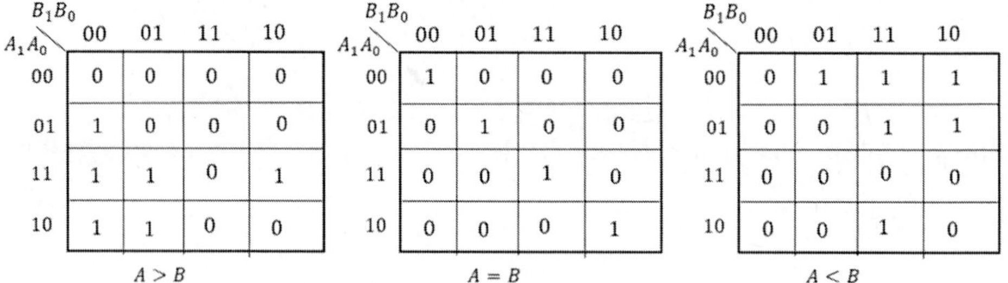

Fig. 4.62: K-maps for each output for a 2-bit comparator

From the above K-maps logical expressions for each output can be expressed as follows:

$$A > B: A_1\overline{B_1} + A_0\overline{B_1}\overline{B_0} + A_1A_0\overline{B_0}$$
$$A = B: (A_0 \odot B_0)\ (A_1 \odot B_1)$$
$$A < B: \overline{A_1}B_1 + \overline{A_0}B_1B_0 + \overline{A_1}\ \overline{A_0}B_0$$

By using these Boolean expressions, we can implement a logic circuit for this comparator as shown in Fig. 4.63.

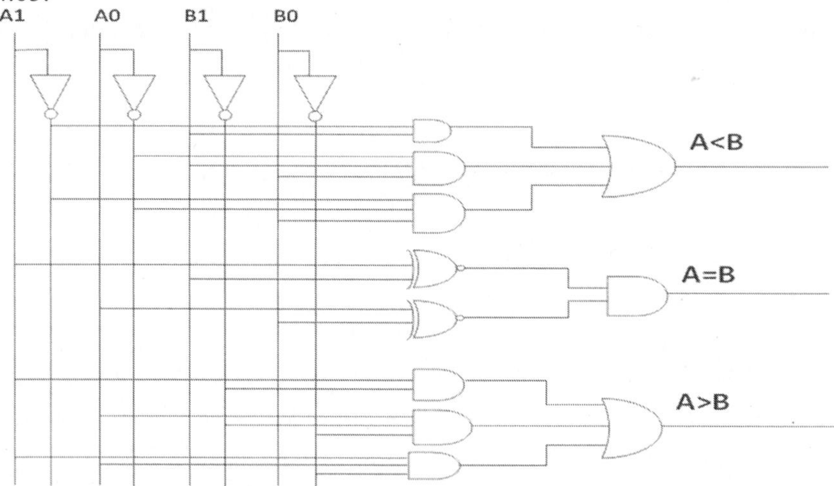

Fig. 4.63: Logic circuit for 2-bit comparator

4-Bit Magnitude Comparator:

A comparator is a logic circuit that used to compare two binary numbers. In case of 4-bit magnitude comparator, it consists of eight inputs (as both the numbers are 4-bit numbers, as given by Table 4.17) and three outputs (to generate three different result less than, equal to, and greater than between two binary numbers).

Table 4.17: Compare two binary numbers of 4 bit each

	Inputs	Outputs
A	$A_3A_2A_1A_0$	$A > B$
B	$B_3B_2B_1B_0$	$A = B$
		$A < B$

The 4-bit combinational logic circuit truth table (Refer Table 4.18) is given below and its logic diagram is shown in

Table 4.18: Truth table for a 4-bit magnitude comparator

Comparing Inputs				Outputs		
$\boldsymbol{A_3, B_3}$	A_2, B_2	A_1, B_1	A_0, B_0	$A > B$	$A = B$	$A < B$
$\boldsymbol{A_3 > B_3}$	d	d	d	1	0	0
$\boldsymbol{A_3 < B_3}$	d	d	d	0	1	0
$\boldsymbol{A_3 = B_3}$	$A_2 > B_2$	d	d	1	0	0
$\boldsymbol{A_3 = B_3}$	$A_2 < B_2$	d	d	0	1	0
$\boldsymbol{A_3 = B_3}$	$A_2 = B_2$	$A_1 > B_1$	d	1	0	0
$\boldsymbol{A_3 = B_3}$	$A_2 = B_2$	$A_1 < B_1$	d	0	1	0
$\boldsymbol{A_3 = B_3}$	$A_2 = B_2$	$A_1 = B_1$	$A_0 > B_0$	1	0	0
$\boldsymbol{A_3 = B_3}$	$A_2 = B_2$	$A_1 = B_1$	$A_0 < B_0$	0	1	0
$\boldsymbol{A_3 = B_3}$	$A_2 = B_2$	$A_1 = B_1$	$A_0 = B_0$	1	0	0
$\boldsymbol{A_3 = B_3}$	$A_2 = B_2$	$A_1 = B_1$	$A_0 = B_0$	0	1	0
$\boldsymbol{A_3 = B_3}$	$A_2 = B_2$	$A_1 = B_1$	$A_0 = B_0$	0	0	1

From Table 4.18 the Boolean functions can be derived for three different result less than, equal to, and greater than between two binary numbers, here $'d'$ represents don't care terms.

Case-1: $A = B$, the two numbers are said to be equal if all pairs of significant digits are equal. If,

$$A_3 = B_3, A_2 = B_2, A_1 = B_1, and\ A_0 = B_0$$

- When the numbers are binary, then the digits are either equal to 1 or 0, and the equality of each pair of bits can be stated logically with an XNOR function as:

  $A = B$: $\quad A_i \odot B_i = 1, \quad i.e., (A_0 \odot B_0).(A_1 \odot B_1).(A_2 \odot B_2).(A_3 \odot B_3) = 1$

Hence

$$A = B; when\ x_0.x_1.x_2.x_3 = 1$$

- The equation is represented in Fig. by using four X-NOR gates and one AND gate. The binary variable represents $A = B$, is equal to 1 if the given input numbers, A and B are same, otherwise is equal to 0.

Case-2: $A > B, or\ A < B$

- To illustrate whether A is greater than or less than B, we check the relative magnitudes of pairs of significant digits, starting from the most significant digit. If the two digits of a pair are same, we compare the next lower significant pair of digits. The comparison continues till a pair of unequal digits is reached.
- if the corresponding digit of A is equal to 1 and that of B is equal to 0, we conclude that A > B. The sequential comparison can be expressed logically by the Boolean function:

  $A > B$: $\quad A_3B_3' + x_3A_2B_2' + x_3x_2A_1B_1' + x_3x_2x_1A_0B_0'$

- If the corresponding digit of A is equal to 0 and that of B is equal to 1, we have A < B. The sequential comparison can be expressed logically by the Boolean function:

$$A < B:\ A_3'B_3 + x_3A_2'B_2 + x_3x_2A_1'B_1 + x_3x_2x_1A_0'B_0$$

- The terms (A > B) and (A < B) are binary output variables and they are equal to 1 when A > B and A < B, respectively.

The logic diagram of the 4-bit magnitude comparator is shown in the below diagram (Refer Fig. 4.64). The four x outputs ($x_0, x_1, x_2, and\ x_3$ are created with XNOR circuits and are applied to an AND gate to give the output binary variable (A = B) The other two outputs that represents $A > B$ and $A < B$ may generate according as per the discussion given above. This implementation is a multilevel implementation and it has a regular pattern.

By using these Boolean expressions, we can implement a logic circuit for this comparator as given below:

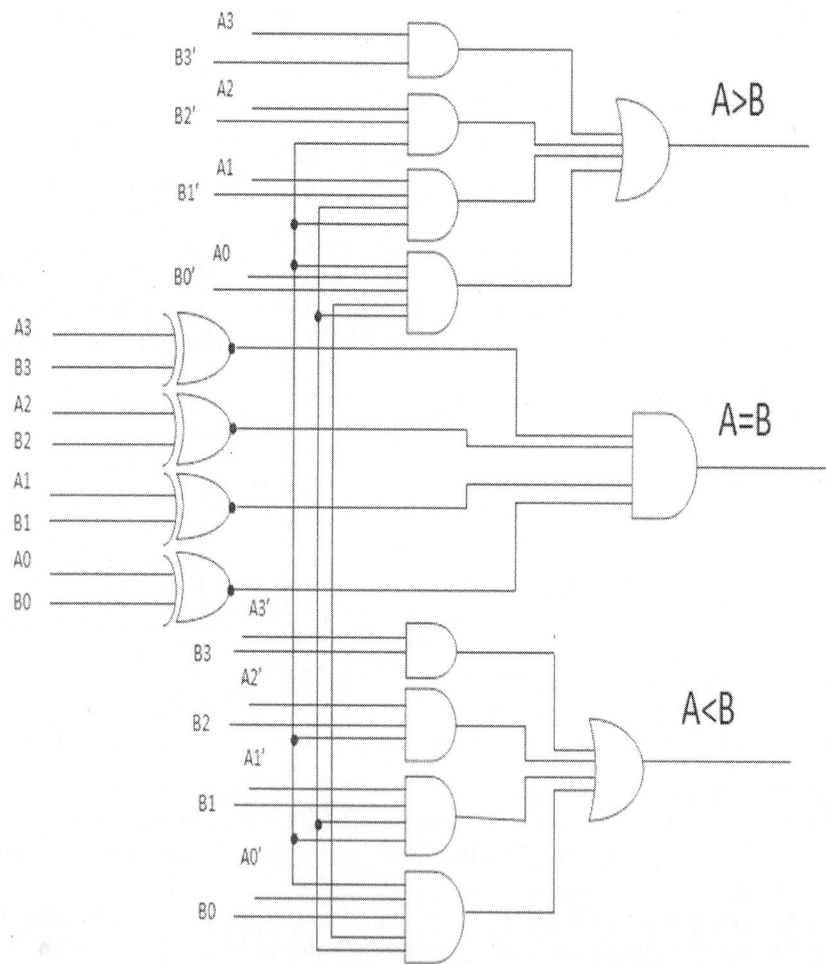

Fig. 4.64: Logic circuit for 4-bit comparator

Cascading Comparator:

A comparator performing the comparison operation to more than four bits by cascading two or more 4-bit comparators is called a cascading comparator. When two comparators are to be cascaded, the outputs of the lower-order comparator are connected to corresponding inputs of the higher-order comparator.

IC 74LS85 can be used to compare two 4-bit binary digits by grounding $I(A > B) and\ I(A < B)$, whereas $I(A = B)$ connector input is given to V_{CC} terminal as shown in Fig. 4.65. In addition to the ordinary comparator, this IC is provided with cascading inputs in order to enable the cascading several comparators. We can compare any number of bits by cascading several of these comparator ICs.

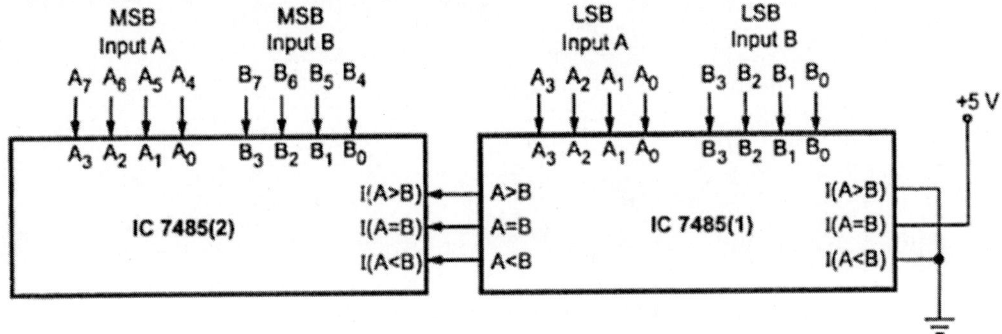

Fig. 4.65: 4-bit comparator IC

Applications of 4-bit Comparator:

- 4-bit comparators are mostly used in electronic devices such as central processing units (CPUs) and microcontrollers (MCUs).
- Comparators are commonly used in control applications in which the physical variables are representing with binary numbers such as temperature and position, etc. and they are compared with some standard value.
- Comparators are widely used as process controllers and for Servo motor control.
- 4-bit comparators are used in password verification and biometric applications.

NOTE: For $n - bit$ comparator, the number of combinations for which, $A = B\ is\ 2^n$ and for $A > B, or\ A < b\ is\ (2^{2n} - 2^n)/2$

Example 4.24 Design a two-bit magnitude comparator. Also, write relevant Boolean Expression.

Solution:

Let $A(A_1A_0)$ and $B(B_1B_0)$ are the two numbers. If $X, Y\ and\ Z$ represents the condition $A = B, A > B\ and\ A < B$ respectively. Hence, expression for $X, Y\ and\ Z$ can be written as follows:

Say

$$x_0 = A_0 \odot B_0,\quad and\ x_1 = A_1 \odot B_1$$

Hence,

$$A = B\ represent\ by\ X = x_1.x_0$$
$$A > B\ represent\ by\ Y = A_1.\overline{B_1} + x_1.A_0.\overline{B_0}$$
$$A < B\ represent\ by\ Z = \overline{A_1}B_1 + x_1.\overline{A_0}.B_0$$

Fig. 4.66 shows the logic diagram of the two-bit comparator.

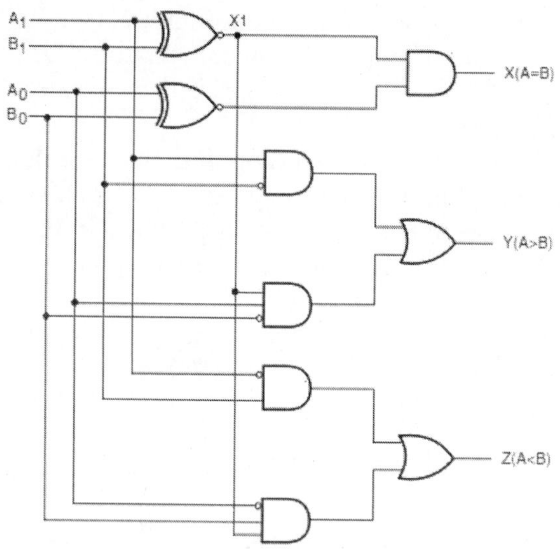

Fig. 4.66: Solution to example 4.24.

Example 4.25 Implement a 3-bit magnitude comparator having only one output that goes HIGH when the two three-bit numbers are equal (Use NAND gates only).

Solution:

For a 3-bit Comparator

$$x_0 = A_0 \odot B_0, \qquad x_1 = A_1 \odot B_1, \qquad x_2 = A_2 \odot B_2$$

Then

$$A = B \; represent \; by \; X = x_2.x_1.x_0$$

For two three-bit numbers to be equal given by the equation X and Fig. 4.67 shows the logic Diagram

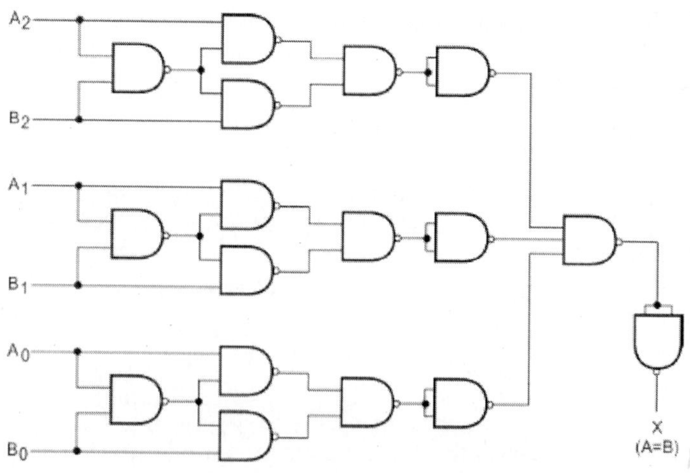

Fig. 4.67: Solution to example 4.25

4.6 Decoders

A decoder is a combinational circuit that converts binary information from n input lines to a maximum of 2^n unique output lines. If the n-bit decoded information has unused or don't-care combinations, the decoder output will have less than 2^n outputs.

The decoder presented here are called n-to-m line decoders where $m < 2^n$. their purpose is to generate the 2^n (or less) minterms of n input variables. As an example, consider let 2 to 4 Decoder has two inputs A_1 & A_0 and four outputs Y_3, Y_2, Y_1 & Y_0. The **block diagram** of 2 to 4 decoder is shown in Fig. 4.68 (a).

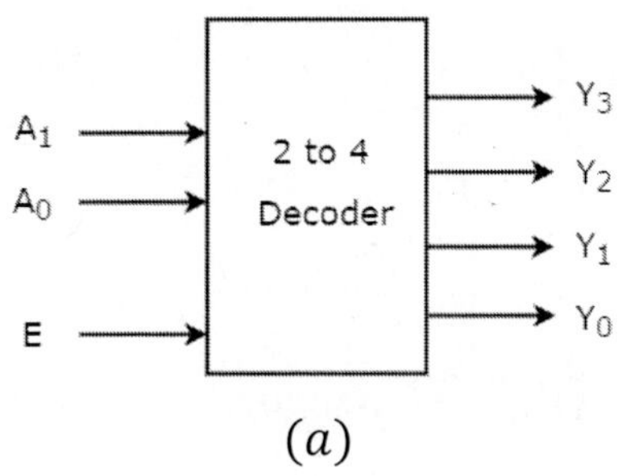

(a)

Enable	Inputs		Outputs			
E	A_1	A_0	Y_3	Y_2	Y_1	Y_0
0	x	x	0	0	0	0
1	0	0	0	0	0	1
1	0	1	0	0	1	0
1	1	0	0	1	0	0
1	1	1	1	0	0	0

(b)

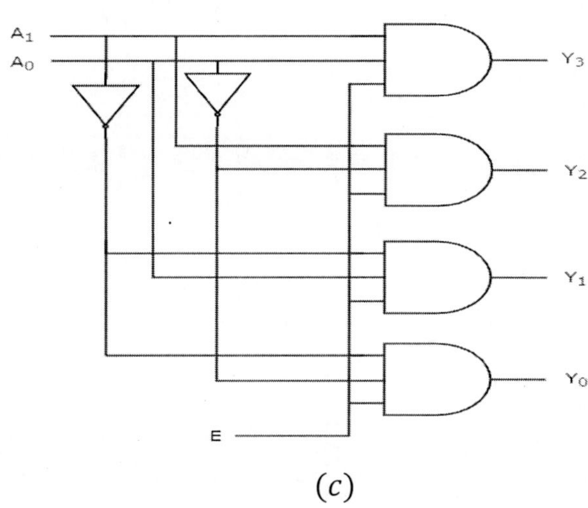

(c)

Fig. 4.68: 2 to 4 decoder

One of these four outputs will be '1' for each combination of inputs when enable, E is '1'. The **Truth table** of 2 to 4 decoder is shown in Fig. 4.68 (b). From Truth table, we can write the **Boolean functions** for each output as Each output is having one product term. So, there are four product terms in total. We can implement these four product terms by using four AND gates having three inputs each & two inverters. The **circuit diagram** of 2 to 4 decoder is shown in Fig. 4.68 (c).

Therefore, the outputs of 2 to 4 decoder are nothing but the **min terms** of two input variables A_1 & A_0, when enable, E is equal to one. If enable, E is zero, then all the outputs of decoder will be equal to zero. Similarly, 3 to 8 decoder produces eight min terms of three input variables A_2, A_1 & A_0 and 4 to 16 decoder produces sixteen min terms of four input variables A_3, A_2, A_1 & A_0.

Example 4.26: Design a 3-to-8-line decoder circuit using two 2-to-4-line decoder.

Solution:

- The three inputs are decoded into eight outputs, each output representing one of the minterms of the 3-input variables.
- A particular application of this decoder would be a binary-to-octal conversion. The input variables may represent a binary number, and the outputs will then represent the eight digits in the octal number system.
- However, a 3-to-8 Line decoder can be used for decoding any 3-bit code to provide eight outputs, one for each element of the code.
- Truth table shows only one output can be equal to 1 at any one time.
- The output line whose value is equal to 1 represents the minterms equivalent of the binary number presently available in the input lines shows in Table 4.19.

Table 4.19: Truth Table for 3-to-8-line decoder

Inputs			**Outputs**							
x	y	z	D_0	D_1	D_2	D_3	D_4	D_5	D_6	D_7
0	0	0	1	0	0	0	0	0	0	0
0	0	1	0	1	0	0	0	0	0	0
0	1	0	0	0	1	0	0	0	0	0
0	1	1	0	0	0	1	0	0	0	0
1	0	0	0	0	0	0	1	0	0	0
1	0	1	0	0	0	0	0	1	0	0
1	1	0	0	0	0	0	0	0	1	0
1	1	1	0	0	0	0	0	0	0	1

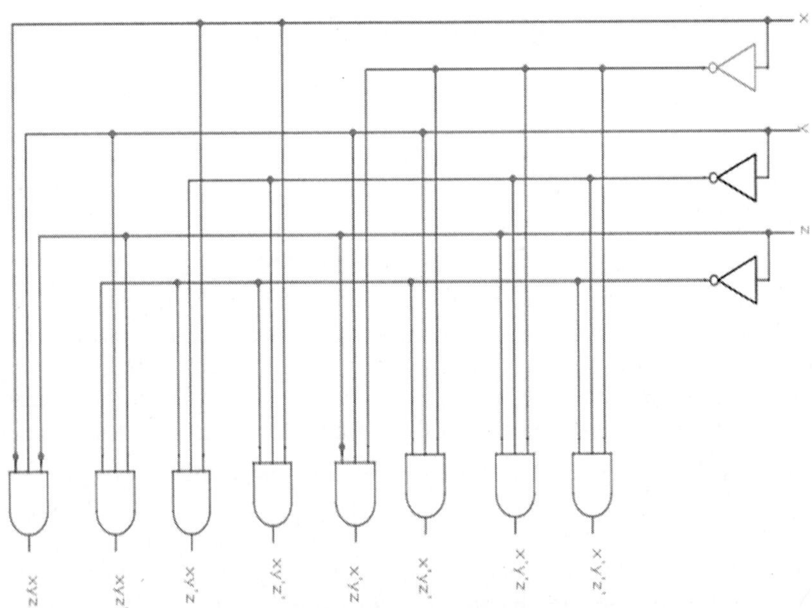

Fig. 4.69: A 3-to-8-line decoder.

A decoder provides the 2^n minterm of n input variables. Since any Boolean function can be expressed in sum of minterms canonical form, one can use a decoder to generate the minterms and an external OR gate to form the sum.so, any combinational circuit with n inputs and m outputs can be implemented with an n-to-2^n line decoder and m OR gates.

The procedure for implementing a combinational circuit by means of a decoder and OR gates requires that the Boolean functions for the circuit be expressed in sum of minterms. This form can be easily obtained from the truth table or by expanding the functions to their sum of minterms. A decoder is then chosen which generates all the minterm of the n inputs variables. The inputs to each OR gate are selected from the decoder outputs according to the minterm list in each function.

From above mentioned truth table 3 to 8 Decoder circuit can be designed as shown in **Fig. 4.69.**

Let us implement **3 to 8 Decoder** circuit using **2 to 4 decoders**. We know that 2 to 4 Decoder has two inputs, A_1 & A_0 and four outputs, Y_3 to Y_0. Whereas, 3 to 8 Decoder has three inputs A_2, A_1 & A_0 and eight outputs, Y_7 to Y_0. Therefore, we require two 2 to 4 decoders for implementing one 3 to 8 Decoder. The **block diagram** of 3 to 8 decoder using 2 to 4 decoders is shown in Fig. 4.70.

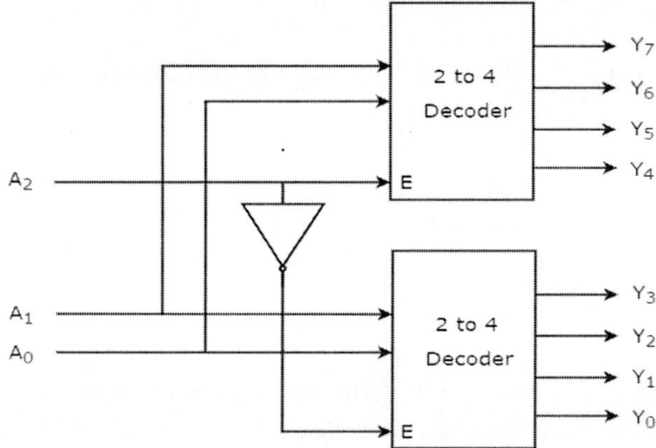

Fig. 4.70: A 3-to-8 line decoder using 2 to 4 decoders.

The parallel inputs A_1 & A_0 are applied to each 2 to 4 decoders. The complement of input A_2 is connected to Enable, E of lower 2 to 4 decoder in order to get the outputs, Y_3 to Y_0. These are the **lower four min terms**. The input, A_2 is directly connected to Enable, E of upper 2 to 4 decoder in order to get the outputs, Y_7 to Y_4. These are the **higher four min terms**.

Example 4.27 Implement a full-adder circuit with a decoder and two OR gates.

Solution:

From the truth table of the full-adder (section 4.2), we obtain the functions for this combinational circuit in sum of minterms:

$$S(x, y, z) = \sum(1, 2, 4, 7)$$

$$C(x, y, z) = \sum(3, 5, 6, 7)$$

Since there are three inputs and a total of eight minterms, we need a 3-to-8-line decoder.the implementation is shown in *Fig.* . The decoder generates the eight minterms for x, y, z. The

OR gate for output S forms the sum of minterms 1,2,4, and 7. The OR gate for output C forms the sum of minterms 3, 5, 6, and 7. A function with a long list of minterms requires an OR gate with a large number of inputs.

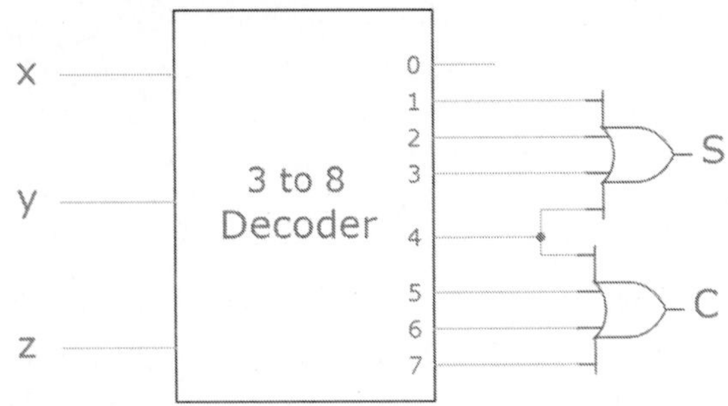

Fig. 4.71: Implementation of a full-adder with a decoder.

The decoder method can be used to implement any combinational circuit. However, its implementation must be compared with all other possible implementations to determine the best solution, especially if the combinational circuit has many outputs and if each output functions (or its complement) is expressed with a small number of minterms. A function F having a list of K minterms can be expressed in its complemented form F' with $2^n - k$ minterms. If the number of minterms in a function is greater than $2^n/2$, then F' can be expressed with fewer minterms than required for F.

4.7 Encoders

An **Encoder** is a combinational circuit that performs the reverse operation of Decoder. An Encoder

- Has a maximum of 2^n input lines and 'n' output lines.
- Will produce a binary code equivalent to the input, which is active High. Therefore, the encoder
- Encodes 2^n input lines with 'n' bits.
- It is optional to represent the enable signal in encoders. A basic Block Diagram of 4 to 2 Encoder is as shown in Fig. 4.72 (a).

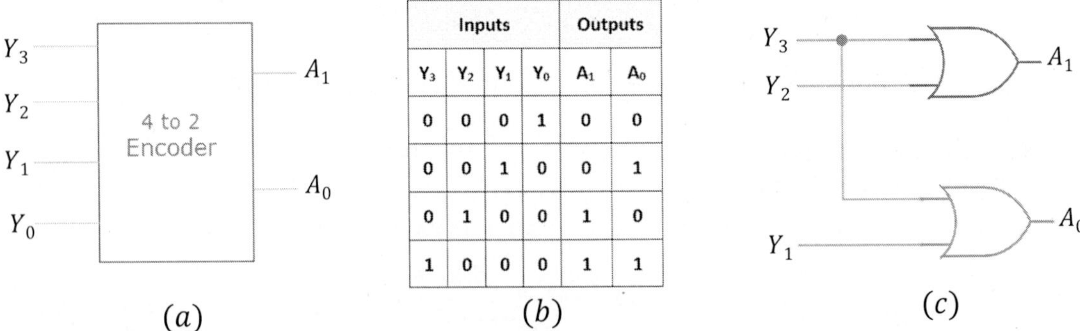

Inputs				Outputs	
Y_3	Y_2	Y_1	Y_0	A_1	A_0
0	0	0	1	0	0
0	0	1	0	0	1
0	1	0	0	1	0
1	0	0	0	1	1

Fig. 4.72: 4 to 2 Encoder (a) Block Diagram (b) Truth table (c) Circuit Diagram

Let 4 to 2 Encoder has four inputs Y_3, Y_2, Y_1 & Y_0 and two outputs A_1 & A_0. At any time, only one of these 4 inputs can be '1' in order to get the respective binary code at the output. The **Truth table** of 4 to 2 encoder is shown in Fig. 4.72 (b).

From Truth table, we can write the Boolean functions for each output as

$$A_1 = Y_3 + Y_2$$
$$A_0 = Y_3 + Y_1$$

We can implement the above two Boolean functions by using two input OR gates. The **circuit diagram** of 4 to 2 encoder is shown in the Fig. 4.72 (c). The circuit diagram contains two OR gates. These OR gates encode the four inputs with two bits.

Example 4.28 Design an Octal to binary Encoder circuit.

Solution:

The octal-to-binary encoder consists of eight inputs, one for each of the eight digits, and three Outputs that generate the corresponding binary number. Octal to binary Encoder has eight inputs, Y_7 to Y_0 and three outputs A_2, A_1 & A_0. Octal to binary encoder is nothing but 8 to 3 Encoder. The block diagram of octal to binary Encoder is shown in Fig. 4.73 (a).

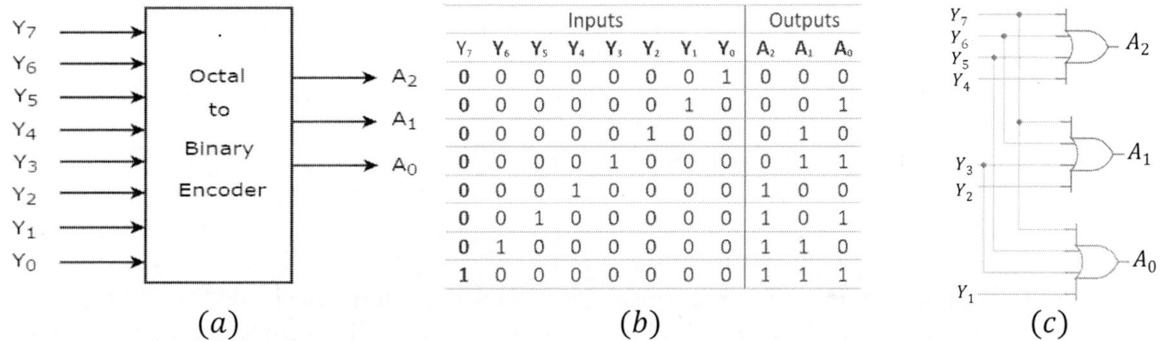

Inputs								Outputs		
Y_7	Y_6	Y_5	Y_4	Y_3	Y_2	Y_1	Y_0	A_2	A_1	A_0
0	0	0	0	0	0	0	1	0	0	0
0	0	0	0	0	0	1	0	0	0	1
0	0	0	0	0	1	0	0	0	1	0
0	0	0	0	1	0	0	0	0	1	1
0	0	0	1	0	0	0	0	1	0	0
0	0	1	0	0	0	0	0	1	0	1
0	1	0	0	0	0	0	0	1	1	0
1	0	0	0	0	0	0	0	1	1	1

Fig. 4.73: Octal to binary Encoder (a) Block diagram (b) Truth-Table (c) Circuit Diagram

At any time, only one of these eight inputs can be '1' in order to get the respective binary code. The **Truth table** of octal to binary encoder is shown in Fig. 4.73 (b). From Truth table, we can write the **Boolean functions** for each output as

$$A_2 = Y_7 + Y_6 + Y_5 + Y_4$$
$$A_1 = Y_7 + Y_6 + Y_3 + Y_2$$
$$A_0 = Y_7 + Y_5 + Y_3 + Y_1$$

We can implement the above Boolean functions by using four input OR gates. The **circuit diagram** of octal to binary encoder is shown in the Fig. 4.73 (c). The above circuit diagram contains three 4-input OR gates. These OR gates encode the eight inputs with three bits. The encoder in Fig. 73 assumes that only one input line can be equal to 1 at any time; otherwise, the circuit has no meaning.

Note: The circuit has eight inputs and could have $2^8 = 256$ possible input combinations. Only eight of these combinations have any meaning. The other input combination are don't-care conditions. The type of encoder available in IC form is called a priority encoder. These encoders establish an input priority to ensure that only the highest-priority input line is encoded.

Drawbacks of Encoder: Following are the drawbacks of normal encoder.

- There is an ambiguity, when all outputs of encoder are equal to zero. Because, it could be the code corresponding to the inputs, when only least significant input is one or when all inputs are zero.
- If more than one input is active High, then the encoder produces an output, which may not be the correct code. For example, if both Y3 and Y6 are '1', then the encoder produces 111 at the output. This is neither equivalent code corresponding to Y3, when it is '1' nor the equivalent code corresponding to Y6, when it is '1'.

So, to overcome these difficulties, we should assign priorities to each input of encoder. Then, the output of encoder will be the binary code corresponding to the active High inputs, which has higher priority. This encoder is called as priority encoder.

4.7.1 Priority Encoder: (Ordinary encoder with a priority function)

In Digital Electronics, the binary encoders are the multi-input combinational logic circuits, which consider all the input lines simultaneously and then convert them into an equivalent single encoded output. An n-bit digital encoder contains 2^n *input lines and* n *output lines*.

- To overcome the disadvantages of binary encoders, priority encoders were developed that work based on the highest priority input.
- It is used to solve the issues in binary encoders, which generate wrong output when more than one input line is active high. If more than one input line is active high (1) at the same time, then this encoder prioritizes every input level and allocates the priority level to each input.
- The output of this encoder corresponds to the input that has the highest priority. To obtain the output, only the input with the highest priority is considered by ignoring all other input lines. This is a type of binary encoder or an ordinary encoder with a priority function.
- The input that has the larger magnitude or highest priority is encoded first rather than other input lines. Hence, the generated output is based on the priority assigned to the inputs.
- In most digital applications, these encoders are used to select the inputs, which have the highest priority level. This process of selecting the input is called arbitration.

For example: A 4 to 2 priority encoder has four inputs Y3, Y2, Y1 & Y0 and two outputs A1 & A0. Here, the input, Y3 has the highest priority, whereas the input, Y0 has the lowest priority. In this case, even if more than one input is '1' at the same time, the output will be the binary code corresponding to the input, which is having **higher priority**. We considered one more **output, V** in order to know, whether the code available at outputs is valid or not.

- If at least one input of the encoder is '1', then the code available at outputs is a valid one. In this case, the output, V will be equal to 1.
- If all the inputs of encoder are '0', then the code available at outputs is not a valid one. In this case, the output, V will be equal to 0.

The **Truth table** and corresponding K-map for 4 to 2 priority Encoder is shown in Fig. 4.74.

Inputs				Outputs		
Y_3	Y_2	Y_1	Y_0	A_1	A_0	V
0	0	0	0	0	0	0
0	0	0	1	0	0	1
0	0	1	x	0	1	1
0	1	x	x	1	0	1
1	x	x	x	1	1	1

(*a*)

$Y_3Y_2 \backslash Y_1Y_0$	00	01	11	10
00	0	0	0	0
01	1	1	1	1
11	1	1	1	1
10	1	1	1	1

(*b*)

$Y_3Y_2 \backslash Y_1Y_0$	00	01	11	10
00	0	0	1	1
01	0	0	0	0
11	1	1	1	1
10	1	1	1	1

(*c*)

Fig. 4.74: 4 to 2 priority encoder (a) Truth table (b) K-map for A_1 (c) K-map for A_0

The simplified **Boolean functions** are

$$A_1 = Y_3 + Y_2$$
$$A_0 = Y_3 + \overline{Y_2}Y_1$$

Similarly, we will get the Boolean function of output, V as

$$V = Y_3 + Y_2 + Y_1 + Y_0$$

We can implement the above Boolean functions using logic gates. The **circuit diagram** of 4 to 2 priority encoder is shown below.

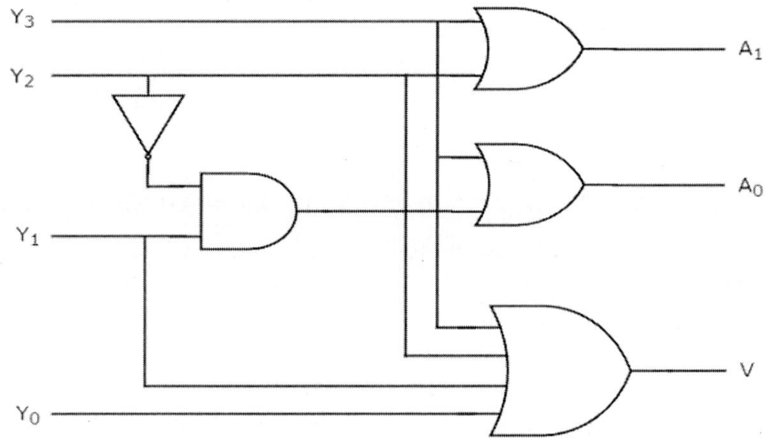

Fig. 4.74 (d): Circuit diagram of 4 to 2 priority encoder

The above circuit diagram contains two 2-input OR gates, one 4-input OR gate, one 2-input AND gate & an inverter. Here AND gate & inverter combination are used for producing a valid code at the outputs, even when multiple inputs are equal to '1' at the same time. Hence, this circuit encodes the four inputs with two bits based on the **priority** assigned to each input.

Example 4.29: Design 8 to 3 Priority Encoder.

Solution:

This kind of encoder is also named an 8-bit or Octal to Binary priority encoder. This type of encoder consists of 8 inputs and 3 outputs. When multiple inputs are active high at the same time, the input with the highest priority is considered to represent the output.

For example, if $D_1, D_2, and\ D_3$ inputs are active high or logic 1 regardless of other input bits, then the encoded output of the priority encoder will be D_3 i.e.,111. Here, the D_1, and D_2 input bits are either irrelevant or don't care conditions. The 8 to 3 priority encoder truth table is shown in Fig. 4.76.

Input								Output		
D7	D6	D5	D4	D3	D2	D1	D0	A	B	C
0	0	0	0	0	0	0	1	0	0	0
0	0	0	0	0	0	1	X	0	0	1
0	0	0	0	0	1	X	X	0	1	0
0	0	0	0	1	X	X	X	0	1	1
0	0	0	1	X	X	X	X	1	0	0
0	0	1	X	X	X	X	X	1	0	1
0	1	X	X	X	X	X	X	1	1	0
1	X	X	X	X	X	X	X	1	1	1

(a)

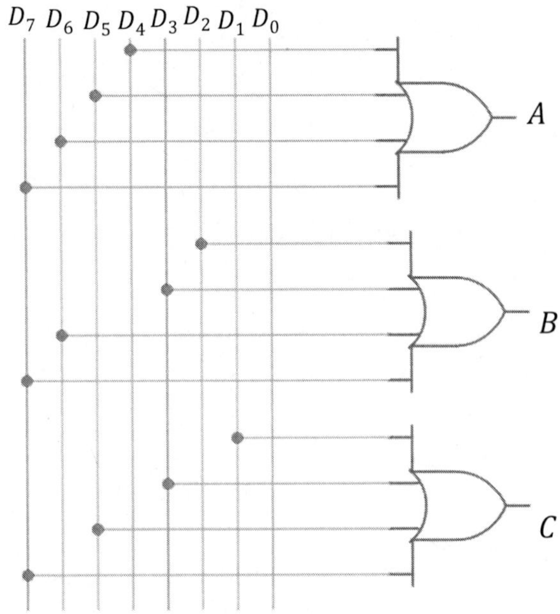

(b)

Fig. 4.75: 8 to 3 Priority Encoder (a) Truth Table (b) Circuit Diagram

From the above truth table, we can observe that D0, D1, D2, D3, D4, D5, D6, D7 are the inputs, and A, B, C are the outputs of an 8 to 3 priority encoder. The output expressions for A, B, and C are written either from Karnaugh map (K-map) simplification or truth table. The output expressions are obtained as shown below,

$$A = D_4 + D_5 + D_6 + D_7$$
$$B = D_2 + D_3 + D_6 + D_7$$
$$C = D_1 + D_3 + D_5 + D_7$$

From these simplified expressions, the 8 to 3 priority encoder circuit diagram is drawn as illustrated with logic gates as shown in Fig. 4.76 (b).

The output 'A' of a priority encoder is represented as active high or logic '1' only when the inputs $D_{4,}\ D_5, D_6\ and\ D_7$ are active high. The output 'B' of an encoder is at logic 1 only when the inputs $D_{2,}\ D_3, D_6\ and\ D_7$ are active high. Similarly, the output 'C' is represented as logic '1' only when the inputs $D_{1,}\ D_3, D_5\ and\ D_7$ are active high.

Some of the ***applications of priority encoder*** are,

- It is used to reduce the no. of wires and connections required for electronic circuit designing that have multiple input lines. Example keypads and keyboards.

- Used in controlling the position in the ship's navigation and robotics arm position.
- Used in the detection of highest priority input in various applications of microprocessor interrupt controllers.
- Used to protect the entire network from hackers by transmitting the binary code over the network.
- Used to encode the analog to digital converter's output.
- Used in synchronization of the speed of motors in industries.
- Used robotic vehicles
- Used in applications of home automation systems with RF
- Used in hospitals for health monitoring systems
- Used in secure communication systems with RF technology to enable secret code.

Note: 8 to 3 types are available in the standard IC 74LS148, which consists of 8 active low or logic 0 inputs and 3 active high or logic 1 output bits. The different properties of this type of encoder include cascading of n bits for priority encoding, encoding of highest priority inputs, code conversions, decimal to BCD conversion, to enable output line with active low when all the input lines are active high.

4.8 De-Multiplexer *(1 to many)*

De-multiplexer is a combinational circuit that performs the reverse operation of Multiplexer. It has single input, $'n'$ selection lines and maximum of 2^n(possible combinations of zeros and ones) outputs. The input will be connected to one of these outputs based on the values of selection lines. De-Multiplexer is also called as **De-Mux**.

A demultiplexer can be seems as a decoder with an enable input (Refer Fig. 4.76 (a)). A demultiplexer is a circuit that receives information on a single line and transmits this information on one of 2^n possible output lines. The selection of a specific output line is controled by the bit values of $'n'$ selection lines as shown in Fig. 4.76 (b).

The decoder of Fig. 4.76 has a single input variable E, that has been connected to all the four outputs, but the input information is directed to only one of the output lines, as specified by the binary value of the two selection lines A and B.

This can be verified from the truth table For example if the selection lines AB = 10 , output D_2 will be the same as the input value E. A decoder with an enable input is referred to as a demultiplexer.

A 1x4 De-Multiplexer (Refer Fig. 4.76 (b)) has one input I, two selection lines, s_1 & s_0 and four outputs Y_3, Y_2, Y_1 &Y_0. The single input '*I*' will be connected to one of the four outputs, Y_3 to Y_0 based on the values of selection lines S_1 & S_0. The **Truth table** of 1x4 De-Multiplexer is shown in Fig. 4.76 (c).

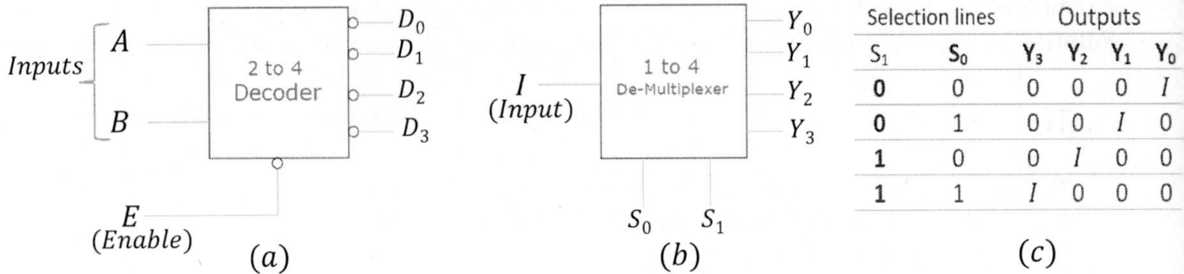

Selection lines		Outputs			
S_1	S_0	Y_3	Y_2	Y_1	Y_0
0	0	0	0	0	*I*
0	1	0	0	*I*	0
1	0	0	*I*	0	0
1	1	*I*	0	0	0

Fig. 4.76: Block diagram for (a) Decoder with Enable (b) Demultiplexer (c) Truth Table for De-mux

From the above Truth table, we can directly write the **Boolean functions** for each output as

$$Y_3 = S_1.S_0.I$$
$$Y_2 = S_1.S_0.I$$
$$Y_1 = S_1.S_0.I$$
$$Y_0 = S_1.S_0.I$$

We can implement these Boolean functions using Inverters & 3-input AND gates. The **circuit diagram** of 1x4 De-Multiplexer is shown in Fig. 4.77.

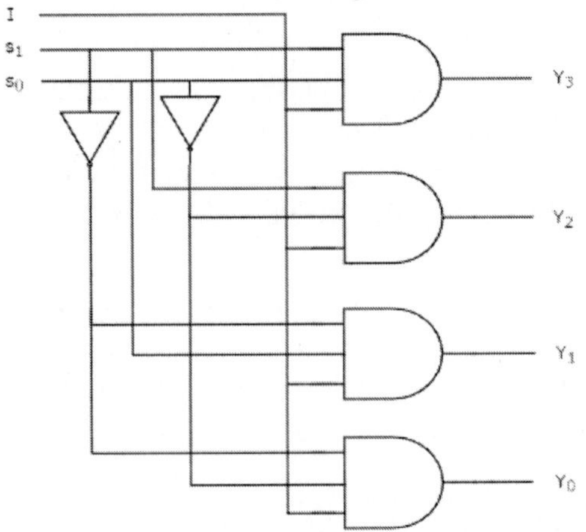

Fig. 4.77: Circuit diagram of 1x4 De-Multiplexer

We can easily understand the operation of the above circuit. Similarly, 1x8 De-Multiplexer and 1x16 De-Multiplexer can be implemented by following the same procedure. However, higher-order De-Multiplexers can be Implemented using lower-order De-Multiplexers.

We know that 1x4 De-Multiplexer has single input, two selection lines and four outputs. Whereas, 1x8 De-Multiplexer has single input, three selection lines and eight outputs. So, in order to construct 1x8 De-Multiplexer two **1x4 De-Multiplexers** are required in second stage in order to get the final eight outputs. Since, the number of inputs in second stage is two, we require **1x2 De-Multiplexer** in first stage so that the outputs of first stage will be the inputs of second stage. Input of this 1x2 De-Multiplexer will be the overall input of 1x8 De-Multiplexer. The same is described below using Truth table and Circuit Diagram below.

Example 4.30: Implement a 1x8 De-Multiplexer using 1x4 and 1x2 De-Multiplexer.

Solution:

Selection Lines			Outputs							
S2	S1	S0	Y7	Y6	Y5	Y4	Y3	Y2	Y1	Y0
0	0	0	0	0	0	0	0	0	0	D
0	0	1	0	0	0	0	0	0	D	0
0	1	0	0	0	0	0	0	D	0	0
0	1	1	0	0	0	0	D	0	0	0
1	0	0	0	0	0	D	0	0	0	0
1	0	1	0	0	D	0	0	0	0	0
1	1	0	0	D	0	0	0	0	0	0
1	1	1	D	0	0	0	0	0	0	0

(a)

(b)

Fig. 4.78: 1x8 De-Multiplexer (a) Truth table (b) Circuit diagram

A 1x8 De-Multiplexer has one input I, three selection lines S_2, S_1, S_0 and eight outputs Y_7 to Y_0. The **Truth table** of 1x8 De-Multiplexer is shown in Fig. 4.78 (a). The input 'D' is connected with one of the eight outputs from Y_0 to Y_7 based on the select lines S_2, S_1 and S_0. For example, if $S_2\ S_1\ S_0 = 0\ 0\ 0$, then the input D is connected to the output Y_0 and so on. From this truth table, the Boolean expressions for all the outputs can be written as follows

$$Y_0 = \bar{S}_2.\bar{S}_1.\bar{S}_0.D$$
$$Y_1 = \bar{S}_2.\bar{S}_1.S_0.D$$
$$Y_2 = \bar{S}_2.S_1.\bar{S}_0.D$$
$$Y_3 = \bar{S}_2.S_1.S_0.D$$
$$Y_4 = S_2.\bar{S}_1.\bar{S}_0.D$$
$$Y_5 = S_2.\bar{S}_1.S_0.D$$
$$Y_6 = S_2.S_1.\bar{S}_0.D$$
$$Y_7 = S_2.S_1.S_0.D$$

From these obtained equations, the logic diagram of this demultiplexer can be implemented by using eight 4-input AND gates and three NOT gates as shown in Fig. 4.78 (b). Different combinations of the select lines activates one AND gate at given time, such that data input will appear at the corresponding output.

Note: There are two popular 1-to-8 demultiplexer integrated circuits. One is the IC 74237, and the other one is IC 74138.

A 1:8 DEMUX can be construct using two 1:4 Demultiplexers and one 1:2 Demultiplexer with a proper cascading as shown in Fig. 4.79. In the figure below, the highest significant bit A of the selection inputs are connected to the enable inputs such that it is complemented before connecting to one DEMUX and to the other it is directly connected.

By this configuration, when A is set to zero, one of the output lines from Y0 to Y3 is selected based on the combination of select lines B and C. Similarly, when A is set to one, based on the select lines one of the output lines from Y4 to Y7 will be selected.

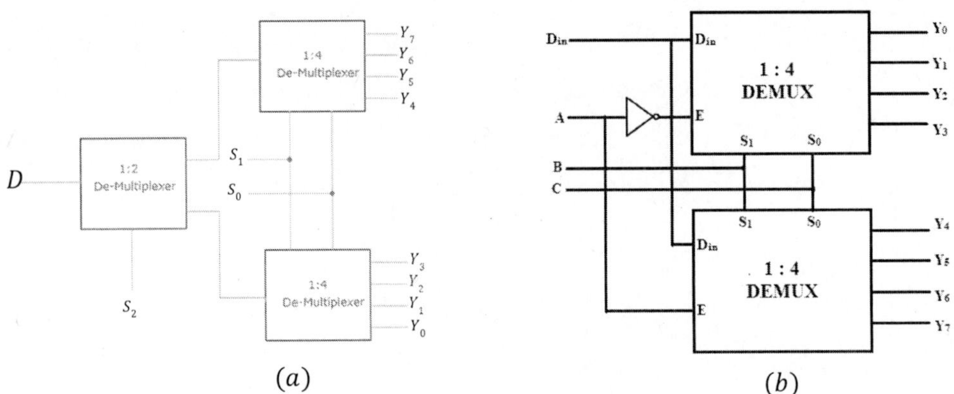

Fig. 4.79: 1:8 demultiplexer implemented by using two 1:4 De-mux and one 1:2 De-mux.

Example 4.31: Implementation a Full Subtractor Using 1:8 DEMUX

Solution:

Demultiplexers are also used for Boolean function implementation as well as combinational circuit design or to produce any truth table output by properly controlling the select lines. Consider the case for implementing a demultiplexer circuit in order to produce the full subtractor output. Fig. 4.80 (a) shows the truth table of a full subtractor.

Input			Output	
A	B	BIN	D	BOUT
0	0	0	0	0
0	0	1	1	1
0	1	0	1	1
0	1	1	0	1
1	0	0	1	0
1	0	1	0	0
1	1	0	0	0
1	1	1	1	1
0	0	0	0	0

(a)

D_{in}, 1:8 De-Multiplexer, Y_0–Y_7, A B B_{in}, D, B_{out}

(b)

Fig. 4.80: Full Subtractor Using 1-to-8 DEMUX (a)Truth table (b)Circuit Diagram

From the above table, the full subtractor the difference output (D) can be written as

$$D = f(A, B, B_{in}) = \sum m(1,2,4,7)$$

Similarly, the Borrow output (B_{out})can be written as

$$B_{out} = f(A, B, B_{in}) = \sum m(1,2,3,7)$$

From these Boolean expressions, a demultiplexer for producing full subtractor output can be built by properly configuring the 1-to-8 DEMUX, such that with input D = 1, it gives the minterms at the output further logically ORing these minterms, the outputs of difference and borrow can be obtained as shown in Fig. 4.80 (b).

Applications of Demultiplexer:

Since the demultiplexers are used to select or enable the one signal out of many, these are extensively used in microprocessor or computer control systems such as, Selecting different IO devices for data transfer (Data Routing), Choosing different banks of memory (Memory Decoding), Depends on the address, enabling different rows of memory chips, Enabling different functional units., Synchronous data transmission systems, Boolean function implementation (as we discussed full subtractor function above), Data acquisition systems, Combinational circuit design, Automatic test equipment systems, Security monitoring systems (for selecting a particular surveillance camera at a time), etc.

4.9 Multiplexer *(Many to 1)*

Multiplexer is a combinational circuit that has maximum of 2^n data inputs, $'n'$ selection lines and single output line. One of these data inputs will be connected to the output based on the values of selection lines. Since there are $'n'$ selection lines, there will be 2^n possible combinations of zeros and ones. So, each combination will select only one data input. Multiplexer is also called as **Mux,** and refereed as 2^n-to-1 multiplexer or $2^n \times 1$ or $2^n\!:1$ multiplexer.

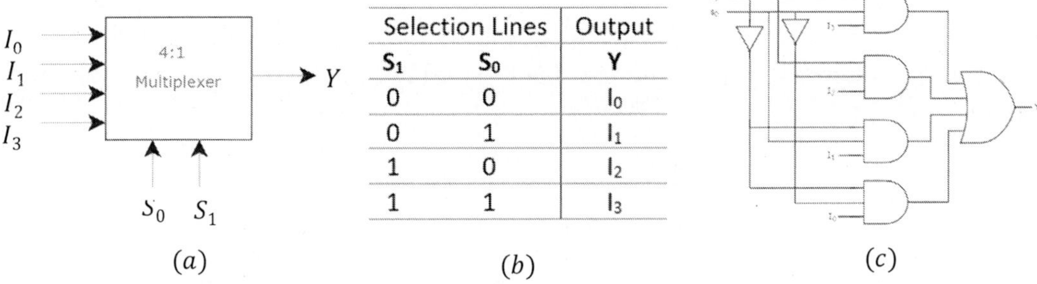

Selection Lines		Output
S_1	S_0	Y
0	0	I_0
0	1	I_1
1	0	I_2
1	1	I_3

Fig. 4.81: 4x1 Multiplexer (a) block diagram (b) Truth Table (c) Circuit Diagram

Consider 4:1 multiplexer, a 4:1 Multiplexer has four data inputs I_3, I_2, I_1 & I_0, two selection lines s_1 & s_0 and one output Y. The **block diagram** of 4x1 Multiplexer is shown in Fig. 4.81 (a). One of these 4 inputs will be connected to the output based on the combination of inputs present at these two selection lines. **Truth table** of 4x1 Multiplexer is Fig. 4.81 (b). From Truth table, we can directly write the **Boolean function** for output, *Y* as

$$Y = \overline{S_1}\,\overline{S_0}I_0 + \overline{S_1}S_0I_1 + S_1\overline{S_0}I_2 + S_1S_0I_3$$

We can implement this Boolean function using Inverters, AND gates & OR gate. The **circuit diagram** of 4x1 multiplexer is shown in Fig. 4.81 (c). Similarly, an 8:1 Multiplexer can also be implemented shows in Fig. 4.82. IC 74151 is an 8-to-1 multiplexer of the TTL family. It has an active LOW ENABLE input and provides complementary outputs. This ENABLE input can be used to control the multiplexing function. When the ENABLE input is active HIGH or active LOW respectively, the output is enabled, and the multiplexer function normally. When the ENABLE input is inactive, the output is disabled.

Input				Output
Selection Line			Enable	
A	B	C	$\bar{E}$	Y
×	×	×	H	L
L	L	L	L	D_0
L	L	H	L	D_1
L	H	L	L	D_2
L	H	H	L	D_3
H	L	L	L	D_4
H	L	H	L	D_5
H	H	L	L	D_6
H	H	H	L	D_7

Y $-D_7$: *Data Inputs*

$\bar{E}$: *Enable Line (Active low)*; Y: *Output*; A, B, C: *Selection Lines*

(*a*)

$\bar{E}$ A B C D_0 D_1 D_2 D_3 D_4 D_5 D_6 D_7

8:1 Multiplexer

Y

(*b*)

Fig. 4.82: 8-to-1 multiplexer (a) Truth table (b) circuit representation

Example 4.32: Design a 16×1 multiplexer using 8×1 and 2×1 multiplexer.

Solution:

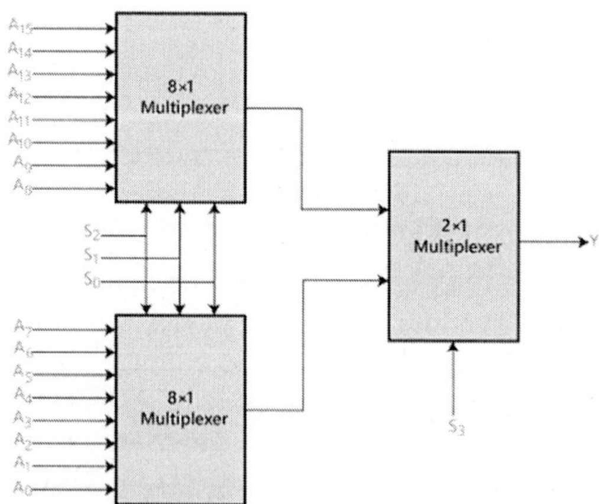

Fig. 4.83: 16×1 multiplexer using 8×1 and 2×1 multiplexer.

To implement the 16×1 multiplexer, two 8×1 multiplexer and one 2×1 multiplexer is needed. The 8×1 multiplexer has 3 selection lines, 4 inputs, and 1 output, whereas, 2×1 multiplexer has only 1 selection line.

At the input end to accommodate 16 data inputs, two 8 ×1 multiplexer needed. As each 8×1 multiplexer produces one output. So, in order to get the final output, a 2×1 multiplexer. The block diagram of 16×1 multiplexer using 8×1 and 2×1 multiplexer is shown in Fig. 4.83..

The same selection lines, s_2, s_1 & s_0 are applied to both 8x1 Multiplexers. The data inputs of upper 8x1 Multiplexer are A_{15} to A_8 and the data inputs of lower 8x1 Multiplexer are A_7 to A_0. Therefore, each 8x1 Multiplexer produces an output based on the values of selection lines, s_2, s_1 & s_0. The outputs of first stage 8x1 Multiplexers are applied as inputs of 2x1 Multiplexer that is present in second stage. The other **selection line,** s_3 is applied to 2x1 Multiplexer.

- If s_3 is zero, then the output of 2x1 Multiplexer will be one of the 8 inputs A_7 to A_0 based on the values of selection lines s_2, s_1 & s_0.
- If s_3 is one, then the output of 2x1 Multiplexer will be one of the 8 inputs A_{15} to A_8 based on the values of selection lines s_2, s_1 & s_0.

Therefore, the overall combination of two 8x1 Multiplexers and one 2x1 Multiplexer performs as one 16x1 Multiplexer.

Example 4.33: Implement the following Boolean function using 8:1 MUX

$$F(A,B,C,D) = \sum m\ (0,1,3,4,8,9,15)$$

Solution:

Following are the steps to implement the given Boolean Function using 8:1 MUX:

Step 1: To find number of select lines and input lines of the MUX

- For $'n'$ variable Boolean function, the number of select lines would be $n - 1$. In this case there are four variables A, B, C & D (As highest value in the expression is 15), hence, $n =$ 4.
- As we know that for an 8:1 MUX the number of select lines would be 3. Therefore, Number of select lines would ben-1 = 3. The variables B, C and D would be used as select lines. And the remaining variable i.e., A, which is the MSB, would be taken as the input variable.

Step 2: Formation of Implementation Table (Refer Table 4.20)

- Implementation Table: Write the MSB i.e. A and A/ at the left side of the table column wise and the other variables i.e., B, C, D at the top of the table row wise sequentially as shown below.
    - Write numbers from 0 to 15 in the cells of the Implementation table.
    - Encircle the numbers or minterms given in the question.
- If both the numbers in a column are encircled, then put '1' against the corresponding input line 'I'.

- If both the numbers in a column are not encircled, then put '0' against the corresponding input line 'I'.
- If only one number is encircled in a particular column, then write its corresponding MSB i.e., A or A/ against its input line 'I'.

Table 4.20: Implementation Table for example 4.33

	$\bar{B}.\bar{C}.\bar{D}$	$\bar{B}.\bar{C}.D$	$\bar{B}.C.\bar{D}$	$\bar{B}.C.D$	$B.\bar{C}.\bar{D}$	$B.\bar{C}.D$	$B.C.\bar{D}$	$B.C.D$
	I_0	I_1	I_2	I_3	I_4	I_5	I_6	I_7
$\bar{A}$	(0)	(1)	2	(3)	(4)	5	6	7
A	(8)	(9)	10	11	12	13	14	(15)
	1	1	0	$\bar{A}$	$\bar{A}$	0	0	A

- Implementation Table if B will be taken as input variable

	$\bar{A}.\bar{C}.\bar{D}$	$\bar{A}.\bar{C}.D$	$\bar{A}.C.\bar{D}$	$\bar{A}.C.D$	$A.\bar{C}.\bar{D}$	$A.\bar{C}.D$	$A.C.\bar{D}$	$A.C.D$
	I_0	I_1	I_2	I_3	I_4	I_5	I_6	I_7
$\bar{B}$	0	1	2	3	8	9	10	11
B	4	5	6	7	12	13	14	15

- Similarly, Implementation Table can be drawn for C and D as well.

Step 3: Draw the circuit to implement the given Boolean Function using 8:1 MUX(Refer Fig. 4.84)

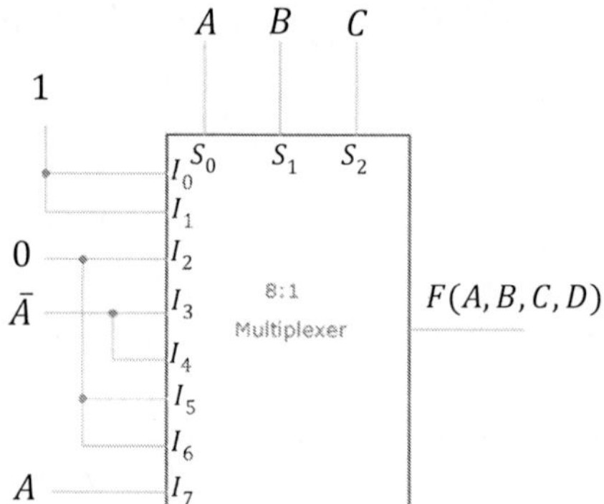

Fig. 4.84: An 8:1 multiplexer for example 4.33

Note: In a 2^n:1 MUX can be used to implement a Boolean function with n+1 variable, and out of n+1 variable, n variables are connected to the n selection lines.

Multiplexers for parallel-to-serial Data Conversion

Although data are processed in parallel in many digital systems to achieve faster processing speeds, when it comes to transmitting these data relatively large distances, this is done serially. The parallel arrangement in this case is highly undesirable as it would require a large number of transmission lines. Multiplexers can possibly be used for parallel-to-serial conversion.

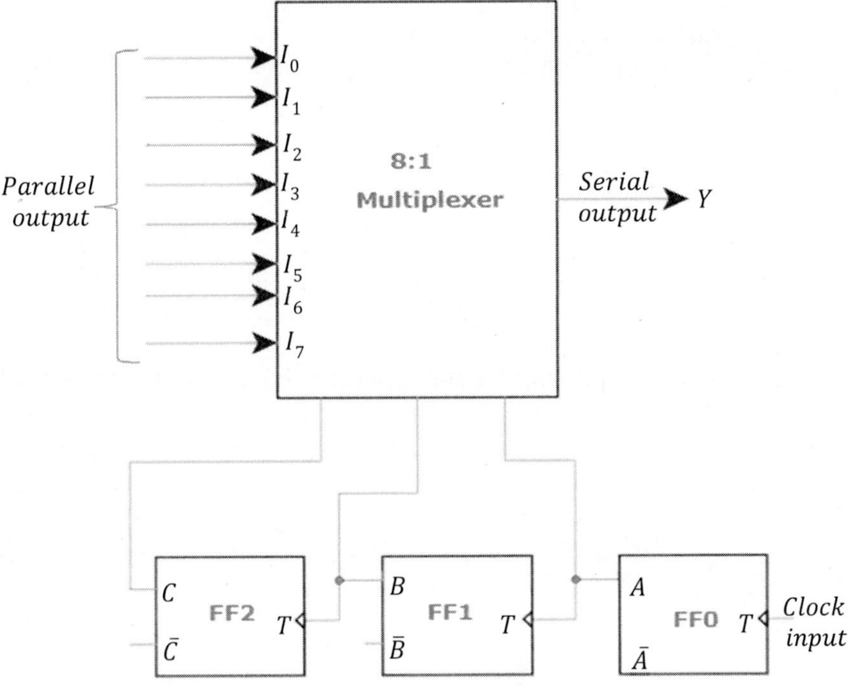

Fig. 4.85: Multiplexer for parallel-to-serial Conversion

Fig. 4.85 shows one such arrangement where an 8:1 multiplexer is used to convert eight-bit parallel binary data to serial form. A three-bit counter controls the selection inputs. As the counter goes through 000 to 111, the multiplexer output goes through I_0 to I_7. The conversion process takes a total of eight clock cycles. the three-bit counter has been constructed with the help of three toggle flip-flops. A variety of counter circuits of various types and complexities are, however, available in IC form. Flip-flops and counters are discussed in detail in upcoming chapter.

Example 4.34: Implement the product-of-sum Boolean function expression using a suitable multiplexer.

$$f(A,B,C) = \prod 1,2,5$$

Solution:

Let the Boolean function be

$$f(A,B,C) = \prod 1,2,5.$$

The equivalent sum-of-products expression can be written as

$$f(A,B,C) = \sum 0,3,4,6,7.$$

The truth table for the given Boolean function is given in Fig. 4.86 (a). The given function can be implemented with a 4-to-1 multiplexer with two selection lines. Variables A and B are chosen for the selection lines.the implementation table as drawn with the help of the truth table is given in Fig. 4.86 (b). Fig. 4.86 (c) shows the hardware implementation.

C	B	A	$f(A,B,C)$
0	0	0	1
0	0	1	0
0	1	0	0
0	1	1	1
1	0	0	1
1	0	1	0
1	1	0	1
1	1	1	1

	I_0	I_1	I_2	I_3
$\overline{C}$	0	1	2	3
C	4	5	6	7
	1	0	C	1

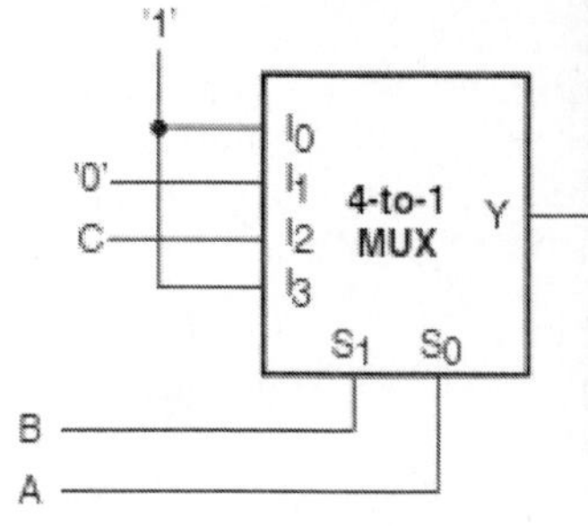

Fig. 4.86: (a) Truth Table (b) Implementation table (c) hardware implementation

Example 4.35 Fig. 4.87 shows the use of an 8-to-1 multiplexer to implement a certain four-variable Boolean function. From the given logic circuit arrangement, derive the Boolean expression implemented by the given circuit.

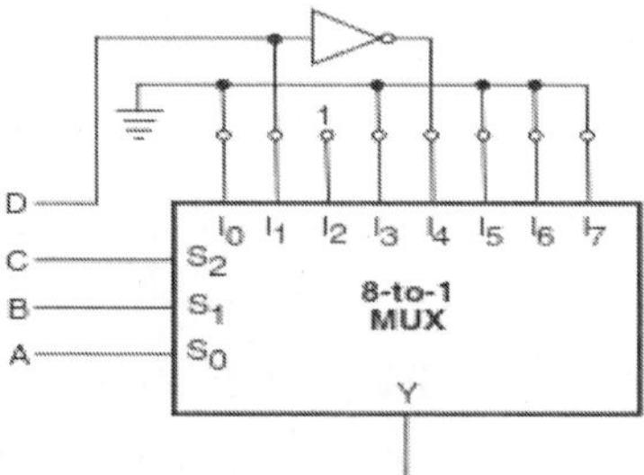

Fig. 4.87: Example 4.28

Solution:

This problem can be solved by simply working backwards in the procedure outlined earlier for designing the multiplexer-based logic circuit for a given Boolean function. Here, the hardware implementation is known and the objective is to determine the corresponding Boolean expression. From the given logic circuit, we can draw the implementation table as given in Table 4.21.

Table 4.21: Implementation table for example 4.35

	$\overline{A}.\overline{B}.\overline{C}$	$\overline{A}.\overline{B}.C$	$\overline{A}.B.\overline{C}$	$\overline{A}.B.C$	$A.\overline{B}.\overline{C}$	$A.\overline{B}.C$	$A.B.\overline{C}$	$A.B.C$
	I_0	I_1	I_2	I_3	I_4	I_5	I_6	I_7
$\overline{D}$	0	2	(4)	6	(8)	10	12	14
D	1	(3)	(5)	7	9	11	13	15
	0	D	1	0	$\overline{D}$	0	0	0

The entries in the first row (0, 2, 4, 6, 8, 10, 12, 14) and the second row (1, 3, 5, 7, 9, 11, 13, 15) are so because the selection variable chosen for application to the inputs is the MSB variable D. Entries in the first row Include all those minterms that contain $\overline{D}$, and entries in the second row include all those minterms that contain D. After writing the entries in the first two rows, the entries in the third row can be filled in by examining the logic status of different input lines in the given logic circuit diagram. Having completed the third row, relevant entries in the first and second rows are highlighted. The Boolean expression can now be written as follows:

$$Y = \sum m\,(3,4,5,8) = A.B.\bar{C}.\bar{D} + A.\bar{B}.C.D + A.\bar{B}.C.\bar{D} + \bar{A}.B.C.D$$

$$Y = A.\bar{B}.C\,(D + \bar{D}) + A.B.\bar{C}.\bar{D} + \bar{A}.B.C.D$$

Hence,

$$Y = A.\bar{B}.C\; + A.B.\bar{C}.\bar{D} + \bar{A}.B.C.D$$

Cascading Multiplexer Circuits

There can possibly be a situation where the desired number of input channels is not available in IC multiplexers. A multiple number of devices of a given size can be used to construct multiplexers that can handle a larger number of input channels. For instance, 8:1 multiplexer can be used to construct 16:1 or 32:1 or even larger multiplexer circuits. The basic steps to be followed to carry out the design are as follows:

1. If 2^n is the number of input lines in the available multiplexer and 2^N is the number of input lines in the desired multiplexer, then the number of individual multiplexers required to construct the desired multiplexer circuit would be 2^{N-n}.
2. From the knowledge of the number of selection inputs of the available multiplexer and that of the desired multiplexer, connect the less significant bits of the selection inputs of the desired multiplexer to the selection inputs of the available multiplexer.
3. The left-over bits of the selection inputs of the desired multiplexer circuit are used to enable or disable the individual multiplexers so that their outputs when ORed produce the final output.

4.10 Multiplier

Multiplication of binary numbers is usually implemented in microprocessors and microcomputers by USING *repeated addition* and *shift* operations. Since the binary adders are designed to add only two binary numbers at a time, instead of adding all the partial products at the end, they are added two at a time and their sum is accumulated in a register called the accumulator register. Also, when the multiplier bit is '0', that very partial product is ignored, as an all '0' line does not affect the final result.

The basic hardware arrangement of such a binary multiplier would comprise shift registers for the multiplicand and multiplier bits, an accumulator register for storing partial products, a binary parallel adder and a clock pulse generator to time various operations. Binary multipliers are also available in IC form. Some of the popular type numbers in the TTL family include 74261 which is a 2 x 4 bit multiplier (a four-bit multiplicand designated as $B_0, B_1, B_2, B_3, and\ B_4$ and a two-bit multiplier designated as $M_0, M_1, and\ M_2$).

The MSBs B_4, and M_2, are used to represent signs. 74284 and 74285 are 4 by 4 bit multipliers. They can be used together to perform high-speed multiplication of two four-bit numbers.

Fig. shows the arrangement. The result of multiplication is often required to be stored in a register. The size of this register (accumulator) depends upon the number of bits in the result, which at the most can be equal to the sum of the number of bits in the multiplier and multiplicand. Some multiplier ICs have an in-built register.

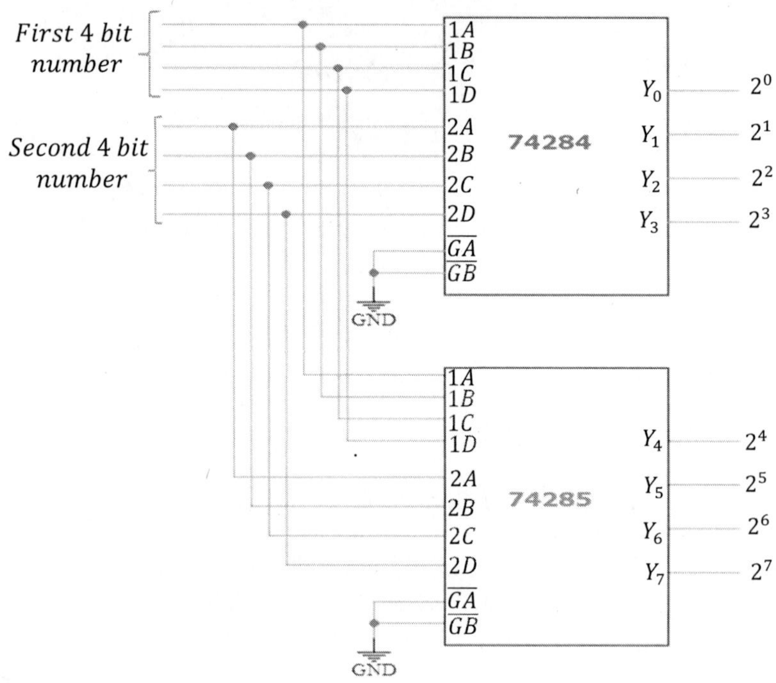

Fig. 4.88: A 4 by 4 Multiplier

Many microprocessors do not have in their ALU the hardware that can perform multiplication or other complex anthmetic operations such as division, determining the square root, trigonometric functions, etc. These operations in these microprocessors are executed through software. For example, a multiplication operation may be accomplished by using a software program that does multiplication through repeated execution of addition and shift instructions.

Other complex operations mentioned above can also be executed with similar programs. Although the use of software reduces the hardware needed in the microprocessor, the computation time in general is higher in the case of software-executed operations when compared with the use of hardware to perform those operations.

4.11 Application-Relevant Information

Application-relevant information such as pin connection diagrams, truth tables, etc., in respect of the more popular of these type numbers is given on the companion website. Table 4. And 4. lists commonly used IC type numbers used as multiplexers, encoders, demultiplexers and decoders.

Table 4.22: Commonly used IC type numbers used as arithmetic operations, multiplexers, encoders, demultiplexers and decoders.

IC No.	Function	Logic Family
7483	Four-bit full adder	TTL
7485	Four-bit magnitude comparator	TTL
74181	Four-Bit ALU and function generator	TTL
74182	Look-ahead carry generator	TTL
74183	Dual carry full adder	TTL
74283	Four-bit full binary adder	TTL
74885	Eight-bit magnitude comparator	TTL
4008	Four-bit binary full adder	CMOS
4527	BCD rate multiplier	CMOS
4585	Four-bit magnitude comparator	CMOS
40181	Four-bit arithmetic logic unit	CMOS
40182	Look-ahead carry generator	CMOS
10179	a Look-ahead carry block	ECL
10180	a Dual high-speed two-bit adder/subtractor	ECL
10182	Four-bit arithmetic logic unit/function generator	ECL
10183	4 by 2 multiplier	ECL
7442	1-of-10 decoder	TTL
74138	1-of-8 decoder/demultiplexer	TTL
74139	Dual 1-of-4 decoder/demultiplexer	TTL
74145	1-of-10 decoder/driver (open collector)	TTL
74147	10-line to four-line priority encoder	TTL
74148	Eight-line to three-line priority encoder	TTL
74150	16-input multiplexer	TTL
74151	Eight-input multiplexer	TTL
74152	Eight-input multiplexer	TTL
74153	Dual four-input multiplexer	TTL
74154	4-of-16 decoder/demultiplexer	TTL
74155	Dual 1-of-4 decoder/demultiplexer	TTL
74156	Dual 1-of-4 decoder/demultiplexer (open collector)	TTL
74157	Quad two-input noninverting multiplexer	TTL
74158	Quad two-input inverting multiplexer	TTL
74247	BCD to seven-segment decoder/driver (open collector)	TTL
74248	BCD to seven-segment decoder/driver with Pull-ups	TTL
74251	Eight-input three-state multiplexer	TTL
74253	Dual four-input three-state multiplexer	TTL
74256	Dual four-bit addressable latch	TTL
74257	Quad two-input non-inverting three-state multiplexer	TTL
74258	Quad two-input inverting three-state multiplexer	TTL

IC No.	Function	Logic Family
74259	Eight-bit addressable latch	TTL
74298	Dual two-input multiplexer with output latches	TTL
74348	Eight-line to three-line priority encoder (three-state)	TTL
74353	Dual four-input multiplexer	TTL
74398	Quad two-input multiplexer with output register	TTL
74399	Quad two-input multiplexer with output register	TTL
4019	Quad two-input multiplexer	CMOS
40147	10-line to four-line BCD priority encoder	CMOS
4511	BCD to seven-segment latch/decoden/driver	CMOS
4512	Eight-input three-state multiplexer	CMOS
4514	1-of-16 decoder/demultiplexer with input latch	CMOS
4515	1-of-16 decoder/demultiplexer with input latch	CMOS
4532	Eight-line to three-line priority encoder	CMOS
4539	Dual four-input multiplexer	CMOS
4543	BCD to seven-segment latch/decoder driver for LCD displays	CMOS
4555	Dual 1-of-4 decoder/demultiplexers	CMOS
4556	Dual 1-of-4 decoder/demultiplexers	CMOS
4723	Dual four-bit addressable latch	CMOS
4724	Fight-bit addressable latch	CMOS
10132	Dual two-input multiplexer with latch and common reset	ECL
10134	Dual multiplexer with latch	ECL
10158	Quad two-input multiplexer (non-inverting)	ECL
10159	Quad two-input multiplexer (inverting)	ECL
10161	3-to-8-line decoder (LOW)	ECL
10162	3-to-8-line decoder (HIGH)	ECL
10164	Eight-line multiplexer	ECL
10165	Eight-input priority encoder	ECL
10171	Dual 2-to-4 line decoder (LOW)	ECL
10172	Dual 2-to-4 line decoder (HIGH)	ECL
10173	Quad two-input multiplexer/latch	ECL
10174	Dual 4-to-1 multiplexer	ECL

Unit summary

This chapter, described combinational logic circuits that can be used to perform anthmetic and related operations in detail. This chapter takes a comprehensive look at yet another class of building blocks used to design more complex combinational circuits, and covers building blocks such as multiplexers and demultiplexers and other derived devices such as encoders and decoders. Particular emphasis is given to the operational basics and use of these devices to design more complex combinational circuits. Application-relevant information in terms of the list of commonly used integrated circuits available in this category, along with their functional description is given towards the end of the chapter. The text has been adequately illustrated with the help of a large number of solved examples.

Review Questions

1. Give the truth table for 4-bit priority encoder.
2. Draw the Truth Table of Full Adder.
3. Realize f= A'B + AB' using minimum universal gates.
4. Draw a tristate inverter and draw its truth table.
5. What is combinational circuit? Give an example.
6. Obtain 3 level NOR – NOR implementation of f (a, b, c, d, e, f) = [ab + cd] ef.
7. What is demux?
   a. Design a 4-bit binary adder/ subtractor circuit.
   b. Basic equations.
   c. Comparison of equations.
8. Design using twos complement Circuit diagram.
9. Design a half adder using NAND – NAND logic.
10. Explain how a full adder can be built using two half adders.
11. Design a half adder using at most three NOR gates.
12. Using 8 to 1 multiplexer, realize the Boolean function
    $T = f(w, x, y, z) = \Sigma (0,1,2,4,5,7,8,9,12,13)$
13. Design an 8421 to gray code converter.
14. Draw the logic diagram of full subtractor and explain its operation.
15. Draw the circuit diagram of NMOS NAND gate and explain its operation.
16. Design a full adder circuit using only NOR gates.
17. How do you characterize or define a combinational circuit? How does it differ from a sequential circuit? Give two examples each of combinational and sequential logic devices.
18. Beginning with the statement of the problem, outline different steps involved in the design of a suitable combinational logic circuit to implement the hardware required to solve the given problem.
19. Write down Boolean expressions representing the SUM and CARRY outputs in terms of three input binary variables to be added. Design a suitable combinational circuit to hardware-implement the design using NAND gates only.
20. Draw the truth table of a full subtractor circuit. Write a minterm Boolean expression for DIFFERENCE and BORROW outputs in terms of minuend variable, subtrahend variable and BORROW-IN. Minimize the expressions and implement them in hardware.
21. Draw the logic diagram of a three-digit BCD adder and briefly describe its functional principle.
22. Briefly describe the concept of look-ahead carry generation with respect to its use in adder circuits. What is its significance while implementing hardware for addition of binary numbers of longer lengths?
23. With the help of a block schematic of the logic circuit, briefly describe how individual four-bit magnitude comparators can be used in a cascade arrangement to perform magnitude comparison of binary numbers of longer lengths.
24. What is a multiplexer circuit? Briefly describe one or two applications of a multiplexer?

25. Is it possible to enhance the capability of an available multiplexer in terms of the number of input lines it can handle by using more than one device? If yes, briefly describe the procedure to do so, with the help of an example.
26. What is an encoder? How does a priority encoder differ from a conventional encoder? With the help of a truth table, briefly describe the functioning of a 10-line to four-line priority encoder with active LOW inputs and outputs and priority assigned to the higher-order inputs.
27. What is a demultiplexer and how does it differ from a decoder? Can a decoder be used as a demultiplexer? If yes, from where do we get the required input line?
28. Briefly describe how we can use a decoder optimally to implement a given Boolean function? Illustrate your answer with the help of an example.
29. Prove that the logic diagram below performs the function of a half-subtractor provided that y represents the DIFFERENCE output and X represents the BORROW output.

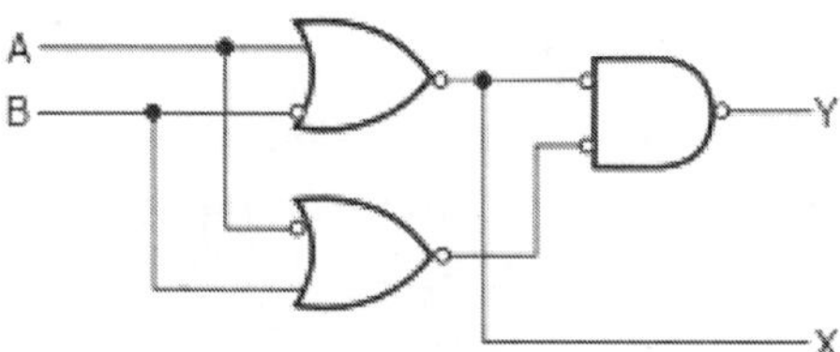

30. Determine the number of 7483s (four-bit binary adders) and 7486s (quad two-input EX-OR gates) required to design a 16-bit adder—subtractor circuit.

Objective Questions

1. A Karnaugh map (K-map) is an abstract form of ____________ diagram organized as a matrix of squares.
   a. Venn diagram b. cycle diagram c. block diagram d. triangular diagram
2. There are ________ cells in 4-variable K-map.
   a. 12 b. 16 c. 18 d.8
3. K- map based Boolean reduction is based on the following Unifying Theorem: A + A' = 1.
   a. Impact b. non-Impact c. Force d. Complementarity.
4. Don't care conditions can be used for simplifying Boolean expressions in ____
   a. Registers b. terms c. K-map d. latches
5. It should be kept in mind that don't care terms should be used along with the terms that are present in
   a. Minterms b. expressions c. K-map d. Latches
6. These logic gates are widely used in ______________ design and therefore are available in IC form.
   a. sampling b. digital c. analog d. systems
7. Which statement below best describes a Karnaugh map?
   a. It is simply a rearranged truth table
   b. The Karnaugh map eliminates the need for using NAND and NOR gates

   c. Variable complements can be eliminated by using Karnaugh maps
   d. A Karnaugh map can be used to replace Boolean rules
8. Each "1" entry in a K-map square represents:
   a. A HIGH for each input truth table condition that produces a HIGH output
   b. A HIGH output on the truth table for all LOW input combinations
   c. A LOW output for all possible HIGH input conditions
   d. A DON'T CARE condition for all possible input truth table combinations
9. Each "0" entry in a K-map square represents:
   a. A HIGH for each input truth table condition that produces a HIGH output
   b. A HIGH output on the truth table for all LOW input combinations
   c. A LOW output for all possible HIGH input conditions
   d. A DON'T CARE condition for all possible input truth table combinations
10. Looping on a K-map always results in the elimination of __________
   a. Variables within the loop that appear only in their complemented form
   b. Variables that remain unchanged within the loop
   c. Variables within the loop that appear in both complemented and uncomplemented form
   d. Variables within the loop that appear only in their uncomplemented form
11. The minimisation of logic expression is done due to
   a. Reduce space b. Reduce number of gates c. Reduce cost d. All of these
12. A Karnaugh map is used for
   a. Minimising Boolean expressions
   b. Develop digital circuits
   c. Computer interface
   d. None of these
13. There are ....cells in a 3 input K map
   a. 2 b. 5 c. 9 d. 8
14. For the given K-map What is the simplified expression?

CD \ AB	00	01	11	10
00	1			1
01	1	1	1	1
11	1	1	1	1
10	1			1

   a. B' + D b. B'D c. B + D' d. AB + CD
15. For the given K-map What is the simplified expression?

C \ AB				
	1			1
	1			1

a. B'
b. AB
c. A'B'+AB'
d. AC+AB

16. For the given K-map What is the simplified expression?

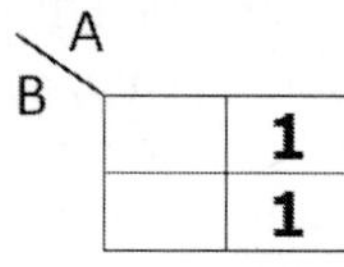

a. AB'+AB
b. A
c. B
d. 1

17. A, B, B_{in}, D and B_{out} are respectively the minuend, the subtrahend, the BORROW-IN, the DIFFERENCE output and the BORROW-OUT in the case of a full subtractor. Determine the bit status of D and B_{out}, for the following values of A, B and B_{in}:
a. A=0, B=1, B_{in} =1
b. A=1, B=1, B_{in}=0
c. A=1, B=1, B_{in}=1
d. A=0, B=0, B_{in}= 1

Reference and Suggested Readings

- *A. Anand Kumar, Fundamentals of digital circuits Second Edition, PHI Learning Private Limited, ISBN (978-81-203-3679-7)*
- *Cook, N. P. (2003) Practical Digital Electronics, Prentice-Hall, NJ, USA,*
- *Floyd, T. L. (2005) Digital Fundamentals, Prentice-Hall Inc., USA.*
- *Jain, R. P. Modern digital electronics. Vol. 1. No. 10. Tata McGraw-Hill Education, New Delhi, 2003.*
- *Morris Mano, M. and Kime, C. R. (2003) Logic and Computer Design Fundamentals, Prentice-Hall Inc., USA.*
- *Malvino, A. P. and Leach, D. P. (1994) Digital Principles and Applications, McGraw-Hill Book Company. USA.*
- *Tocci, R. J. and N. S. Widmer. 2004. Digital Systems Principles and Applications, 9th ed. Upper Saddle River, NJ: Prentice Hall.*
- *Tokheim, R. L. (1994) Schaum's Outline Series of Digital Principles, McGraw-Hill Companies Inc., USA.*

5 Sequential Circuit and System

UNIT SPECIFICS

Through this unit we have discussed the following aspects:

- *To study and understand the operation and application of Multivibrators (Astable, Monostable & Bistable).*
- *To Apply the concepts of system design and Boolean algebra for designing various sequential circuits.*
- *To understand the utility of Flip flops as 1-bit storage device.*
- *To understand different type of Shift Registers.*
- *To study and analyse various application of Counters.*
- *To identify various application relevant information for Finite State Machines.*

RATIONALE

A sequential circuit is a type of digital circuit that has memory and can store data over time. It is composed of a combinational logic circuit, which performs the logical operations on the inputs, and a memory element, which stores the output of the combinational logic circuit and provides it as an input for the next cycle of operation.

Sequential circuits can be synchronous or asynchronous. In synchronous circuits, the memory element is triggered by a clock signal, which ensures that all the circuit elements are synchronized and operate at the same frequency. In asynchronous circuits, the memory element is triggered by the input signals, and the circuit operates at different speeds.

A sequential system is a broader concept that encompasses not only digital circuits but also other systems that have memory and can store data over time. Sequential systems can be found in many areas of science and engineering, including control systems, communication systems, and signal processing systems.

In summary, a sequential circuit is a type of digital circuit that has memory and can store data over time, while a sequential system is a broader concept that encompasses systems with memory and data storage capabilities, including digital circuits.

PRE-REQUISITES

Combinational logic circuit design, Boolean Algebra

UNIT OUTCOMES

List of outcomes of this unit is as follows:
U5-O1: Describe various types of Multivibrator and their uses.
U5-O2: Identify the difference between Combinational and Sequential circuits.
U5-O3: Understand the concept of various Flip Flops.
U5-O4: Analyse and understand the data transfer using Shift Registers.
U5-O5: Implement and realization of counters using Flip Flops.
U5-O6: Identify various application of Finite state Machines.

Unit-5 Outcomes	**EXPECTED MAPPING WITH COURSE OUTCOMES** (1- Weak Correlation; 2- Medium correlation; 3- Strong Correlation)					
	CO-1	**CO-2**	**CO-3**	**CO-4**	**CO-5**	**CO-6**
U5-O1	3	3	3	-	3	1
U5-O2	1	1	2	2	1	-
U5-O3	2	1	3	1	2	1
U5-O4	-	-	3	1	2	2
U5-O5	3	3	3	-	3	1
U5-O6	3	3	3	-	3	1

➢ Scan the below QR codes for more information on the topic written below the QR code.

Multivibrators

Waveform Generators

Sequential Circuit

Latches and Flip flop

Shift Registers

Digital Counter

Unit outcomes

On completion of this lesson students will be able to learn about various types of sequential circuit, and also how sequential circuits are different from combinational circuits. Students will also able to understand the utility of the sequential circuits in digital systems. This unit will develop an understanding about the Flip Flops, Multivibrator, Shift registers, Counters etc. Students will get to know that sequential systems can be modelled using sequential circuits, which can be implemented using various digital logic components such as flip-flops, registers, and counters. Also, a detailed study has been given about synchronous and asynchronous systems.

5.1 Multivibrator

A *Multivibrator* circuit oscillates between a "HIGH" state and a "LOW" state producing a continuous output. Hence, an electronic device that produces a non-sinusoidal waveform as its output, is known as a **Multivibrator**. The generated non-sinusoidal waveforms may be a square wave, rectangular wave, a triangular wave, sawtooth wave, or ramp wave etc. The Multivibrator is a switching circuit and its basic configuration is shown below in Fig. 5.1:

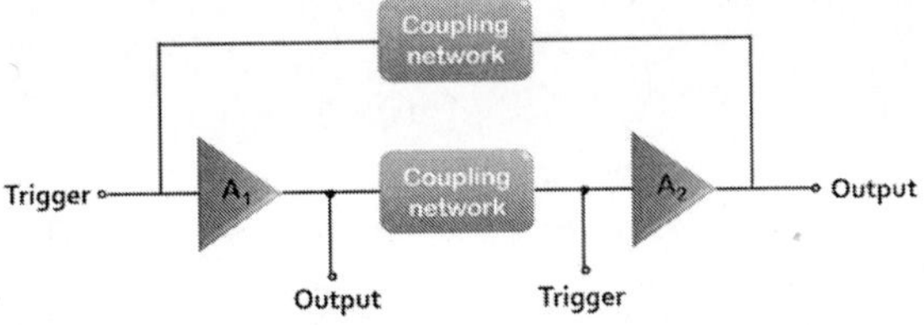

Fig. 5.1: Basic Configuration of a Multivibrator

Here, two amplifiers are employed in which the output from the first stage (A_1) acts as input to the second stage (A_2). The output of the A_2 is then through a feedback path is provided to the input of the A_1. Here, the operation of the circuit is controlled by two conditions *on* and *off* of the circuit. It performs oscillations between high and low state i.e., 0 and 1 thereby generating a continuous output.

A Multivibrator has been used for three main functions in digital systems

- To store binary number,
- To count pulses
- To provide synchronization between the arithmetic operations

Sequential logic circuits which use a clock signal for synchronization are dependent upon the frequency and therefore the clock pulse width to activate their switching action. Sequential circuits can also change their switching state using either the rising edge, falling edge, or both edges of the clock signal. The following key terms are commonly used in association to a timing pulse or in general clock pulse waveform as shown in Fig. 5.2.

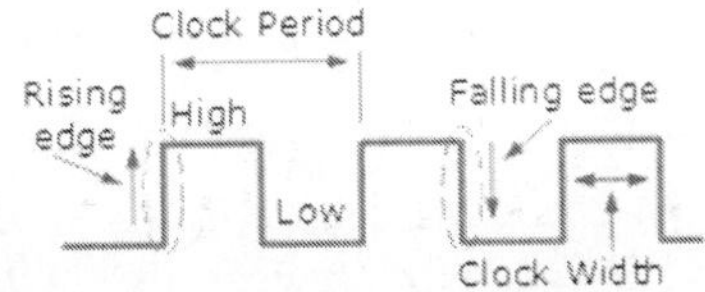

Fig. 5.2: Timing/Clock Waveform

- Active HIGH – If the transition or state change occurs from a "LOW" to a "HIGH" on the arriving of clock's pulse rising edge or during the clock width.
- Active LOW – If the transition or state change occurs from a "HIGH" to a "LOW" on the arriving of clock's pulses falling edge.
- Clock Width – This is the time during which the value of the clock signal is equal to a logic "1", or HIGH.
- Clock Period – This is the time between successive transitions in the same direction, ie, between two rising or two falling edges.
- Duty Cycle – This is the ratio of the clock width to the clock period.
- Clock Frequency – The clock frequency is the reciprocal of the clock period,

$$Frequency\ (f) = \frac{1}{Clock\ Period\ (T)}$$

Hence, the duty cycle is determined from the period of these states; a wave having the same ON and OFF time means a 50% duty cycle i.e., $T_{ON} = T_{OFF}$. The duty cycle can be represented by the equation given below

$$T_{Duty} = \frac{T_{ON}}{T_{ON} + T_{OFF}}$$

Note: Astable Multivibrator generally have an even 50% duty cycle, that is that 50% of the cycle time the output is "HIGH" and the remaining 50% of the cycle time the output is "OFF". In other words, the duty cycle for an Astable timing pulse is 1:1.

5.2 Types of Multivibrators

Clock pulse generation circuits can be a combination of analog and digital circuits that produce a continuous series of pulses (called Astable Multivibrator) or a pulse of a specific duration (called Monostable Multivibrator). Combining two or more multivibrator circuit provides generation of a desired pattern of pulses (including pulse width, time between pulses and frequency of pulses). The various types of coupling network are basically categorized by the type of coupling network that is employed in the circuit of the multivibrator. There are basically three types of clock pulse generation circuits as shown in Fig. 5.3

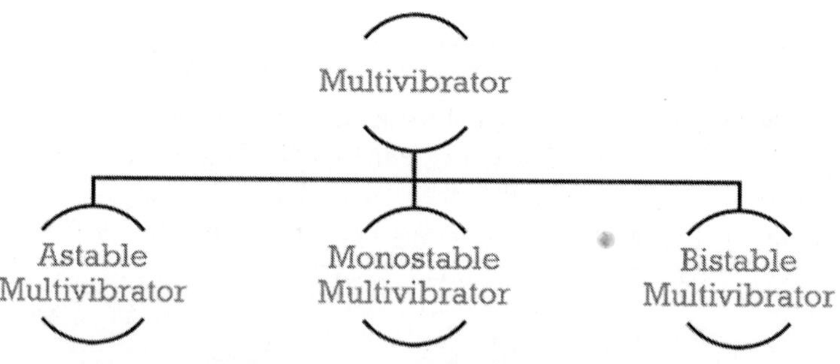

Fig.5.3: Types of Multivibrators

- Astable – It is also called a *free-running Multivibrator.* It means that, **NONE** of the states are stable. Moreover, switches continuously between two states. This action produces a train of square wave pulses at a fixed known frequency (Like pendulum of a clock).

- Monostable – A *one-shot Multivibrator* that has only **ONE** stable state as once externally triggered, it returns back to its first stable state (like our door bell).
- Bistable – A *flip-flop* that has **TWO** stable states producing a single pulse either HIGH or LOW in value (switch of a fan).

The input and output waveforms for various multivibrators is shown in Fig. 5.4.

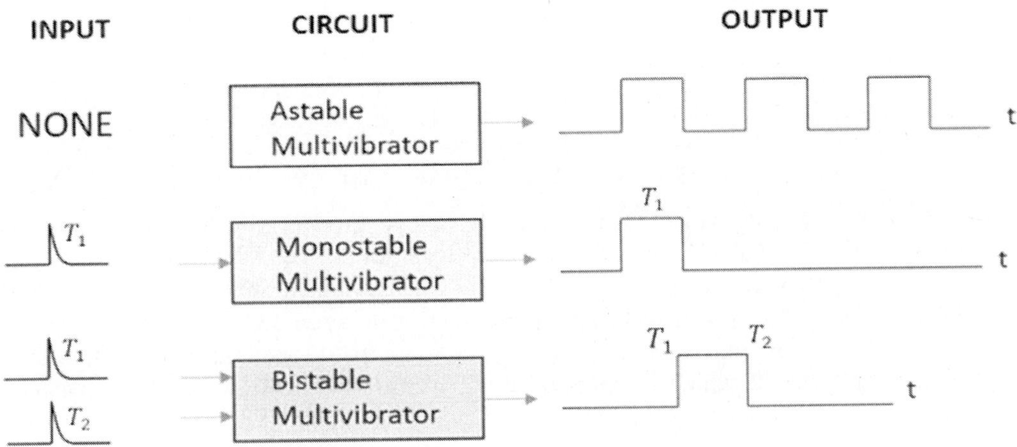

Fig. 5.4: Input and output waveform for Multivibrators

One way of producing a very simple clock signal (or pulse) is by the interconnection of digital logic gates. As NAND gates contain current amplification, they can also be used to provide a suitable clock signal or timing pulse with the aid of a single *Capacitor* and *Resistor* to provide the required feedback and timing functions. These timing circuits are often used because of their simplicity and are also useful if a logic circuit once designed has some unused gates which can be utilised to create a monostable or Astable oscillator.

The trigger requirements of all three types of multivibrators are-

- Astable Multivibrators - Do not require any external triggering
- Monostable Multivibrators - One trigger pulse is enough for each disturbance.
- Bistable Multivibrators - One trigger pulse is required whenever the current state needs to be changed.

5.2.1 Monostable Multivibrator

Monostable Multivibrators or "one-shot" pulse generators are generally used to convert short sharp pulses into much wider ones for timing applications. Monostable multivibrators generate a single output pulse, either "HIGH" or "LOW", when a suitable external trigger signal or trigger pulse 'T' is applied. It has a single stable state and a quasi-state. Here, out of the two coupling networks, one provides ac coupling and the other provides dc coupling. Thus, providing one stable and one quasi-state. A triggering pulse is required in order to have transition from the stable state to the quasi-state. However, in order to have transition again from quasi-state to a stable state no any triggering pulse is provided. Thus, after a certain time period decided by the time constant, the circuit comes back to its initial state i.e., the stable state without the need of external signal as shown in Fig. 5.5.

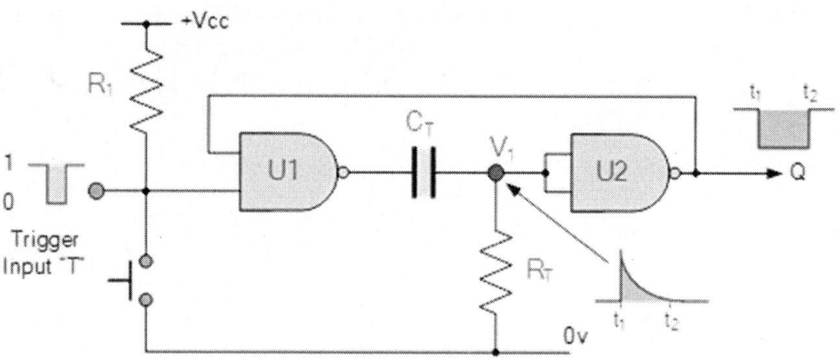

Fig. 5.5: Simple NAND Gate Monostable Circuit

Suppose that initially the trigger input '*T*' is held HIGH at logic level *"1"* by the resistor R_1 so that the output from the first NAND gate U1 is LOW at logic level "0", (NAND gate principals). The timing resistor, R_T is connected to a voltage level equal to logic level "0", which will cause the capacitor, C_T to be fully discharged. The output of U1 is therefore LOW. As the timing capacitor is completely discharged, junction V_1 will also be equal to "0" resulting in the output from the second NAND gate U2, which is connected as an inverting NOT gate, to be HIGH (logic-1). The output from the second NAND gate, (U2) is fed back to one input of U1 to provide the necessary positive feedback. Since the junction V_1 and the output of U1 are both at logic "0" no current flows in the timing capacitor C_T. This results in the circuit being Stable and it will remain in this stable state until a trigger input *T* is applied.

If a negative pulse is now applied either externally or by the action of the push-button to the trigger input of the NAND gate U1, the output of U1 will go HIGH to logic "1" (NAND gate principles). Since the voltage across the capacitor cannot change instantaneously (capacitor charging principals) this will cause the junction at $V_1$and also the input to U2 to go HIGH, which in turn will make the output of the NAND gate U2 change state to logic-0. The circuit will now remain in this second timing state even if the trigger input pulse *T* is removed. This is known as the *Meta-stable state*.

The voltage across the capacitor will now increase as the capacitor C_T starts to charge up from the HIGH output of U1 at a time constant determined by the resistor/capacitor combination. This charging process continues until the charging current is unable to hold the input of U2 and therefore junction V_1 HIGH.

When this happens, the output of U2 switches HIGH again, logic-1, which in turn causes the output of U1 to go LOW and the capacitor discharges into the output of U1 under the influence of resistor R_T. The circuit has now switched back to its original stable state. Thus, for each negative going trigger pulse, the monostable multivibrator circuit produces a LOW going output pulse. The length of the output time period is determined by the capacitor/resistor combination (RC Network) and is given as the Time Constant $T = 0.69RC$ of the circuit in seconds. Since the input impedance of the NAND gates is very high, large timing periods can be achieved. One main disadvantage of Monostable Multivibrators is that the time between the application of the next trigger pulse *T* has to be greater than the *RC* time constant of the circuit.

Like NAND gate monostable type circuit above, it is also possible to build simple monostable timing circuits. The following circuit (Refer Fig. 5.6) shows how a basic monostable multivibrator circuit can be constructed using just two 2-input Logic "NOR" Gates.

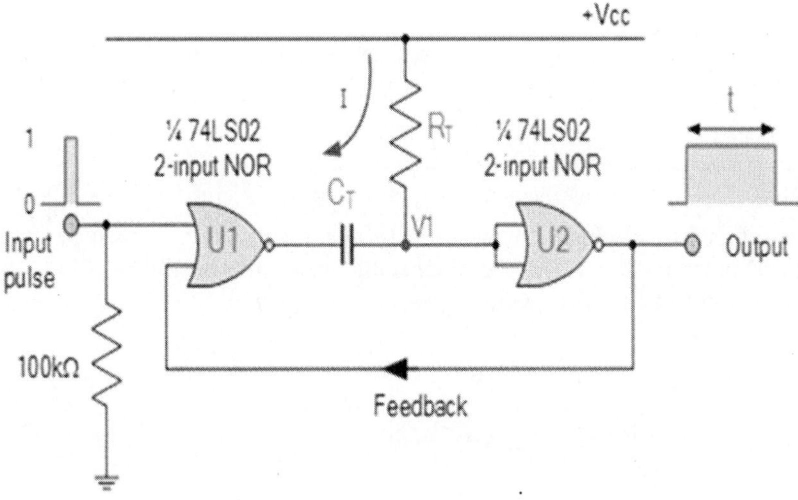

(a) NOR Gate Monostable Multivibrator

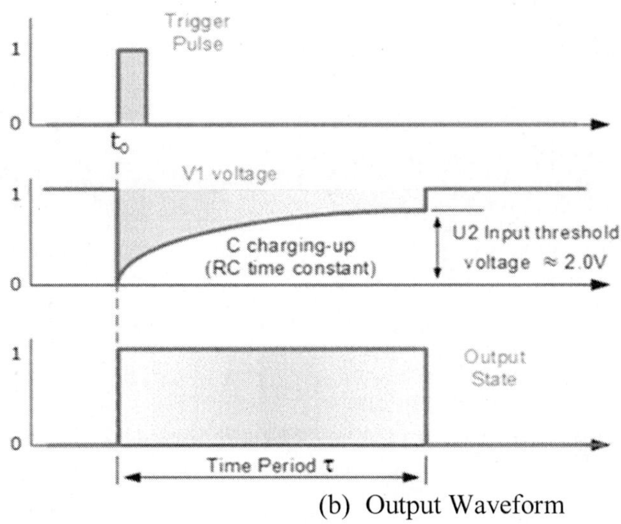

(b) Output Waveform

Fig. 5.6: NOR Gate Monostable Multivibrator

Suppose initially that the trigger input is LOW at a logic level "0" so that the output from the first NOR gate U1 is HIGH at logic level "1", (NOR gate principals). The resistor, R_T is connected to the supply voltage so is also equal to logic level *"1"*, which means that the capacitor, C_T has the same charge on both of its plates. Junction V1 is therefore equal to this voltage so the output from the second NOR gate U2 will be LOW at logic level "0". This then represents the circuits "Stable State" with zero output.

When a positive trigger pulse is applied to the input at time t_0, the output of the first NOR gate U1 goes LOW taking with it the left-hand plate of capacitor C_T thereby discharging the capacitor. As both plates of the capacitor are now at logic level "0", so too is the input to the second NOR gate, U2 resulting in an output equal to logic level "1". This then represents the circuits second state, the "Unstable State" with an output voltage equal to +Vcc.

The second NOR gate, U2 will maintain this second unstable state until the timing capacitor now charging up through resistor, R_T reaches the minimum input threshold voltage of U2 (approx. 2.0V). This causes U2 to change state as a logic level "1" value has now appeared on its inputs.

The NOR gates output resets to logic "0" which in turn is fed back (feedback loop) to one input of U1. This action automatically returns the monostable back to its original stable state and awaiting a second trigger pulse to restart the timing process once again.

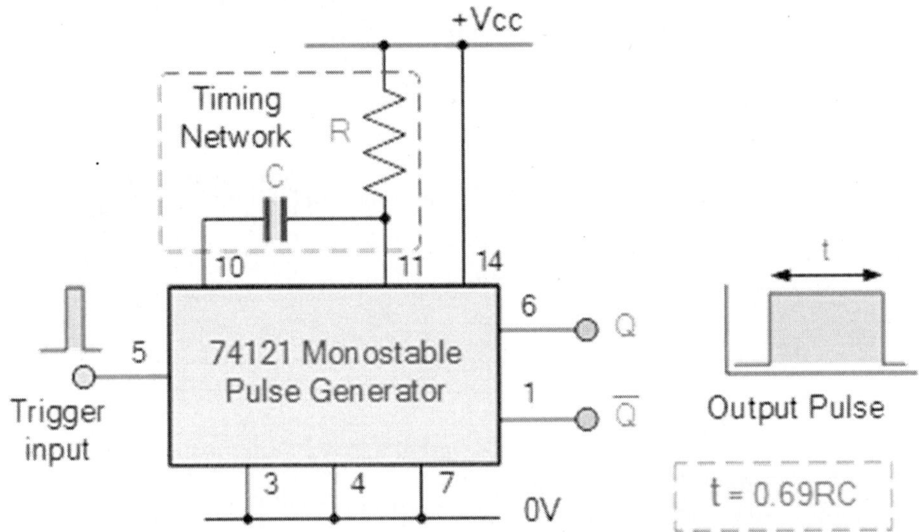

Fig. 5.7: 74LS121 Monostable Multivibrator

This monostable pulse generator (Refer Fig. 5.7) IC can be configured to produce an output pulse on either a rising-edge trigger pulse or a falling-edge trigger pulse. The 74LS121 can produce pulse widths from about 10ns to about 10ms width a maximum timing resistor of 40kΩ and a maximum timing capacitor of 1000uF.

5.2.2 Astable Multivibrator Circuits

Astable Multivibrators are the most commonly used type of multivibrator circuit. An astable multivibrator have no permanent "meta" or "steady" state but is continually changing its output from one state (LOW) to another state (HIGH) and then back again. This continual switching action from "HIGH" to "LOW" and "LOW" to "HIGH" produces a continuous and stable square wave output which switches abruptly between the two logic levels making it ideal for timing and clock pulse applications. It is a type of multivibrator also termed as a **free running multivibrator**. It is called so because here the state changes on its own after some predetermined time interval and thus does not require a triggering pulse. Here, the output of the circuit simply oscillates between high and low state freely. Hence is just an oscillator.

The coupling network employed in astable multivibrator with the help of a coupling capacitor provides ac coupling to the amplifier (Refer Fig. 5.8). A phase shift of 180° is provided by each amplifier stage in the mid-band, thereby generating a total phase shift of 0° or 360°. Hence providing positive feedback. Thus, the circuit has no stable state, and the two states are merely temporary ones termed as quasi-state. So, a continuous output is generated by performing successive transitions from one state to another after a fixed time duration. It is to be noted here that the time duration of switching between one quasi-state to the other depends on the time constant and other parameters of the circuit

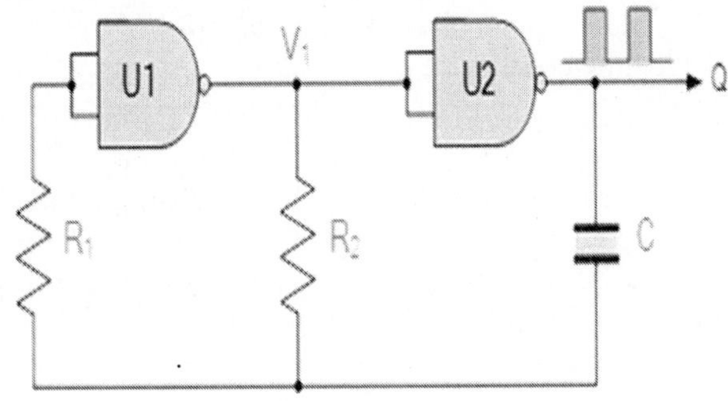

Circuit aiagram using NAND Gate

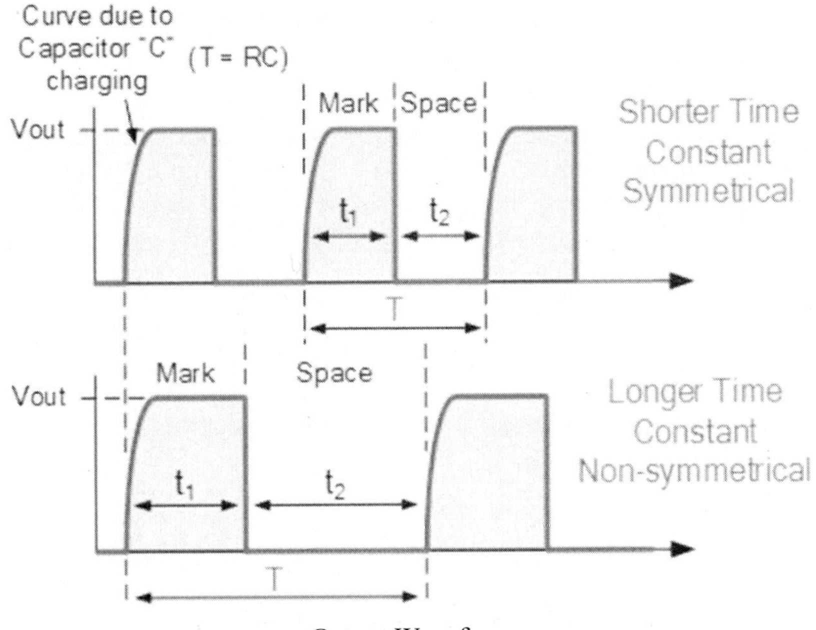

Output Waveform

Fig. 5.8: Astable Multivibrator

As with the previous monostable multivibrator circuit above, the timing cycle is determined by the RC time constant of the resistor-capacitor, RC Network. Then the output frequency can be varied by simply changing the value(s) of the resistors and capacitor in the circuit.

The **astable multivibrator** circuit uses two CMOS NOT gates such as the CD4069 or the 74HC04 hex inverter ICs, or as in our simple circuit below a pair of CMOS NAND gates such as the CD4011 or the 74LS132 as well as a RC timing network. The two NAND gates are connected as inverting NOT gates.

Suppose that initially the output from the NAND gate U2 is HIGH at logic level "1", then the input must therefore be LOW at logic level "0" (NAND gate principles) as will be the output from the first NAND gate U1. Capacitor, *C* is connected between the output of the second NAND gate U2 and its input via the timing resistor, R_2. The capacitor now charges up at a rate determined by the time constant of R_2 and *C*.

As the capacitor, *C* charges up, the junction between the resistor R_2 and the capacitor, *C*, which is also connected to the input of the NAND gate U1 via the stabilising resistor, R_2 decreases until the lower threshold value of U1 is reached. At this point U1 changes state and the output of U1 now becomes HIGH. This change causes NAND gate U2 to also change state as its input has now changed from a logic "0" to a logic "1" condition resulting in the output of NAND gate U2 becoming LOW.

Capacitor *C* is now becoming reverse biased so starts to discharge itself through the input of U1. Capacitor, *C* charges up again in the opposite direction determined by the time constant of both R_2 and *C* as before until it reaches the upper threshold value of NAND gate U1. This causes U1 to change state and the cycle repeats itself over again.

Then, the time constant for a NAND gate **Astable Multivibrator** is given as $T = 2.2RC$ in seconds with the output frequency given as $f = 1/T$.

Example 5.1: For an Astable Multivibrator, resistor $R_2 = 10k\Omega$ and the capacitor $C = 45nF$, find the oscillation frequency of the circuit.

Solution:

$$f = \frac{1}{T} = \frac{1}{2.2RC} = \frac{1}{2.2 \times 10k\Omega \times 45nF} = 1KHZ$$

Then the output frequency is calculated as being 1kHz, which equates to a time constant of 1ms. Thus, the output waveform would look something like Fig. 5.9.

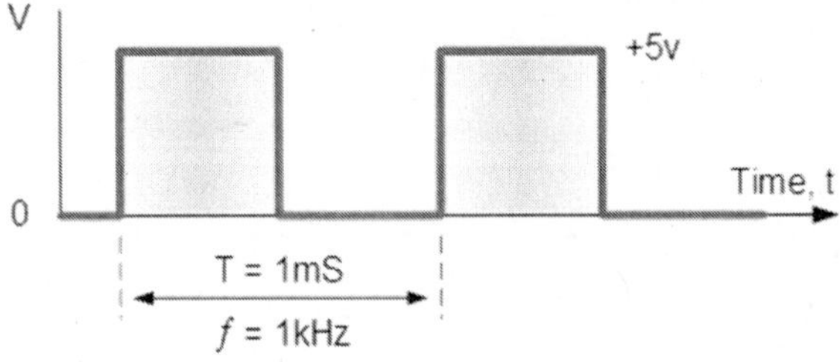

Fig. 5.9: output frequency

5.2.3 Bistable Multivibrator Circuits

The **Bistable Multivibrators** circuit is basically a SR flip-flop that we look at in the previous tutorials with the addition of an inverter or NOT gate to provide the necessary switching function. As with flip-flops, both states of a bistable multivibrator are stable, and the circuit will remain in either stable state indefinitely. This type of multivibrator circuit passes from one state to the other "only" when a suitable external trigger pulse T is applied and therefore to switch through a full "SET-RESET" cycle **two** triggering pulses are required. This type of circuit is also known as a "Bistable Latch", "Toggle Latch" or simply "T-latch". A bistable multivibrator has 2 stable states. Here, a separate trigger pulse is required in order to have transition from one stable state to another stable state. Here, only dc coupling is provided by the coupling networks and hence the energy storing element is not required.

When the trigger pulse is first applied, the transistor in the circuit gets cut-off thus showing an off stable state. However, as another triggering pulse is applied, the transistor again starts its conduction. Thereby changing its state from one state to another. As the overall operation depends on two triggers, these circuits are also termed as **flip-flops** or **trigger circuits**. For the purpose of counting and storing binary elements, bistable multivibrators are used.

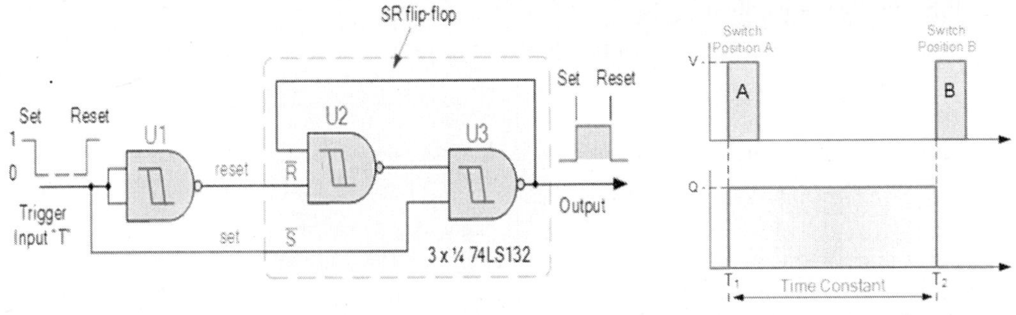

using NAND Gate Output Waveform

Fig. 5.10: Bistable Multivibrators

The simplest way to make a **Bistable Latch** is to connect together a pair of Schmitt NAND gates to form a SR latch as shown in Fig. 5.10 above. The two NAND gates, U2 and U3 form the bistable which is triggered by the input NAND gate, U1. This U1 NAND gate can be omitted and replaced by a single toggle switch to make a switch debounce circuit which we saw previously in the SR Flip-flop tutorial.

When the input pulse goes "LOW" the bistable latches into its "SET" state, with its output at logic level "1", until the input goes "HIGH" causing the bistable to latch into its "RESET" state, with its output at logic level "0". The output of a bistable multivibrator will stay in this "RESET" state until another input pulse is applied and the whole sequence will start again.

Then a **Bistable Latch** or "Toggle Latch" is a two-state device in which both states either HIGH or LOW, (logic "1" or logic "0") are stable. **Bistable Multivibrators** have many applications such as frequency dividers, counters or as a storage device in computer memories but they are best used in circuits such as *Latches* and *Counters*.

So **Bistable Multivibrators** can produce a very short output pulse or a much longer rectangular shaped output whose leading edge rises in time with the externally applied trigger pulse and whose trailing edge is dependent upon a second trigger pulse as shown.

5.3 Waveform Generators

In the previous section various multivibrator circuits have been discussed which can be used as relaxation oscillators to produce either a square or rectangular wave at their outputs for use as clock and timing signals. But it is also possible to construct basic This waveform generation discussion would be incomplete without some examples of digital regenerative switching circuits, since it illustrates both the switching action and operation of waveform generators used for generating square waves for use as timing or sequential waveforms.

We know that regenerative switching circuits such as Astable Multivibrators are the most commonly used type of relaxation oscillator as they produce a constant square wave output, making them ideal as a digital Waveform Generator. Astable multivibrators make excellent oscillators because they switch continuously between their two unstable states at a constant repetition rate thereby producing a continuous square wave output with a 1:1 mark-space ratio ("ON" and "OFF" times the same) from its output.

In this section we will look at some of the different ways we can construct waveform generators using just standard TTL and CMOS logic circuits along with some additional discrete timing components.

5.3.1 Schmitt Waveform Generators

Simple Waveform Generators can be constructed using basic Schmitt trigger action inverters such as the TTL 74LS14. This method is by far the easiest way to make a basic astable waveform generator. When used to produce clock or timing signals, the astable multivibrator must produce a stable waveform that switches quickly between its "HIGH" and "LOW" states without any distortion or noise, and Schmitt inverters do just that.

A **Schmitt trigger** is a comparator (not exclusively) circuit that makes use of positive feedback (small changes in the input lead to large changes in the output in the same phase) to implement hysteresis (a fancy word for delayed action) and is used to remove noise from an analog signal while converting it to a digital one (Refer Fig. 5.11). It was invented way back in 1937 by Otto H. Schmitt (whose legacy is somewhat understated) who called it a 'thermionic trigger'. Comparators are especially very sensitive to inputs, as because of their very high gain even tiny changes in the input can cause instant change of state on the output. A Schmitt trigger makes use of ***positive feedback***, i.e., it takes a sample of the output and feeds it back into the input so as to 'reinforce', so to speak, the output which is the exact opposite to negative feedback, which tries to nullify any changes to the output. This reinforcing property is useful as it makes the comparator decide the state of the output it wants, and makes it stay there, even within what would normally be the dead zone.

We know that the output state of a Schmitt inverter is the opposite or inverse to that of its input state, (NOT gate principles) and that it can change state at different voltage levels giving it "hysteresis". Schmitt inverters use a Schmitt trigger action that changes state between an upper and a lower threshold level as the input voltage signal increases and decreases about the input terminal as shown in Fig. 5.12. This upper threshold level "sets" the output and the lower threshold level "resets" the output which equates to a logic "0" and a logic "1" respectively for an inverter. Consider the circuit below.

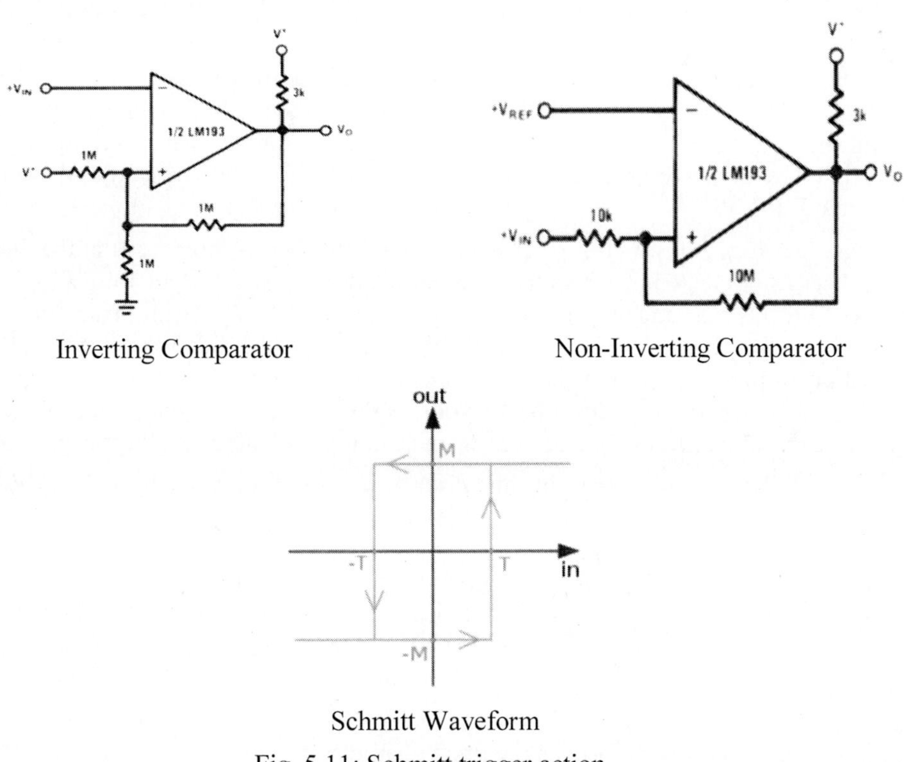

Inverting Comparator

Non-Inverting Comparator

Schmitt Waveform

Fig. 5.11: Schmitt trigger action

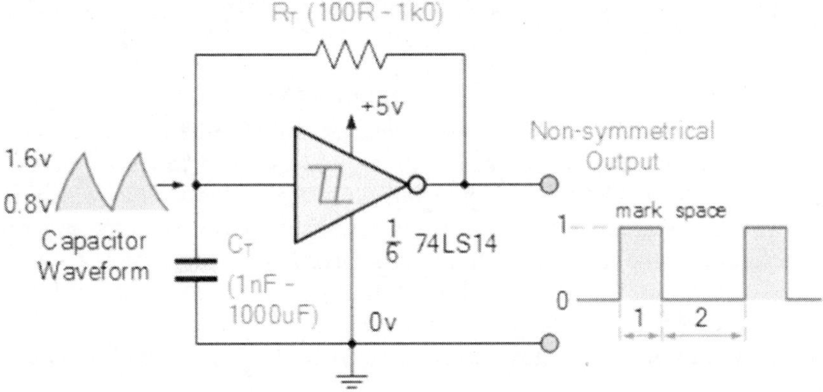

Fig. 5.12: Schmitt Inverter Waveform Generator

This simple waveform generator circuit consists of a single TTL 74LS14 Schmitt inverter logic gate with a capacitor, *C* connected between its input terminal and ground, (0V) and the positive feedback required for the circuit to oscillate being provided by the feedback resistor, *R*.

Assume that the charge across the capacitors plates is below the Schmitt's lower threshold level of 0.8 volt (Datasheet value). This therefore makes the input to the inverter at a logic "0" level resulting in a logic "1" output level (inverter principals). One side of the resistor R is now connected to the logic "1" level (+5V) output while the other side of the resistor is connected to the capacitor, *C* which is at a logic "0" level (0.8v or below). The capacitor now starts to charge up in a positive direction through the resistor at a rate determined by the *RC* time constant of the combination as shown in Fig. 5.12.

When the charge across the capacitor reaches the 1.6-volt upper threshold level of the Schmitt trigger (datasheet value) the output from the Schmitt inverter rapidly changes from a logic level "1" to a logic level "0" state and the current flowing through the resistor changes direction. This change now causes the capacitor that was originally charging up through the resistor, *R* to begin to discharge itself back through the same resistor until the charge across the capacitors plates reaches the lower threshold level of 0.8 volts and the inverters output switches state again with the cycle repeating itself over and over again as long as the supply voltage is present.

So, the capacitor, *C* is constantly charging and discharging itself during each cycle between the inputs upper and lower threshold levels of the Schmitt inverter producing a logic level "1" or a logic level "0" at the inverters output. However, the output waveform is not symmetrical producing a duty cycle of about 33% or 1/3 as the mark-to-space ratio between "HIGH" and "LOW" is 1:2 respectively due to the input gate characteristics of the TTL inverter. The value of the feedback resistor, (*R*) MUST also be kept low to below *1kΩ* for the circuit to oscillate correctly, 220R to 470R is good, and by varying the value of the capacitor, *C* to vary the frequency.

Also, at high frequency levels the output waveform changes shape from a square shaped waveform to a trapezoidal shaped waveform as the input characteristics of the TTL gate are affected by the rapid charging and discharging of the capacitor. The frequency of oscillation for Schmitt Waveform Generators is therefore given as:

$$f = \frac{1}{1.2\,RC}$$

With a resistor value between: 100R to *1kΩ*, and a capacitor value of between: *1nF* to *1000uF*. This would give a frequency range of between *1Hz* to *1MHz*, (high frequencies produce waveform distortion). Generally, standard TTL logic gates do not work too well as waveform generators due to their average input and output characteristics, distortion of the output waveform and low value of feedback resistor required, resulting in a large high value capacitor for low frequency operation. Also, TTL oscillators may not oscillate if the value of the feedback capacitor is too small. However, we can also make Astable Multivibrators using better CMOS logic technology that operate from a 3V to 15V supply such as the CMOS 40106B Schmitt Inverter.

5.3.2 CMOS Schmitt Waveform Generator

The CMOS 40106 is a single input inverter with the same Schmitt-trigger action as the TTL 74LS14 but with very good noise immunity, high bandwidth, high gain and excellent input/output characteristics to produce a more "squarer" output waveform as shown in Fig. 5.13 below.

The Schmitt waveform generators circuit for the CMOS 40106 is basically the same as that for the previous TTL 74LS14 inverter, except for the addition of the 10kΩ resistor which is used to prevent the capacitor from damaging the sensitive MOSFET input transistors as it discharges rapidly at higher frequencies. The mark-space ratio is more evenly matched at about 1:1 with the feedback resistor value increased to below 100kΩ resulting in a smaller and cheaper timing capacitor, C.

The frequency of oscillation may not be the same as: (1/1.2RC) as CMOS input characteristics are different to TTL. With a resistor value between: 1kΩ and 100kΩ, and a capacitor value of between: 1pF to 100uF. This would give a frequency range of between 0.1Hz to 100kHz. Schmitt Inverter Waveform Generators can also be made from a variety of different logic gates connected to form an inverter circuit. The basic Schmitt astable multivibrator circuit can be easily modified with some additional components to produce different outputs or frequencies. For example, two inverse waveforms or multiple frequencies and by changing the fixed feedback resistor to a potentiometer the output frequency can be varied as shown below.

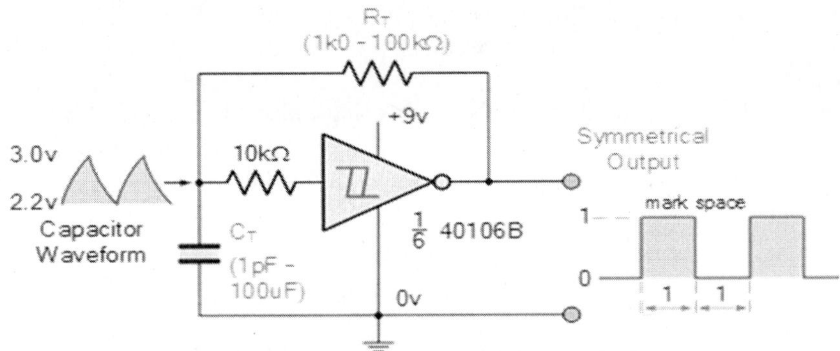

Fig.5.13: CMOS Schmitt Waveform Generator

5.3.3 Clock Waveform Generators

In the circuit given below (Fig. 1), an additional Schmitt Inverter has been added to the output of the Schmitt waveform generator to produce a second waveform that is the inverse or mirror image of the first producing two complementary output waveforms, so when one output is "HIGH" the other is "LOW". This second Schmitt inverter also improves the shape of the inverse output waveform but adds a small "gate delay" to it so it is not exactly in synch with the first.

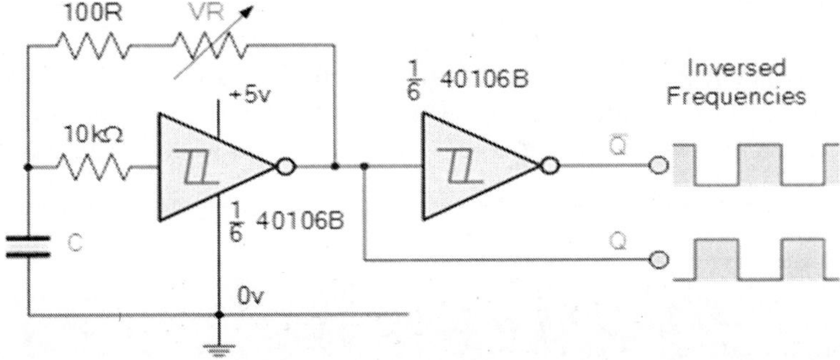

Fig.5.14: Schmitt inverter to improves the shape of the inverse output waveform

Also, the output frequency of the oscillator circuit can be varied by changing the fixed resistor, R into a potentiometer but a smaller feedback resistor is still required to prevent the potentiometer from shorting out the inverter when its at its minimum value, 0Ω. We can also use the two complementary outputs, Q and Q of the first circuit to alternatively flash two sets of lights or LED's by connecting their outputs directly to the bases of two switching transistors. In order to generate a very low frequency output of a few Hertz to flash the LED's, Schmitt waveform generators use high value timing capacitors which themselves can be physically large and expensive.

One alternative solution is to use a smaller value capacitor to generate a much higher frequency, say 1kHz or 10kHz, and then divide this main clock frequency down into individual smaller ones until the required low frequency value is achieved, and the second circuit above does just that.

The circuit as shown in Fig. 5.15 below shows the oscillator being used to drive the clock input of a ripple counter. Ripple counters are basically a number of divide-by-2, D-type flip-flops cascaded together to form a single divide-by-N counter, where N is equal to the counters bit-count such as the CMOS 4024 7-bit Ripple Counter or the CMOS 4040 12-bit Ripple Counter. The fixed clock frequency produce by the Schmitt astable clock pulse circuit is divided into a number of different sub-frequencies such as, *f*÷2, *f*÷4, *f*÷8, *f*÷256, etc, up to the maximum "Divide-by-n" value of the ripple counter being used. The process of using either "Flip-flops", "Binary Counters" or "Ripple Counters" to divide a main fixed clock frequency into different sub-frequencies is known as Frequency Division and we can use it to obtain a number of frequency values from a single waveform generator

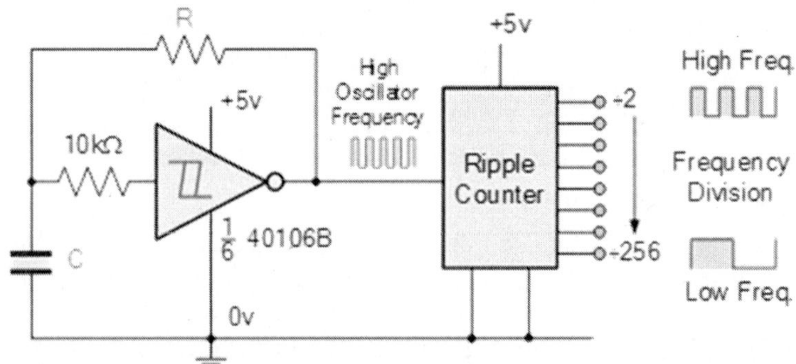

Fig. 5.15: Oscillator being used to drive the clock input

5.3.4 NAND Gate Waveform Generators

Schmitt Waveform Generators can also be made using standard CMOS Logic NAND Gates connected to produce an inverter circuit. Here, two NAND gates are connected together to produce another type of RC relaxation oscillator circuit that will generate a square wave shaped output waveform as shown in Fig. 5.16 below.

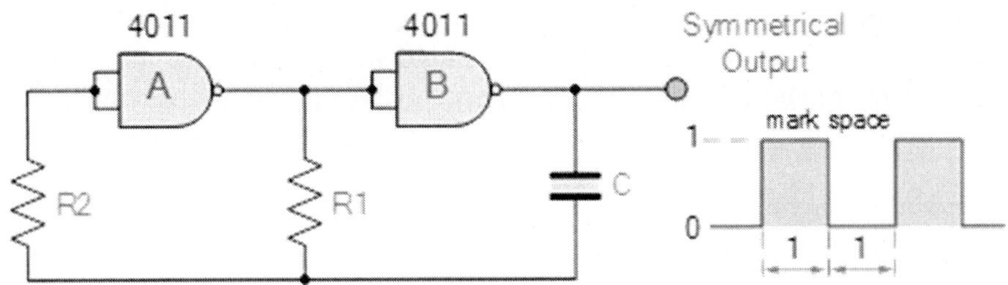

Fig. 5.16: NAND Gate Waveform Generator

In this type of waveform generator circuit, the *RC* network is formed from resistor, R_1 and the capacitor, *C* with this *RC* network being controlled by the output of the first NAND gate. The output from this R_1C network is fed back to the input of the first NAND gate via resistor, R_2 and when the charging voltage across the capacitor reaches the upper threshold level of the first NAND gate, the NAND gate changes state causing the second NAND gate to follow it, thereby change state and producing a change in the output level. The voltage across the R_1C network is now reversed and the capacitor begins to discharge through the resistor until it reaches the lower threshold level of the first NAND gate causing the two gates to change state once again.

Like the previous Schmitt waveform generators circuit above, the frequency of oscillation is determined by the R_1C time constant which is given as

$$\tau = \frac{1}{2.2\ RC}$$

Generally, R_2 is given a value that is 10 times the value of resistor R_1.When high stability or guaranteed self-starting is required, CMOS Waveform Generators can be made using three inverting NAND gates or any three logic inverters for that matter, connected together as shown below producing a circuit that is sometimes called "the ring of three" waveform generator. The frequency of oscillation is determined again by the R1C time constant, the same as for the two gate oscillator above, and which is given as: 1/2.2R1C when R2 has a value that is 10 times the value of resistor, R1.

5.3.5 Stable NAND Gate Waveform Generator

The addition of the extra NAND gate guarantees that the oscillator will start even with very low capacitor values. Also, the stability of the waveform generator is greatly improved as it is less susceptible to power supply variations due to its threshold triggering level being nearly half of the supply voltage.

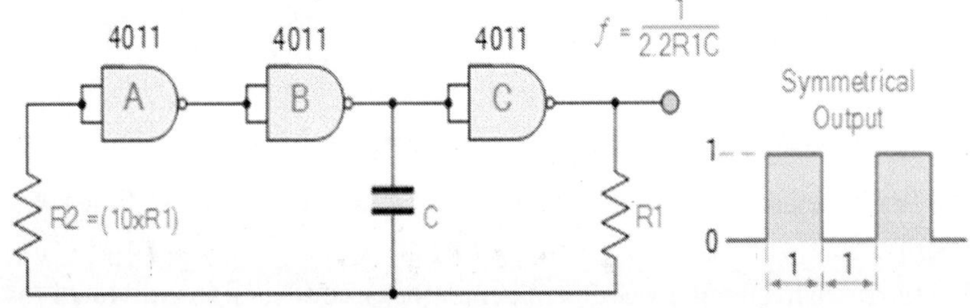

Fig. 5.17: Stable NAND Gate Waveform Generator

The amount of stability is mainly determined by the frequency of oscillation and generally speaking, the lower the frequency the more stable the oscillator becomes. As this type of waveform generator operates at nearly half or 50% of the supply voltage the resultant output waveform has very nearly a 50% duty cycle, 1:1 mark-space ratio. The three-gate waveform generator has many advantages over the previous two gate oscillator above but it's one big disadvantage is that it uses an additional logic gate as shown in Fig. 5.17.

5.3.6 Ring Type Waveform Generator

We have seen above that Waveform Generators can be made using both TTL and the better CMOS logic technology with an RC network producing a time delay within the circuit when connected across either one, two or even three logic gates to form a simple RC Relaxation Oscillator. But we can also make waveform generators using just Logic NOT Gates or in other words Inverters without any additional passive components connected to them as shown in Fig. 5.18.

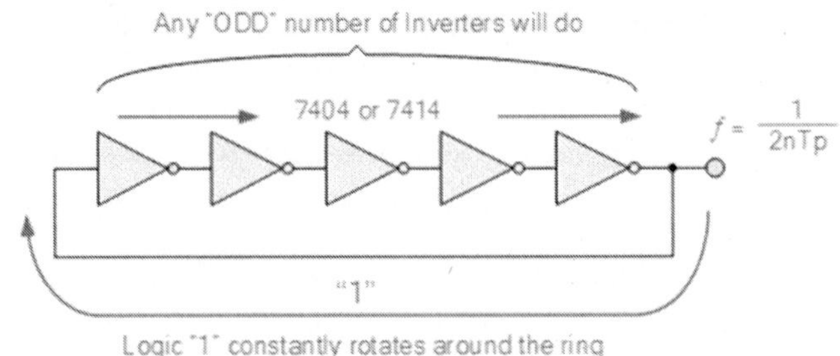

Fig. 5.18: Ring waveform generator circuit

By connecting together any ODD number (3, 5, 7, 9 etc) of NOT gates to form a "ring" circuit, so that the output of the ring is connected straight back to the input of the ring the circuit will continue to oscillate as a logic level "1" constantly rotates around the network producing an output frequency that is determined by the propagation delays of the inverters used.

The frequency of oscillation is determined by the total propagation delay of the Inverters used within the ring and which itself is determined by the type of gate technology, TTL, CMOS, BiCMOS that the inverter is made from. Propagation delay or propagation time, is the total time required (usually in Nanoseconds) for a signal to pass straight through the Inverter from a logic "0" arriving at the input to it producing a logic "1" at its output. Also, for this type of ring waveform generator circuit variations in the supply voltage, temperature and load capacitance all affect the propagation delay of logic gates. Generally, an average propagation delay time will be given in the manufacturers data sheets for the type of digital logic gates being used with the frequency of oscillation given as:

$$f = \frac{1}{2nT_p}$$

Where: f is the frequency of Oscillation, n is the number of gates used and T_p is the propagation delay time for each gate.

For example. Assume that a simple waveform generator circuit has five individual Inverters connected together in series chain to form a Ring Oscillator. If the propagation delay time for each Inverter is given as being 8 Nano-seconds (8ns). Then the frequency of oscillation of the circuit would be given as:

$$f = \frac{1}{2nT_p} = \frac{1}{2 \times 5 \times 8 \times 10^{-9}} = 12.5MHz$$

Of course, this is not really a practical oscillator mainly due to its instability and very high oscillation frequency, 10's of Megahertz depending upon the type of logic gate technology used, and in our simple example it was calculated as 12.5MHz!!

The ring oscillator output frequency can be "tuned" a little by varying the number of Inverters used within the ring but it is much better to use a more stable RC waveform generator like the ones we have discussed above. Nevertheless, it does show that logic gates can be connected together to produce logic-based waveform generators and badly designed digital circuits with lots of gates, signal paths and feedback loops have been known to oscillate unintentionally. By using a RC network across the Inverter circuit, the frequency of oscillation can be accurately controlled producing a more practical astable relaxation oscillator circuit for use in many general electronic applications.

5.4 Sequential Circuit

We already know that the combinational circuits implement the essential functions of a digital computer. "A circuit known as combinational as long as its steady state outputs depend only on its current inputs". In these circuits, there is no ability to retain the information regarding the state of the circuit and any prior input level conditions have no effect on the present outputs because they provide no memory. So, for the later purposes, sequential circuit is used. In sequential logic circuit, the present values of outputs are dependent on both present values of the inputs and the past values of inputs. A sequential logic circuit consist of two parts as shown in Fig. 5.19.

- The *memory elements* i.e. flip-flop which is made up of an assembly of logic gates.
- The *combinational logic circuits* needed to produce the excitation inputs to the memory elements and to produce the required outputs. Sequential circuits find wide application in digital systems as counters, registers, control logic, memories and other complex functions.

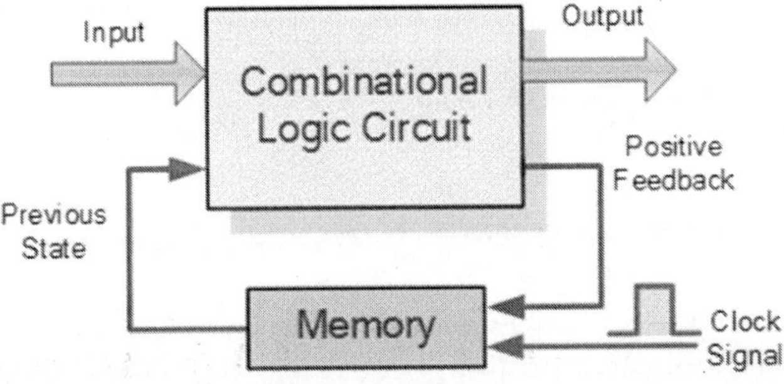

Fig. 5.19: Block diagram of a sequential logic circuit

Unlike Combinational Logic circuits that change state depending upon the actual signals being applied to their inputs at that time, Sequential Logic circuits have some form of inherent "Memory" built in. This means that sequential logic circuits are able to consideration of their previous input state as well as those actually present, a sort of "before" and "after" effect is involved with sequential circuits.

In other words, the output state of a "sequential logic circuit" is a function of the following three states, the "present input", the "past input" and/or the "past output". Sequential Logic circuits remember these conditions and stay fixed in their current state until the next clock signal changes one of the states, giving sequential logic circuits "Memory".

Sequential logic circuits are generally termed as two state or Bistable devices which can have their output or outputs set in one of two basic states, a logic level "1" or a logic level "0" and will remain "latched" (hence the name latch) indefinitely in this current state or condition until some other input trigger pulse or signal is applied which will cause the bistable to change its state once again.

The word "Sequential" means that things happen in a "sequence", one after another and in Sequential Logic circuits, the actual clock signal determines when things will happen next. Simple sequential logic circuits can be constructed from standard Bistable circuits such as: Flipflops, Latches and Counters and which themselves can be made by simply connecting together universal NAND Gates and/or NOR Gates in a particular combinational way to produce the required sequential circuit. Table below shows the Difference Between Combinational and Sequential Circuit

Difference Between Combinational and Sequential Circuit

Parameters	Combinational Circuit	Sequential Circuit
Meaning and Definition	It is a type of circuit that generates an output by relying on the input it receives at that instant, and it stays independent of time.	It is a type of circuit in which the output does not only rely on the current input. It also relies on the previous ones.
Feedback	A Combinational Circuit requires no feedback for generating the next output. It is because its output has no dependency on the time instance.	The output of a Sequential Circuit, on the other hand, relies on both- the previous feedback and the current input. So, the output generated from the previous inputs gets transferred in the form of feedback. The circuit uses it (along with inputs) for generating the next output.
Performance	We require the input of only the current state for a Combinational Circuit. Thus, it performs much faster and better in comparison with the Sequential Circuit.	In the case of a Sequential Circuit, the performance is very slow and also comparatively lower. Its dependency on the previous inputs makes the process much more complex.
Complexity	It is very less complex in comparison. It is because it basically lacks implementation of feedback.	This type of circuit is always more complex in its nature and functionality. It is because it implements the feedback, depends on previous inputs and also on clocks.

Parameters	Combinational Circuit	Sequential Circuit
Elementary Blocks	Logic gates form the building/ elementary blocks of a Combinational Circuit.	Flip-flops form the building/ elementary blocks of a Sequential Circuit.
Operation	One can use these types of circuits for both- Boolean as well as Arithmetic operations.	You can mainly make use of these types of circuits for storing data.

5.4.1 Classification of Sequential Logic

As standard logic gates are the building blocks of combinational circuits, bistable latches and flipflops are the basic building blocks of sequential logic circuits. Sequential logic circuits can be constructed to produce either simple edge-triggered flip-flops or more complex sequential circuits such as storage registers, shift registers, memory devices or counters. Either way sequential logic circuits can be divided into the following three main categories as shown in Fig. 5.20.

- Event Driven – asynchronous circuits that change state immediately when enabled.
- Clock Driven – synchronous circuits that are synchronised to a specific clock signal.
- Pulse Driven – which is a combination of the two that responds to triggering pulses.

Asynchronous sequential circuits

If some or all the outputs of a sequential circuit do not change affect with respect to active transition of clock signal, then that sequential circuit is called as Asynchronous sequential circuit. That means, all the outputs of asynchronous sequential circuits do not change affect at the same time. Therefore, most of the outputs of asynchronous sequential circuits are not in synchronous with either only positive edges or only negative edges of clock signal.

Synchronous sequential circuits

If all the outputs of a sequential circuit change affect with respect to active transition of clock signal, then that sequential circuit is called as Synchronous sequential circuit. That means, all the outputs of synchronous sequential circuits change affect at the same time. Therefore, the outputs of synchronous sequential circuits are in synchronous with either only positive edges or only negative edges of clock signal.

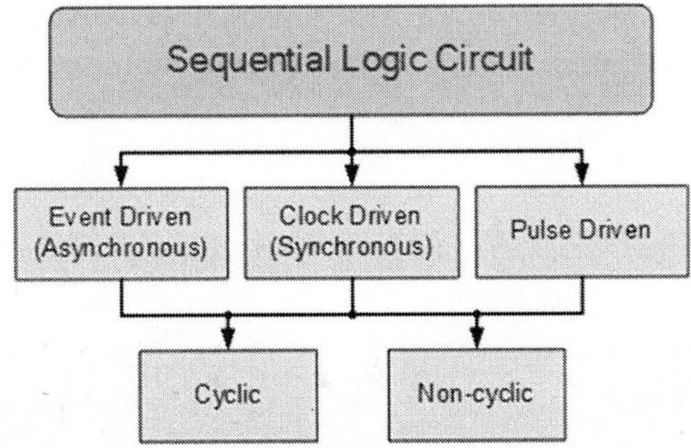

Fig. 5.20: Type of sequential Logic circuits

5.4.2 Clock signal

Clock signal is a periodic signal and its ON time and OFF time need not be the same. We can represent the clock signal as a square wave, when both its ON time and OFF time are same. This clock signal is shown in Fig. 5.21.

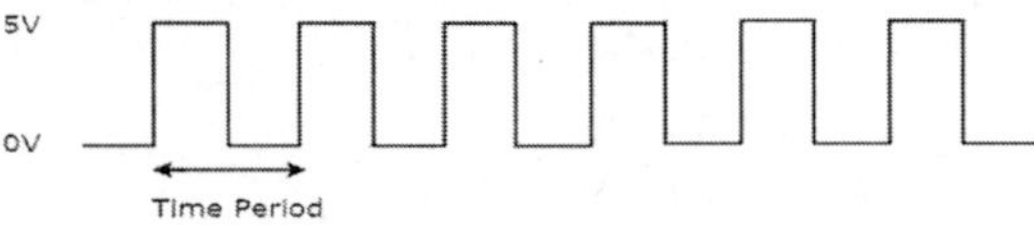

Fig. 5.21: A basic clock signal

In the above figure, square wave is considered as clock signal. This signal stays at logic High 5V for some time and stays at logic Low 0V for equal amount of time. This pattern repeats with some time period. In this case, the time period will be equal to either twice of ON time or twice of OFF time. We can represent the clock signal as train of pulses, when ON time and OFF time are not same. This clock signal is shown in Fig. 5.22.

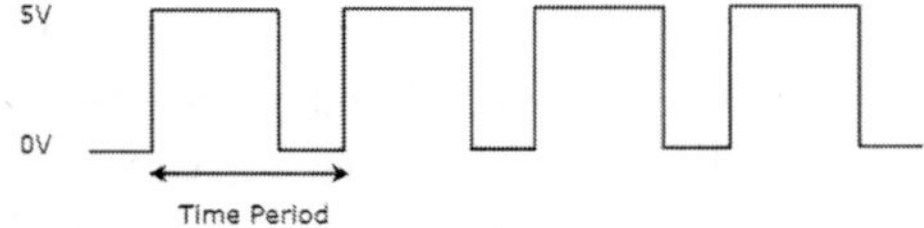

Fig.5.22: Clock signal with ON and OFF time

In the above figure, train of pulses is considered as clock signal. This signal stays at logic High 5V for some time and stays at logic Low 0V for some other time. This pattern repeats with some time period. In this case, the time period will be equal to sum of ON time and OFF time. The reciprocal of the time period of clock signal is known as the frequency of the clock signal. All sequential circuits are operated with clock signal. So, the frequency at which the sequential circuits can be operated accordingly the clock signal frequency has to be chosen.

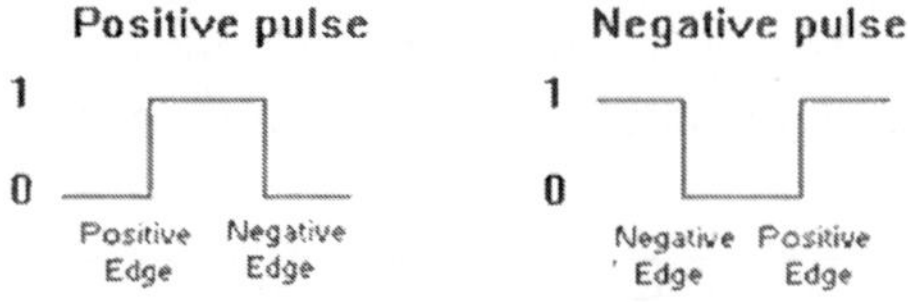

Fig. 5.23: The clock pulse transition

Following are the two possible types of triggering that are used in sequential circuits as shown in Fig. 5.23.

- Level triggering
- Edge triggering

Level triggering: There are two levels, namely logic High and logic Low in clock signal. Following are the two types of level triggering.

- Positive level triggering
- Negative level triggering

If the sequential circuit is operated with the clock signal when it is in Logic High, then that type of triggering is known as Positive level triggering. It is highlighted in Fig. 5.24.

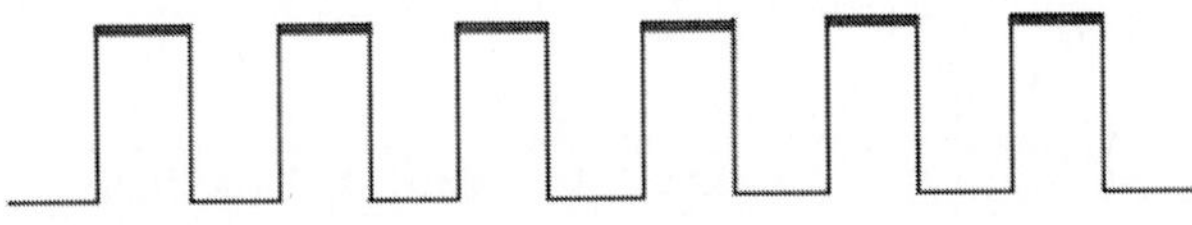

Fig.5.24: The level triggering (Positive Level)

If the sequential circuit is operated with the clock signal when it is in Logic Low, then that type of triggering is known as Negative level triggering. It is highlighted in Fig. 5.25.

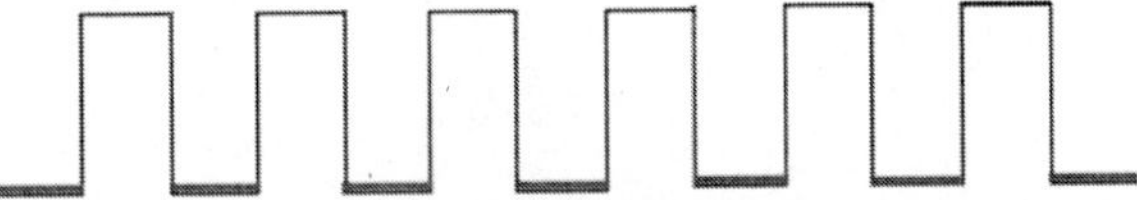

Fig.5.25: The level triggering (Negative Level)

Edge triggering: There are two types of transitions that occur in clock signal. That means, the clock signal transitions either from Logic Low to Logic High or Logic High to Logic Low. Following are the two types of edge triggering based on the transitions of clock signal.

- Positive edge triggering
- Negative edge triggering

If the sequential circuit is operated with the clock signal that is transitioning from Logic Low to Logic High, then that type of triggering is known as Positive edge triggering. It is also called as rising edge triggering. It is shown in Fig. 5.26.

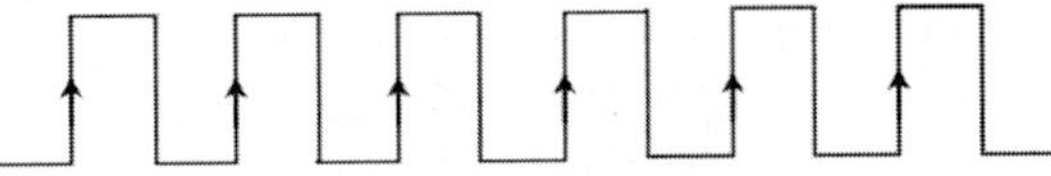

Fig. 5.26: Positive Edge Triggering

If the sequential circuit is operated with the clock signal that is transitioning from Logic High to Logic Low, then that type of triggering is known as Negative edge triggering. It is also called as falling edge triggering. It is shown in Fig. 5.27.

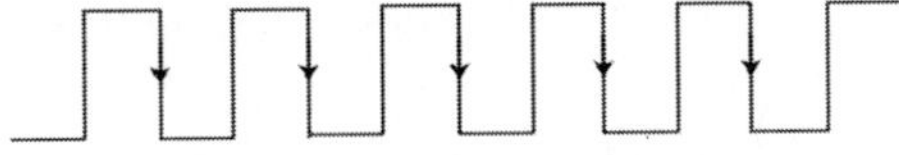

Fig. 5.27: Negative Edge Triggering

5.5 Latches

There are two types of memory elements based on the type of triggering that is suitable to operate it.

- Latches
- Flip-flops

Latches operate with enable signal, which is **level sensitive**. Whereas, flip-flops are edge sensitive. We will discuss about flip-flops in next chapter. Now, let us discuss about SR Latch & D Latch one by one.

5.5.1 SR Latch

SR Latch is also called as **Set Reset Latch**. This latch affects the outputs as long as the enable, E is maintained at '1'. The **circuit diagram** of SR Latch along with its State Table is shown in Fig. 5.28.

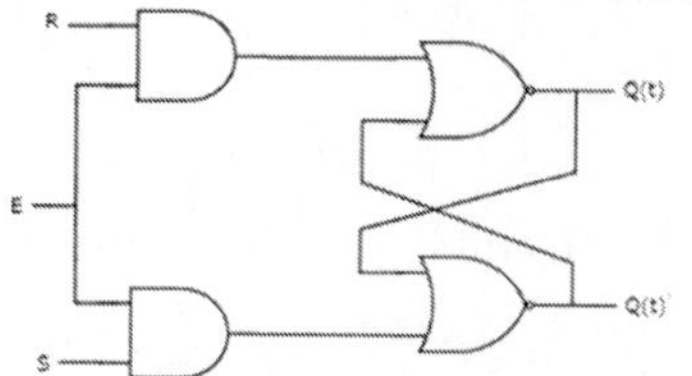

Circuit Diagram for SR Latch

Inputs		Outputs
S	*R*	Q_{t+1}
0	0	Q_t *(Hold)*
0	1	0 *(Reset)*
1	0	1 *(Set)*
1	1	Undefined

State Table for SR Latch

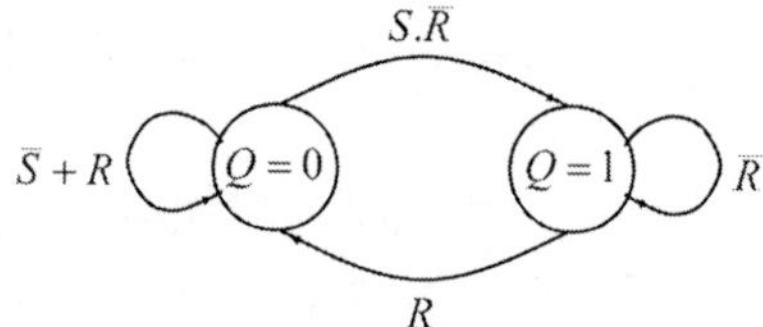

State Diagram

Fig.5.28: SR Latch (a) Circuit diagram (b) State table (c) state diagram

This circuit has two inputs *S* & *R* and two outputs Q_t & Q_t' The upper NOR gate has two inputs *R* & complement of present state, Q_t' and produces next state, Q_{t+1} when enable, E is '1'.

Similarly, the lower NOR gate has two inputs *S* & present state, Q_t and produces complement of next state, $Q_{t+1}$when enable, *E* is '1'. We know that a 2-input NOR gate produces an output, which is the complement of another input when one of the inputs is '0'. Similarly, it produces '0' output, when one of the inputs is '1'.

- If *S* = *1*, then next state Q_{t+1} will be equal to '1' irrespective of present state, Q_t values.
- If *R* = *1*, then next state Q_{t+1} will be equal to '0' irrespective of present state, Q_t values.
- If both inputs *(S & R)* are '1', then the next state Q_{t+1} value is undefined.
- If both inputs *(S & R)* are '', then the next state $Q_{t+1} = Q_t$ value is undefined.

5.5.2 D Latch

There is one drawback of SR Latch. That is the next state value can't be predicted when both the inputs S & R are one. So, we can overcome this difficulty by D Latch. It is also called as Data Latch. The circuit diagram of D Latch is shown in Fig. 5.29.

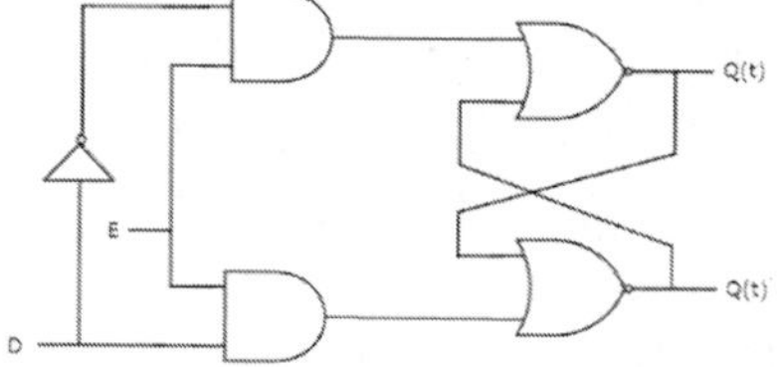

Circuit Diagram for D Latch

Inputs	Outputs
D	Q_{t+1}
0	0 *(Reset)*
1	1 *(Set)*

State Table for D Latch

Fig. 5.29: Circuit diagram and truth table for D Latch

This circuit has single input D and two outputs Q_t & Q_t'. *D* Latch is obtained from *SR* Latch by placing an inverter between *S* & *R* inputs and connect *D* input to *S*. That means we eliminated the combinations of *S* & *R* are of same value.

- If D = 0, i.e., S = 0 then due to invert operation R = 1, then next state Q_{t+1} will be equal to '0' irrespective of present state, Q_t values. This is corresponding to the second row of *SR* Latch state table.
- If D = 1, i.e., S = 1 then due to invert operation R = 0, then next state Q_{t+1} will be equal to '1' irrespective of present state, Q_t values. This is corresponding to the third row of *SR* Latch state table.
- Therefore, D Latch Hold the information that is available on data input, D. That means the output of D Latch is sensitive to the changes in the input, D as long as the enable is High. Here, various Latches have been implemented using cross coupling between NOR gates. Similarly, these Latches may be implemented using NAND gates.

5.5.3 Difference between Latch and Flip Flop

Difference between Latch and Flip-Flop

Latch	Flip-Flop
The latch is transparent, because the input is directly connected to the output when enable is high. It means Latch is sensitive to pulse duration (also called soft barrier)	Flip-flop is a pair of latches (master and slave flop). Flip-flop is sensitive to pulse transition. The signal only propagates through on the rising/falling edge (also called hard barrier)
Less Area/Power (less gates)	More Area/ Power (more gates) because the flip-flop contains two latches.
Fast :(The longer combinational path can be compensated by shorter path delays in the subsequent logic stages. That's why, for higher performance, circuits designer are turning to latched-based design.)	Slow: (The delay of a combinational logic path of a design using edge-triggered flip-flops is always less than the clock period except for those specified as false paths and multiple-cycle paths. Hence the longest path of a design limits the circuit performance.)
Require more tool manipulation and more hand calculations to verify that they meet the timing	Easy to check design timing using Static Timing Analysis (STA) tools
Cycle-borrowing to gain more setup time on the next register stage, as long as each loop completes in one cycle. To meet the timing in the design, Designers consider latches to adjust timing mismatch.	Data launches on one rising edge, so it must be set up before the next rising edge. If it arrives late, the system fails. If it arrives early, time is wasted due to hard edges in Flops
For ASICs with large clock skew, latches have substantial benefits for reducing the clock period	Even for the high-speed pulsed flip-flops with zero setup time, as they are not transparent, the impact of the clock skew is not reduced
Level-sensitive latches reduce the impact of the inaccuracy of wire load models and process variation.	Flip-flops demand the highly accurate wire load model and process
In DFT, Latches are needed as a lockup state at the clock domain crossings in the scan chain to avoid unpredictable behavior	In DFT, use flops that can be scanned (controllable and observable)

Latch	Flip-Flop
In FPGA, level-sensitive transparent latches should be avoided in FPGAs	In FPGA, edge-sensitive flip-flops are used exclusively. Timings analysis is more appropriate with flops for FPGA tools
Circuit analysis is complex. You may see last minutes timing mismatch surprises at the implantation stage.	Circuit analysis is easy
High-speed microprocessor designs typically use master-slave latches instead of flip-flops so that logic can be added between the rising and falling clock edges. Most of these companies have written their own specialized STA tools to verify latch-based designs.	The most commonly used flop in the design world is D type flip-flop. FSM implementation mostly involves D Flip-flops due to a minimum number of logic gates and lesser cost as compared to other types of flip-flops.
For non-timing-critical configuration registers, latches work great, duc to fewer gates and less power consumption	For non-power aware design, Flip flops are preferred over Latches
The latch is an asynchronous block. Therefore, you must ensure that the combinational functions, which generate input signals for the latch, are race-free. Otherwise, they may generate glitches, which may be latched, causing hazards in your system.	A flip-flop, on the other hand, is edge-triggered and only changes state when a control signal goes from high to low or low to high
Latch-based design is noisy because any noise in the enable signal disrupts the latch output easily.	Flip-flop-based design is robust

5.6 Flip Flops

We can implement flip-flops in two methods. In first method, cascade two latches in such a way that the first latch is enabled for every positive clock pulse and second latch is enabled for every negative clock pulse. So that the combination of these two latches become a flip-flop. In the second method, we can directly implement the flip-flop, which is edge sensitive. In this section, let us discuss the following flip-flops using second method.

- SR Flip-Flop
- D Flip-Flop
- JK Flip-Flop
- T Flip-Flop

5.6.1 SR Flip-Flop

The SR flip-flop, can be considered as one of the most basic sequential logic circuit possible. This simple flip-flop is basically a one-bit memory bistable device that has two inputs, one which will "SET" the device (meaning the output = "1"), and is labelled *S* and one which will "RESET" the device (meaning the output = "0"), labelled *R*. Then the *SR* description stands for "Set-Reset".

The reset input resets the flip-flop back to its original state with an output Q that will be either at a logic level "1" or logic "0" depending upon this set/reset condition.

A basic NAND gate SR flip-flop circuit provides feedback from both of its outputs back to its opposing inputs and is commonly used in memory circuits to store a single data bit. Then the SR flip-flop actually has three inputs, Set, Reset and its current output Q relating to it's current state or history. *The term "Flip-flop" relates to the actual operation of the device, as it can be "flipped" into one logic Set state or "flopped" back into the opposing logic Reset state (Refer Fig. 5.30).*

The NAND Gate SR Flip-Flop is the simplest way to make any basic single bit set-reset SR flip-flop is to connect together a pair of cross-coupled 2-input NAND gates as shown, to form a Set-Reset Bistable also known as an *active LOW SR NAND Gate*, so that there is feedback from each output to one of the other NAND gate inputs. This device consists of two inputs, one called the Set, *S* and the other called the Reset, *R* with two corresponding outputs *Q* and its inverse or $\bar{Q}$ (not-Q) as shown in Fig. 5.31.

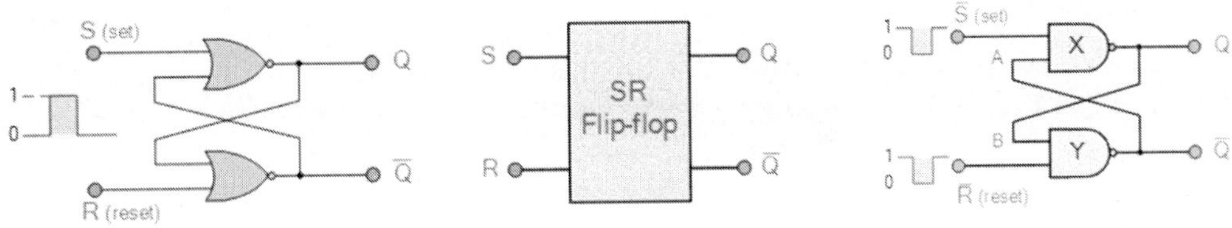

(a) SR Flip Flop using NOR (b)Symbol SR Flip – Flop (c) SR Flip Flop using NAND

Fig. 5.30: Circuit Diagram of SR Flip Flop and its symbol (Using NOR gates)

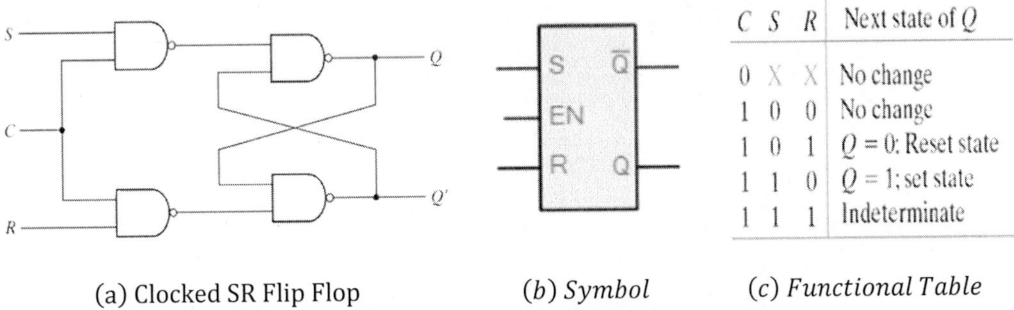

C	S	R	Next state of Q
0	X	X	No change
1	0	0	No change
1	0	1	Q = 0; Reset state
1	1	0	Q = 1; set state
1	1	1	Indeterminate

(a) Clocked SR Flip Flop (*b*) *Symbol* (*c*) *Functional Table*

Fig. 5.31: Circuit Diagram of SR Flip Flop and its symbol (Using NAND gates)

The Set State (Refer Fig. 5.30 (c))

Consider the circuit shown above If the input *R* is at logic level *"0" (R = 0)* and input *S* is at logic level *"1" (S = 1),* the NAND gate *Y* has at least one of its inputs at logic *"0"* therefore, its output $\bar{Q}$ must be at a logic level *"1"* (NAND Gate principles). Output *Q* is also fed back to input *"A"* and so both inputs to NAND gate *X* are at logic level *"1"*, and therefore its output *Q* must be at logic level *"0"*.

Again, as per the NAND gate principals. If the reset input *R* changes state, and goes HIGH to logic *"1"* with *S* remaining HIGH also at logic level *"1"*, NAND gate *Y* inputs are now *R* = *"1"* and *B* = *"0"*. Since one of its inputs is still at logic level *"0"* the output at *Q* still remains HIGH at logic level *"1"* and there is no change of state. Therefore, the flip-flop circuit is said to be "Set" with $\bar{Q}$ = *"1"* and *Q* = *"0"*.

Reset State (Refer Fig. 5.30(c))

In this second stable state, Q is at logic level "0", ($\bar{Q}$ = 0) its inverse output i.e., Q is at logic level "1", (Q = 1), and is given by R = "1" and S = "0".

As gate *X* has one of its inputs at logic "0" its output Q must equal logic level "1" (again NAND gate principles). Output Q is fed back to input "B", so both inputs to NAND gate *Y* are at logic "1", therefore, Q = "0".

If the set input, 'S' now changes state to logic "1" with input R remaining at logic "1", output Q still remains LOW at logic level "0" and there is no change of state. Therefore, the flip-flop circuits "Reset" state has also been latched and we can define this "set/reset" action in the following truth table.

Truth Table for this Set-Reset Function

State	S	R	Q	$\bar{Q}$	Description
Set	1	0	0	1	Set $\bar{Q}$ » 1
	1	1	0	1	no change
Reset	0	1	1	0	Reset $\bar{Q}$ » 0
	1	1	1	0	no change
Invalid	0	0	1	1	Invalid Condition

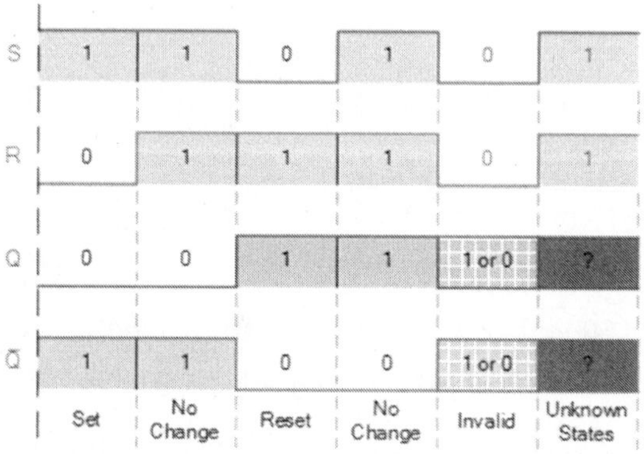

Fig. 5.32: Timing Diagram for SR FF

- It can be seen that when both inputs S = "1" and R = "1" the outputs Q and Q can be at either logic level "1" or "0", depending upon the state of the inputs S or R BEFORE this input condition existed. Therefore, the condition of S = R = "1" does not change the state of the outputs Q and Q.

- However, the input state of S = "0" and R = "0" is an undesirable or invalid condition and must be avoided. The condition of S = R = "0" causes both outputs Q and Q to be HIGH together at logic level "1" when we would normally want Q to be the inverse of Q.

The result is that the flip-flop loses control of Q and Q, and if the two inputs are now switched "HIGH" again after this condition to logic "1", the flip-flop becomes unstable and switches to an unknown data state based upon the unbalance as shown in the following switching diagram (Refer Fig. 5.32).

S-R Flip-flop Switching Diagram

This unbalance can cause one of the outputs to switch faster than the other resulting in the flip-flop switching to one state or the other which may not be the required state and data corruption will exist. This unstable condition is generally known as its **Meta-stable** state.

Then, a simple NAND gate SR flip-flop or NAND gate SR latch can be set by applying a logic "0", (LOW) condition to its Set input and reset again by then applying a logic "0" to its Reset input. The SR flip-flop is said to be in an "invalid" condition (Meta-stable) if both the set and reset inputs are activated simultaneously.

As we have seen above, the basic NAND gate SR flip-flop requires logic "0" inputs to flip or change state from Q to Q and vice versa. We can however, change this basic flip-flop circuit to one that changes state by the application of positive going input signals with the addition of two extra NAND gates connected as inverters to the S and R inputs as shown.

Excitation Table of the SR Latch

During the design process we usually know the transition from present state to next state and wish to find the latch input conditions that will cause the required transition. For this reason, we need a table that lists the required inputs for a given change of state. Such a table is called an excitation table. and it specifies the excitation behaviour of the sequential circuits. These are used in the synthesis (design) of sequential circuits, which we shall see later. The excitation of the SR latch is shown in Fig. 5.33.

Q_t	S	R	Q_{t+1}
0	0	0	0
0	0	1	0
0	1	0	1
0	1	1	Not Used
1	0	0	1
1	0	1	0
1	1	0	1
1	1	1	Not Used

Excitation Table for SR Flip Flop

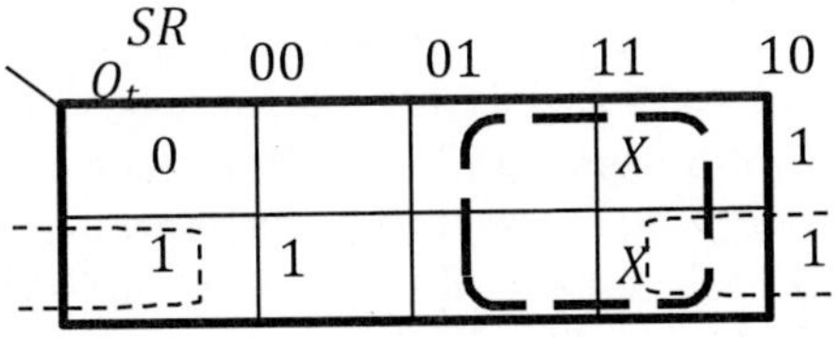

Q_t \ SR	00	01	11	10
0			X	1
1	1		X	1

K-Map for Q_{t+1} in SR FF

Fig. 5.33: SR Flip Flop (a) Excitation Table (b) K-map for Excitation table

Hence the excitation equation (Sometime called Characteristic equation) of the SR flip flop is given by

$$Q_{t+1} = S + \bar{R}Q_t$$

5.6.2 CLOCKED D FLIP-FLOP

One way to eliminate the undesirable condition of the indeterminate state in the SR latch is to ensure that inputs S and R both should never equal to 1 at the same time. This is done in the D latch. The D flip-flop has only one input referred to as the D (data) input & two outputs as usual Q and Q'. It transfers the data at the input after the delay of one clock pulse at the output Q. So, in some cases the input is referred to as a delay input and the flip-flop gets the name delay (D) flip-flop. It can be easily constructed from an S-R flip-flop by simply incorporating an inverter between S and R such that the input of the inverter is at the S end & the output of the inverter is at the R end.

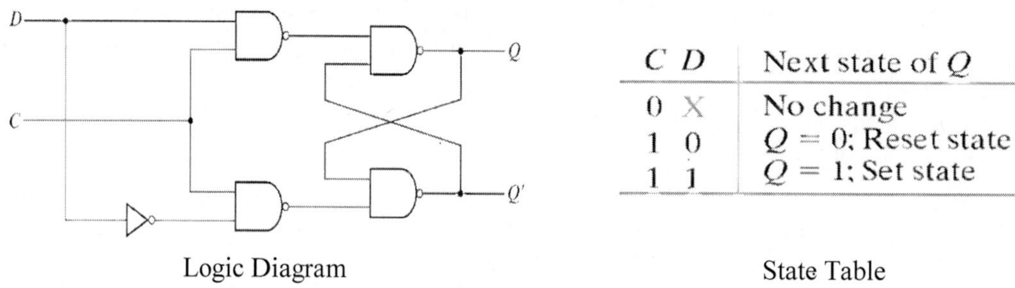

C D	Next state of Q
0 X	No change
1 0	Q = 0; Reset state
1 1	Q = 1; Set state

Logic Diagram — State Table

Fig. 5.34: D flip-flop (a) Logic Diagram (b) State Table

We can get rid of the undefined condition, i.e., S = R = 1 condition, of the S-R flip-flop in the D flip flop. The D flip-flop is either used as a delay device or as a latch to store one bit of binary information. The truth table of D flip-flop is given in the table below. The structure of the D flip-flop is shown in Figure below, which is being constructed using NAND gates. The same structure can be constructed using only NOR gates. Logic diagram and the state table is shown in Fig. 5.34.

- ***Case 1.*** If the CLK input is low, the value of the D input has no effect, since the S and R inputs of the basic NAND flip-flop are kept as 1.
- ***Case 2.*** If the CLK = 1 and D = 1, the NAND gate 1 produces 0, which forces the output of NAND gate 3 as 1. On the other hand, both the inputs of NAND gate 2 are 1, which gives the output of gate 2 as 0. Hence, output of NAND gate 4 is forced to be 1, i.e., Q = 1, whereas both the inputs of gate 5 are 1 and the output is 0, i.e., Q'=0. Hence, we find that when D = 1, after one clock pulse passes Q = 1, which means the output follows D.
- ***Case 3.*** If the CLK = 1, and D = 0, the NAND gate 1 produces 1. Hence both the inputs of NAND gate 3 are 1, which gives the output of gate 3 as 0. On the other hand, D = 0 forces the output of NAND gate 2 to be 1. Hence the output of NAND gate 5 is forced to be 1, i.e., Q' = 1, whereas both the inputs of gate 4 are 1 and the output is 0, i.e, Q = 0. Hence, we find that when D = 0, after one clock pulse passes Q = 0, which means the output again follows D.

A simple way to construct a D flip-flop using an S-R flip-flop is shown in Fig. 5.35. The logic symbol of a D flip-flop is also shown below. A D flip-flop is most often used in the construction of sequential circuits like registers.

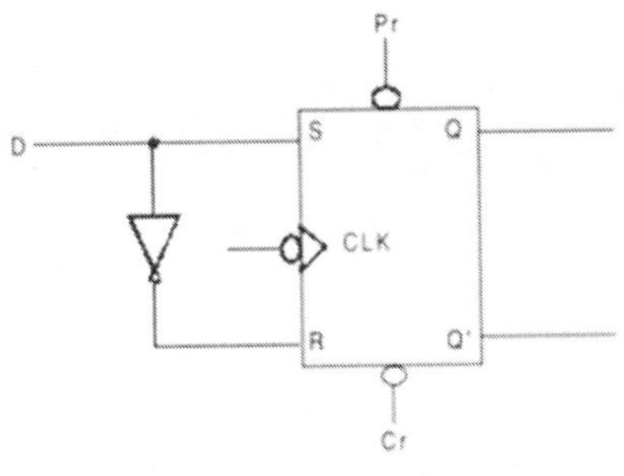

(a) D FF using RS FF

Logic symbol of D FF

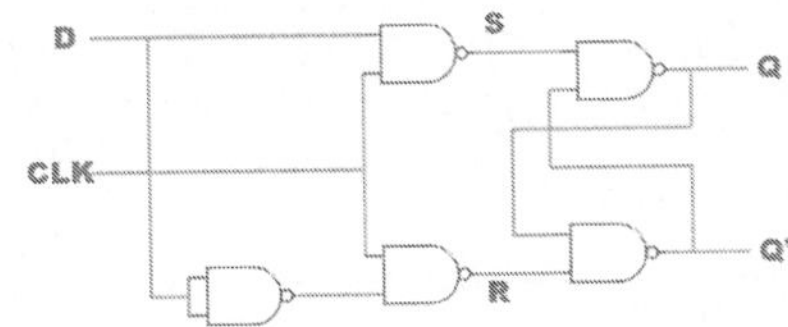

(c) Conversion of RS FF to D FF

Input D_n	*Output* Q_{n+1}
0	0
1	1

State Table

Fig. 5.35: D Flip Flop

Excitation Table of a D Flip-flop

As we have already discussed the characteristic equation of an S-R flip-flop, we can similarly find out the characteristic equation of a D flip-flop. The characteristic table of a D flip-flop is shown in Fig. 5.36. From the characteristic table we have to find out the characteristic equation of the D flip-flop.

Q_t	D	Q_{t+1}
0	0	0
0	1	1
1	0	0
1	1	1

Excitation Table for D Flip Flop

Q_t \ D	0	1
0		1
1		1

K-Map for Q_{t+1} in D FF

Fig. 5.36: D Flip Flop (a) Excitation Table (b) K-Map for $\boldsymbol{Q_{t+1}}$

Hence the excitation equation (Sometime called Characteristic equation) of the D flip flop is given by

$$Q_{t+1} = D$$

5.6.3 J-K Flip-Flop

A J-K flip-flop (As shown in Fig. 5.37) has very similar characteristics to an S-R flip-flop. The only difference is that the undefined condition for an S-R flip-flop, i.e., S=R=1 condition, is also included in this case. Inputs J and K behave like inputs S and R to set and reset the flip-flop respectively. When J = K = 1, the flip-flop is said to be in a toggle state, which means the output switches to its complementary state every time a clock pass.

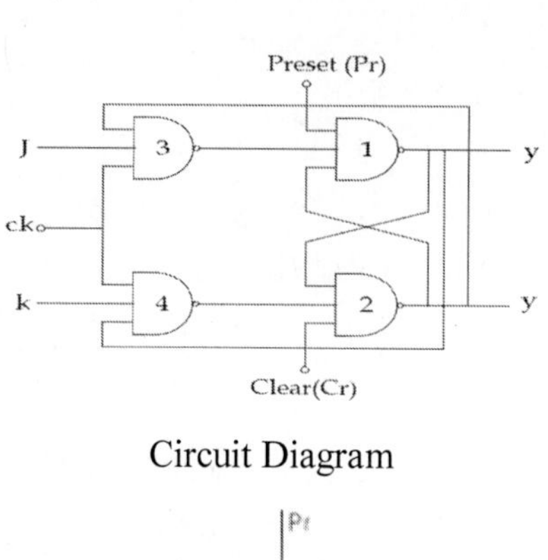

Circuit Diagram

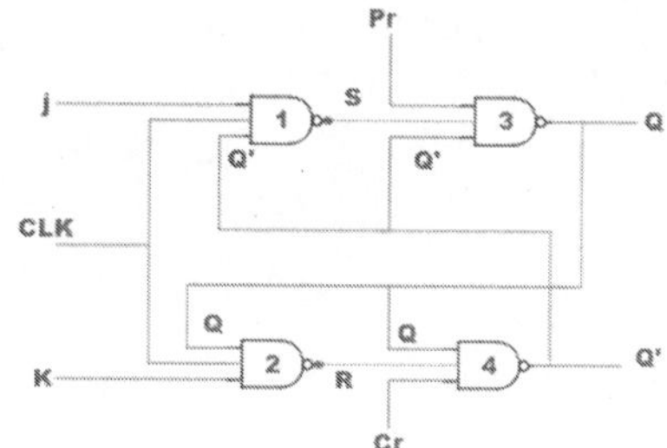

Logic Diagram

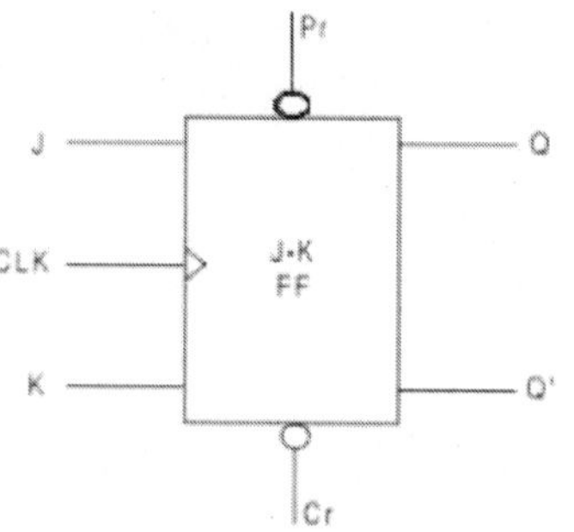

Logic Symbol

Clk	J	K	Qn+1
0	X	X	Memory
1	0	0	Memory
1	0	1	0
1	1	0	1 (Set)
1	1	1	Toggle

State table

Fig. 5.37: JK Flip Flop

An S-R flip-flop converted into a J-K flip-flop (Refer Fig. 5.38)

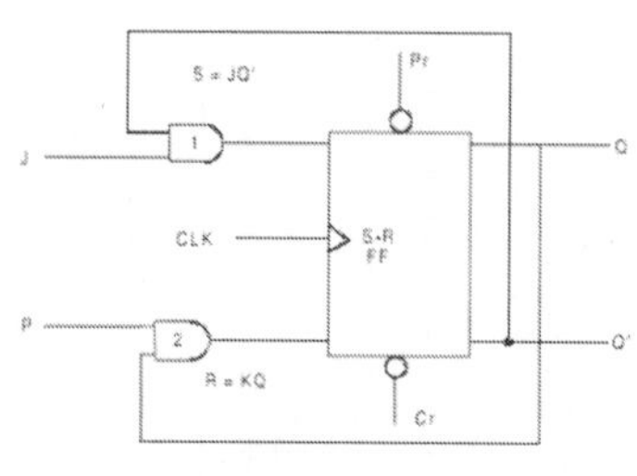

JK Using SR

Data inputs		Outputs		Inputs to S-R FF		Output
J_n	K_n	Q_n	Q'_n	S_n	R_n	Q_{n+1}
0	0	0	1	0	0	0
0	0	1	0	0	0	1
0	1	0	1	0	0	0
0	1	1	0	0	1	0
1	0	0	1	1	0	1
1	0	1	0	0	0	1
1	1	0	1	1	0	1
1	1	1	0	0	1	0

Excitation Table

Fig. 5.38: JK Flip Flop (a) Using SR FF and (b) Excitation Table

- ***Case 1.*** When the clock is applied and J = 0, whatever the value of Q', (0 or 1), the output of NAND gate 1 is 1. Similarly, when K = 0, whatever the value of Q (0 or 1), the output of gate 2 is also 1. Therefore, when J = 0 and K = 0, the inputs to the basic flip-flop are S = 1 and R = 1. This condition forces the flip-flop to remain in the same state (Refer Fig. 5.40).
- ***Case 2.*** When the clock is applied and J = 0 and K = 1 & the previous state of the flip-flop is reset (i.e., Q =0 and Q' = 1), then S=1and R= 1. Since S =1 and R = 1, the basic flip-flop does not alter the state and remains in the reset state. But if the flip-flop is in set condition (i.e., Q=1& Q' =0), then S=1 and R =0. Since S = 1 and R = 0, the basic flip-flop changes its state and resets.

- ***Case 3.*** When the clock is applied and J = 1 and K = 0 and the previous state of the flip-flop is reset (i.e., Q=0 and Q'= 1), then S=0 and R = 1. Since S = 0 and R = 1, the basic flip-flop changes its state and goes to the set state. But if the flip-flop is already in set condition (i.e., Q=1 and Q=0), then S=1 and R = 1. Since S = 1 and R = 1, the basic flip-flop does not alter its state and remains in the set state.
- ***Case 4.*** When the clock is applied and J = 1 and K = 1 and the previous state of the flip-flop is reset (i.e., Q=0 and Q'=1), then S=0 and R = 1. Since S = 0 and R = 1, the basic flip-flop changes its state and goes to the set state. But if the flip-flop 1s already in set condition (i.e., Q, = 1 and Q' = 0), then S=1 and R = 0. Since S = 1 and R = 0, the basic flip-flop changes its state and goes to the reset state. So, we find that for J=1 and K = 1, the flip-flop toggles its state from set to reset and vice versa. Toggle means to switch to the opposite state.

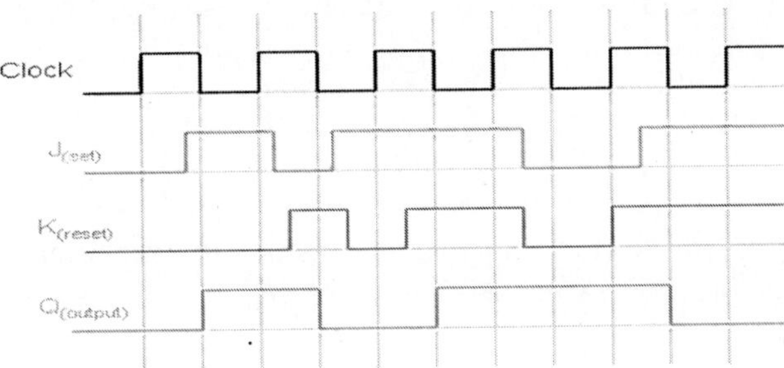

Fig. 5.39: Timing Diagram

Excitation Table of a J-K Flip-flop

As we have already discussed the characteristic equation of an S-R flip-flop, we can similarly find out the characteristic equation of a J-K flip-flop. The characteristic table of a J-K flip-flop is given in the table below. From the characteristic table we have to find out the characteristic equation of the J-K flip-flop as shown in Fig. 5.40.

Flip Flop Inputs		**Present State**	**Next State**
J	K	Q_t	Q_{t+1}
0	0	0	0
0	0	1	1
0	1	0	0
0	1	1	0
1	0	0	1
1	0	1	1
1	1	0	1
1	1	1	0

Excitation Table for JK Flip Flop

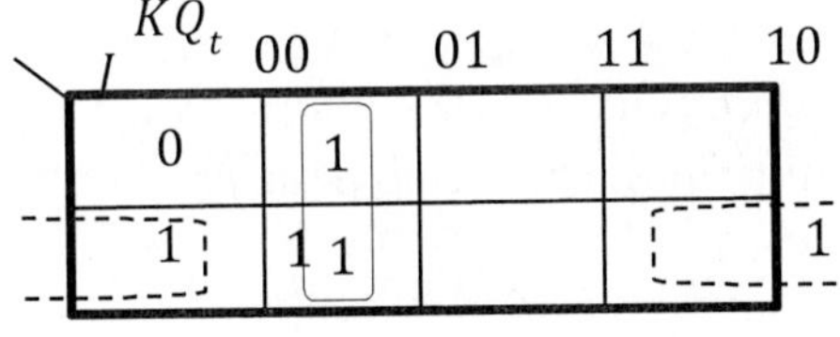

K-Map for Q_{t+1}

Fig. 5.40: JK Flip Flop (a) Excitation Table (b) K-Map for $\boldsymbol{Q_{t+1}}$

Hence the excitation equation (Sometime called Characteristic equation) of the SR flip flop is given by

$$Q_{t+1} = J\overline{Q_t} + \bar{K}Q_t$$

Race-around Condition of a J-K Flip-flop

The inherent difficulty of an S-R flip-flop (i.e., S = R = 1) is eliminated by using the feedback connections from the outputs to the inputs of gate 1 and gate 2 as discussed in JK flip-flop. Truth tables JK flip-flop were formed with the assumption that the inputs do not change during the clock pulse (CLK = 1). But the consideration is not true because of the feedback connections. Consider, for example, that the inputs are J=K =1and Q=1, and a pulse as shown in Fig. 5.41 below, is applied at the clock input.

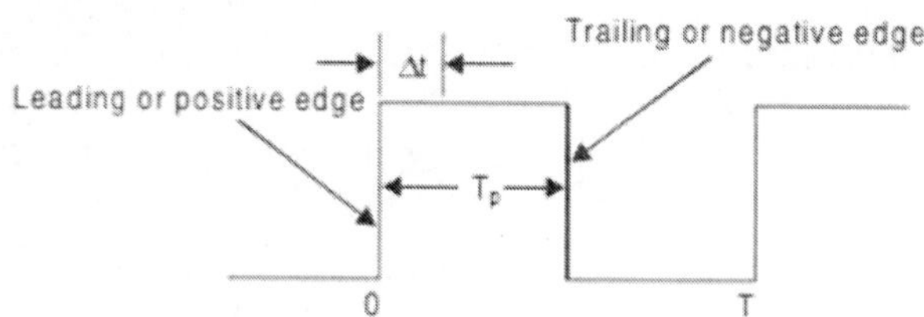

Fig. 5.41: The clock input while considering Race-around condition

Consider, for example, that the inputs are J =K = 1 and Q = 1, and a pulse as shown above is applied at the lock input. After a time, interval at equal to the propagation delay through two NAND gates in series, the outputs will change to Q = 0. So now we have J =K = 1 and Q = 0. After another time interval of at the output will change back to Q = 1. Hence, we conclude that for the time duration of T_p, of the clock pulse, the output will oscillate between 0 and 1. Hence, at the end of the clock pulse, the value of the output is not certain. This situation is referred to as a race-around condition.

Generally, the propagation delay of TTL gates is of the order of nanoseconds. So, if the clock pulse is of the order of microseconds, then the output will change thousands of times within the clock pulse. This race-around condition can be avoided if $t_p < \Delta t < T$. Due to the small propagation delay of the ICs it may be difficult to satisfy the above condition. A more practical way to avoid the problem is to use the master-slave (M-S) configuration as discussed below.

5.6.4 Master-Slave J-K Flip-flop

A master-slave (M-S) flip-flop is shown in Figure below. Basically, a master-slave flip-flop is a system of two flip-flops. One being designated as master and the other is the slave. From the figure below, we see that a clock pulse is applied to the master and the inverted form of the same clock pulse is applied to the slave.

When CLK = 1, the first flip-flop (i.e., the master) is enabled and the outputs Q, and Q', respond to the inputs J and K according to the table shown in Figure 7.13. At this time the second flip-flop (i.e., the slave) is disabled because the CLK is LOW to the second flip-flop. Similarly, when CLK becomes LOW, the master becomes disabled and the slave becomes active, since now the CLK to it is HIGH. Therefore, the outputs Q and Q' follow the outputs Q, and Q', respectively. Since the second flip-flop just follows the first one, it is referred to as a slave and the first one is called the master. Hence, the configuration is referred to as a master-slave (M-S) flip-flop as shown in Fig. 5.42.

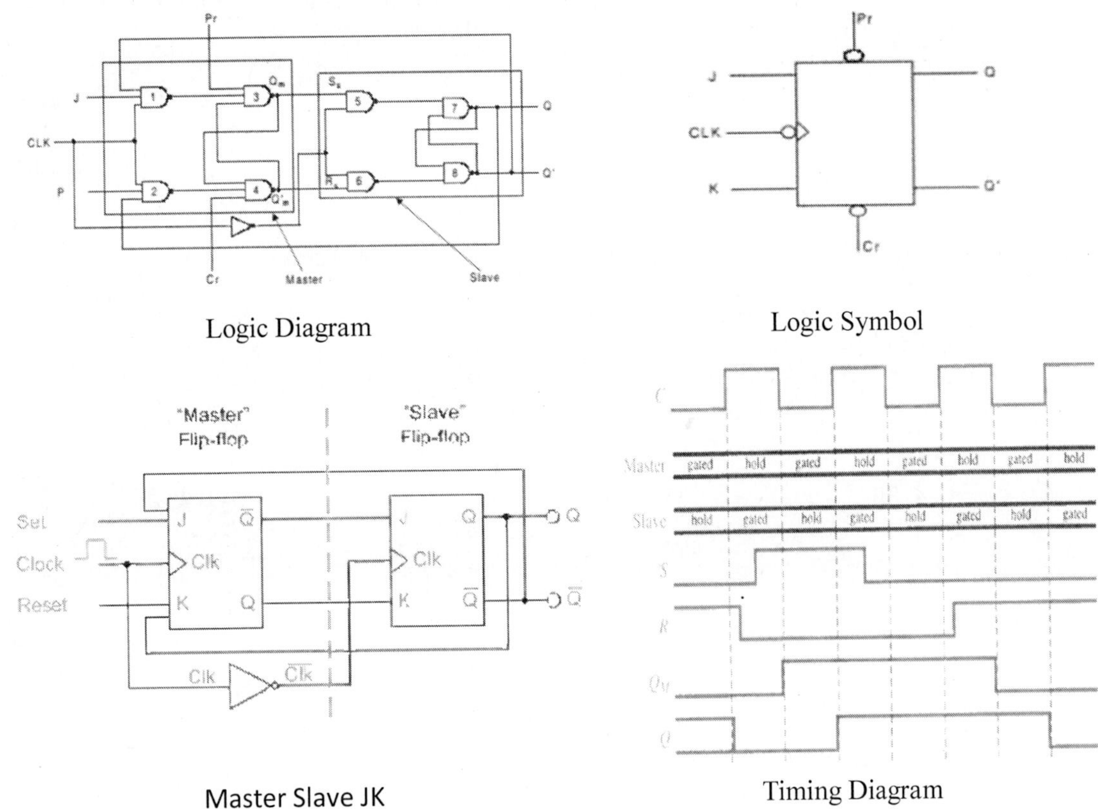

Logic Diagram

Logic Symbol

Master Slave JK

Timing Diagram

Fig. 5.42: Master slave JK FF (a) Logic Diagram (b) Logic Symbol (c) using JK FF (d) Timing Diagram

In this type of circuit configuration, the inputs to the gates 5 and 6 do not change at the time of application of the clock pulse. Hence the race-around condition does not exist. The state of the master-slave flip-flop, shown in above Figure, changes at the negative transition (trailing edge) of the clock pulse. Hence, it becomes negative triggering a master-slave flip-flop. This can be changed to a positive edge triggering flip-flop by adding two inverters to the system, one before the clock pulse 1s applied to the master and an additional one in between the master and the slave. The logic symbol of a negative edge master-slave is shown in Fig. 5.42. The system of master-slave flip-flops is not restricted to J-K master-slave only. There may be an S-R master-slave or a D master-slave, etc.,

5.6.5 T Flip-flop

With a slight modification of a J-K flip-flop, we can construct a new flip-flop called a T flip-flop. If the two inputs J and K of a J-K flip-flop are tied together it is referred to as a T flip-flop. Hence, a T flip-flop has only one input T and two outputs Q and Q'. The name T flip-flop actually indicates the fact that the flip-flop has the ability to toggle. It has actually only two states, i.e., toggle state and memory state. Since there are only two states, a T flip-flop is a very good option to use in counter design and in sequential circuits design where switching an operation is required. The truth table of a T flip-flop is shown in Fig. 5.43 below.

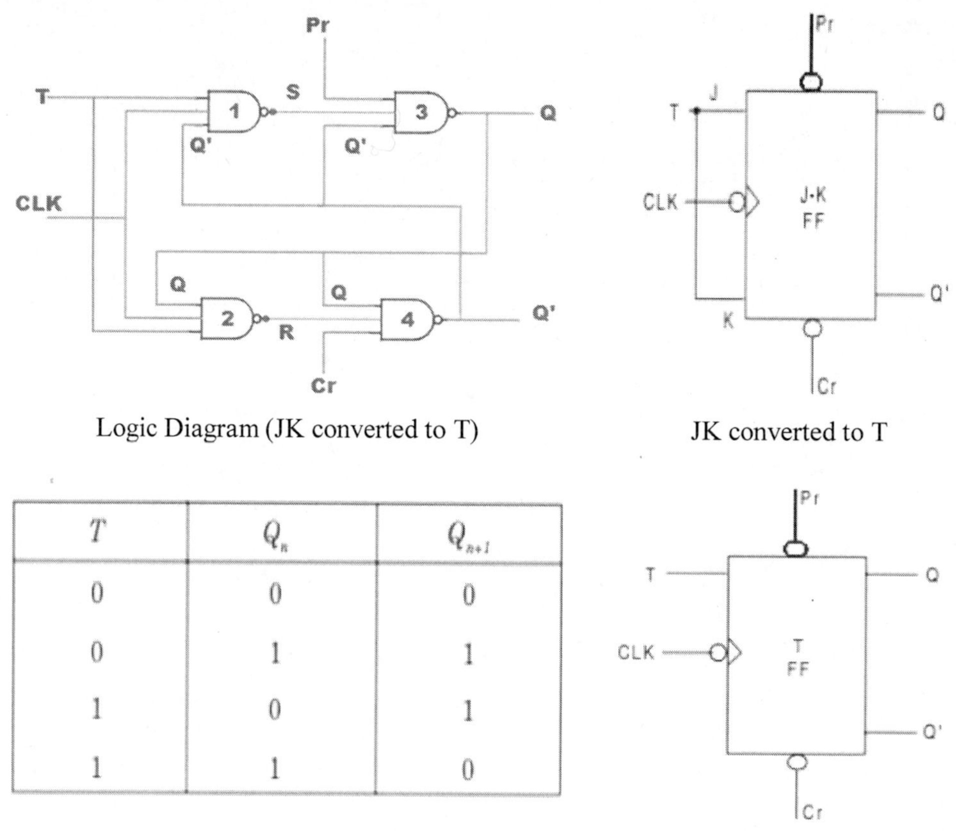

Logic Diagram (JK converted to T)

JK converted to T

T	Q_n	Q_{n+1}
0	0	0
0	1	1
1	0	1
1	1	0

Excitation table

Logic Symbol

Fig. 5.43: T FF (a and b) JK converted to T (c) Excitation Table (d) Logic Symbol

If the T input is in '*0*' state (i.e., $J = K = 0$) prior to a clock pulse, the Q output will not change with the clock pulse. On the other hand, if the T input is in '*1*' state (i.e., $J = K = 1$) prior to a clock pulse, the Q output will change to Q' with the clock pulse. In other words, we may say that, if T = 1 and the device is clocked, then the output toggles its state.

The truth table shows that when T = 0, then $Q_{t+1} = Q_t$, i.e, the next state is the same as the present state and no change occurs. When T = 1, then $Q_{t+1} = Q'_t$, i.e., the state of the flip-flop is complemented. The circuit diagram of a T flip-flop and the block diagram of the T flip-flop is shown in Fig. 5.43.

Input	Present State	Next State
T	Q_t	Q_{t+1}
0	0	0
0	1	1
1	0	0
1	1	0

Excitation table

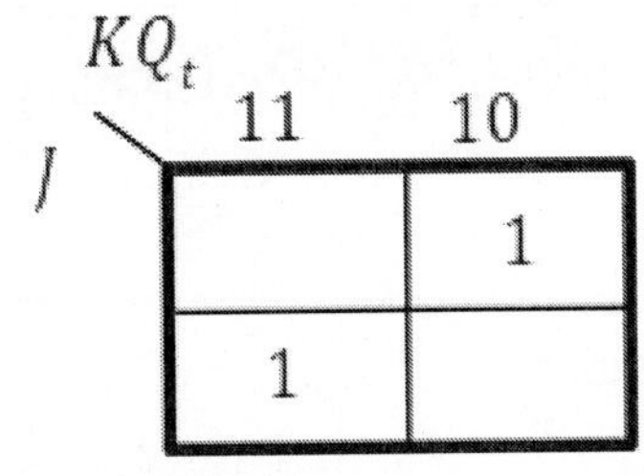

K-map for Q $_{t+1}$

Fig. 5.44: T FF: (a) Excitation Table (b) K-Map for Q_{t+1}

5.7 Flip Flop Conversion

For the conversion of one flip flop to another, a combinational circuit has to be designed first. If a JK Flip Flop is required, the inputs are given to the combinational circuit and the output of the combinational circuit is connected to the inputs of the actual flip flop. Thus, the output of the actual flip flop is the output of the required flip flop. The following flip flop conversions will be explained.

- SR Flip Flop to JK Flip Flop
- JK Flip Flop to SR Flip Flop
- SR Flip Flop to D Flip Flop
- D Flip Flop to SR Flip Flop
- JK Flip Flop to T Flip Flop
- JK Flip Flop to D Flip Flop
- D Flip Flop to JK Flip Flop

5.7.1 SR Flip Flop to JK Flip Flop

As told earlier, J and K will be given as external inputs to S and R. As shown in the logic diagram below, S and R will be the outputs of the combinational circuit. The truth tables for the flip flop conversion are given below. The present state is represented by Q_t and $Q_{t+1}$is the next state to be obtained when the J and K inputs are applied.

For two inputs J and K, there will be eight possible combinations. For each combination of J, K and Q_t, the corresponding Q_{t+1} states are found Q_t simply suggests the future values to be obtained by the JK flip flop after the value of Q_t. The table is then completed by writing the values of S and R required getting each Q_{t+1} from the corresponding Q_t. That is, the values of S and R that are required to change the state of the flip flop from Q_t to Q_{t+1} are written.

JK Inputs		Outputs		SR Inputs	
J	K	Q_t	Q_{t+1}	S	R
0	0	0	0	0	X
0	0	1	1	X	0
0	1	0	0	0	X
0	1	1	0	0	1
1	0	0	1	1	0
1	0	1	1	X	0
1	1	0	1	1	0
1	1	1	0	0	1

(a) Conversion Map

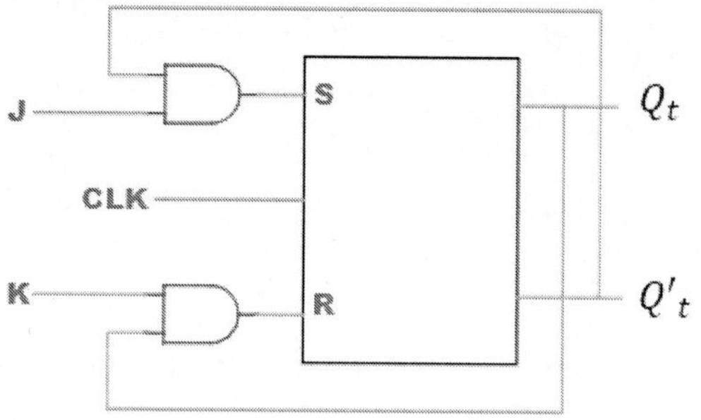

(b) Logic Diagram

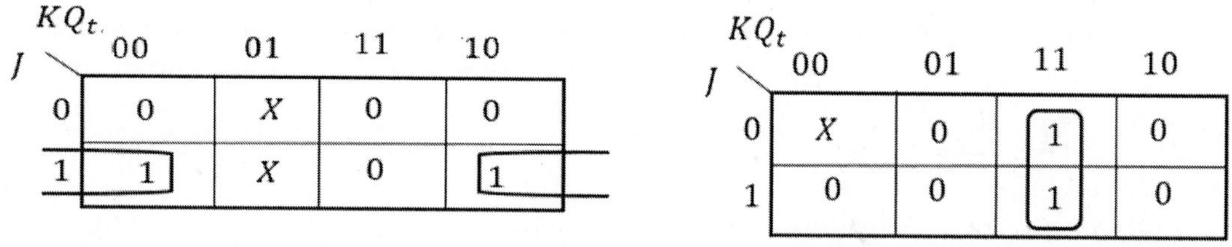

J \ KQ_t	00	01	11	10
0	0	X	0	0
1	1	X	0	1

$S = \bar{J}Q_t$

J \ KQ_t	00	01	11	10
0	X	0	1	0
1	0	0	1	0

$R = KQ_t$

(c) K map

Fig. 5.45: SR Flip Flop to JK Flip Flop (a) Conversion Map (b) Logic Diagram (c) K-Map

5.7.2 JK Flip Flop to SR Flip Flop

This will be the reverse process of the above explained conversion. S and R will be the external inputs to J and K. As shown in the logic diagram below, J and K will be the outputs of the combinational circuit. Thus, the values of J and K have to be obtained in terms of S, R and Qp. The logic diagram is shown below.

A conversion table is to be written using S, R, Q_t and Q_{t+1}. J and K. For two inputs, S and R, eight combinations are made. For each combination, the corresponding, Q_t and Q_{t+1} outputs are found. The outputs for the combinations of S=1 and R=1 are not permitted for an SR flip flop. Thus, the outputs are considered invalid and the J and K values are taken as “don’t cares”.

SR Inputs		Outputs		JK Inputs	
S	R	Q_t	Q_{t+1}	J	K
0	0	0	0	0	X
0	0	1	1	X	0
0	1	0	0	0	X
0	1	1	0	X	1
1	0	0	1	1	X
1	0	1	1	X	0
1	1	Invalid		Don't Care	
1	1	Invalid		Don't Care	

Conversion Map

J — S; CLK; K — R; Q_t; Q'_t

Logic Diagram

$S \backslash RQ_t$	00	01	11	10
0	0	X	X	0
1	1	X	X	X

$J = S$

$S \backslash RQ_t$	00	01	11	10
0	X	0	1	X
1	X	0	X	X

$K = R$

K-map

Fig. 5.46: JK Flip Flop to SR Flip Flop (a) Conversion Map (b) Logic Diagram (c) K-Map

5.7.3 SR Flip Flop to D Flip Flop

As shown in the figure, S and R are the actual inputs of the flip flop and D is the external input of the flip flop. The four combinations, the logic diagram, conversion table, and the K-map for S and R in terms of D and Q_t are shown below.

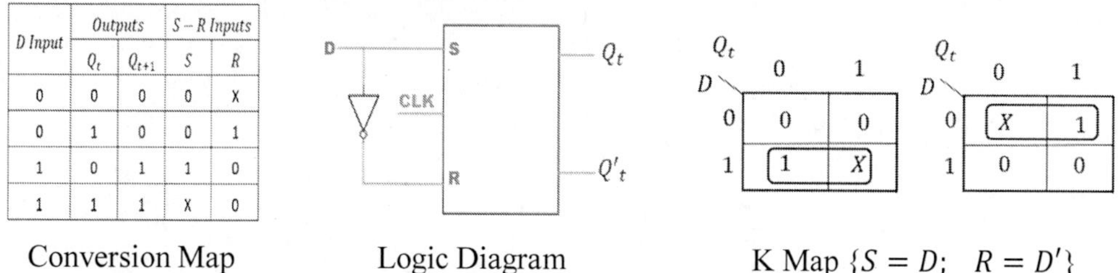

D Input	Outputs		S – R Inputs	
	Q_t	Q_{t+1}	S	R
0	0	0	0	X
0	1	0	0	1
1	0	1	1	0
1	1	1	X	0

$D \backslash Q_t$	0	1
0	0	0
1	1	X

$D \backslash Q_t$	0	1
0	X	1
1	0	0

Conversion Map Logic Diagram K Map $\{S = D; \ \ R = D'\}$

Fig. 5.47: SR Flip Flop to D Flip Flop (a) Conversion Map (b) Logic Diagram (c) K-Map

5.7.4 D Flip Flop to SR Flip Flop

D is the actual input of the flip flop and S and R are the external inputs. Eight possible combinations are achieved from the external inputs S, R and Q_t. But, since the combination of S=1 and R=1 are invalid, the values of Q_{t+1} and D are considered as "don't cares". The logic diagram showing the conversion from D to SR, and the K-map for D in terms of S, R and Q_t are shown below.

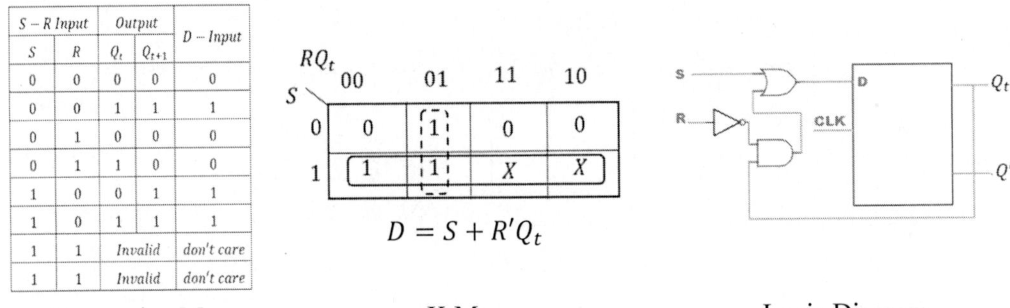

S − R Input		Output		D − Input
S	R	Q_t	Q_{t+1}	
0	0	0	0	0
0	0	1	1	1
0	1	0	0	0
0	1	1	0	0
1	0	0	1	1
1	0	1	1	1
1	1	Invalid		don't care
1	1	Invalid		don't care

Conversion Map K Map Logic Diagram

Fig. 5.48: D Flip Flop to SR Flip Flop (a) Conversion Map (b) Logic Diagram (c) K-Map

5.7.5 JK Flip Flop to T Flip Flop

J and K are the actual inputs of the flip flop and T is taken as the external input for conversion. Four combinations are produced with T and Q_t. J and K are expressed in terms of T and Q_t. The conversion table, K-maps, and the logic diagram are given below.

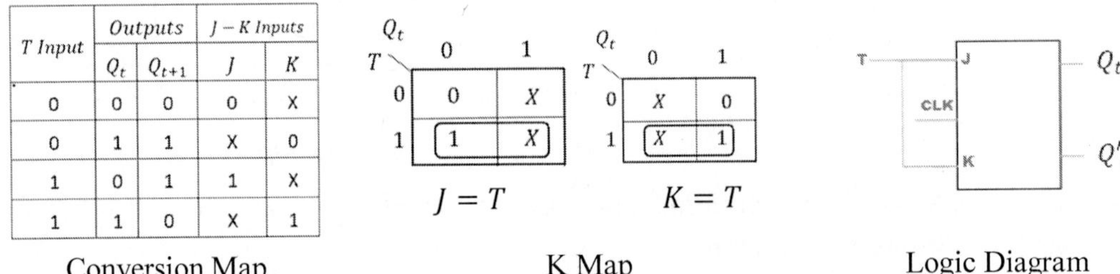

T Input	Outputs		J − K Inputs	
	Q_t	Q_{t+1}	J	K
0	0	0	0	X
0	1	1	X	0
1	0	1	1	X
1	1	0	X	1

Conversion Map K Map Logic Diagram

Fig. 5.49: JK Flip Flop to T Flip Flop (a) Conversion Map (b) Logic Diagram (c) K-Map

5.7.6 JK Flip Flop to D Flip Flop

D is the external input and J and K are the actual inputs of the flip flop. D and Q_t make four combinations. J and K are expressed in terms of D and Q_t. The four-combination conversion table, the K-maps for J and K in terms of D and Q_t, and the logic diagram showing the conversion from JK to D are given below.

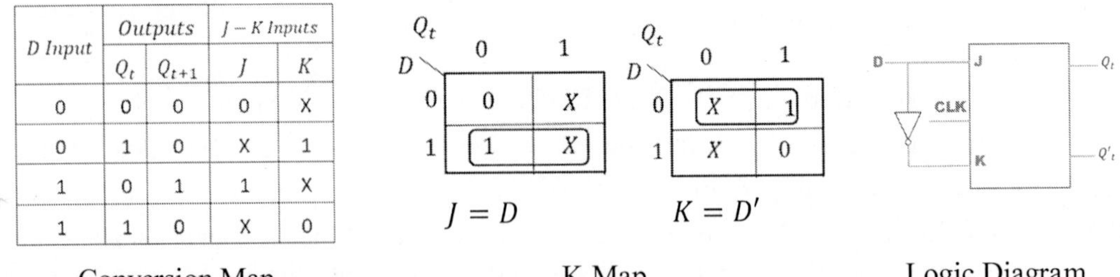

D Input	Outputs		J − K Inputs	
	Q_t	Q_{t+1}	J	K
0	0	0	0	X
0	1	0	X	1
1	0	1	1	X
1	1	0	X	0

Conversion Map K Map Logic Diagram

Fig.5.50: JK Flip Flop to D Flip Flop (a) Conversion Map (b) Logic Diagram (c) K-Map

5.7.7 D Flip Flop to JK Flip Flop

In this conversion, D is the actual input to the flip flop and J and K are the external inputs. J, K and Q_t make eight possible combinations, as shown in the conversion table below. D is expressed in terms of J, K and Q_t. The conversion table, the K-map for D in terms of J, K and Q_t and the logic diagram showing the conversion from D to JK are given in the figure below.

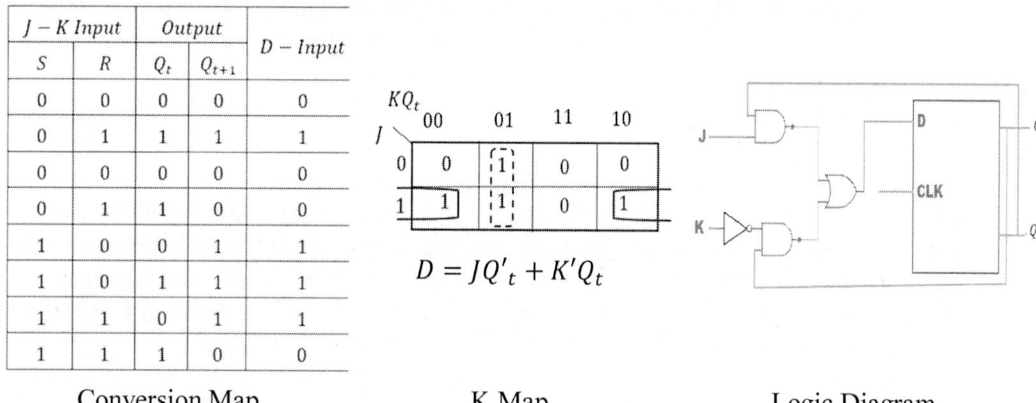

$J-K$ Input		Output		$D-$ Input
S	R	Q_t	Q_{t+1}	
0	0	0	0	0
0	1	1	1	1
0	0	0	0	0
0	1	1	0	0
1	0	0	1	1
1	0	1	1	1
1	1	0	1	1
1	1	1	0	0

Conversion Map K Map Logic Diagram

Fig. 5.51: D Flip Flop to JK Flip Flop (a) Conversion Map (b) Logic Diagram (c) K-Map

5.8 Applications of Flip-Flops

As we know Flip-flops are fundamental building blocks in digital electronics, and they are used in a wide range of applications, including:

- Registers: Flip-flops are often used as building blocks for registers, which are circuits that store data for later use.
- Counters: A counter is a circuit that counts the number of clock pulses it receives. Flip-flops are the key component of most counters, and they are used to store the count value.
- Memory: Flip-flops are used in memory circuits to store digital data. Memory circuits can be volatile (RAM) or non-volatile (ROM, EEPROM).
- Oscillators: Flip-flops can be used to create oscillators, which are circuits that produce periodic waveforms. By connecting the output of a flip-flop to its input, it can produce a clock signal.
- Digital Timers: Flip-flops are used in digital timers to keep track of time. A timer circuit can be created using a combination of flip-flops and other digital logic gates.
- Frequency Dividers: A flip-flop can be used to create a frequency divider circuit that divides the input clock frequency by a fixed factor.
- Multiplexers/Demultiplexers: Flip-flops are used in multiplexers and demultiplexers, which are circuits that allow multiple inputs to be switched to a single output or a single input to be switched to multiple outputs.
- Sequential Logic Circuits: Flip-flops are used in sequential logic circuits to store and manipulate digital signals. These circuits can be used in many applications, including control systems, digital signal processing, and communication systems.

Overall, flip-flops are versatile building blocks that can be used in a wide range of digital circuits, making them an essential component of modern electronics

5.9 Shift Registers

Shift registers are a type of sequential logic circuit, mainly for storage of digital data. They are a group of flip-flops connected in a chain so that the output from one flip-flop becomes the input of the next flip-flop. Most of the registers possess no characteristic internal sequence of states. All flip-flop is driven by a common clock, and all are set or reset simultaneously. The Shift Register is another type of sequential logic circuit that can be used for the storage or the transfer of binary data.

A shift register basically consists of several single bit "D-Type Data Latches", one for each data bit, either a logic "0" or a "1", connected together in a serial arrangement so that the output from one data latch becomes the input of the next latch and so on. Data bits may be fed in or out of a shift register serially, that is one after the other from either the left or the right direction, or all together at the same time in a parallel configuration. The individual data latches that make up a single shift register are all driven by a common clock (Clk) signal making them synchronous devices. Data inputs are given at the *'D'* inputs of the flipflops. The four-bit data is transferred to the outputs at the falling edge of the clock pulse

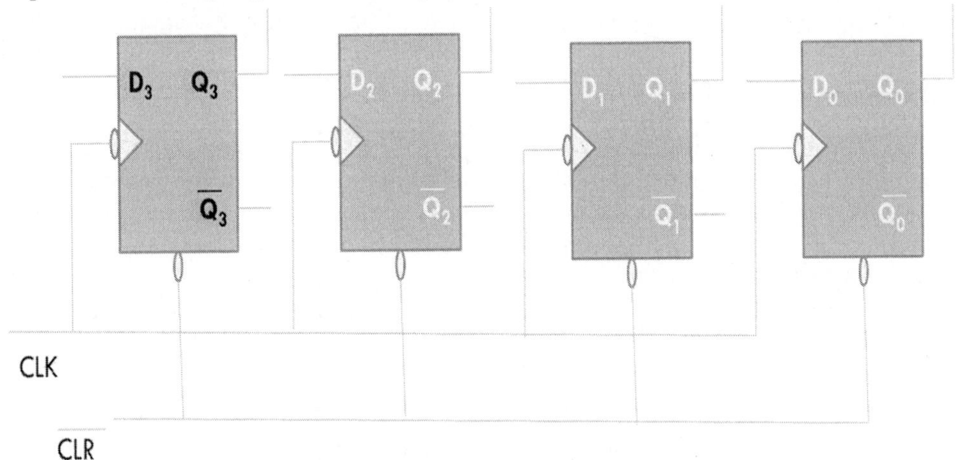

Fig. 5.52: A shift register using D FF

Generally, shift registers operate in one of four different modes with the basic movement of data through a shift register being:

- Serial-in to Parallel-out (SIPO) - the register is loaded with serial data, one bit at a time, with the stored data being available at the output in parallel form.
- Serial-in to Serial-out (SISO) - the data is shifted serially "IN" and "OUT" of the register, one bit at a time in either a left or right direction under clock control.
- Parallel-in to Serial-out (PISO) - the parallel data is loaded into the register simultaneously and is shifted out of the register serially one bit at a time under clock control.
- Parallel-in to Parallel-out (PIPO) - the parallel data is loaded simultaneously into the register, and transferred together to their respective outputs by the same clock pulse.

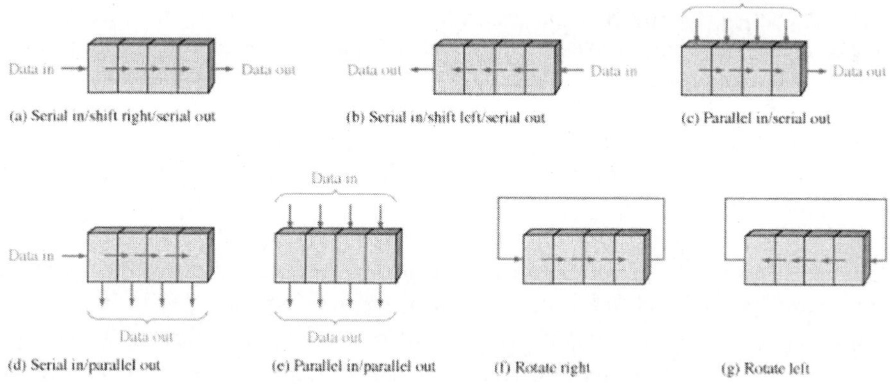

Fig. 5.53: Different Modes of Shift operations

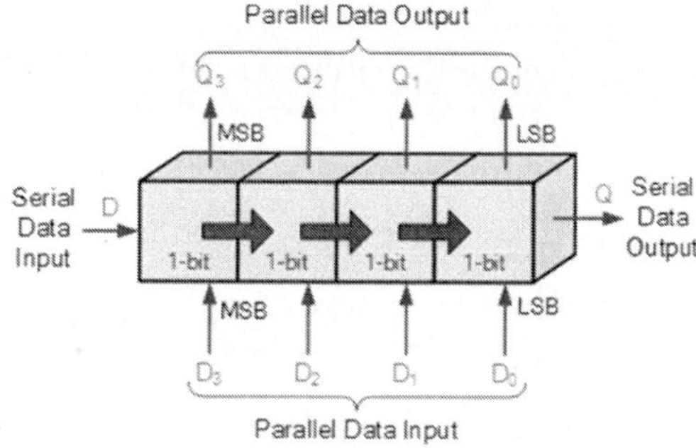

Fig. 5.54: Mode of data shifting (Serial/Parallel)

5.9.1 Serial-in/Serial-out shift shift register

Serial-in, serial-out shift registers delay data by one clock time for each stage. They will store a bit of data for each register. A serial-in, serial-out shift register may be one to 64 bits in length, longer if registers or packages are cascaded. Below is a single stage shift register receiving data which is not synchronized to the register clock. The "data in" at the D pin of the type D FF(Flip-Flop) does not change levels when the clock changes for low to high. We may want to synchronize the data to a system wide clock in a circuit board to improve the reliability of a digital logic circuit.

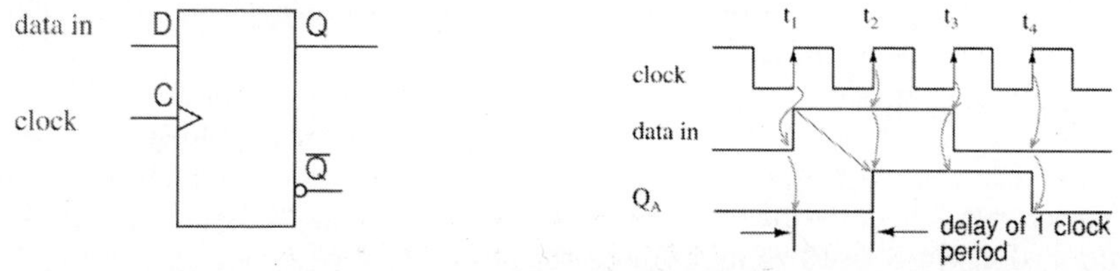

Fig. 5.55: Operation of D FF used in shift register

The figure above illustrated that whatever" data in" is present at the D pin of a type D FF is transferred from D to output Q at clock time. Since in the present example shift register uses positive edge sensitive storage elements, hence the output Q follows the D input when the clock transitions from low to high as shown by the up arrows on the diagram above. There is no doubt what logic level is present at clock time because the data is stable well before and after the clock edge. This is seldom the case in multi-stage shift registers. But, this was an easy example to start with. We are only concerned with the positive, low to high, clock edge. The falling edge can be ignored. It is very easy to see Q follow D at clock time above.

Since data, above, present at D is clocked to Q at clock time, and Q cannot change until the next clock time, the D FF delays data by one clock period, provided that the data is already synchronized to the clock. The QA waveform is the same as" data in" with a one clock period delay.

A more detailed look at what the input of the type D Flip-Flop sees at clock time follows. Refer to the figure below. Since" data in" appears to changes at clock time (above), we need further information to determine what the D FF sees. If the" data in" is from another shift register stage, another same type D FF, we can draw some conclusions based on data sheet information. Manufacturers of digital logic make available information about their parts in data sheets, formerly only available in a collection called a data book. Data books are still available; though, the manufacturer's web site is the modern source. Data must be present (t_S) before the clock and after (t_H) the clock. Data is delayed from D to Q by propagation delay (t_p).

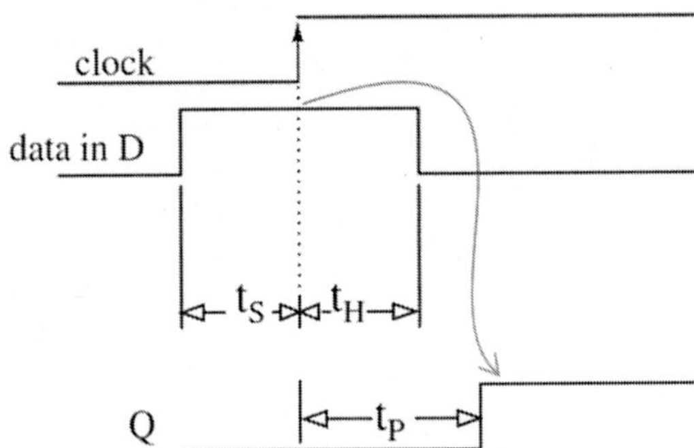

Fig. 5.56: D Flip-Flop as sees at clock time

t_S, is the setup time, i.e., the time data must be present before clock time. In this case data must be present at D for 100ns prior to the clock. Furthermore, the data must be held for hold time t_H =60ns after clock time. These two conditions must be met to reliably clock data from D to Q of the Flip-Flop.

There is no problem meeting the setup time of 60ns as the data at D has been there for the whole previous clock period if it comes from another shift register stage. For example, at a clock frequency of 1 Mhz, the clock period is 1000 µs, plenty of time. Data will actually be present for 1000µs prior to the clock, which is much greater than the minimum required t_S of 60ns.

The hold time t_H =60ns is met because D connected to Q of another stage cannot change any faster than the propagation delay of the previous stage t_p=200ns. Hold time is met as long as the propagation delay of the previous D FF is greater than the hold time. Data at D driven by another stage Q will not change any faster than 200ns for the CD4006b. To summarize, output Q follows input D at nearly clock time if Flip-Flops are cascaded into a multi-stage shift register.

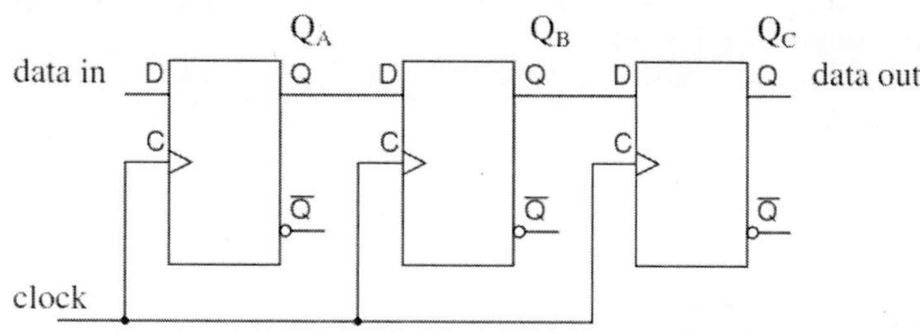

Fig. 5.57: Serial-in/serial-out shift register using D flip-flop

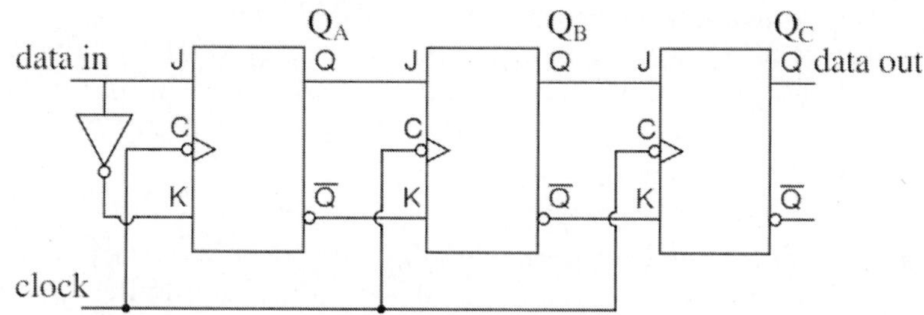

Fig. 5.58: Serial-in/serial-out shift register using JK flip-flop

Type JK FFs cascaded Q to J, Q' to K with clocks in parallel to yield an alternate form of the shift register above. A serial-in/serial-out shift register has a clock input, a data input, and a data output from the last stage. In general, the other stage outputs are not available Otherwise, it would be a serial-in, parallel-out shift register. The waveforms below are applicable to either one of the preceding two versions of the serial-in, serial-out shift register. The three pairs of arrows show that a three-stage shift register temporarily stores 3-bits of data and delays it by three clock periods from input to output.

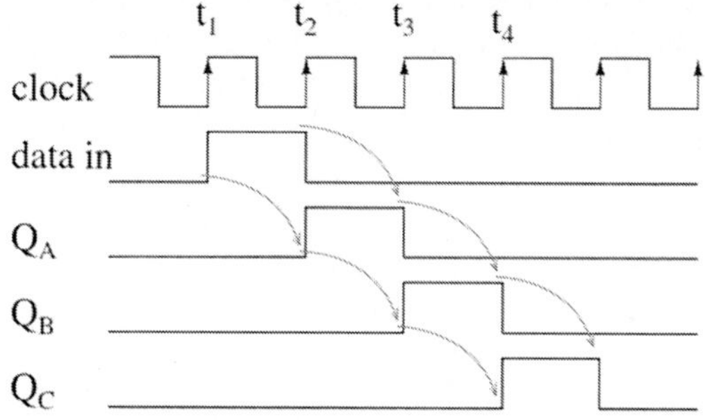

Fig. 5.59: Clock Waveform for Serial in Serial out shift register

- At clock time t_1 a "data in" of 0 is clocked from D to Q of all three stages. In particular, D of stage A sees a logic 0, which is clocked to Q_A where it remains until time t_2.
- At clock time t_2 a "data in" of 1 is clocked from D to Q_A. At stages B and C, a 0, fed from preceding stages is clocked to Q_B and Q_C.
- At clock time t_3 a "data in" of 0 is clocked from D to Q_A. Q_A goes low and stays low for the remaining clocks due to "data in" being 0. Q_B goes high at t_3 due to a 1 from the previous stage. Q_C is still low after t_3 due to a low from the previous stage.

- Q_C finally goes high at clock t_4 due to the high fed to D from the previous stage Q_B. All earlier stages have 0s shifted into them.
- And, after the next clock pulse at t_5, all logic 1s will have been shifted out, replaced by 0s

Serial-in/serial-out devices

We will take a closer look at the following parts available as integrated circuits, courtesy of Texas Instruments. For complete device data sheets follow the links.

- CD4006b 18-bit serial-in/ serial-out shift register
  - (http://focus.ti.com/docs/prod/folders/print/cd4006b.html)
- CD4031b 64-bit serial-in/ serial-out shift register
  - (http://focus.ti.com/docs/prod/folders/print/cd4031b.html)
- CD4517b dual 64-bit serial-in/ serial-out shift register
  - (http://focus.ti.com/docs/prod/folders/print/cd4517b.html)

The following serial-in/ serial-out shift registers are 4000 series CMOS (Complementary Metal Oxide Semiconductor) family parts. As such, they will accept a VDD, positive power supply of 3-Volts to 15-Volts. The VSS pin is grounded. The maximum frequency of the shift clock, which varies with VDD, is a few megahertz. See the full data sheet for details.

5.9.2 Parallel-in, Serial-out shift register

Parallel-in/ serial-out shift registers do everything that the previous serial-in/ serial-out shift registers do plus input data to all stages simultaneously. The parallel-in/ serial-out shift register stores data, shifts it on a clock by clock basis, and delays it by the number of stages times the clock period. In addition, parallel-in/ serial-out really means that we can load data in parallel into all stages before any shifting ever begins. This is a way to convert data from a parallel format to a serial format. By parallel format we mean that the data bits are present simultaneously on individual wires, one for each data bit as shown below. By serial format we mean that the data bits are presented sequentially in time on a single wire or circuit as in the case of the "data out".

The diagram below we take a close look at the internal details of a 3-stage parallel-in/ serial-out shift register. A stage consists of a type D Flip-Flop for storage, and an AND-OR selector to determine whether data will load in parallel, or shift stored data to the right. In general, these elements will be replicated for the number of stages required. We show three stages due to space limitations. Four, eight or sixteen bits is normal for real parts.

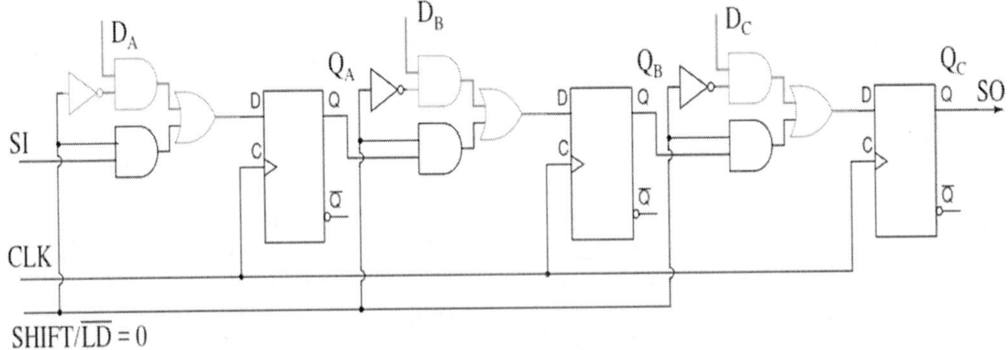

Fig. 5.60: Parallel-in, serial-out shift register showing parallel data input

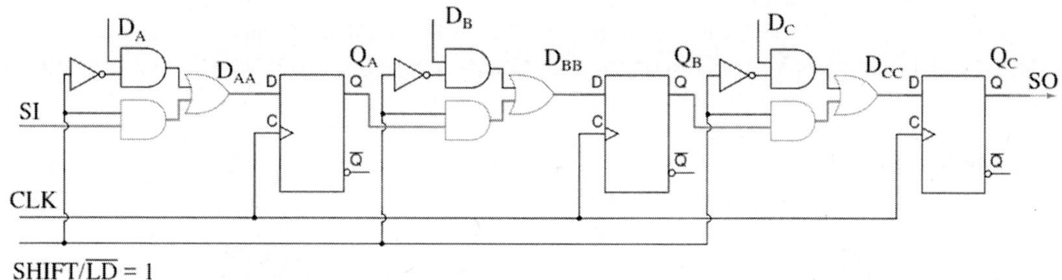

Fig. 5.61: Parallel-in, serial-out shift register showing serial data output/shift

Above we show the parallel load path when SHIFT/LD' is logic low. The upper NAND gates serving D_A, D_B and D_C are enabled, passing data to the D inputs of type D Flip-Flops Q_A Q_B D_C respectively. At the next positive going clock edge, the data will be clocked from D to Q of the three FFs. Three bits of data will load into Q_A Q_B D_C at the same time.

The type of parallel load just described, where the data loads on a clock pulse is known as synchronous load because the loading of data is synchronized to the clock. This needs to be differentiated from asynchronous load where loading is controlled by the preset and clear pins of the Flip-Flops which does not require the clock. Only one of these load methods is used within an individual device, the synchronous load being more common in newer devices.

The shift path is shown above when SHIFT/LD' is logic high. The lower AND gates of the pairs feeding the OR gate are enabled giving us a shift register connection of SI to D_A, Q_A to D_B, Q_B to D_C, Q_C to SO. Clock pulses will cause data to be right shifted out to SO on successive pulses. The waveforms below show both parallel loading of three bits of data and serial shifting of this data. Parallel data at D_A D_B D_C is converted to serial data at SO

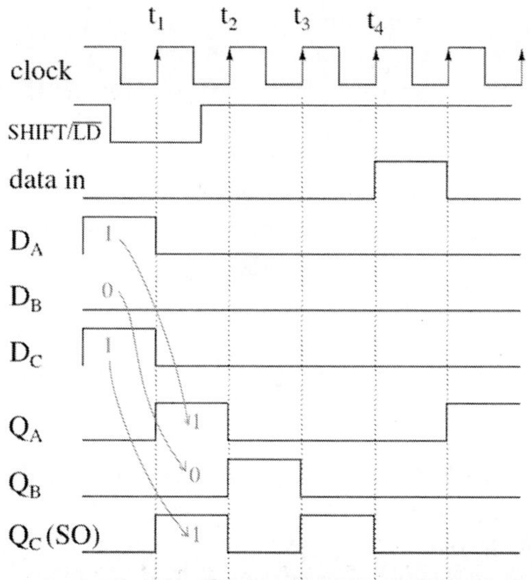

Fig. 5.62: Clock diagram for parallel-in, serial-out shift register

What we previously described with words for parallel loading and shifting is now set down as waveforms above. As an example we present 101 to the parallel inputs D_{AA} D_{BB} and D_{CC} . Next, the SHIFT/LD' goes low enabling loading of data as opposed to shifting of data. It needs to be low a short time before and after the clock pulse due to setup and hold requirements. It is considerably wider than it has to be. Though, with synchronous logic it is convenient to make it wide. We could have made the active low SHIFT/LD' almost two clocks wide, low almost a clock before t_1 and back high just before t_3. The important factor is that it needs to be low around clock time t1 to enable parallel loading of the data by the clock.

Note that at t_1 the data 101 at D_A D_B D_C is clocked from D to Q of the Flip-Flops as shown at Q_A Q_B Q_C at time t_1. This is the parallel loading of the data synchronous with the clock. Now that the data is loaded, we may shift it provided that SHIFT/LD' is high to enable shifting, which it is prior to t_2. At t_2 the data 0 at Q_C is shifted out of SO which is the same as the Q_C waveform. It is either shifted into another integrated circuit, or lost if there is nothing connected to SO. The data at Q_B, a 0 is shifted to Q_C. The 1 at Q_A is shifted into Q_B. With "data in" a 0, Q_A becomes 0. After t_2, Q_A Q_B Q_C = 010.

After t_3, Q_A Q_B Q_C = 001. This 1, which was originally present at Q_A after t_1, is now present at SO and Q_C. The last data bit is shifted out to an external integrated circuit if it exists. After t_4 all data from the parallel load is gone. At clock t_5 we show the shifting in of a data 1 present on the SI, i.e., serial input.

Why provide SI and SO pins on a shift register? These connections allow us to cascade shift register stages to provide large shifters than available in a single IC (Integrated Circuit) package. They also allow serial connections to and from other ICs like microprocessors.

Parallel-in/serial-out devices

Let's take a closer look at parallel-in/ serial-out shift registers available as integrated circuits, cour- tesy of Texas Instruments. For complete device data sheets follow these the links.

- SN74ALS166 parallel-in/ serial-out 8-bit shift register, synchronous load
  - (http://www-s.ti.com/sc/ds/sn74als166.pdf)
- SN74ALS165 parallel-in/ serial-out 8-bit shift register, asynchronous load
  - (http://www-s.ti.com/sc/ds/sn74als165.pdf)
- CD4014B parallel-in/ serial-out 8-bit shift register, synchronous load
  - (http://www-s.ti.com/sc/ds/cd4014b.pdf)
- SN74LS647 parallel-in/ serial-out 16-bit shift register, synchronous load
  - (http://www-s.ti.com/sc/ds/sn74ls674.pdf)

5.9.3 Serial-in, parallel-out shift register

A serial-in/parallel-out shift register is similar to the serial-in/ serial-out shift register in that it shifts data into internal storage elements and shifts data out at the serial-out, data-out, pin. It is different in that it makes all the internal stages available as outputs. Therefore, a serial-in/parallel-out shift register converts data from serial format to parallel format. If four data bits are shifted in by four clock pulses via a single wire at data-in, the data becomes available simultaneously on the four Outputs QA to QD after the fourth clock pulse. The practical application of the serial-in/parallel-out shift register is to convert data from serial format on a single wire to parallel format on multiple wires. Perhaps, we will illuminate four LEDs (Light Emitting Diodes) with the four outputs (Q_A Q_B Q_C Q_D).

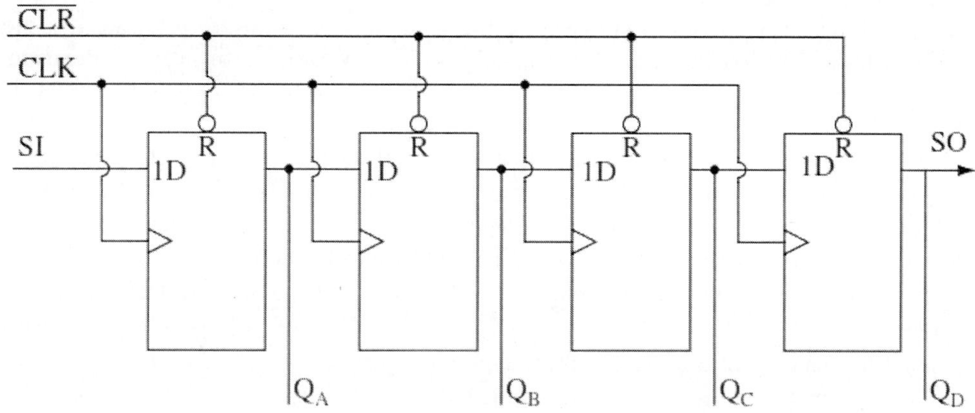

Fig. 5.63: Serial-in/ Parallel out shift register details

The above details of the serial-in/parallel-out shift register are fairly simple. It looks like a serial-in/ serial-out shift register with taps added to each stage output. Serial data shifts in at SI (Serial Input). After a number of clocks equal to the number of stages, the first data bit in appears at SO (Q_D) in the above figure. In general, there is no SO pin. The last stage (Q_D above) serves as SO and is cascaded to the next package if it exists. If a serial-in/parallel-out shift register is so similar to a serial-in/ serial-out shift register, why do manufacturers bother to offer both types? Why not just offer the serial-in/parallel-out shift register? They actually only offer the serial-in/parallel-out shift register, as long as it has no more than 8-bits. Note that serial-in/ serial-out shift registers come in gigger than 8-bit lengths of 18 to to 64-bits. It is not practical to offer a 64-bit serial-in/parallel-out shift register requiring that many output pins. See waveforms below for above shift register.

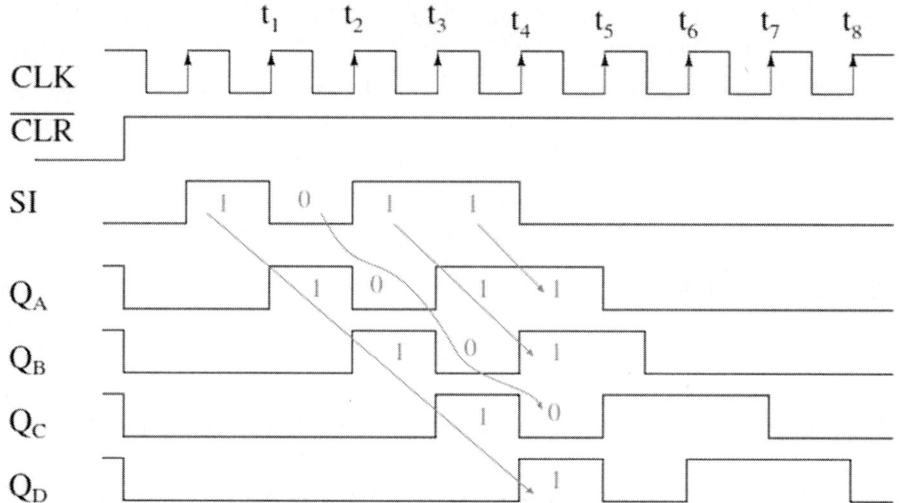

Fig. 5.64: Timing Diagram for Serial-in/ Parallel out shift register

The shift register has been cleared prior to any data by CLR', an active low signal, which clears all type D Flip-Flops within the shift register. Note the serial data 1011 pattern presented at the SI input. This data is synchronized with the clock CLK. This would be the case if it is being shifted in from something like another shift register, for example, a parallel-in/ serial-out shift register (not shown here).

On the first clock at t1, the data 1 at SI is shifted from D to Q of the first shift register stage. After t2 this first data bit is at Q_B. After t3 it is at Q_C. After t4 it is at Q_D. Four clock pulses have shifted the first data bit all the way to the last stage Q_D. The second data bit a 0 is at Q_C after the 4th clock. The third data bit a 1 is at Q_B. The fourth data bit another 1 is at Q_A. Thus, the serial data input pattern 1011 is contained in (Q_D Q_C Q_B Q_A). It is now available on the four outputs.

It will available on the four outputs from just after clock t_4 to just before t_5. This parallel data must be used or stored between these two times, or it will be lost due to shifting out the Q_D stage on following clocks t_5 to t_8 as shown above.

Serial-in/ parallel-out devices

Let's take a closer look at Serial-in/ parallel-out shift registers available as integrated circuits, cour- tesy of Texas Instruments. For complete device data sheets follow the links

- SN74ALS164A serial-in/ parallel-out 8-bit shift register
  - (http://www-s.ti.com/sc/ds/sn74als164a.pdf)
- SN74AHC594 serial-in/ parallel-out 8-bit shift register with output register
  - (http://www-s.ti.com/sc/ds/sn74ahct594.pdf)
- SN74AHC595 serial-in/ parallel-out 8-bit shift register with output register
  - (http://www-s.ti.com/sc/ds/sn74ahct595.pdf)
- CD4094 serial-in/ parallel-out 8-bit shift register with output register
  - (http://www-s.ti.com/sc/ds/cd4094b.pdf)
  - (http://www.st.com/stonline/books/pdf/docs/2069.pdf)

5.9.4 Parallel-in, parallel-out, universal shift register

The purpose of the parallel-in/ parallel-out shift register is to take in parallel data, shift it, then output it as shown below. A universal shift register is a do-everything device in addition to the parallel-in/ parallel-out function.

Above we apply four bits of data to a parallel-in/ parallel-out shift registers at D_A D_B D_C D_D. The mode control, which may be multiple inputs, controls parallel loading vs shifting. The mode control may also control the direction of shifting in some real devices. The data will be shifted one bit position for each clock pulse. The shifted data is available at the outputs Q_A Q_B Q_C Q_D. The "data in" and "data out" are provided for cascading of multiple stages. Though, above, we can only cascade data for right shifting. We could accommodate cascading of left-shift data by adding a pair of left pointing signals," data in" and" data out", above.

The internal details of a right shifting parallel-in/ parallel-out shift register are shown below. The tri-state buffers are not strictly necessary to the parallel-in/ parallel-out shift register, but are part of the real-world device shown below

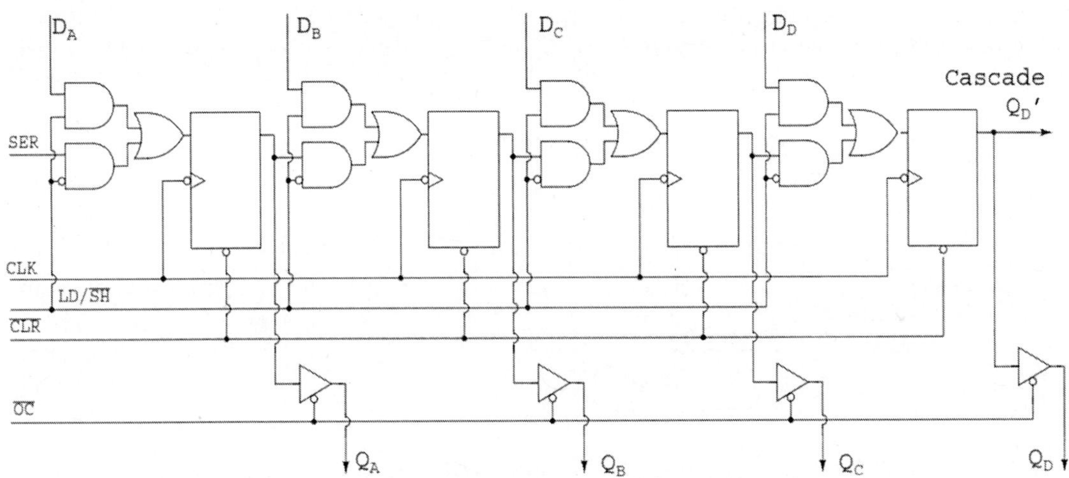

Fig. 5.65: 74LS395 parallel-in/ parallel-out shift register with tri-state output

The 74LS395 so closely matches our concept of a hypothetical right shifting parallel-in/ parallel-out shift register that we use an overly simplified version of the data sheet details above. See the link to the full data sheet more more details, later in this chapter.

LD/SH' controls the AND-OR multiplexer at the data input to the FF's. If LD/SH'=1, the upper four AND gates are enabled allowing application of parallel inputs D_A D_B D_C D_D to the four FF data inputs. Note the inverter bubble at the clock input of the four FFs. This indicates that the 74LS395 clocks data on the negative going clock, which is the high to low transition. The four bits of data will be clocked in parallel from D_A D_B D_C D_D to Q_A Q_B Q_C Q_D at the next negative going clock. In this "real part", OC' must be low if the data needs to be available at the actual output pins as opposed to only on the internal FFs.

The previously loaded data may be shifted right by one-bit position if LD/SH'=0 for the succeeding negative going clock edges. Four clocks would shift the data entirely out of our 4-bit shift register. The data would be lost unless our device was cascaded from Q_D' to SER of another device.

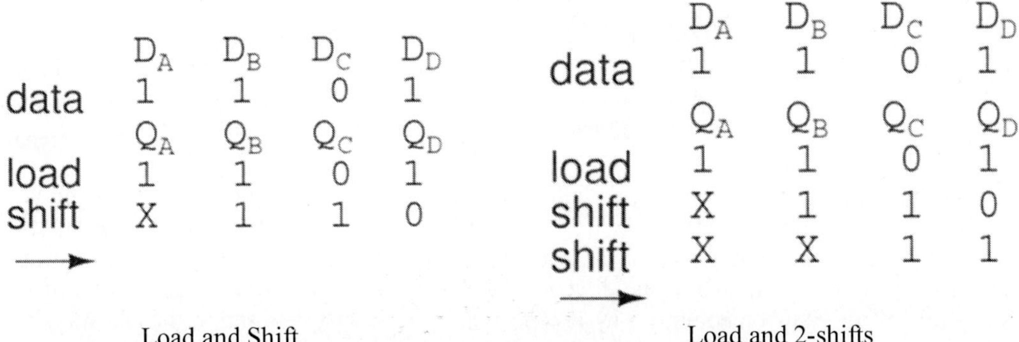

Fig. 5.66: Parallel in Parallel out shift register

Above, a data pattern is presented to inputs D_A D_B D_C D_D. The pattern is loaded to Q_A Q_B Q_C Q_D. Then it is shifted one bit to the right. The incoming data is indicated by X, meaning the we do no know what it is. If the input (SER) were grounded, for example, we would know what data (0) was shifted in. Also shown, is right shifting by two positions, requiring two clocks.

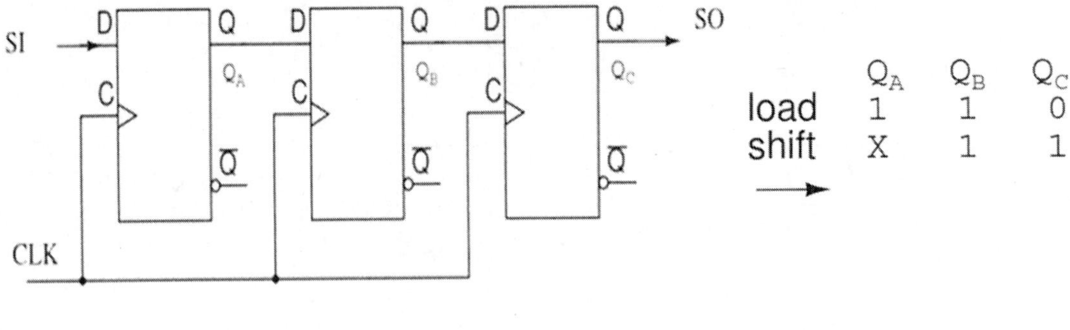

(a) Circuit Diagram (b) Load and Right Shift

Fig. 5.67: Shift Right in Parallel in Parallel out

The above figure serves as a reference for the hardware involved in right shifting of data. It is too simple to even bother with this figure, except for comparison to more complex figures to follows

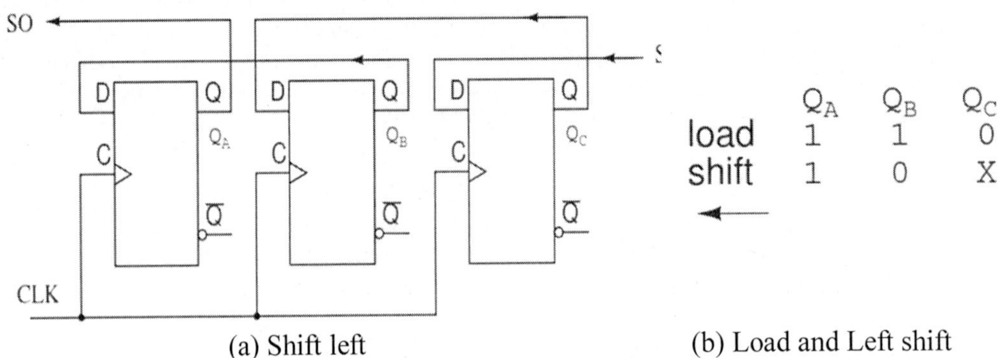

(a) Shift left (b) Load and Left shift

Fig. 5.68: Shift Right in Parallel in Parallel out

If we need to shift left, the FFs need to be rewired. Compare to the previous right shifter. Also, SI and SO have been reversed. SI shifts to Q_C. Q_C shifts to Q_B. Q_B shifts to Q_A. Q_A leaves on the SO connection, where it could cascade to another shifter SI. This left shift sequence is backwards from the right shift sequence

There is one problem with the "shift left" figure above. There is no market for it. Nobody manufactures a shift-left part. A "real device" which shifts one direction can be wired externally to shift the other direction. Or, should we say there is no left or right in the context of a device which shifts in only one direction. However, there is a market for a device which will shift left or right on command by a control line. Of course, left and right are valid in that context.

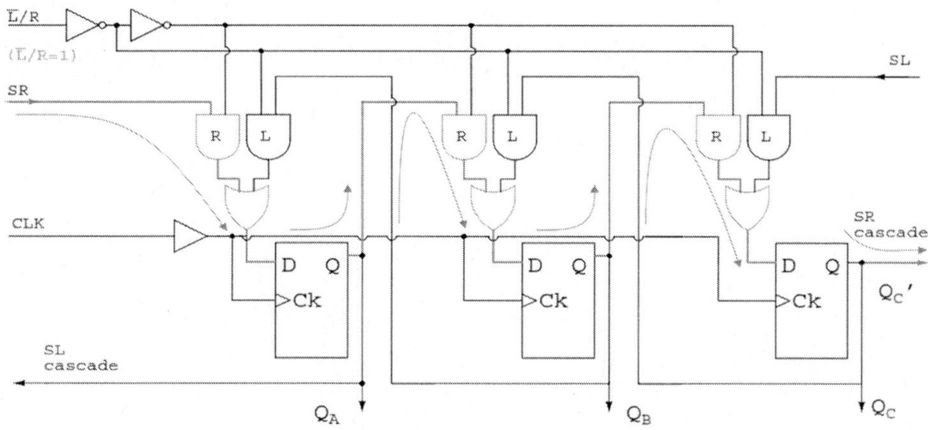

Fig. 5.69: Shift left/ right, right action

What we have above is a hypothetical shift register capable of shifting either direction under the control of L'/R. It is setup with L'/R=1 to shift the normal direction, right. L'/R=1 enables the multiplexer AND gates labeled R. This allows data to follow the path illustrated by the arrows, when a clock is applied. The connection path is the same as the "too simple" "shift right" figure above.

Data shifts in at SR, to Q_A, to Q_B, to Q_C, where it leaves at SR cascade. This pin could drive SR of another device to the right. What if we change L'/R to L'/R=0?

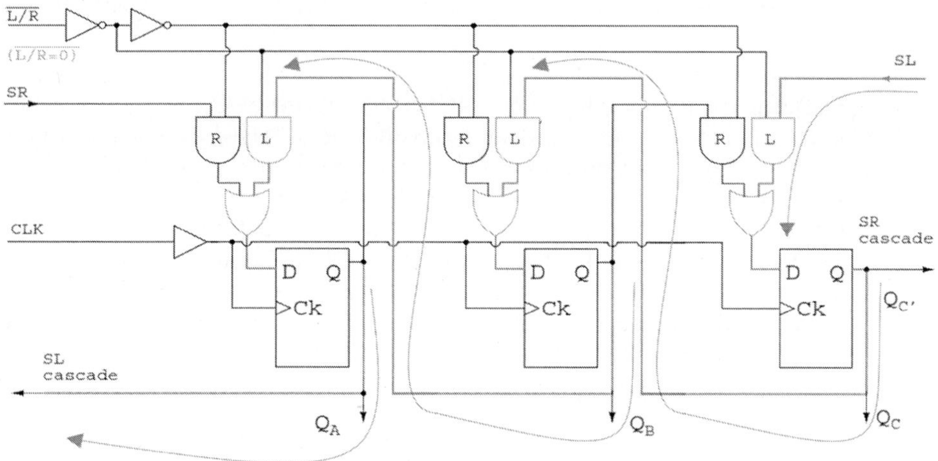

Fig.5.70: Shift left/ right register, left action

With L'/R=0, the multiplexer AND gates labeled L are enabled, yielding a path, shown by the arrows, the same as the above "shift left" figure. Data shifts in at SL, to Q_C, to Q_B, to QA, where it leaves at SL cascade. This pin could drive SL of another device to the left. The prime virtue of the above two figures illustrating the "shift left/ right register" is simplicity. The operation of the left right control L'/R=0 is easy to follow. A commercial part needs the parallel data loading implied by the section title. This appears in the figure below.

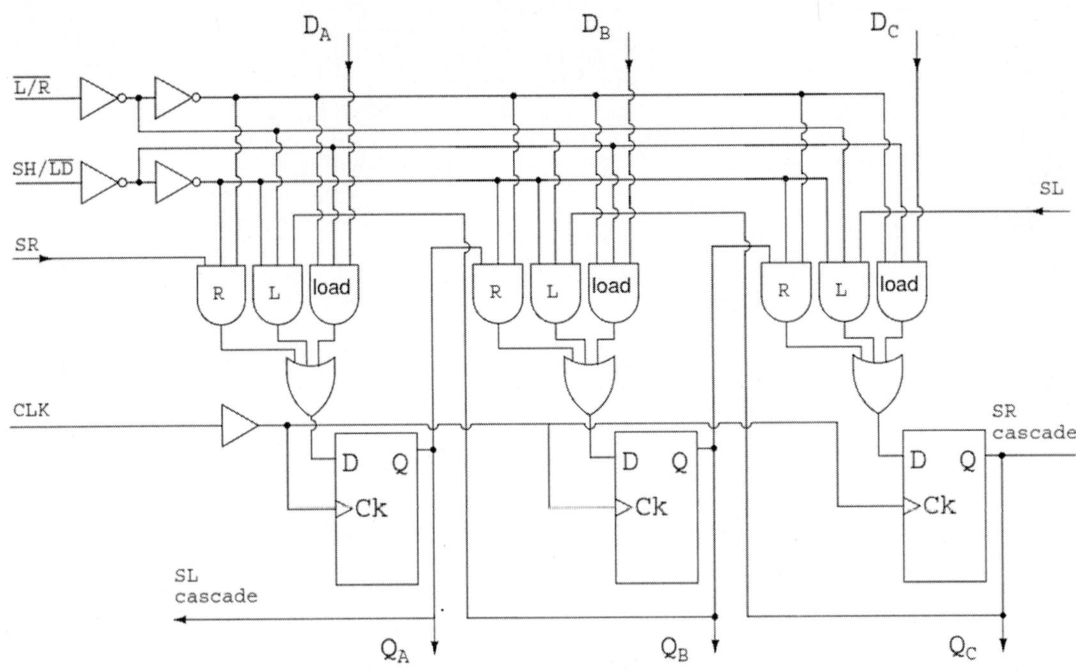

Fig. 5.71: Shift left/ right/ load

Now that we can shift both left and right via L'/R, let us add SH/LD', shift/ load, and the AND gates labeled "load" to provide for parallel loading of data from inputs D_A D_B D_C. When SH/LD'=0, AND gates R and L are disabled, AND gates "load" are enabled to pass data D_A D_B D_C to the FF data inputs. the next clock CLK will clock the data to Q_A Q_B Q_C. As long as the same data is present it will be re-loaded on succeeding clocks. However, data present for only one clock will be lost from the outputs when it is no longer present on the data inputs. One solution is to load the data on one clock, then proceed to shift on the next four clocks. This problem is remedied in the 74ALS299 by the addition of another AND gate to the multiplexer.

If SH/LD' is changed to SH/LD'=1, the AND gates labeled "load" are disabled, allowing the left/ right control L'/R to set the direction of shift on the L or R AND gates. Shifting is as in the previous figures. The only thing needed to produce a viable integrated device is to add the fourth AND gate to the multiplexer as alluded for the 74ALS299. This is shown in the next section for that part.

Parallel-in/ parallel-out and universal devices

Let's take a closer look at Serial-in/ parallel-out shift registers available as integrated circuits, courtesy of Texas Instruments. For complete device data sheets, follow the links.

- SN74LS395A parallel-in/ parallel-out 4-bit shift register
  - (http://www-s.ti.com/sc/ds/sn74ls395a.pdf)
- SN74ALS299 parallel-in/ parallel-out 8-bit universal shift register
  - (http://www-s.ti.com/sc/ds/sn74als299.pdf)

5.9.5 Shift Register Counters

Two of the most common types of shift register counters are introduced here: the Ring counter and the Johnson counter. They are basically shift registers with the serial outputs connected back to the serial inputs in order to produce particular sequences. These registers are classified as counters because they exhibit a specified sequence of states

Ring counters

If the output of a shift register is fed back to the input. a ring counter results. Ring Counters are also known as re-circulating shift registers. The data pattern contained within the shift register will recirculate as long as clock pulses are applied. For example, the data pattern will repeat every four clock pulses in the figure below. However, we must load a data pattern. All 0's or all 1's doesn't count. Is a continuous logic level from such a condition useful?

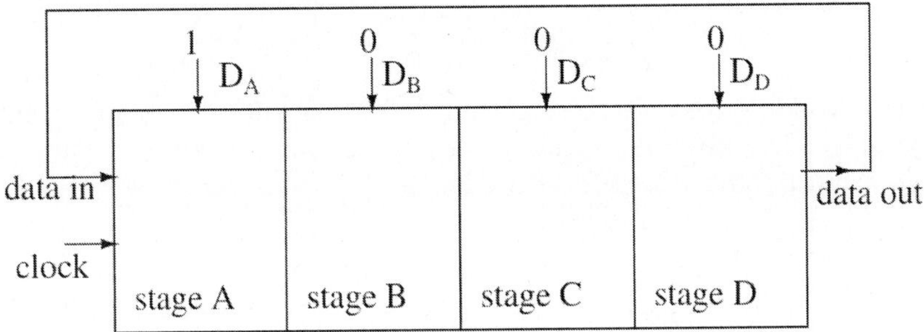

Fig. 5.72: Ring Counter, PISO shift register where output fed back to input

We make provisions for loading data into the parallel-in/ serial-out shift register configured as a ring counter below. Any random pattern may be loaded. The most generally useful pattern is a single 1. Loading binary 1000 into the ring counter, above, prior to shifting yields a viewable pattern. The data pattern for a single stage repeats every four clock pulses in our 4-stage example. The waveforms for all four stages look the same, except for the one clock time delay from one stage to the next. See figure below

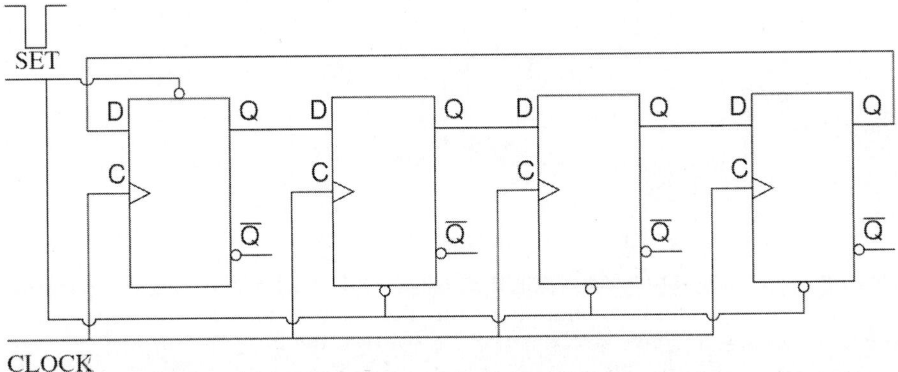

Fig. 5.73: A ring counter with 1000 input (Set one stage and clear other three stages)

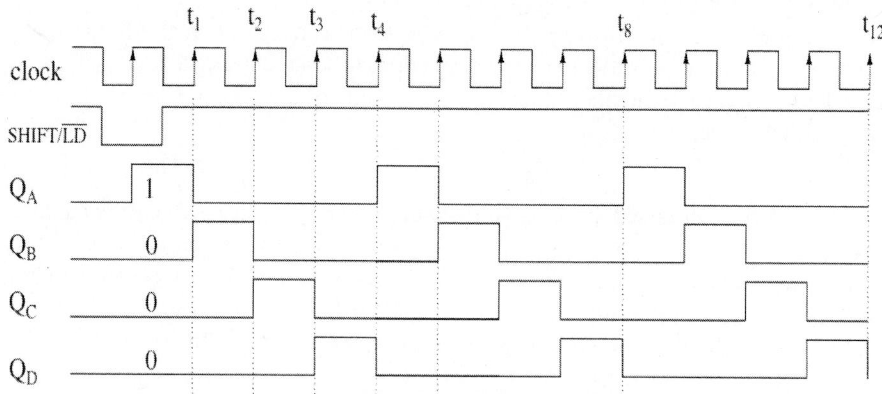

Fig. 5.74: Timing Diagram

The circuit above is a divide by 4 counter. Comparing the clock input to any one of the outputs, shows a frequency ratio of 4:1. How may stages would we need for a divide by 10 ring counter? Ten stages would recirculate the 1 every 10 clock pulses.

The requirement for initialization is a disadvantage of the ring counter over a conventional counter. At a minimum, it must be initialized at power-up since there is no way to predict what state flip-flops will power up in. In theory, initialization should never be required again. In actual practice, the flip-flops could eventually be corrupted by noise, destroying the data pattern. A "self-correcting" counter, like a conventional synchronous binary counter would be more reliable

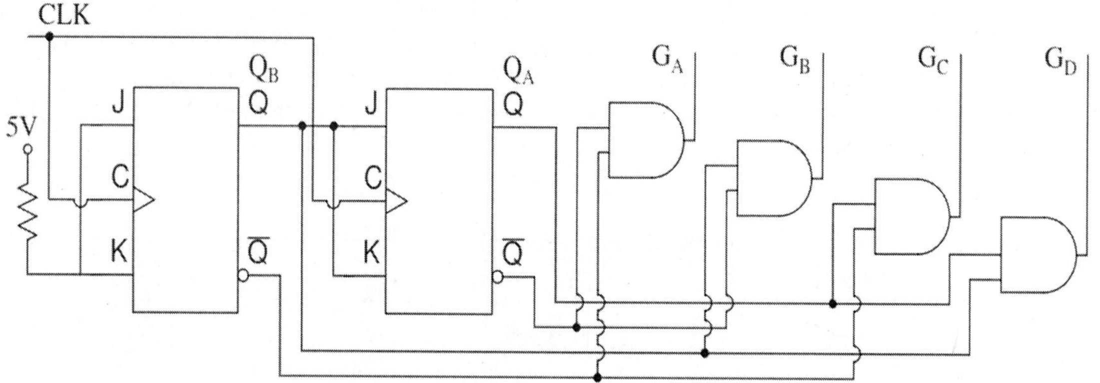

Fig.5.75: Binary synchronous counter with decoder gates

The above binary synchronous counter needs only two stages, but requires decoder gates. The ring counter had more stages, but was self- decoding, saving the decode gates above. Another disadvantage of the ring counter is that it is not "self-starting". If we need the decoded outputs, the ring counter looks attractive, in particular, if most of the logic is in a single shift register package. If not, the conventional binary counter is less complex without the decoder.

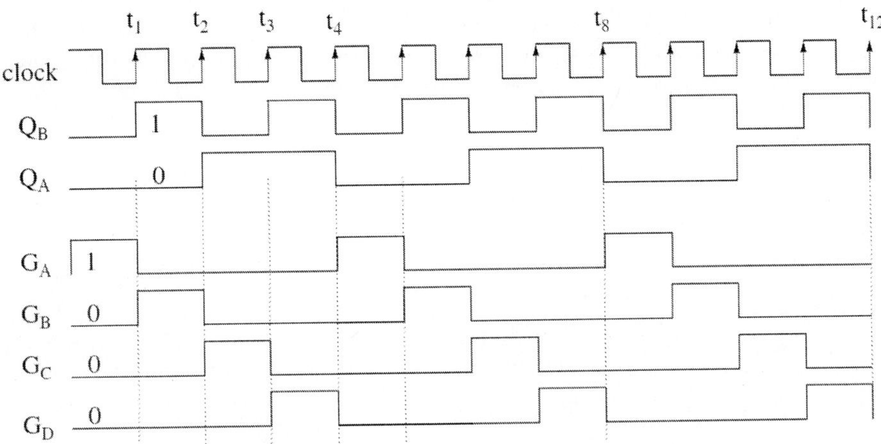

Fig. 5.76: Timing Diagram for binary synchronous counter with decoder

The waveforms decoded from the synchronous binary counter are identical to the previous ring counter waveforms. The counter sequence is (Q_A Q_B) = (00 01 10 11).

Johnson counters

The switch-tail ring counter, also known as the Johnson counter, overcomes some of the limitations of the ring counter. Like a ring counter a Johnson counter is a shift register fed back on its' self. It requires half the stages of a comparable ring counter for a given division ratio. If the complement output of a ring counter is fed back to the input instead of the true output, a Johnson counter results. The difference between a ring counter and a Johnson counter is which output of the last stage is fed back (Q or Q'). Carefully compare the feedback connection below to the previous ring counter. A Johnson Counter re-circulates the last flipflop Q(inverted) output back to the input of the first Flip-Flop. It doesn't Require an initialization value, and will provide a predictable output state sequence. A 4-bit Johnson counter has a modulus of 8, meaning there are 8 unique output states.

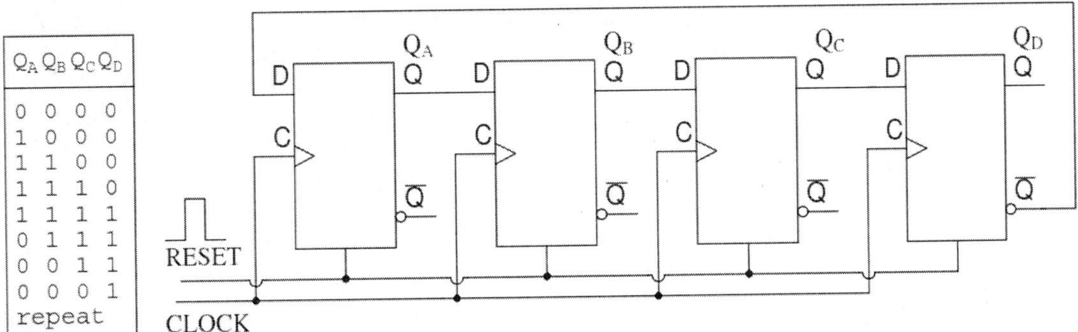

Fig. 5.77: Johnson counter (note the Q'D to DA feedback connection)

This "reversed" feedback connection has a profound effect upon the behavior of the otherwise similar circuits. Recirculating a single 1 around a ring counter divides the input clock by a factor equal to the number of stages. Whereas, a Johnson counter divides by a factor equal to twice the number of stages. For example, a 4-stage ring counter divides by 4. A 4-stage Johnson counter divides by 8. Start a Johnson counter by clearing all stages to 0s before the first clock. This is often done at power-up time. Referring to the figure below, the first clock shifts three 0s from (Q_A Q_B Q_C) to the right into (Q_B Q_C Q_D).

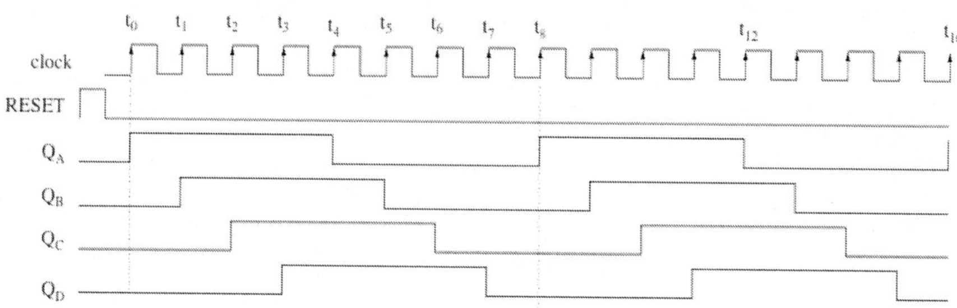

Fig. 5.78: Timing Diagram

The 1 at Q'_D (the complement of Q) is shifted back into Q_A. Thus, we start shifting 1s to the right, replacing the 0s. Where a ring counter recirculated a single 1, the 4-stage Johnson counter recirculates four 0s then four 1s for an 8-bit pattern, then repeats. The above waveforms illustrate that multi-phase square waves are generated by a Johnson counter. The 4-stage unit above generates four overlapping phases of 50% duty cycle. How many stages would be required to generate a set of three phase waveforms? For example, a three stage Johnson counter, driven by a 360 Hertz clock would generate three 120o phased square waves at 60 Hertz.

5.10 Digital Counter

A **Counter** is a device which stores (and sometimes displays) the number of times a particular event or process has occurred, often in relationship to a clock signal. Counters are used in digital electronics for counting purpose, they can count specific event happening in the circuit. For example, in UP counter a counter increases count for every rising edge of clock. Not only counting, a counter can follow the certain sequence based on our design like any random sequence 0,1,3,2…. They can also be designed with the help of flip flops. They are used as frequency dividers where the frequency of given pulse waveform is divided. Counters are sequential circuit that count the number of pulses can be either in binary code or BCD form. The main properties of a counter are timing, sequencing, and counting. Counter works in two modes i.e., up counter and down counter.

The counter is one of the widest applications of the flip flop. Based on the clock pulse, the output of the counter contains a predefined state. The number of the pulse can be counted using the output of the counter.

5.10.1 Application of Counters in Digital Electronics

Here are the applications of the counters in digital electronics:

- **Frequency Measurement and Division**

  The counter is used to measure the frequency of a signal, simply by counting the no of cycles in a particular given time period and the counter is also used to divide the input clock frequency by a fixed integer value.

- **Timing**
  The counter is also used to generate timing signals like pulse-width modulated (PWM) signals. These signals are commonly used in power electronics to control the speed of motors and regulate the brightness of LEDs.
- **Binary Arithmetic**
  Binary arithmetic operations like addition, subtraction, multiplication, and division are used in digital systems by counters.
- **Data Storage**
  Counters can also be used as memory elements in a digital circuit if take an example of a binary operator. The binary counter can be used to store a binary value that represents a state in a digital system.
- **Digital Signal Processing**
  Counters are also used in digital signal processing applications like filtering and signal analysis.

Counter can be broadly divided into synchronous and asynchronous types. Synchronous counter has its flip-flops clocked at the same time, whilst asynchronous counter is not. The clock of the preceding flip-flop of the asynchronous flip-flop is fed from the output of the previous flip-flop. Asynchronous counter suffers delay problem whilst, synchronous counter will not. Asynchronous Counter is also referred as ripple counter for the reason of delay feeding of the clock pulse from one flip-flop to another. Counters are broadly divided into two categories

1. Asynchronous counter
2. Synchronous counter

5.10.2 Asynchronous Counter (Ripple Counter)

The asynchronous counter is also called as Ripple counter and is made up of a series of flip-flops. In asynchronous counter we don't use universal clock, only first flip flop is driven by main clock and the clock input of rest of the following flip flop is driven by output of previous flip flops. Below is a diagram of the 2-bit **Asynchronous counter** in which we used two T flip-flops. Apart from the T flip flop, we can also use the JK flip flop by setting both of the inputs to 1 permanently. The external clock pass to the clock input of the first flip flop, i.e., FF-A and its output, i.e., is passed to clock input of the next flip flop, i.e., FF-B.

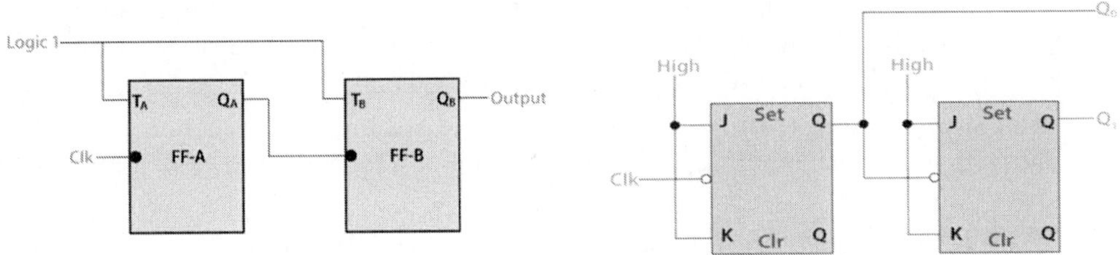

2-bit Ripple counter using T Flip-Flop   2-bit Ripple counter using JK Flip-Flop

Fig. 5.79: Basic Idea of Ripple counter

We can understand it by following diagram for a 4-bit Ripple counter. Ripple counter is a special type of **Asynchronous** counter in which the clock pulse ripples through the circuit. The n-MOD ripple counter forms by combining n number of flip-flops. The n-MOD ripple counter can count 2^n states, and then the counter resets to its initial value. It is evident from timing diagram that Q_0 is changing as soon as the rising edge of clock pulse is encountered, Q_1 is changing when rising edge of Q_0 is encountered (because Q_0 is like clock pulse for second flip flop) and so on. In this way ripples are generated through Q_0, Q_1, Q_2, Q_3 hence it is also called **RIPPLE counter and serial counter.** A ripple counter is a cascaded arrangement of flip flops where the output of one flip flop drives the clock input of the following flip flop

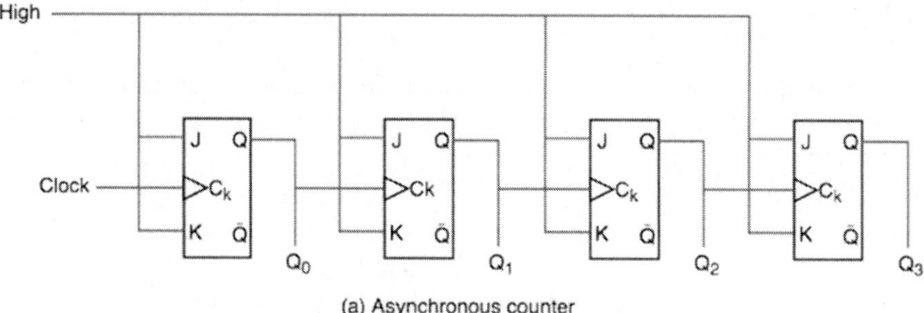

(a) Asynchronous counter

Fig. 5.80: Asynchronous Counter

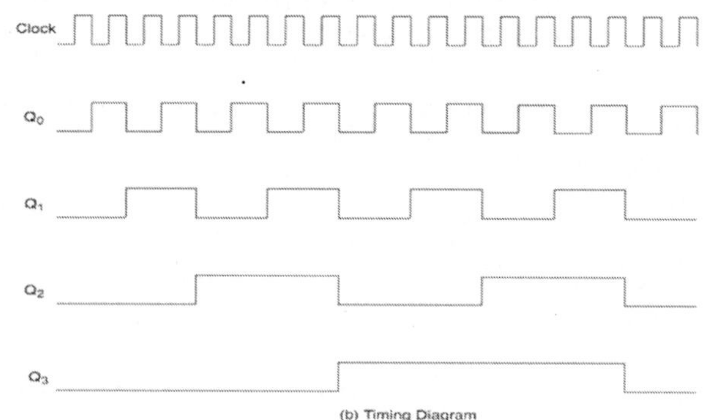

(b) Timing Diagram

Fig. 5.81: Timing Diagram

UP/DOWN Ripple Counters

In the UP/DOWN ripple counter all the FFs operate in the toggle mode. So, either T flip-flops or JK flip-flops are to be used. The LSB flip-flop receives clock directly. But the clock to every other FF is obtained from ($Q = \overline{Q}$) output of the previous FF.

- **UP counting mode (M=0)**: The Q output of the preceding FF is connected to the clock of the next stage if up counting is to be achieved. For this mode, the mode select input M is at logic 0 (M=0).
- **DOWN counting mode (M=1)**: If M = 1, then the $\overline{Q}$output of the preceding FF is connected to the next FF. This will operate the counter in the counting mode.

Example: Design a 3-bit binary up/down ripple counter.

Solution:

- 3-bit – hence three FFs are required.
- UP/DOWN – So a mode control input is essential.
- For a ripple up counter, the Q output of preceding FF is connected to the clock input of the next one.

- For a ripple up counter, the Q output of preceding FF is connected to the clock input of the next one.
- For a ripple down counter, the Q bar output of preceding FF is connected to the clock input of the next one.
- Let the selection of Q and $\overline{Q}$ output of the preceding FF be controlled by the mode control input M such that, If M = 0, UP counting. So connect Q to CLK. If M = 1, DOWN counting. So, connect $\overline{Q}$ to CLK.

Block Diagram

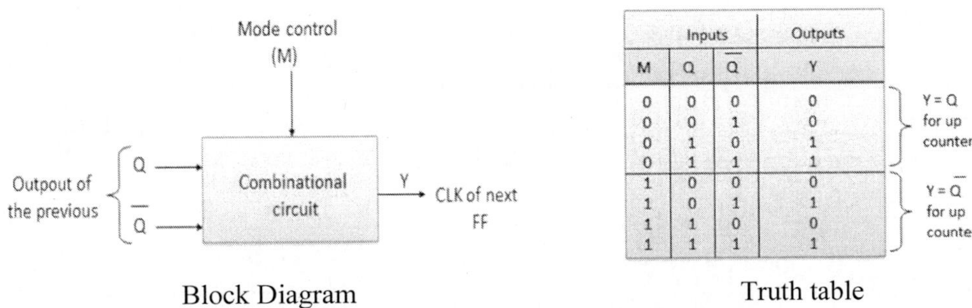

Inputs			Outputs
M	Q	$\overline{Q}$	Y
0	0	0	0
0	0	1	0
0	1	0	1
0	1	1	1
1	0	0	0
1	0	1	1
1	1	0	0
1	1	1	1

Block Diagram Truth table

Fig. 5.82: Block Diagram and Truth Table for Example

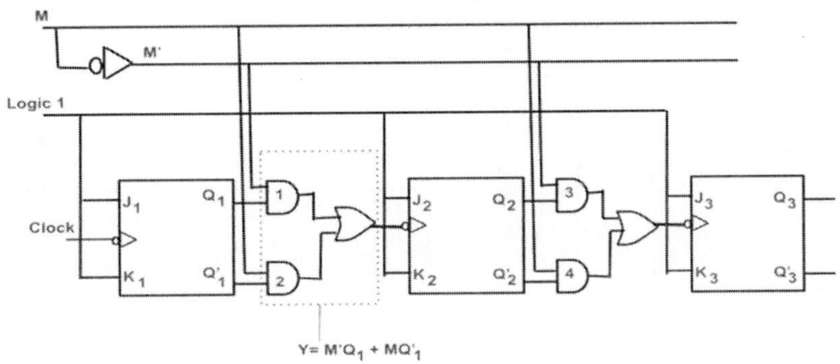

Fig. 5.83: Circuit diagram3-bit binary up/down ripple counter

Operation

S.N.	Condition	Operation
1	**Case 1 – With M = 0 (Up counter)**	• If M = 0 and M bar = 1, then the AND gates 1 and 3 in fig. will be enabled whereas the AND gates 2 and 4 will be disabled. • Hence Q_A gets connected to the clock input of FF-B and Q_B gets connected to the clock input of FF-C. • These connections are same as those for the normal up counter. Thus, with M = 0 the circuit work as an up counter.

2	**Case 2:** **With M = 1** **(Down counter)**	• If M = 1, then AND gates 2 and 4 in fig. are enabled whereas the AND gates 1 and 3 are disabled. • Hence Q_A bar gets connected to the clock input of FF-B and Q_B bar gets connected to the clock input of FF-C. • These connections will produce a down counter. Thus, with M = 1 the circuit works as a down counter.

Timing Diagram (Initially $Q_3 = 0$, $Q_2 = 0$, $Q_1 = 0$)

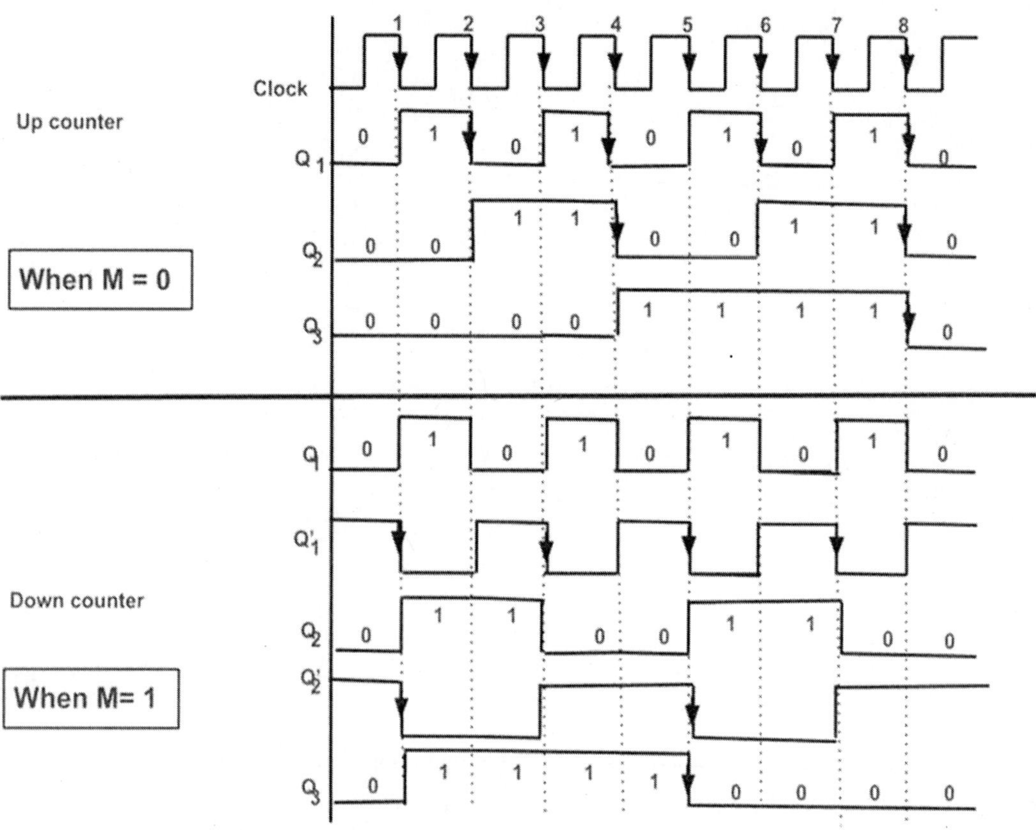

Fig. 5.84: Timing diagram for 3-bit binary up/down ripple counter

The 2-bit ripple counter is called as MOD-4 counter and 3-bit ripple counter is called as MOD-8 counter. So, in general, an n-bit ripple counter is called as modulo-N counter. Where, MOD number = 2^n.

Type of modulus

- 2-bit up or down (MOD-4)
- 3-bit up or down (MOD-8)
- 4-bit up or down (MOD-16)

5.10.3 Synchronous Counter

Unlike the asynchronous counter, synchronous counter has one global clock which drives each flip flop so output changes in parallel. The one advantage of synchronous counter over asynchronous counter is, it can operate on higher frequency than asynchronous counter as it does not have cumulative delay because of same clock is given to each flip flop. It is also called as parallel counter.

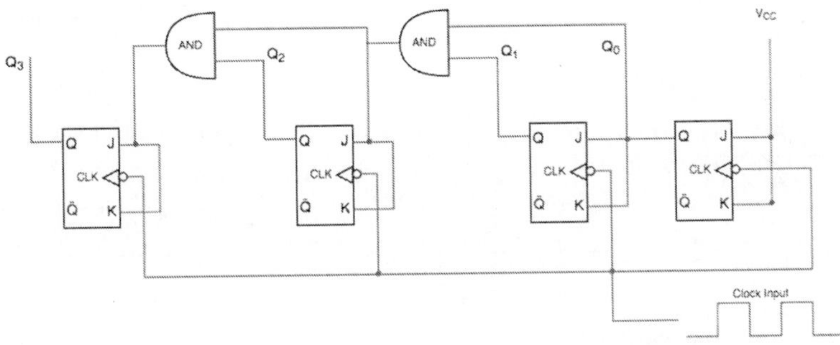

Fig. 5.85: A basic Synchronous Counter

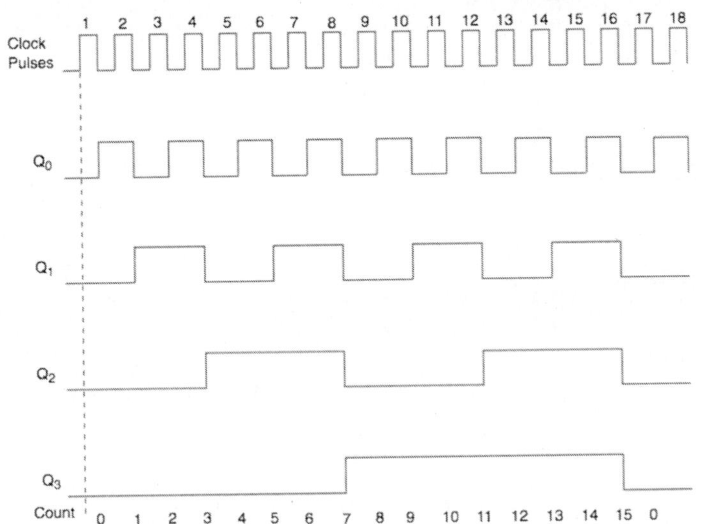

Fig. 5.86: Timing diagram for a Synchronous Counter

From circuit diagram we see that Q0 bit gives response to each falling edge of clock while Q1 is dependent on Q0, Q2 is dependent on Q1 and Q0 , Q3 is dependent on Q2,Q1 and Q0.

5.10.4 Modulus Counter (MOD-N Counter)

Procedure to Design Synchronous Counters

- Obtain the truth table of the logic sequence for intended counter to be designed. Alternatively obtain the state diagram of the counter.
- Determine the number and type of flip-flop to be used.

- From the excitation table of the flip-flop, determine the next state logic.
- From the output state, use Karnaugh map for simplification to derive the circuit output functions and the flip-flop output functions.
- Draw the logic circuit diagram.
- Simulate the circuit using software.
- Build the circuit

From the function tables of the flip-flops learnt earlier, the excitation or characteristic table of SR flip-flop, D flip-flip, JK flip-flop, and T flip-flop are shown in Fig. 9.7. and 9.8 respectively. Q_t is denoting the output of the present state and Q_{t+1} denotes the output of next state.

Characteristic Table of (a) SR FF (b) D FF (c) JK FF (d) T FF

Q_t	Q_{t+1}	S	R
0	0	0	X
0	1	1	0
1	0	0	1
1	1	X	0

(a) Characteristic Table of SR FF

Q_t	Q_{t+1}	D
0	0	0
0	1	1
1	0	0
1	1	1

(b) Characteristic Table of D FF

Q_t	Q_{t+1}	J	K
0	0	0	X
0	1	1	X
1	0	X	1
1	1	X	0

(c) Characteristic Table of JK FF

Q_t	Q_{t+1}	T
0	0	0
0	1	1
1	0	1
1	1	0

(d) Characteristic Table of T FF

5.10.5 Decade Counter

A decade counter counts ten different states and then reset to its initial states.

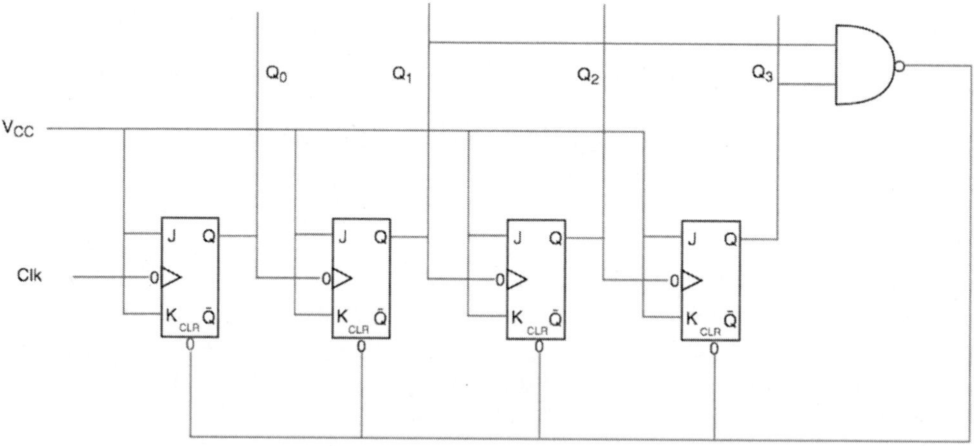

Fig. 5.87: Decade counter circuit diagram

We see from circuit diagram that we have used nand gate for Q3 and Q1 and feeding this to clear input line because binary representation of 10 is— 1010

And we see Q3 and Q1 are 1 here, if we give NAND of these two bits to clear input then counter will be clear at 10 and again start from beginning.

<u>**Important point**</u>: Number of flip flops used in counter are always greater than equal to (**$\log_2 n$**) where n=number of states in counter.

<u>Solved Examples</u>

Example: Design A synchronous decade counter will count from zero to nine and repeat the sequence.

Solution: The state diagram of this counter is shown in figure Below

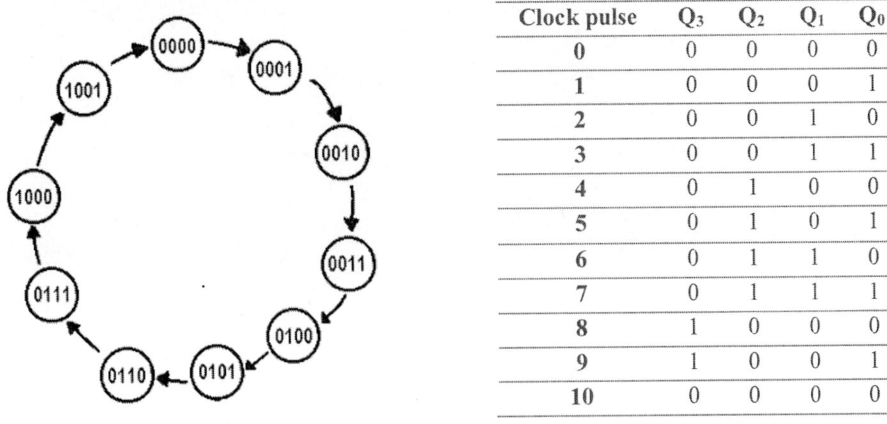

Clock pulse	Q_3	Q_2	Q_1	Q_0
0	0	0	0	0
1	0	0	0	1
2	0	0	1	0
3	0	0	1	1
4	0	1	0	0
5	0	1	0	1
6	0	1	1	0
7	0	1	1	1
8	1	0	0	0
9	1	0	0	1
10	0	0	0	0

Truth table

Fig. 5.88: State diagram and truth table of synchronous decade counter

A simple decade counter will count from 0 to 9 but we can also make the decade counters which can go through any ten states between 0 to 15(for 4 bit counter).

Since there are ten states, four JK flip-flops are required. The truth tables of present and next state for the decade counter are shown in Fig. 9.10. Using the excitation table of JK flip-flop and the outputs of J and K are filled.

Truth table and state table of a synchronous decade counter

Present State				Next State				Output							
Q_3	Q_2	Q_1	Q_0	Q_3	Q_2	Q_1	Q_0	J_3	K_3	J_2	K_2	J_1	K_1	J_0	K_0
0	0	0	0	0	0	0	1	0	X	0	X	0	X	1	X
0	0	0	1	0	0	1	0	0	X	0	X	1	X	X	1
0	0	1	0	0	0	1	1	0	X	0	X	X	0	1	X
0	0	1	1	0	1	0	0	0	X	1	X	X	1	X	1
0	1	0	0	0	1	0	1	0	X	X	0	0	X	1	X
0	1	0	1	0	1	1	0	0	X	X	0	1	X	X	1
0	1	1	0	0	1	1	1	0	X	X	0	X	0	1	X
0	1	1	1	1	0	0	0	1	X	X	1	X	1	X	1
1	0	0	0	1	0	0	1	X	0	0	X	0	X	1	X
1	0	0	1	0	0	0	0	X	1	0	X	0	X	X	1

The Karnaugh maps of the output J_0, K_0, J_1, K_1, J_2, K_2, J_3, and K_3 are shown in Fig. below. The simplified results are at the bottom of the Karnaugh maps.

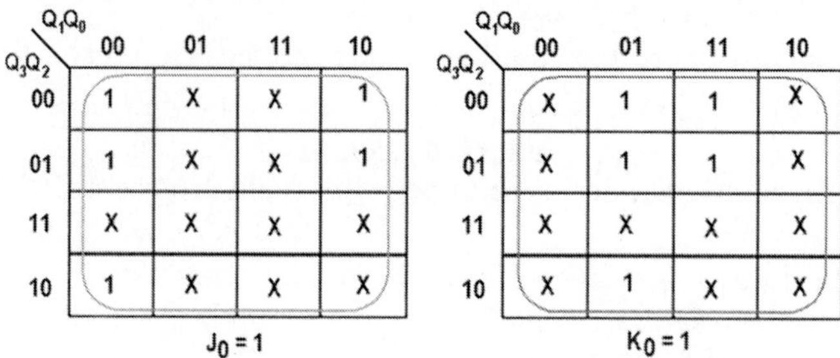

Q_3Q_2 \ Q_1Q_0	00	01	11	10
00	1	X	X	1
01	1	X	X	1
11	X	X	X	X
10	1	X	X	X

$J_0 = 1$

Q_3Q_2 \ Q_1Q_0	00	01	11	10
00	X	1	1	X
01	X	1	1	X
11	X	X	X	X
10	X	1	X	X

$K_0 = 1$

Fig. 5.89: Karnaugh Map for J_0 and K_0

Q_3Q_2 \ Q_1Q_0	00	01	11	10
00		1	X	X
01		1	X	X
11	X	X	X	X
10			X	X

$J_1 = \overline{Q}_3 Q_0$

Q_3Q_2 \ Q_1Q_0	00	01	11	10
00	X	X	1	
01	X	X	1	
11	X	X	X	X
10	X	X	X	X

$K_1 = \overline{Q}_3 Q_0$

Fig. 5.90: Karnaugh Map for J_1 and K_1

Q_3Q_2 \ Q_1Q_0	00	01	11	10
00			1	
01	X	X	X	X
11	X	X	X	X
10			X	X

$J_2 = Q_1Q_0$

Q_3Q_2 \ Q_1Q_0	00	01	11	10
00	X	X	X	X
01			1	
11	X	X	X	X
10			X	X

$K_2 = Q_1Q_0$

Fig. 5.91: Karnaugh Map for J_2 and K_2

Q_3Q_2 \ Q_1Q_0	00	01	11	10
00				
01			1	
11	X	X	X	X
10	X	X	X	X

$J_3 = Q_3Q_0 + Q_2Q_1Q_0$

Q_3Q_2 \ Q_1Q_0	00	01	11	10
00	X	X	X	X
01	X	X	X	X
11	X	X	X	X
10		1	X	X

$K_3 = Q_3Q_0 + Q_2Q_1Q_0$

Fig. 5.92: Karnaugh Map for J_3 and K_3

Based on the results obtained from the Karnaugh maps, the circuit design of synchronous decade counter is shown in Fig. below.

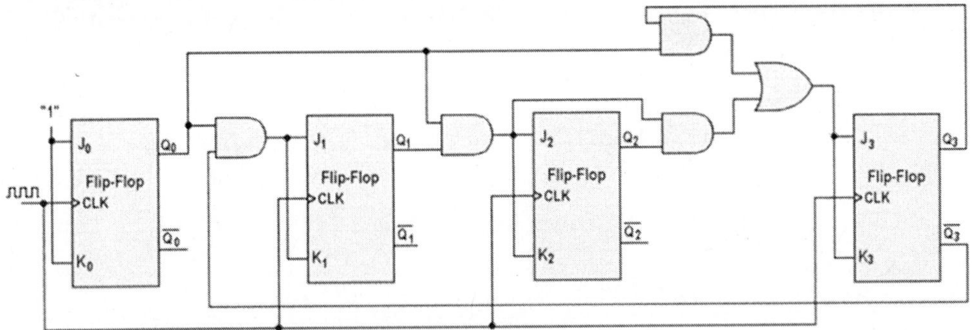

Fig. 5.93: A synchronous decade counter designed using JK flip flop

Example: Design a Modulus-six counter using SR Flip-Flop

Solution:

The modulus six counter will count 0, 2, 3, 6, 5, and 1 and repeat the sequence. The state diagram is shown in Fig. 9.18.

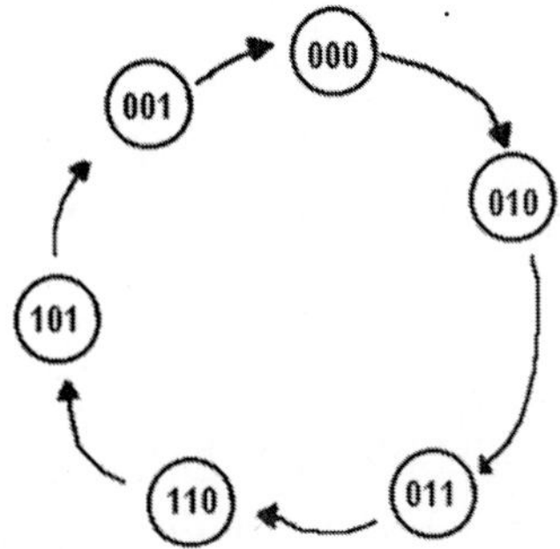

Fig. 5.94: State diagram of a modulus six counter

This modulus six counter requires three SR flip-flops for the design. The truth table of a modulus six counter is shown in Fig. above. From the excitation table of SR flip-flop, the logic of output S_2, R_2, S_1, R_1, S_0, and R_0 are filled and. Truth table and state table of a Modulus-six counter is

[Present State			Next State			Output					
Q_2	Q_1	Q_0	Q_2	Q_1	Q_0	R_2	S_2	R_1	S_1	R_0	S_0
0	0	0	0	1	0	0	X	1	0	0	X
0	1	0	0	1	1	0	X	X	0	1	0
0	1	1	1	1	0	1	0	X	0	0	1
1	1	0	1	0	1	X	0	0	1	1	0
1	0	1	0	0	1	0	1	0	X	X	0
0	0	1	0	0	0	0	X	0	X	0	1

Fig. 5.95: Design of K-maps

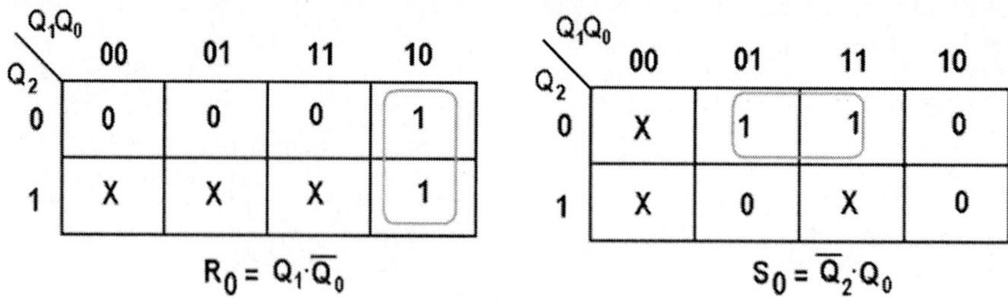

Fig. 5.96: Karnaugh Map for R_0 and S_0

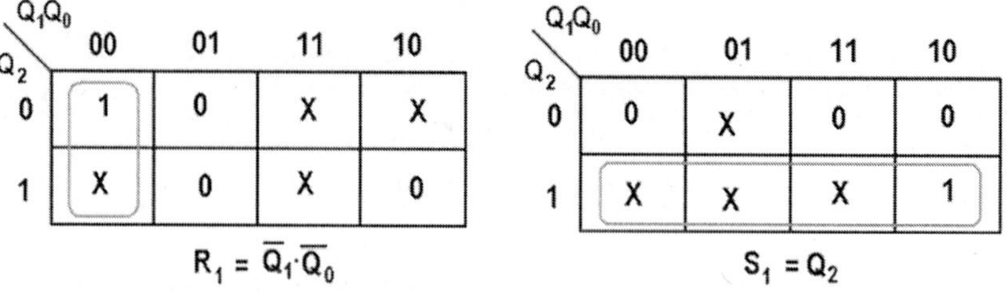

Fig. 5.97: Karnaugh Map for R_1 and S_1

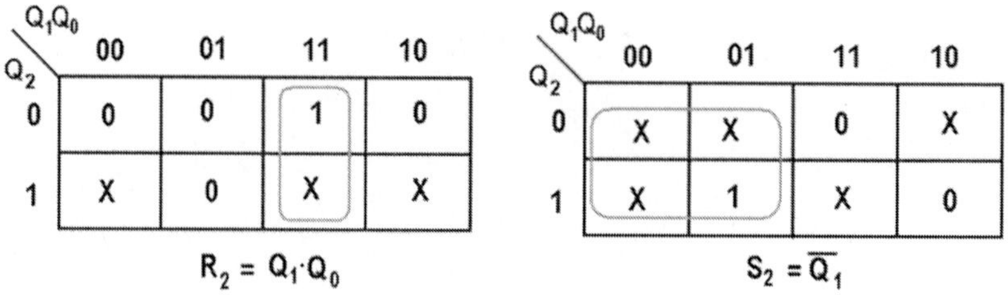

Fig. 5.98: Karnaugh Map for R_2 and S_2

With the known output logic functions, the logic design of the synchronous modulus six counter is shown in Fig. below.

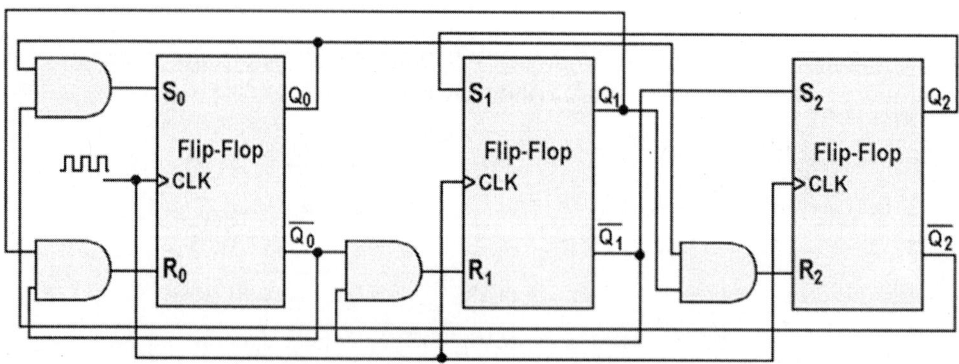

Fig. 5.99: Logic diagram of synchronous modulus six counter

Question: Design a 2 bit up/down counter with an input D which determines the up/down function. Thus, when D=0, the count sequence is 00,01,10,11,00 ... when D=1, the count sequence is 00,11,10,01,00 ...

Solution:

Draw the state diagram for the given sequence.

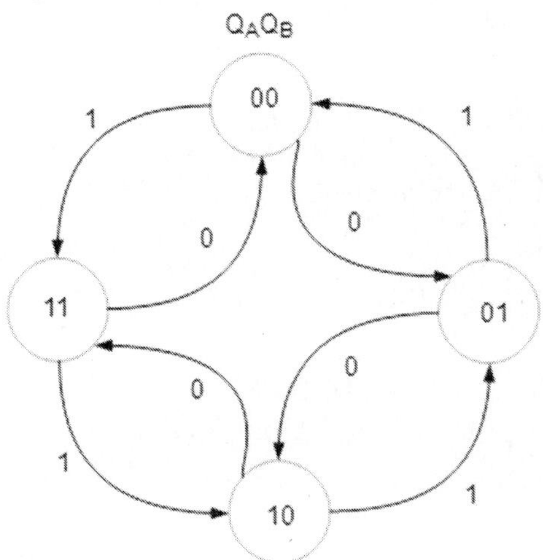

Fig. 5.100: State diagram for the example

Next State table

Develop a next-state table for the specific counter sequence. Using the state diagram as a reference, fill up the present state and next state columns.

Next state and excitation table for example

Present State			Next State		JK flip flop inputs			
D	Q_A	Q_B	Q_A	Q_B	J_A	K_A	J_B	K_B
0	0	0	0	1	0	X	1	X
0	0	1	1	0	1	X	X	1
0	1	0	1	1	X	0	1	X
0	1	1	0	0	X	1	X	1
1	0	0	1	1	1	X	1	X
1	0	1	0	0	0	X	X	1
1	1	0	0	1	X	1	1	X
1	1	1	1	0	X	0	X	1

FF transition table

Next, the FF transition table is completed using data from the respective present state, next state and the JK flip flop transition table below.

Excitation Table and truth Table of JK FF

Q_t	Q_{t+1}	J	K
0	0	0	X
0	1	1	X
1	0	X	1
1	1	X	0

J	K	Q
0	0	Q_0
0	1	0
1	0	1
1	1	$Toggle$

The JK flip flop truth table is not required in the design procedure but has been included to explain how the JK flip flop transition table is obtained. Click on any green cell of the JK flip flop transition table to learn how its value has been derived.

Design K-Map

We have to design 4 K-maps for $J_A, J_B, K_A, and\ K_B$ for the inputs $Q_A,\ Q_B\ and\ D$. And find the values as

$$J_A = \bar{D}Q_B + D\overline{Q_B}$$
$$K_A = \bar{D}Q_B + D\overline{Q_B}$$
$$J_B = 1$$
$$K_B = 1$$

Hence from the above equations a circuit can be designed as

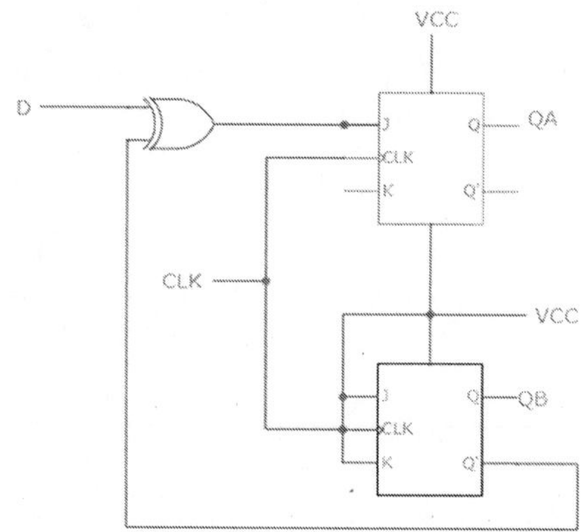

Fig. 5. 101: Circuit diagram for 2 bit up/down counter

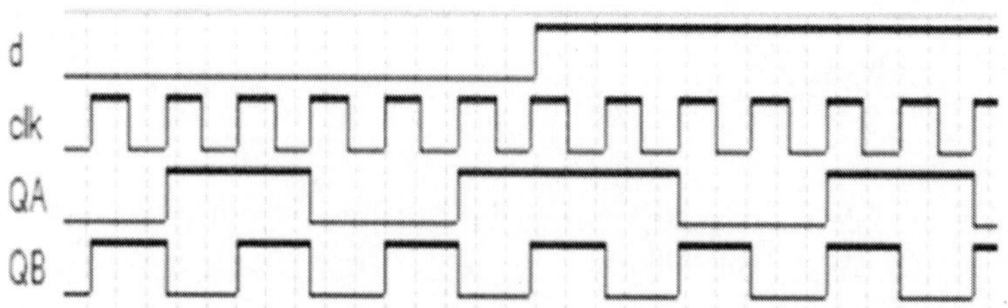

Fig. 5. 102: Timing diagram for 2 bit up/down counter

Example: Design a synchronous Mod-10 Counter using T-FF

Solution:

If we are designing mod *N* counter and n number of flip-flops are required then *n* can be found out by this equation. $N <= 2^n$

Here we are designing Mod-10 counter Therefore, N= 10 and number of Flip flops(n) required is

For n =3, *10<=8, which is false.*

For n= 4,*10<=16, which is true.*

Therefore, number of FF required is 4 for Mod-10 counter.

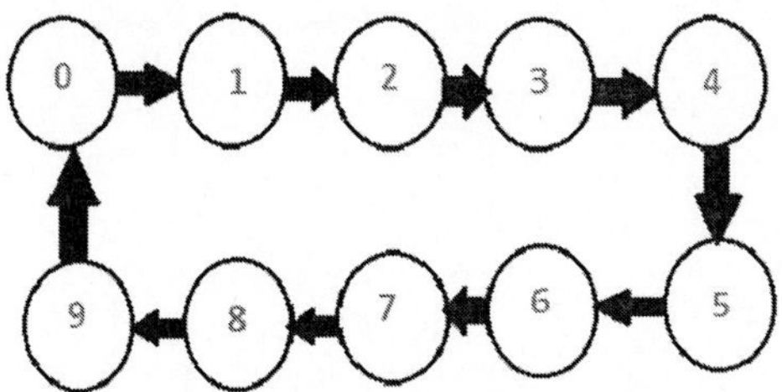

State Diagram:

Q_t	Q_{t+1}	T
0	0	0
0	1	1
1	0	1
1	1	0

Excitation table of T Flip flops

Fig. 5.103: Counting sequence of decade counter and Excitation Table for T-FF

A decade counter is called as mod -10 or divide by 10 counters. It counts from 0 to 9 and again reset to 0. It counts in natural binary sequence. Here 4 T Flip flops are used. It resets after $Q_3\ Q_2\ Q_1\ Q_0$ = 1001.

Circuit excitation table –

Here $Q_3\ Q_2\ Q_1\ Q_0$ are present states of four flip-flops and $Q^*_3\ Q^*_2\ Q^*_1\ Q^*_0$ are next counting state of 4 Flip flops. If there is a transition in current state i.e. if Q3 value changes from 0 to 1 or 1 to 0 then there's corresponding T(toggle) bit is written as 1 otherwise 0.

Circuit Excitation table

Q_3	Q_2	Q_1	Q_0	Q'_3	Q'_2	Q'_1	Q'_0	T_3	T_2	T_1	T_0
0	0	0	0	0	0	0	1	0	0	0	1
0	0	0	1	0	0	1	0	0	0	1	1
0	0	1	0	0	0	1	1	0	0	0	1
0	0	1	1	0	1	0	0	0	1	1	1
0	1	0	0	0	1	0	1	0	0	0	1
0	1	0	1	0	1	1	0	0	0	1	1
0	1	1	0	0	1	1	1	0	0	0	1
0	1	1	1	1	0	0	0	1	1	1	1
1	0	0	0	1	0	0	1	0	0	0	1
1	0	0	1	0	0	0	0	1	0	0	1

Design K-Map

Create Karnaugh map for each FF input in terms of flip-flop outputs as the input variable

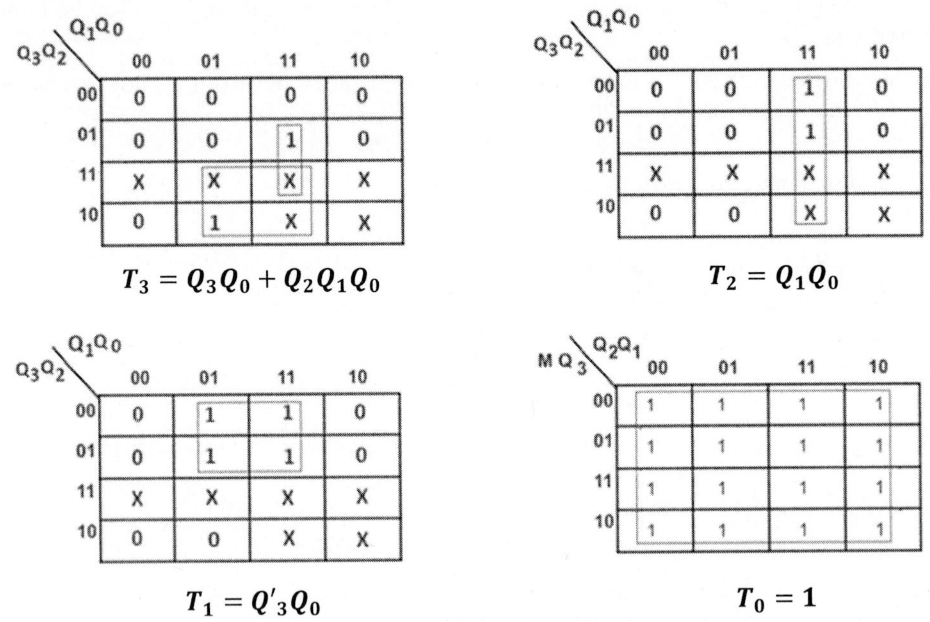

Fig. 5.104: **K map for finding minimal expressions**

Create circuit diagram

Here negative edge triggered clock is used for toggling purpose.

- The clock is provided to every Flip flop at same instant of time.
- The toggle(T) input is provided to every Flip flop according to the simplified equation of K map.

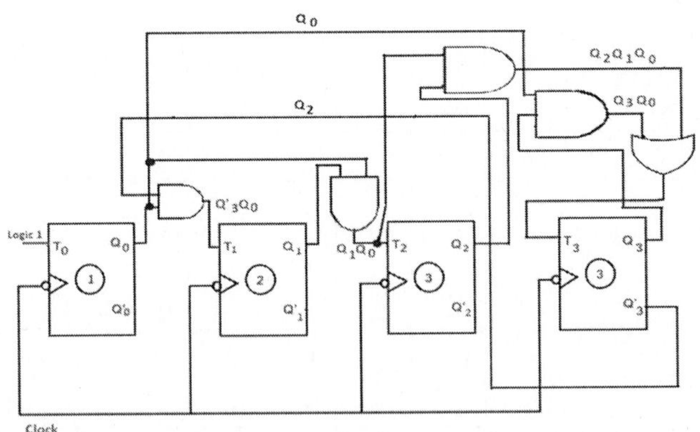

Fig. 5.105: Circuit Diagram for Mod-10 Counter using T-FF

Timing Diagram

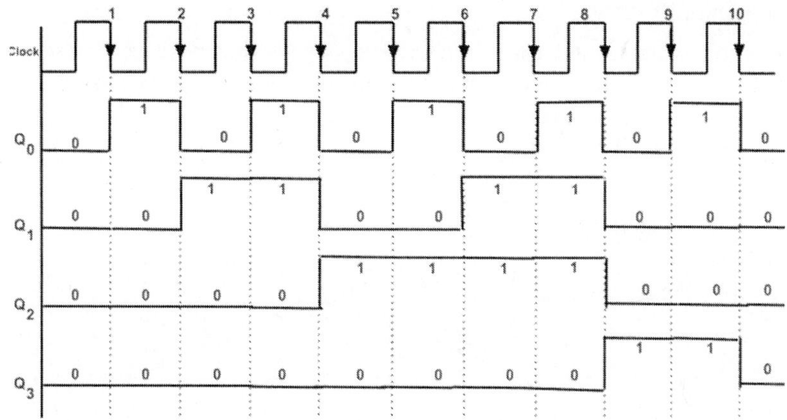

Fig. 5.106: Timing Diagram for Mod-10 Counter using T-FF

Explanation:

- Initially Q_3 Q_2 Q_1 Q_0 are 0 0 0 0.
- The sequence of counter can be verified from the timing diagram. At every falling edge of the clock output Q_0 toggles because T_0 is connected to logic 1.
- T_1 becomes 1 only when expression $T1 = Q'_3Q_0$ becomes 1 also if clock falling edge occurs (because there is negative edge triggering) then the output state of T_1 i.e. Q_1 will change.
- T_2 becomes 1 only when expression $T2 = Q_1Q_0$ becomes 1 also if clock falling edge occurs then the output state Q2 will change.
- T_3 becomes 1 only when expression $T1 = Q_3Q_0 + Q_2Q_1Q_0$ resultant becomes 1 also if clock falling edge occurs (because there is negative edge triggering) then the state of Q3 will change.
- We get Output as Q_3(MSB) Q_2 Q_1 Q_0(LSB).
- After 10th falling edge the output state of all the FFs again becomes 0 0 0 0.

Question: Difference between Straight Ring Counter and Twisted Ring Counter

Solution:

STRAIGHT RING COUNTER	***TWISTED RING COUNTER***
It connects the output of the last shift register to the input of first shift register.	It connects the complement of output of the last shift register to the input of the first register.
It is known as One hot counter.	It is known as Walking ring counter or Johnson's counter.
Number of states = number of flip-flops	Number of states = 2 x number of flip-flops
It circulates a single bit (0 or 1) around the ring.	It circulates stream of 1 followed by stream of 0.
PRESET is used in first shift register.	PRESET is not used in twisted ring counter.
CLEAR is used for last (n-1) flip-flops.	CLEAR is used for all flip-flips in it.
It is used in successive approximation and stepper motor control.	It is used in phase shift or function generator.

Example: Draw the timing diagram of 4-bit SISO shift register (Figure 2) for 1101 data input. Assume that the register is initially clear.

Solution: The left most bit which is a 1 goes into the register on first positive going edge of clock. As remaining bits are entered with the clock, they are shifted from right to left through the shift register. The shift register contains Q3Q2Q1Q0=1101 after four positive going edge of clock. Complete output is available at Q3 in seven clock cycles.

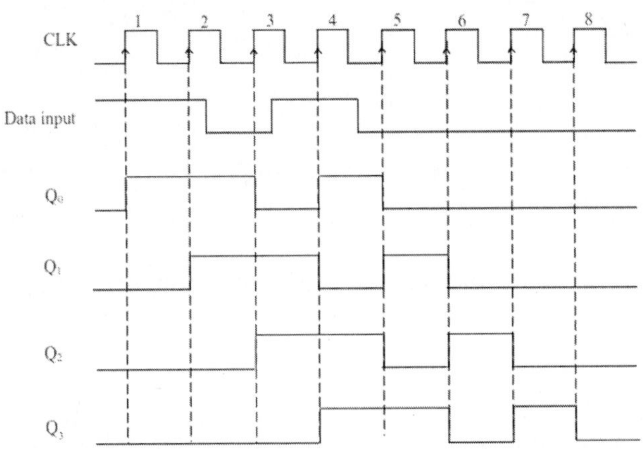

Fig. 5.107: Timing diagram of 4-bit SISO shift register using D flip-flops.

Example: Draw the logic diagram of 4-bit SISO shift register using JK flip-flops instead of D flip-flops.

Solution: At the positive going edge of clock, if J=1 and K=0 is applied then flip-flop stores a 1 and if J=0 and K=1 is applied then flip-flop stores a 0 so a 4-bit SISO shift register using JK flip-flops can be realized as shown in Figure below.

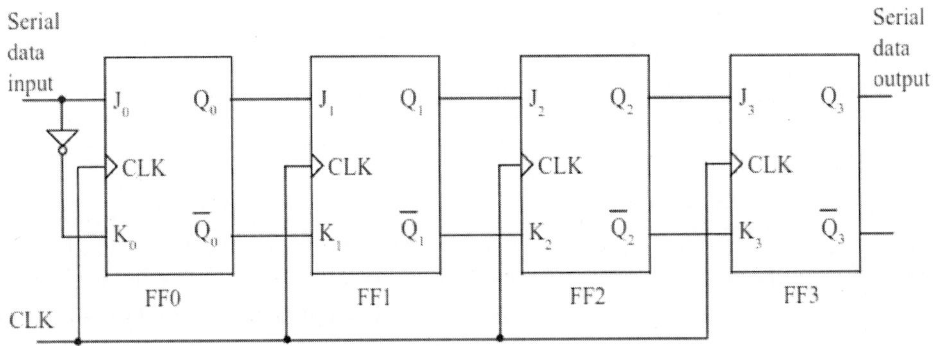

Fig. 5.108: Logic diagram of 4-bit SISO shift register using JK flip-flops

Example: Draw the timing diagram of 4-bit SIPO shift register (Figure 8) for 0111 data input. Assume that the register initially contains all 1's.

Solution: The left most bit which is a 0 goes into the register on first positive going edge of clock. As remaining bits are entered with the clock, they are shifted from right to left through the shift register. The register contains $Q_3Q_2Q_1Q_0$=0111 after four positive going edge of clock. Complete output is available at Q_3, Q_2, Q_1 and Q_0 after four clock cycles.

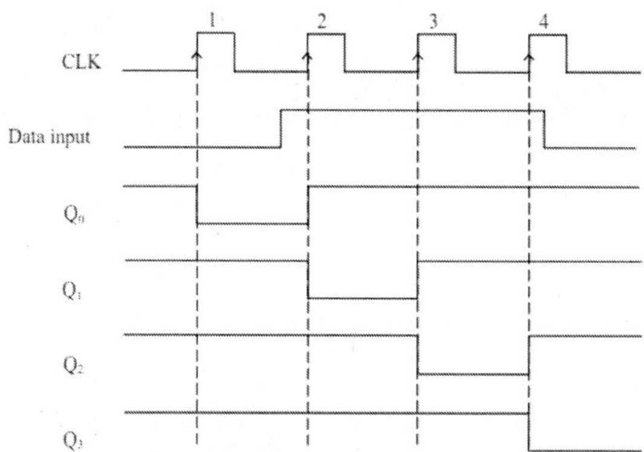

Fig. 5.109: Timing diagram of 4-bit SIPO shift register using D flip-flops

Example: The data input 1101 is entered serially into a 4-bit SIPO shift register which is initially clear. What is the output after four clock cycles? If the data input remains 0 after fourth positive going edge of clock, what is the state of the register after (a) 2 additional clock edges and (b) 4 additional clock edges?

Solution: Sequence of states for 4-bit SIPO shift register is given is table below.

After clock edge	Q_3 (MSB)	Q_2	Q_1	Q_0 (LSB)
0	0	0	0	0
1	0	0	0	**1**
2	0	0	**1**	**1**
3	0	**1**	**1**	**0**
4	**1**	**1**	**0**	**1**
5	**1**	**0**	**1**	0
6	**0**	**1**	0	0
7	**1**	0	0	0
8	0	0	0	0

Example: How to convert 4-bit PIPO shift register as SISO shift register?

Solution: To convert 4-bit PIPO shift register as SISO shift register, the output of each flip flop is connected to the input of next flip-flop. Q0 output of first flip-flop goes to D1 input of second flip-flop, Q1 output of second flip-flop goes to D2 input of third flip-flop, Q2 output of third flip-flop goes to D3 input of fourth flip-flop. Serial data input is applied at D0 and serial output is received at Q3.

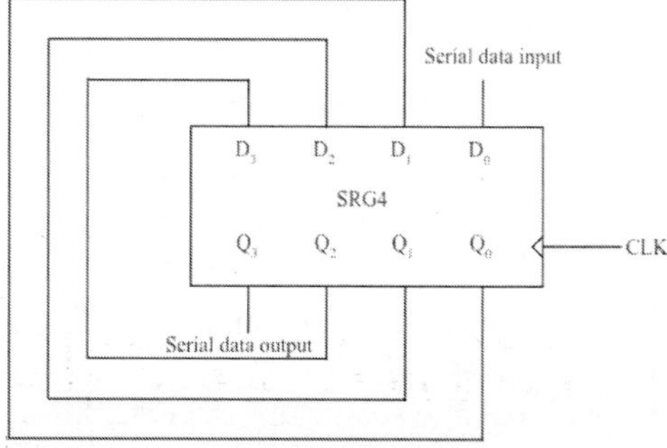

Fig. 5.110: Connection diagram for SISO shift register using PIPO shift register

Example: Draw the timing diagram of 4-bit PIPO shift register (Figure 14) for 0101 data input. Assume that the register is initially clear.

Solution: Level of $Shift/\overline{Load}$ allows four bits of data to load in parallel or shift in the register. When $Shift/\overline{Load} = 0$, data input comes in and the positive going edge of clock sets or resets flip-flop according to the input means data present at D3, D2, D1 and D0 input is loaded in parallel into the register simultaneously. When $Shift/\overline{Load}= 1$, shifting operation takes place and Q0 output of first flip-flop goes to D1 input of second flip-flop (FF1), Q1 output of second flip-flop goes to D2 input of third flip-flop (FF2) and Q2 output of third flip- flop goes to D3 input of fourth flip-flop (FF3).

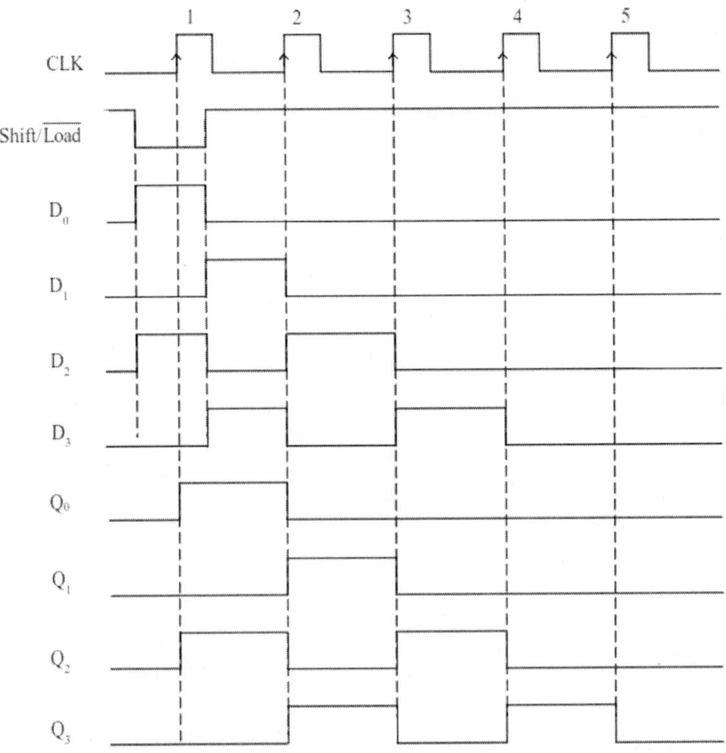

Fig. 5.111: Timing diagram of 4-bit PIPO shift register using D flip-flops.

Review Questions:

1. What do you mean by sequential logic circuit and combinational logic circuits?
2. What do you mean by latch and FF?
3. What are the main differences between synchronous and asynchronous sequential logic circuit?
4. List some of the advantages of synchronous sequential logic circuit and disadvantages of asynchronous sequential logic circuit?
5. Distinguish between combinational and sequential Logic Circuit?
6. What is clock? Describe different types of clock?
7. What is the purpose of clock signal?

8. How many types of sequential logic circuits? Describe briefly.
9. Why is the S-R FF called Set-Reset FF? Deduce the transition table for the S-R FF.
10. What do you mean by characteristic equation? Write K-map of S-R FF and characteristics equation.
11. What are the conditions that must be met in clocked S-R FF?
12. Deduce the restrictions which must be imposed upon the inputs S and R for correct operation of the bistable.
13. Draw S-R FF using NAND gate, explain its operation. What will be the state of y and $\bar{y}$ after FF has been cleared?
14. Draw S-R FF using NOR and explains its operation.
15. Why J-K FF used?
16. What are the modification of S-R FF? Illustrate with diagram.
17. Describe the working principle of J-K FF.
18. Draw the circuit diagram of J-K FF.
19. How can characteristic equation of J-K FF be got from that of S-R FF?
20. What is master slave J-K FF? Illustrate its operation.
21. How can a JK FF be modified to operate as D F-F?
22. Construct the characteristics equation of D FF from that of J-K FF?
23. Why D FF is called Delay FF? Explain the operation of D-FF.
24. Draw the Circuit Diagram of D-FF?
25. Why is T -FF called Trigger FF? Explain T to D and D to T conversion.
26. State the procedure for design a synchronous counter.
27. Draw the timing diagrams of the decade counter using two T-FF.
28. Design a modulus seven synchronous counter that can count 0, 3, 5, 7, 9, 11, and 12 using D flip-flop.
29. Design a 5-bit johnson counter using D Flip-Flop.
30. Design an asynchronous decade counter using JK FF.
31. Which shift register can have a complete binary number loaded into it at a time and have it shifted out one bit at a time? (PISO shift register)
32. Which type of shift register can have only one bit of data entered into it at a time but has all data bits available as output simultaneously? (SIPO shift register)
33. How many bit position shifting of binary data is required so that it multiplies by 4? (as shifting from right to left by one bit position is same as multiplication by 2 so to multiply by 4 we need to shift binary data to the left by two bit positions)
34. How many clock cycles are required to enter or store 2 bytes of data serially into a 16-bit shift register? (one clock cycle is required to store a bit and as a byte of data has 8 bits so to store 2 bytes =16 bits of data, we require 16 clock cycles).
35. In which type of shift register do we have access only to one flip-flop output? (serial output shift register)
36. What is the number of states for a ring counter having five flip-flops?
37. How many bit position shifting of binary data is required so that it multiplies by 4?
38. How many clock cycles are required to enter or store 2 bytes of data serially into a 16-bit shift register?
39. In which type of shift register do we have access only to one flip-flop output?
40. What is the number of states for a ring counter having five flip-flops?

Multiple Choice Questions (MCQs)

1. In sequential logic circuit the present values of outputs are dependent on
   a) past values of input
   b) present values of input
   c) both present values of the inputs and the past values of inputs
   d) none of them.
2. The purpose of clock is to
   a) synchronize the overall action and to prevent the FF from changing state until the right time
   b) synchronize the overall action
   c) respond the clock signal
   d) none of the above.
3. In synchronous sequential logic circuit, contents of the memory can change
   a) only at discrete instants of time
   b) at any instant of time
   c) continuously
   d) none of the above.
4. Which one of following statement in true?
   a) In synchronous S-R FF, the output changes after a brief time delay in response to a change in the input.
   b) In asynchronous S-R FF time delay in response to a change in the input.
   c) The clocked S-R FF outputs should be allowed to change only when CK = 1.
   d) iv) none of the above.
5. Which one is the true statement?
   a) When S = R = 0, output is unchanged
   b) When S = 0, R = 1, then output is 1
   c) When S = 1, R = 1, then output is 1
   d) When S = 1, R = 0, then output is indeterminate condition.
6. Which one of the following statement is true for JK FF?
   a) The K input alone performs Reset function.
   b) The K input alone performs Reset function causing the output to be 0.
   c) The J input alone performs Reset function.
   d) None of the above.
7. The T FF is called
   a) Delay flip-flop b) Trigger flip-flop c) Falling flip-flop d) none of the above.
8. "Irrespective of the present state it complements when clock pulse occurs". Which one of the following is true for the above statement?
   a) T FF b) D FF c) J-K FF d) S-R FF.
9. Storage capacity of a register is determined by
   a) Method of data input
   b) Method of data output
   c) Type of flip-flops used
   d) Number of flip-flops used
10. In SISO shift register
    a) Data input comes in serially and stored output goes out serially
    b) Data input comes in serially and stored output goes out in parallel

c) Data input comes in parallel and stored output goes out serially
d) Data input comes in parallel and stored output goes out in parallel

11. In PISO shift register
    a) Data input comes in serially and stored output goes out serially
    b) Data input comes in serially and stored output goes out in parallel
    c) Data input comes in parallel and stored output goes out serially
    d) Data input comes in parallel and stored output goes out in parallel
12. Data input comes in parallel means
    a) Data output of each stage is available
    b) Bits are entered bit by bit on one line
    c) Bits are entered simultaneously into respective stages
    d) None of the above
13. A 4-bit PISO shift register has
    a) One serial input (D0) and one serial output (Q3)
    b) One serial input (D0) and four parallel output (Q3, Q2, Q1 and Q0)
    c) Four parallel input (D3, D2, D1 and D0) and one serial output (Q3)
    d) Four parallel input (D3, D2, D1 and D0) and four parallel output (Q3, Q2, Q1 and Q0)
14. What is the modulus of a ring counter having 4 flip-flops
    a) 4 b) 8 c) 16 d) None of the above
15. Storage capacity of a register is determined by
    a) Method of data input
    b) Method of data output
    c) Type of flip-flops used
    d) Number of flip-flops used
16. In SISO shift register
    a) Data input comes in serially and stored output goes out serially
    b) Data input comes in serially and stored output goes out in parallel
    c) Data input comes in parallel and stored output goes out serially
    d) Data input comes in parallel and stored output goes out in parallel
17. In PISO shift register
    a) Data input comes in serially and stored output goes out serially
    b) Data input comes in serially and stored output goes out in parallel
    c) Data input comes in parallel and stored output goes out serially
    d) Data input comes in parallel and stored output goes out in parallel
18. Data input comes in parallel means
    a) Data output of each stage is available
    b) Bits are entered bit by bit on one line
    c) Bits are entered simultaneously into respective stages
    d) None of the above
19. 4-bit PISO shift register has
    a) One serial input (D0) and one serial output (Q3)
    b) One serial input (D0) and four parallel output (Q3, Q2, Q1 and Q0)
    c) Four parallel input (D3, D2, D1 and D0) and one serial output (Q3)

d) Four parallel input (D3, D2, D1 and D0) and four parallel output (Q3, Q2, Q1 and Q0)

20. Registers are used mainly to store binary data.
    a) True b) False
21. A flip-flop has 4-bit memory.
    a) True b) False
22. Shift registers shift content of registers once in every clock cycle.
    a) True b) False
23. A 4-bit SISO shift register has 4-bit memory.
    a) True b) False
24. A SIPO shift register can display all input bits at a time.
    a) True b) False
25. A PISO shift register can display all input bits at a time.
    a) True b) False
26. The output of a ring counter is always square wave.
    a) True b) False
27. The output of a ring counter is unique for each state.
    a) True b) False
28. The output of a Johnson counter is always square wave.
    a) True b) False
29. All flip-flops in a shift register have common clock input.
    a) True b) False

References and Suggested Readings

- *A. Anand Kumar, Fundamentals of digital circuits Second Edition, PHI Learning Private Limited, ISBN (978-81-203-3679-7)*
- *Cook, N. P. (2003) Practical Digital Electronics, Prentice-Hall, NJ, USA,*
- *Floyd, T. L. (2005) Digital Fundamentals, Prentice-Hall Inc., USA.*
- *Jain, R. P. Modern digital electronics. Vol. 1. No. 10. Tata McGraw-Hill Education, New Delhi, 2003.*
- *Morris Mano, M. and Kime, C. R. (2003) Logic and Computer Design Fundamentals, Prentice-Hall Inc., USA.*
- *Malvino, A. P. and Leach, D. P. (1994) Digital Principles and Applications, McGraw-Hill Book Company. USA.*
- *Rafiquzzaman, M. (2005) Fundamentals of Digital Logic and Microcomputer Design, Wiley-Interscience, New York, USA.*
- *Tocci, R. J. and N. S. Widmer. 2004. Digital Systems Principles and Applications, 9th ed. Upper Saddle River, NJ: Prentice Hall.*
- *Tokheim, R. L. (1994) Schaum's Outline Series of Digital Principles, McGraw-Hill Companies Inc., USA.*

6 A/D and D/A Convertors

UNIT SPECIFICS

Through this unit we have discussed the following aspects:

- *To understand Analog to digital converter (ADCs) and its various parameters.*
- *To understand Digital to Analog converter (DACs) and its various parameters.*
- *To understand different methods used for ADCs.*
- *To understand different methods used for DACs.*
- *To study and analyse Voltage to frequency and voltage to time-based convertors.*
- *To identify various application relevant information for ADCs and DACs.*

RATIONALE

Most of the real-world physical quantities such as voltage, current, temperature, pressure and time etc. are available in analog form. Even though an analog signal represents a real physical parameter with accuracy, it is difficult to process, store or transmit the analog signal without introducing considerable error because of the involvement of noise.

Therefore, for processing, transmission and storage purposes, it is often convenient to express these variables in digital form. It gives better accuracy and reduces noise. The operation of any digital communication system is based upon analog to digital (A/D) and digital to analog (D/A) conversion.

An analog-to-digital (A/D) converter is defined as a device that receives an analog input and supplies a digital output (usually binary or decimal number). In designing A/D conversion devices, several basic requirements like sampling time, resolution, conversion time must be taken into consideration. Mainly ramp type and successive approximation type convertors are used.

A digital-to-analog (D/A) converter is defined as a device that receives a digital input and outputs an analog signal. Many digital-to-analog converters (DACs) are based upon the principle of the voltage divider or switched resistors.

PRE-REQUISITES

Circuit analysis, KCL, KVL, Quantization concepts, Basic Arithmetic.

UNIT OUTCOMES

List of outcomes of this unit is as follows:

U6-O1: Describe ADCs and DACs along with their different parameters.

U6-O2: Apply various methods for DACs and ADCs implementation.

U6-O3: implement and realization of ADCs and DACs for various type of input signals.

U6-O4: Identify various application relevant information for combinational circuits.

Unit-6 Outcomes	EXPECTED MAPPING WITH COURSE OUTCOMES (1- Weak Correlation; 2- Medium correlation; 3- Strong Correlation)					
	CO-1	CO-2	CO-3	CO-4	CO-5	CO-6
U6-O1	3	3	3	-	3	1
U6-O2	1	1	2	2	1	-
U6-O3	2	1	3	1	2	1
U6-O4	-	-	3	1	2	2

➢ Scan the below QR codes for more information on the topic written below the QR code.

DAC and ADC Converters

R-2R Ladder DAC with example

Slope and Dual slope ADC

Counter type ADC

Unit outcomes

On completion of this lesson students will be able to learn about various terms of A/D and D/A converters. Design different types of D/A converter circuit and describe their operation and understand the advantages, disadvantages, and limitation of several types of digital-to-analog converters (DAC). Understand the operation of various types of analog-to-digital converter circuitry such as the digital ramp A/D circuit. Compare the advantages and disadvantages of different analog-to-digital converter circuits

6.1 Introduction

A digital quantity will have a value that is specified as one of two possibilities such as 0 or LOW or true and 1 or HIGH or false and so on. In practice, the voltage representation a digital quantity such as a value that is anywhere within specified ranges.

Example, for TTL logic:

0V to 0.8V = logic 0

2V to 5V = logic 1

Any voltage falling in the range 0 to 0.8 V is given the digital value 0, and any voltage in the range 2 to 5 V is assigned the digital value 1. The digital circuits respond accordingly to all voltage values within a given range. Most physical variables like temperature, pressure, light intensity, audio signals, position, rotational speed, and flow rate are analog in nature and can take on any value within a continuous range of values. Digital systems perform all of their internal operations using digital circuitry and digital operations. Any information that has to be inputted to a digital system must first be put into digital form. Similarly, the outputs from a digital system are always in digital form.

However, the physical variable is normally a non-electrical quantity. A transducer is a device that converts the physical variable to an electrical variable. Some common transducers include thermistors, photocells, photodiodes, flow meters, pressure transducers, and tachometers. The electrical output of the transducer is an analog current or voltage that is proportional to the physical variable it is monitoring. For example, the physical variable could be the temperature of water.

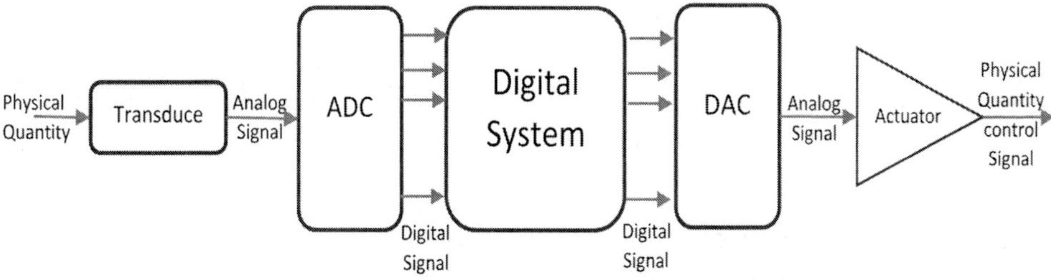

Fig.6.1: Interfacing with the analog world using Analog-to-Digital Converter (ADC) and Digital-to-Analog Converter (DAC).

Analog to Digital Convertor (ADC):

Let's say that the water temperature varies from 80 to 1500 F and that a thermistor and its associated circuitry convert this water temperature to a voltage ranging from 800 to 1500mV. The transducer's output is directly proportional to temperature; such that each 10°F produces a 10mV output. Analog-to-digital converter (ADC) and digital-to-analog converter (DAC) are used to interface a computer to the analog world so that the computer can monitor and control a physical variable Fig..

The Analog to Digital Convertor (ADC) converts this analog input to a digital output. This digital output consists of a number of bits that represent the value of the analog input. For example, the ADC might convert the transducer's 800- to 1500-mV analog values to binary values ranging from 01010000 (80) to 10010110 (150). Note that the binary output from the ADC is proportional to the analog input voltages so that each unit of the digital output represents 10mV. The digital representation of the analog vales is transmitted from the ADC to the digital computer, which stores the digital value and processes it according to a program of instructions that it is executing.

Digital to Analog Convertor (DAC):

This digital output from the computer is connected to a Digital to Analog Convertor (DAC), which converts it to a proportional analog voltage or current. For example, the computer might produce a digital output ranging from 0000000 to 11111111, which the DAC converts to a voltage ranging from 0 to 10V.

Actuator:

The analog signal from the DAC is often connected to some device or circuit that serves as an actuator to control the physical variable. For our water temperature example, the actuator might be an electrically controlled valve that regulates the flow of hot water into the tank in accordance with the analog voltage from the DAC. The flow rate would vary in proportion to this analog voltage, with 0 V producing no flow and 10 V producing the maximum flow rate. Thus we see that ADCs and DACs function as interfaces between a completely digital system, like a computer, and the analog world.

6.2 Quantization and Encoding:

In a digital-to-analog converter, the possible number of inputs is fixed. And in analog-to-digital converter, the input analog voltage can have any value in a range, but the digital output can have only 2N discrete values for an N-bit A/D converter. This analog voltage is represented in 2N intervals this process is known as quantization. Each interval is then assigned a unique N-bit binary code, this process is called encoding. Consider an analog voltage in the range of 0 to V and a 2-bit digital output for any voltage in this range. Let us divide the whole range of analog voltage in 8 intervals (2-bit output) of the size S = V/4. Each interval is assigned a 2-bit binary value. The interval of the analog voltage and their corresponding digital values assigned are shown in Table 6.1.

Table **6.1**: Quantization and Encoding using two-bit binary value

Analog voltage	**V**	**3/4V**	**2/4V**	**1/4V**	**0**
Equivalent digital value	-	11	10	01	00

An analog voltage to be processed using digital techniques may be a fixed (D.C.) voltage or a time varying voltage *v(t)*. The process of converting an analog signal to a digital form involves a sequence of four process, which are Sampling, Holding, Quantizing and Encoding respectively.

Sampling and holding operations are done simultaneously using a circuit known as sample-and-hold (S/H) circuit. These operations are required to be performed for *v(t)* not for D.C. signals. Quantization and Encoding processes are done simultaneously using a circuit referred to as an analog-to-digital converter (A/D converter or ADC.)

The process of conversion of an analog signal to digital signal is referred to as an Analog-to-Digital conversion. The output of the system may be required to be analog form so this digital output has to be converted back to the analog form. This process is referred to as a Digital-to-Analog Converter (D/A converter or DAC). For example, a digital communication system is used to transmit messages which are in the form of analog electrical signals. This requires an A/D converter at the transmitting end and a D/A converter at the receiving end. In microprocessors-based process control system, A/D and D/A converters are often used and are referred to as peripherals or I/O devices. Since D/A converters are used as sub-systems in many A/D converters so, D/A converters will be discussed first.

6.3 Digital to Analog (D/A or DAC) Converters

DAC conversion is the process of taking a value represented in digital code (such as straight binary or BCD) and converting it to a voltage or current which is proportional to the digital value. Fig. 6.2 shows the symbol for a typical 4-bit D/A converter. Now, let us examine the various input/output relationships.

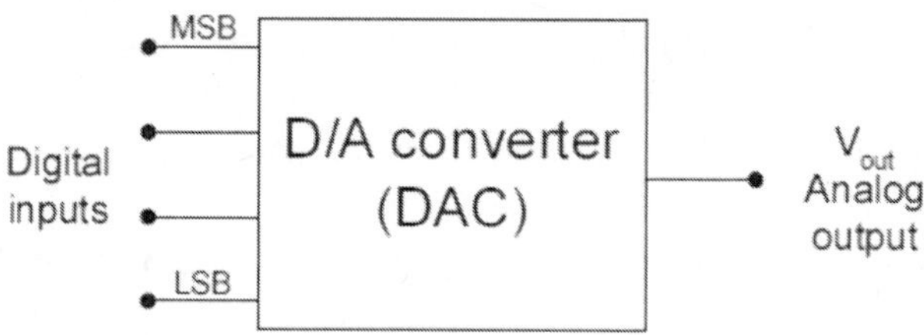

Fig. 6.2: Four-bit DAC with voltage output.

The digital inputs D, C, B, and A are usually derived from the output register of a digital system. The $2^4 = 16$ different binary numbers represented by these 4 bits for each input number, the D/A converter output voltage is a unique value. In fact, for this case, the analog output voltage V_{out} is equal in volts to the binary number.

In general, the analog output voltage V_{out} of an N-bit straight binary D/A converter is related to the digital input by the equation

$$V_{out} = K(2^{N-1}\, b_{N-1} + 2^{N-2} b_{N-2} + \cdots + 2^2 b_2 + 2b_1 + b_0)$$

Where K is a proportionality factor and it is constant value for a given DAC.

$$b_N = 1 \text{ if the nth bit of the digital input is 1.}$$
$$= 0 \text{ if the nth bit of the digital input is 0.}$$

Hence, Analog output = K × digital input

The analog output can of course be a voltage or current. When it is a voltage, K will be in voltage units, and when the output is current, K will be in current units. For the DAC of $K=1\ V$, so that

$$V_{OUT} = (1\text{ V}) \times \text{digital input}$$

We can use this to calculate V_{OUT} for any value of digital input. For example, with a digital input of $1100_2 = 12_{10}$, we obtain

$$V_{OUT} = 1V \times 12 = 12V$$

Example 6.1: A 5-bit DAC has a current output. For a digital input of 101000, an output current of 10mA is produced. What will I_{OUT} be for a digital input of 11101?

Solution: The digital input 101000_2 is equal to decimal 20. Since I_{OUT} = 10 mA for this case, the proportionality factor as 0.5 mA. Thus, we can find for a digital input such as $11101_2 = 29_{10}$

$$\text{Hence, } I_{OUT} = (0.5\text{mA}) \times 29 = 14.5\text{ mA}$$

Note: Remember, the proportionality factor, K, will vary from one DAC to another.

Example 6.2: What is the largest value of output voltage from an 8-bit DAC that produces 1.0V for a digital input of 00110010?

Solution:

$$00110010_2 = 50_{10}$$
$$\text{Hence, } 1.0\text{ V} = K \times 50$$

Therefore, K = 20 mV

The largest output will occur for an input of $11111111_2 = 255_{10}$.

$$V_{OUT}\text{ (max)} = 20\text{mV} \times 255$$
$$= 5.10\text{ V}$$

Note: *The output of a DAC is technically not an analog quantity because it can take on only specific values like the 16 possible voltage levels for V_{out}. Thus, in that sense, it is actually digital. However, the number of different possible output levels can be increased and the difference between successive values can be decreased by increasing the number of input bits. This will allow us to produce an output that is more and more like an analog quantity that varies continuously over a range of values.*

6.3.1 Specifications for D/A converters

Input Weight:

For the DAC of it should be noted that each digital input contributes a different amount to the analog output. This is easily seen if we examine the cases where only one input is HIGH (Refer Table 6.2). The contributions of each digital input are weighted according to their position in the binary number.

Table6.2: Input Weights:

D	C	B	A		V_{OUT} (V)
0	0	0	1	→	1
0	0	1	0	→	2
0	1	0	0	→	4
1	0	0	0	→	8

Thus, A, which is the LSB, has a weight of 1V, B has a weight of 2V, C has a weight of 4 V, and D, the MSB, has the largest weight 8V. The weights are successively doubled for each bit, beginning with the LSB. Thus, we can consider V_{OUT} to be the weighted sum of the digital inputs. For instance, to find V_{OUT} for the digital input 0111 we can add the weights of the C, B, and A bits to obtain 4 V + 2V + 1V=7V.

Example 6.3: A 5-bit D/A converter produces V_{OUT} = 0.2 V for a digital input of 0001. Find the value of V_{OUT} for an input of 11111.

Solution: 0.2 V is the weight of the LSB. Thus, the weights of the other bits must be 0.4 V, 0.8 V, 1.6 V, and 3.2 V respectively. For a digital input of 11111, then, the value of V_{OUT} will be

$$3.2\text{ V} + 1.6\text{ V} + 0.8\text{V} + 0.4\text{V} + 0.2\text{ V} = 6.2\text{ V}.$$

Resolution:

Resolution of a D/A converter is defined as the smallest change that can occur in the analog output as a result of a change in the digital input. As shown in Fig. 6.3 it can be seen that the resolution is 1V, since V_{OUT} can change by no less than 1 V when the digital input value is changed.

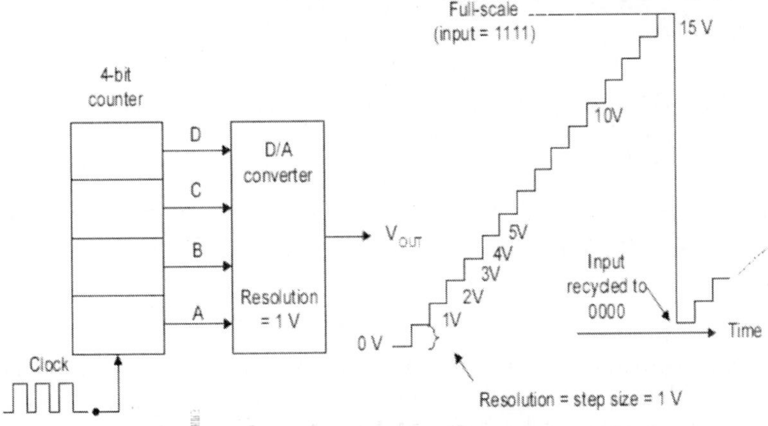

Fig. 6.3: Output wave forms of a four-bit DAC.

The resolution is always equal to the weight of the LSB and is also referred to as the step size. As the counter is being continually cycled through its 16 states by the clock signal, the DAC output is a staircase waveform that goes up 1 V per step. When the counter is at 1111, the DAC output is at its maximum value of 15 V; this is its full-scale output. When the counter recycles to 0000, the DAC output returns to 0V. The resolution or step size of the jumps in the staircase waveform; in this case, each step is 1 V.

Note that the staircase has 16 levels corresponding to the 16 input states, but there are only 15 steps or jumps between the 0-V level and full-scale, so, for an N-bit DAC the number of different levels will be 2^N, and the number of steps will be $2^N - 1$.

The resolution (step size) is the same as the proportionality factor in the DAC input/output relationship:

$$\text{analog output} = K \times \text{digital input}$$

A new interpretation of this expression would be that the digital input is equal to the number of the step, K is the amount of voltage (or current) per step, and the analog output is the product of the two.

Example 6.4: For the DAC of Example 6.3 determine V_{OUT} for a digital input of 10001.
Solution: The step size is 0.2 V, which is the proportionality factor K.
The digital input is $10001 = 17_{10}$. Thus, we have:

$$V_{OUT} = (0.2\text{ V}) \times 17 = 3.4\text{V}$$

Percentage Resolution:

Resolution can be expressed as the amount of voltage or current per step, it is also useful to express it as a percentage of the full-scale output. To illustrate, in Fig. 6.3 the DAC has a maximum full-scale output of 15 V (when the digital input is 1111). The step size is 1V, which gives a percentage resolution

$$\%resolution = \frac{Step\ Size}{Full\ Scale\ (F.S)} \times 100\%$$

Hence for the above example

$$\%resolution = \frac{1V}{15V} \times 100\% = 6.67\%$$

Example 6.5: A 10-bit DAC has a step size of 10 mV. Determine the full-scale output voltage and the percentage resolution.
Solution:

With 10 bits, there will be $2^{10} - 1 = 1023$ steps of 10mV each. The full-scale output will therefore be 10mV × 1023 = 10.23 V and

$$\%resolution = \frac{10\ mV}{10.23\ V} \times 100 \cong 0.1\%$$

Problem 6.4 helps to illustrate the fact that the percentage resolution becomes smaller as the number of input bits is increased. In fact, the percentage resolution can also be calculated from.

$$\%resolution = \frac{1}{Total\ Number\ of\ Steps} \times 100\%$$

For an N-bit binary input code the total number of steps is $2^N - 1$. Thus, for the previous example,

$$\%resolution = \frac{1}{2^{10}-1} \times 100\%$$
$$= \frac{1}{1023} \times 100 \cong 0.1\%$$

Note: This means that it is only the number of bits which determines the percentage resolution. Increase of the number of bits increases the number of steps to reach full scale.

Accuracy:

There are several ways of specifying accuracy. The two most common are called full-scale error and linearity error. which are normally expressed as a percentage of the converter's full-scale output (% F.S.).

Full-scale error is the maximum deviation of the DAC's output from its expected (ideal) value, expressed as a percentage of full scale. For example, assume that the DAC has an accuracy of ± 0.01% F.S. Since this converter has a full-scale output of 9.375 V, this percentage converts to

$$\pm 0.01\% \times 9.375 \text{ V} = \pm 0.9375 \text{ mV}$$

This means that the output of this DAC can, at any time, be off by as much as 0.9375mV from its expected value.

Linearity error is the maximum deviation in step size from the ideal step size. For example, the DAC has an expected step size of 0.625 V. If this converter has a linearity error of ± 0.01F.S., this would mean that the actual step size could be off by as much as 0.9375 mV

Example 6.6: A certain 8-bit DAC has a full-scale output of 2mA and a full-scale error of ± 0.5% F.S. What is the range of possible outputs for an input of 10000000?

Solution:

The *step size* is 2mA/255 = 7.84 μA.

Since 10000000 = 12810, the ideal output should be 128 × 7.84 μA.

The error can be as much as

$$\pm 0.5\% \times 2mA = \pm 10\mu A$$

Thus, the actual output can deviate by this amount from the ideal 1004μA, so the actual output can be anywhere from 994 to 1014 μA.

Offset Error:

Ideally, the output of a DAC will be zero volts when the binary input is all 0's. In practice, however, there will be a very small output voltage for this situation; this is called offset error. This offset error, if not corrected, will be added to the expected DAC output for all input cases. Offset error can be negative as well as positive. Many DACs will have an external offset adjustment that allows you to zero the offset. This is usually accomplished by applying all 0s to the DAC input and monitoring the output while an offset adjustment potentiometer is adjusted until the output is as close to 0 V as required.

Settling Time: The operating speed of a DAC is usually specified by giving its settling time, which is the time required for the DAC output to go from zero to full scale as the binary input is changed from all 0's to all 1's. Typical values for settling time range from 50 ns to 10 μs.

Monotonicity: A DAC is monotonic if its output increases as the binary input is incremented from one value to the next. Another way to describe this is that the staircase output will have no downward steps as the binary input is incremented from zero to full scale.

6.3.2 Weighted resistor D/A converter

The purpose of a digital-to-analog converter is to convert a binary word to a proportional current or voltage. In this conversion methods uses a weighted resistor and a summing amplifier. The binary weighted resistors produce binary-weighted current which are summed up by the op-amp to produce proportional output voltage. Several different binary codes such as straight binary, BCD and offset binary are commonly used as inputs to D/A converters.

Operational Amplifiers such as the inverting op-amp circuit uses negative feedback to both reduce and control its very high open-loop gain, A_{OL}. It does this by feeding back a small fraction of its output signal back to its input terminal.

For an inverting amplifier, the input voltage V_{IN} is connected directly to its inverting input via a suitable input resistor R_{IN} and that the inverting amplifiers closed-loop voltage gain, $A_{V(CL)}$ is determined by the ratio of these two resistors as shown in Fig. 6.4. that represents, an Inverting Operational Amplifier Circuit

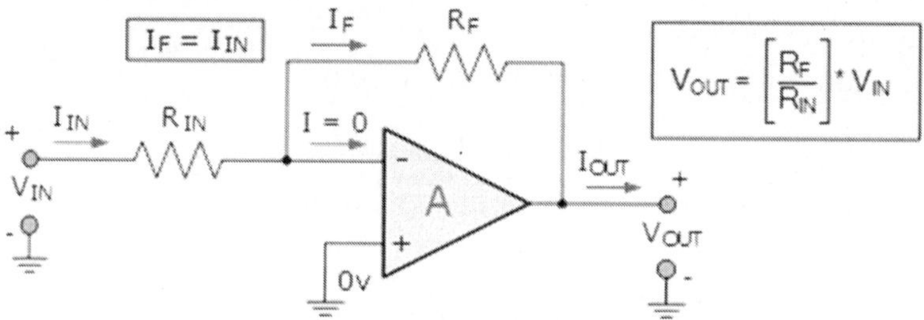

Fig. 6.4: Op-Amp in Inverting Mode

Then we can see that V_{OUT} is given as V_{IN} multiplied by the closed-loop Gain (A_{CL}), which is determined by the ratio of the feedback resistance, R_F to the input resistance, R_{IN}. So, by altering the values of either R_F or R_{IN} we can change the closed-loop gain of the op-amp and therefore the value of V_{OUT} (I_F*R_f) for a given input signal.

Here in this inverting operational amplifier example, we have used a single input voltage signal, but what if we added another input resistor to combine two or more analogue signals into a single output, what would be the effect on the circuit and its gain.

6.3.3 Digital-to-Analogue Converter Summing Amplifier

By connecting multiple inputs to the negative terminal of the operational amplifier, we can convert the single input circuit from above into a *summing amplifier* or to be more precise, a "summing inverting voltage amplifier" circuit.

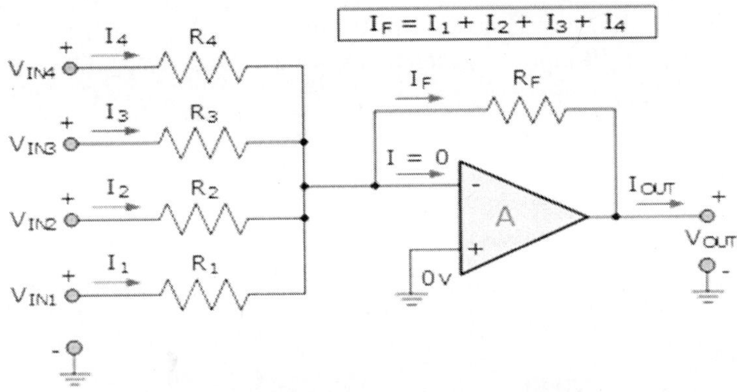

Fig. 6.5: Op-amp as summing amplifier

As the negative feedback created by the feedback resistor, R_F biases the inverting input of the op-amp at zero potential, any input signals are effectively electrically isolated from each other with the output being the inverted sum of all the input signals combined. Thus, a summing amplifier in the inverting mode produces the negative sum of any number of input voltages, whereas a no-inverting summing amplifier would produce the positive sum of any number of input voltages. Consider the circuit below (Refer Fig. 6.5) represents, Inverting Summing Amplifier Circuit

In the summing amplifier circuit above, the output voltage, (V_{OUT}) is proportional to the sum of the four input voltages, V_{IN1}, V_{IN2}, V_{IN3}, and V_{IN4} and we can modify the original equation for the inverting amplifier configuration above to take account of these four new input values as follows:

$$I_F = I_1 + I_2 + I_3 + I_4 = \frac{V_{in1}}{R_1} + \frac{V_{in2}}{R_2} + \frac{V_{in3}}{R_3} + \frac{V_{in4}}{R_4}$$

$$V_{OUT} = \frac{R_F}{I_F} \times V_{in} = -\left(\frac{R_F}{R_1} \times V_{in1} + \frac{R_F}{R_2} \times V_{in2} + \frac{R_F}{R_3} \times V_{in3} + \frac{R_F}{R_4} \times V_{in4}\right)$$

$$If,\ R_1 = R_2 = R_3 = R_4 = R_{in}$$

$$V_{OUT} = -\frac{R_F}{R_{in}}(V_{in1} + V_{in2} + V_{in3} + V_{in4})$$

$$If,\ R_1 = R_2 = R_3 = R_4 = R_F$$

$$V_{OUT} = -(V_{in1} + V_{in2} + V_{in3} + V_{in4})$$

Then we can see that the output voltage is an inverted, scaled sum of the four input voltages as each input voltage is multiplied by its corresponding gain and added to the next to produce the total output. If all the resistances are the same and of an equal value, that is: $R_F = R_1 = R_2 = R_3 = R_4$, then each input channel will have a closed-loop voltage gain of unity (1) so the output voltage is given simply by:

$$V_{OUT} = -(V_{IN1} + V_{IN2} + V_{IN3} + V_{IN4})$$

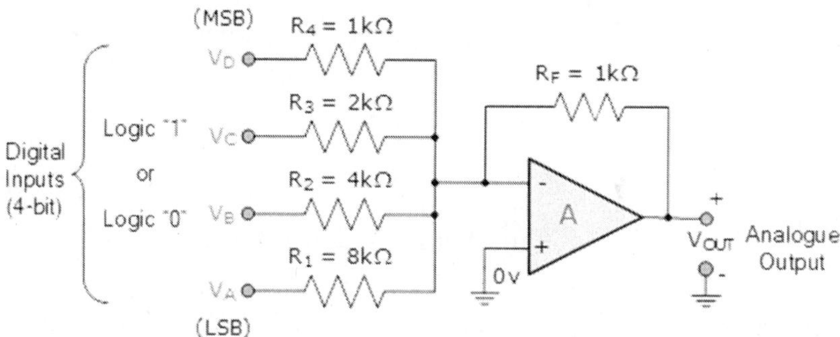

Fig. 6.6: 4-bit Binary Weighted Digital-to-Analogue Converter

So, if we now assume that the four inputs of the summing amplifier are binary inputs with voltage values of either 0 or 5 volts (LOW or HIGH, 0 or 1) and we double the resistive values of each input resistor with regards to the previous one, we can produce an output condition which would be the weighted sum of these four input voltages creating the basic circuit for a 4-bit binary weighted digital-to-analogue converter, or 4-bit weighted D/A converter. Labelling the four summing inputs as A, B, C, D and making $R_F = 1k\Omega$, with the four input resistors ranging from $1k\Omega$ to $8k\Omega$ (or multiples thereof), we can construct a simple 4-bit binary weighted analogue-to-digital converter circuit as shown in Fig. 6.6.

For a 4-bit binary number there are $2^4 = 16$ possible combinations or A, B, C, and D ranging from 0000_2 to 1111_2 which corresponds to decimal 0 to 15 respectively. If we make the weight of each input bit double with respect to the other, we end up with an 8-4-2-1 binary code ratio corresponding to 2^3, 2^2, 2^1 and 2^0.

So, if we set the "D" input resistance at $1k\Omega$, the "C" input resistance at $2k\Omega$ (that is the double of D), the "B" input resistance at $4k\Omega$ (double C), and the "A" input resistance at $8k\Omega$ (double B), with the feedback resistance R_F set again at $1k\Omega$, then the transfer characteristic of the 4-bit binary weighted digital-to-analogue converter would be given by

$$V_{OUT} = -\left[\frac{R_F}{R_4}V_D + \frac{R_F}{R_3}V_C + \frac{R_F}{R_2}V_B + \frac{R_F}{R_1}V_A\right]$$

$$V_{OUT} = -\left[\frac{1K\Omega}{1K\Omega}V_D + \frac{1K\Omega}{2K\Omega}V_C + \frac{1K\Omega}{4K\Omega}V_B + \frac{1K\Omega}{8K\Omega}V_A\right]$$

$$V_{OUT} = -\left[1 * V_D + \frac{1}{2} * V_C + \frac{1}{4} * V_B + \frac{1}{8} * V_A\right]$$

So, we can see that if a TTL voltage of +5 volts (logic 1) is applied to the summing amplifiers input, V_D which represents the most significant bit (MSB), the op-amp's gain will be R_F/R_4 = 1kΩ/1kΩ = 1 (unity). Thus, with a 4-bit binary code of 1000 applied, the output of the digital-to-analogue converter circuit will be -5 volts.

Likewise, if +5 volts (logic 1) is applied to the summing amplifiers input V_C, the op-amp's gain will be R_F/R_3 = 1kΩ/2kΩ = 1/2 (one half). So, the 4-bit binary code of 0100 would produce an analogue output voltage of -2.5 volts.

Again, with a logic "1" applied to the summing amplifiers input V_B, the op-amp's gain will be R_F/R_2 = 1kΩ/4kΩ = 1/4 (one quarter) with the 4-bit binary code of 0010 producing an output voltage of -1.25 volts. Finally, a logic "1" applied to the summing amplifiers input, V_A which represents the least significant bit (LSB), the op-amp's gain will therefore be R_F/R_1 = 1kΩ/8kΩ = 1/8 (one eighth) with the 4-bit binary code of 0001 producing an output voltage of -0.625 volts, (a 12.5% resolution).

Table 6.3: 4-bit Binary Weighted D/A Converter Output

Digital Inputs				**V_{OUT} Expression**	**V_{OUT}**
D	C	B	A	$1*V_D + {}^1/_2*V_C + {}^1/_4*V_B + {}^1/_8*V_A$	**in Volts**
0	**0**	**0**	**0**	**0*5 + 0*5 + 0*5 + 0*5**	**0**
0	**0**	**0**	**1**	**0*5 + 0*5 + 0*5 + ¹/₈*5**	**–0.625**
0	**0**	**1**	**0**	**0*5 + 0*5 + ¹/₄*5 + 0*5**	**–1.25**
0	**0**	**1**	**1**	**0*5 + 0*5 + ¹/₄*5 + ¹/₈*5**	**–1.875**
0	**1**	**0**	**0**	**0*5 + ¹/₂*5 + 0*5 + 0*5**	**–2.50**
0	**1**	**0**	**1**	**0*5 + ¹/₂*5 + 0*5 + ¹/₈*5**	**–3.125**
0	**1**	**1**	**0**	**0*5 + ¹/₂*5 + ¹/₄*5 + 0*5**	**–3.75**
0	**1**	**1**	**1**	**0*5 + ¹/₂*5 + ¹/₄*5 + ¹/₈*5**	**–4.375**
1	**0**	**0**	**0**	**1*5 + 0*5 + 0*5 + 0*5**	**–5.00**
1	**0**	**0**	**1**	**1*5 + 0*5 + 0*5 + ¹/₈*5**	**–5.625**
1	**0**	**1**	**0**	**1*5 + 0*5 + ¹/₄*5 + 0*5**	**–6.25**
1	**0**	**1**	**1**	**1*5 + 0*5 + ¹/₄*5 + ¹/₈*5**	**–6.875**
1	**1**	**0**	**0**	**1*5 + ¹/₂*5 + 0*5 + 0*5**	**–7.50**
1	**1**	**0**	**1**	**1*5 + ¹/₂*5 + 0*5 + ¹/₈*5**	**–8.125**
1	**1**	**1**	**0**	**1*5 + ¹/₂*5 + ¹/₄*5 + 0*5**	**–8.75**
1	**1**	**1**	**1**	**1*5 + ¹/₂*5 + ¹/₄*5 + ¹/₈*5**	**–9.375**

The resolution of this simple 8-4-2-1 binary weighted digital-to-analogue converter will produce an output voltage change of 0.625 volts per 1-bit change in the binary number, and we can express this output voltage change in the following table (Refer Table 6.3).

Where the output voltages are all negative due to the inverting input of the summing amplifier. By increasing the number of binary digits and the resistive summing network so that each resistor has a different weighting, the resolution of the analogue output voltage for a binary weighted digital-to-analogue converter can be increased.

For example, an 8-bit DAC with TTL +5 inputs would produce a resolution of 0.039 (1/128*V) volts, while a 12-bit DAC would be 0.00244 (1/2048*V) volts per step (1 LSB) change of the input binary (or non-binary) code.

Clearly then the disadvantage here is that a binary weighted resistor DAC requires a large range of high precision resistors (one per bit) for an "n"-bit DAC making it impractical (and expensive) for converters with more than a just a few bits of resolution.

But we can expand on this idea of a binary weighted digital-to-analogue circuit configuration which uses different value resistors one step further by converting it into a R-2R resistor ladder DAC which requires only two precision resistance values, namely R and 2R.

Example 6.7: Assume V_{REF} = 10 V and R_F =10 kΩ. Determine the resolution and full-scale output for this DAC. Assume that R_L is much smaller than R_F.
Solution:

$$I_0 = V_{REF}/R_F = 1 \text{ mA}.$$

This is the weight of the MSB. The other three currents will be 0.5, 0.25, and 0.125 mA. The LSB is 0.125 mA, which is also the resolution. The full-scale output will occur when the binary inputs are all HIGH so that each current switch is closed and

$$I_{OUT} = 1 + 0.5 + 0.25 + 0.125 = 1.875mA$$

Note that the output current is proportional to V_{REF}. If V_{REF} is increased or decreased, the resolution and full-scale output will change proportionally.

6.3.4 R-2R Ladder D/A converter

Compared to the R-2R DAC , the binary weighted digital-to-analog converter has some practical limitations like expensive and impractical. The biggest problem is the large difference in resistor values between the LSB and MSB, especially in high-resolution DACs. most widely used DAC circuits that uses resistance's fairly close in value is the R/2R ladder network. Here the resistance values span a range of only 2 to 1. In this conversion methods uses a R-2R resistor ladder network and operational amplifier.

With a much simpler approach is using a R-2R resistive ladder network to construct a **R-2R Digital-to-Analogue Converter** requires only two precision resistances. The R-2R resistive ladder network uses just two resistive values. One resistor has the base value "R", and the second resistor has twice the value of the first resistor, "2R", no matter how many bits are used to make up the ladder network.

So, for example, we could just use a standard 1kΩ resistor for the base resistor "R", and therefore a 2kΩ resistor for "2R" (or multiples there of as the base value of R is not too critical). Thus, the resistive value of 2R will always be twice the value of the base resistor, R. That is 2R = 2*R. This means that it is much easier for us to maintain the required accuracy of the resistors along the ladder network compared to the previous weighted resistor DAC.

6.3.5 R-2R Resistive Ladder Network

A R-2R resistive ladder network provides a simple means of converting digital voltage signals into an equivalent analogue output. Input voltages are applied to the ladder network at various points along its length and the more input points the better the resolution of the R-2R ladder.

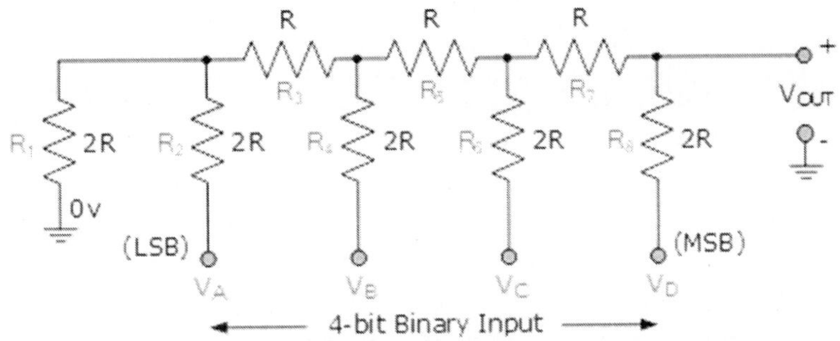

Fig. 6.7: 4-bit R-2R Resistive Ladder Network

The output signal as a result of all these input voltage points is taken from the end of the ladder which is used to drive the inverting input of an operational amplifier. A R-2R resistive ladder network is long strings of parallel and series connected resistors acting as interconnected voltage dividers along its length. Consider the basic 4-bit R-2R ladder network (4-bits because it has four input points) shown in Fig. 6.7.

This 4-bit resistive ladder circuit may look complicated, but it's all about connecting resistors together in parallel and series combinations and working back to the input source using simple circuit laws to find the proportional value of the output. Let's, assume all the binary inputs are grounded at 0 volts, that is: $V_A = V_B = V_C = V_D = 0V$ (LOW). The binary code corresponding to these four inputs will therefore be: **0000.**

Starting from the left-hand side and using the simplified equation for two parallel resistors and series resistors, we can find the equivalent resistance of the ladder network as shown in Fig. 6.8:

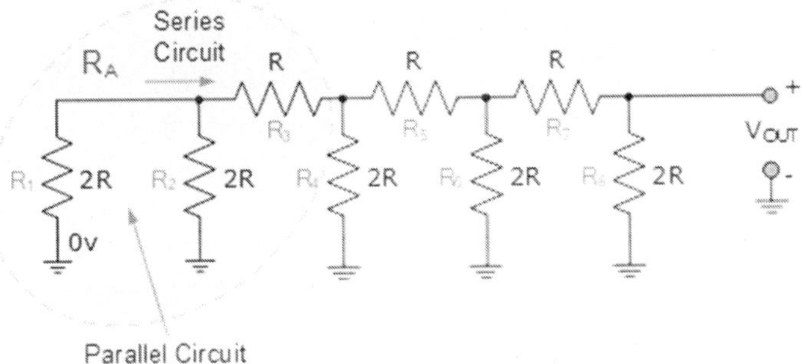

Fig. 6.8: Equivalent resistance of the ladder network

Resistors R_1 and R_2 are in "parallel" with each other but in "series" with resistor R_3. Then we can find the equivalent resistance of these three resistors and call it R_A for simplicity (or any other form of identification you want).

$$R_A = R_3 + \frac{R_1 \times R_2}{R_1 + R_2}$$

$$if, R_1 = R_2 = 2R \ and \ R_3 = R$$

$$R_A = R + \frac{2R \times 2R}{2R + 2R} = R + R = 2R$$

Then R_A is equivalent to "2R". Now we can see (Refer Fig. 6.9) that the equivalent resistance "R_A" is in parallel with R_4 with the parallel combination in series with R_5. Again, we can find the equivalent resistance of this combination and call it R_B.

$$R_A = R_5 + \frac{R_A \times R_4}{R_B + R_4} = R + \frac{2R \times 2R}{2R + 2R} = R + R = 2R$$

So, R_B combination is equivalent to "2R". Hopefully we can see that this equivalent resistance R_B is in parallel with R_6 with the parallel combination in series with R_7 as shown in Fig. 6.10.

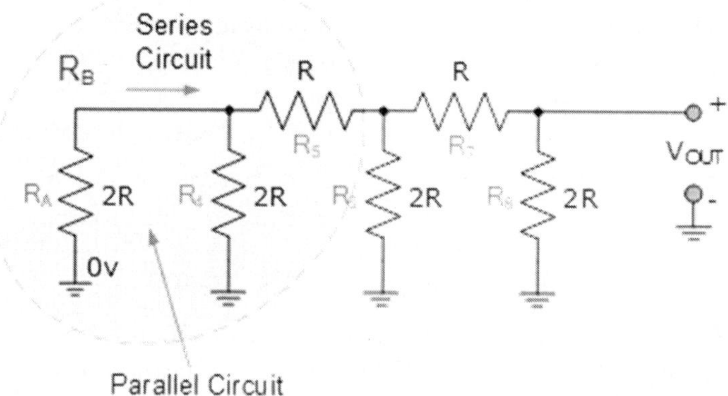

Fig. 6.9: Calculation for RB

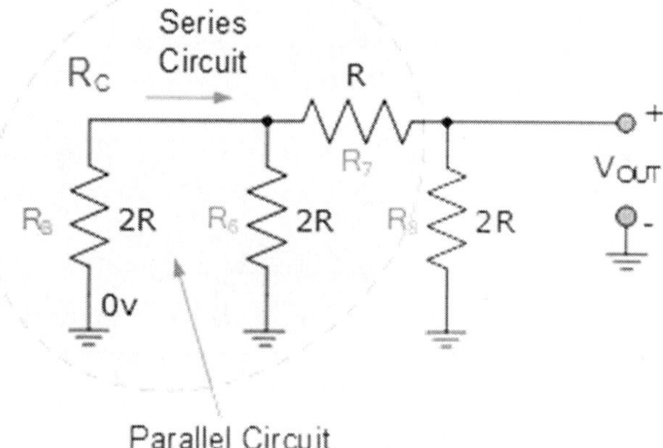

Fig. 6.10: Calculation for Rc

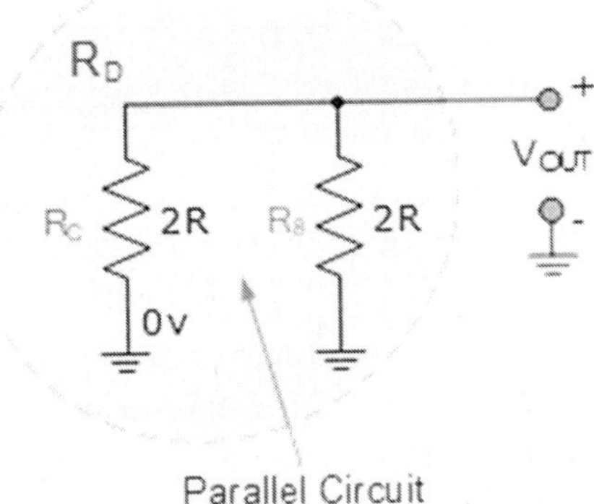

Fig. 6.11: Calculation for RD

As before we find the equivalent resistance and call it R_C.

$$R_A = R_7 + \frac{R_B \times R_6}{R_B + R_6} = R + \frac{2R \times 2R}{2R + 2R} = R + R = 2R$$

Again, resistor combination R_C is equivalent to "2R" which is in parallel with R_8 as shown in Fig. 6.11.

As shown above, when two equal resistor values are parallel together, the resulting value is one-half, so 2R in parallel with 2R equals an equivalent resistance of R. So, the whole 4-bit R-2R resistive ladder network comprising of individual resistors connected together in parallel and series combinations has an equivalent resistance (R_{EQ}) of "R" when a binary code of "0000" is applied to its four inputs.

Therefore, with a binary code of "0000" applied as inputs, our basic 4-bit R-2R digital-to-analogue converter circuit would look something like this as shown in Fig. 6.12:

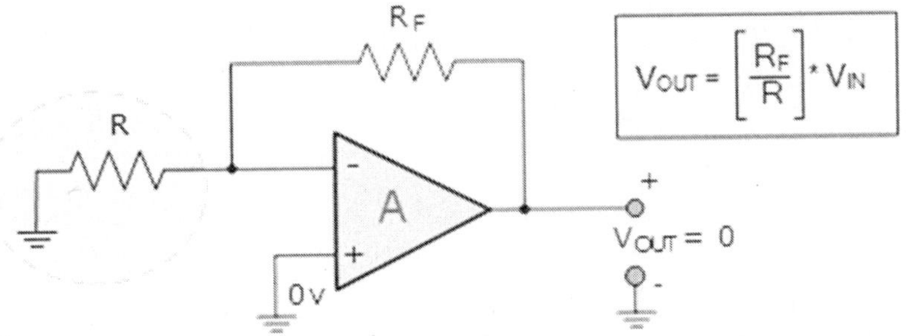

Fig. 6.12: R-2R DAC Circuit with Four Zero (LOW) Inputs

As shown in Fig. 6.12, the output voltage for an inverting operational amplifier is given as:

$$V_{OUT} = \left[\frac{R_F}{R}\right] * V_{IN}$$

If we make R_F equal to R, that is $R_F = R = 1$, and as R is terminated to ground (0V), then there is no V_{IN} voltage value, ($V_{IN} = 0$) so the output voltage would be: (1/1)*0 = 0 volts. So for a 4-bit R-2R DAC with four grounded inputs (LOW), the output voltage will be "zero" volts, thus a 4-bit digital input of **0000** produces an analogue output of 0 volts.

So, what happens now if we connect input bit V_A HIGH to +5 volts. What would be the equivalent resistive value of the R-2R ladder network and the output voltage from the op-amp.

Refer from Fig. 6.13, input V_A is HIGH and logic level "1" and all the other inputs grounded at logic level "0". As the R/2R ladder network is a linear circuit we can find Thevenin's equivalent resistance using the same parallel and series resistance calculations as above to calculate the expected output voltage. The output voltage, V_{OUT} is therefore calculated at 312.5 milli-volts (312.5 mV).

As we have a 4-bit R-2R resistive ladder network, this 312.5 mV voltage change is one-sixteenth the value of the +5V input (5/0.3125 = 16) voltage so is classed as the Least Significant Bit, (LSB). Being the least significant bit, input V_A will therefore determine the "resolution" of our simple 4-bit digital-to-analogue converter, as the smallest voltage change in the analogue output corresponds to a single step change of the digital inputs. Thus, for our 4-bit DAC this will be 312.5mV (1/16th) for a +5V input

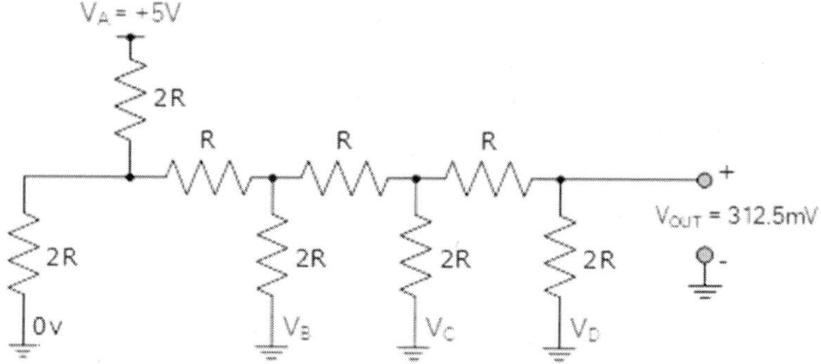

Fig. 6.13: R-2R DAC with Input VA

Now let's see what happens to the output voltage if we connect input bit V_B HIGH to +5 volts.

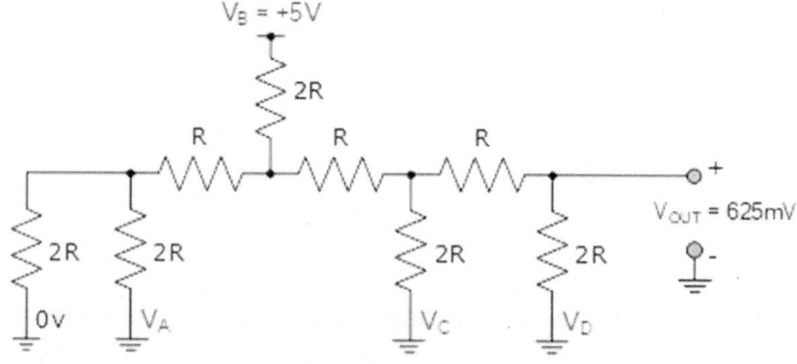

Fig. 6.14: R-2R DAC with Input VB

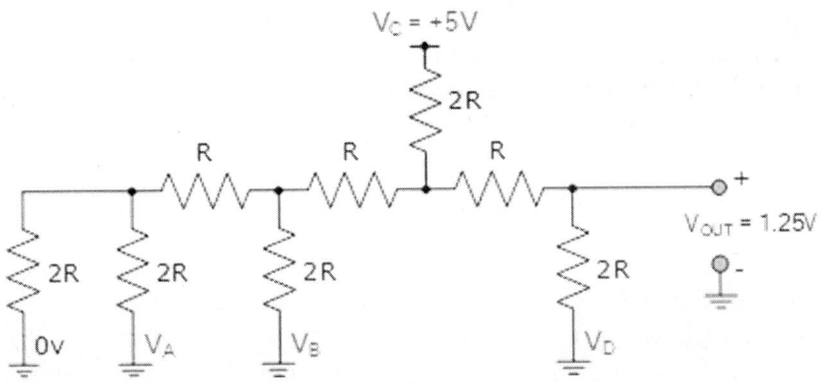

Fig. 6.15: R-2R DAC with Input VC

Refer from Fig. 6.14, input V_B is HIGH i.e., at logic level "1" and all the other inputs grounded at logic level "0", the output voltage, V_{OUT} is calculated at 625mV, and which is one-eighth (1/8th) the value of the +5V input (5/0.625 = 8) voltage. We can also see that it is double the output voltage when only input bit V_A was applied, and we would expect this as its the 2_{nd} bit (input) so has double the weighting of the 1_{st} bit. Now let's see what happens to the output voltage if we connect input bit V_C HIGH to +5 volts.

Refer from Fig. 6.15, if input V_C is HIGH i.e., at logic level "1" and the other input bits at logic level "0", the output voltage, V_{OUT} is calculated at 1.25 volts, and which is one-quarter (1/4) the value of the +5V input (5/1.25 = 4) voltage. Again we can see that this voltage is double the output of input bit V_B but also 4 times the value of bit V_A. This is because input V_C is the 3_{rd} bit so has double the weighting of the 2_{nd} bit and four times the weighting of the 1_{st} bit.

Finally let's see what happens to the output voltage if we connect input V_D HIGH to +5 volts.

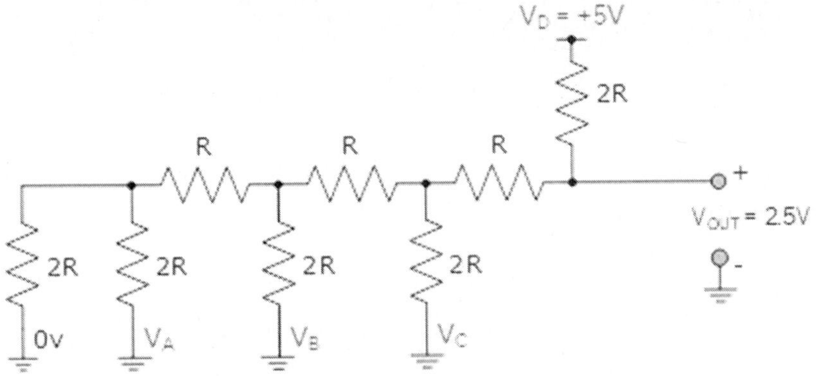

Fig. 6.16: R-2R DAC with Input VD

Refer from Fig. 6.16, if only input V_D is HIGH i.e., at logic level "1" and the other inputs at logic level "0", the output voltage, V_{OUT} is calculated at 2.5 volts. This is on-half (1/2) the value of the +5V input (5/2.5 = 2) voltage. Again, we can see that this voltage is double the output of input bit V_C, 4 times the value of bit V_B and 8 times the value of input bit V_A as it is the 4_{th} bit and therefore classed as the Most Significant Bit, (MSB).

Then we can see that if input V_A represents the LSB and therefore controls the DAC's resolution, and input V_B is double V_A, input V_C is four times greater than V_A, and input V_D is eight times greater than V_A, we can obtain a relationship for the analogue output voltage of our 4-bit digital-to-analogue converter with the following Digital-to-Analogue Output Voltage equation

$$V_{OUT} = \frac{V_A + 2V_B + 4V_C + 8V_D}{16}$$

Where the denominator value of 16 corresponds to the 16 (2^4) possible combinations of inputs to the 4-bit R-2R ladder network of the DAC. We can expand this equation further to obtain a generalised R-2R DAC equation for any number of digital inputs for a R-2R D/A converter as the weighting of each input bit will always be referenced to the least significant bit (LSB), giving us a generalised R-2R DAC equation given as,

$$V_{OUT} = \frac{V_A + 2V_B + 4V_C + 8V_D + 16V_E + 32V_F + \dots\dots\dots etc.}{2^n}$$

Where: "n" represents the number of digital inputs within the R-2R resistive ladder network of the DAC producing a resolution of: $V_{LSB} = V_{IN}/2^n$.

Table 6.4: 4-bit R-2R D/A Converter Voltage Output

Digital Inputs				**V_{OUT} Expression**	**V_{OUT}**
D	C	B	A	$(8{*}V_D + 4{*}V_C + 2{*}V_B + 1{*}V_A)/2^4$	in Volts
0	0	0	0	(0*5 + 0*5 + 0*5 + 0*5)/16	0
0	0	0	1	(0*5 + 0*5 + 0*5 + 1*5)/16	0.3125
0	0	1	0	(0*5 + 0*5 + 2*5 + 0*5)/16	0.6250
0	0	1	1	(0*5 + 0*5 + 2*5 + 1*5)/16	0.9375
0	1	0	0	(0*5 + 4*5 + 0*5 + 0*5)/16	1.2500
0	1	0	1	(0*5 + 4*5 + 0*5 + 1*5)/16	1.5625
0	1	1	0	(0*5 + 4*5 + 2*5 + 0*5)/16	1.8750
0	1	1	1	(0*5 + 4*5 + 2*5 + 1*5)/16	2.1875
1	0	0	0	(8*5 + 0*5 + 0*5 + 0*5)/16	2.5000
1	0	0	1	(8*5 + 0*5 + 0*5 + 1*5)/16	2.8125
1	0	1	0	(8*5 + 0*5 + 2*5 + 0*5)/16	3.1250
1	0	1	1	(8*5 + 0*5 + 2*5 + 1*5)/16	3.4375
1	1	0	0	(8*5 + 4*5 + 0*5 + 0*5)/16	3.7500
1	1	0	1	(8*5 + 4*5 + 0*5 + 1*5)/16	4.0625
1	1	1	0	(8*5 + 4*5 + 2*5 + 0*5)/16	4.3750
1	1	1	1	(8*5 + 4*5 + 2*5 + 1*5)/16	4.6875

Clearly then input bit V_A when HIGH will cause the smallest change in the output voltage, while input bit V_D when HIGH will cause the greatest change in the output voltage. The expected output voltage is therefore calculated by summing the effect of all the individual input bits which are connected HIGH.

Ideally, the ladder network should produce a linear relationship between the input voltages and the analogue output as each input will have a step increase equal to the LSB, we can create a table of expected output voltage values for all 16 combinations of the 4 inputs with +5V representing a logic "1" condition as shown in Table 6.4.

You may have noticed that the full-scale analogue output voltage for a binary code of **1111** never reaches the same value as the digital input voltage (+5V) as it is always less by the equivalent of one LSB bit, (312.5mV in this example).

However, the higher the number of digital input bits (resolution) the nearer the analogue output voltage reaches full-scale when all the input bits are HIGH. Likewise, when all the input bits are LOW, the resulting lower resolution of LSB makes V_{OUT} closer to zero volts.

6.3.6 R-2R Digital-to-Analogue Converter

Now that we understand what a *R-2R resistive ladder network* is and how it works, we can use it to produce a R-2R Digital-to-Analogue Converter. Again, using our 4-bit R-2R resistive ladder network from above and adding it to an inverting operational amplifier circuit, we can create a simple R-2R digital-to-analogue converter as shown in Fig. 6.17

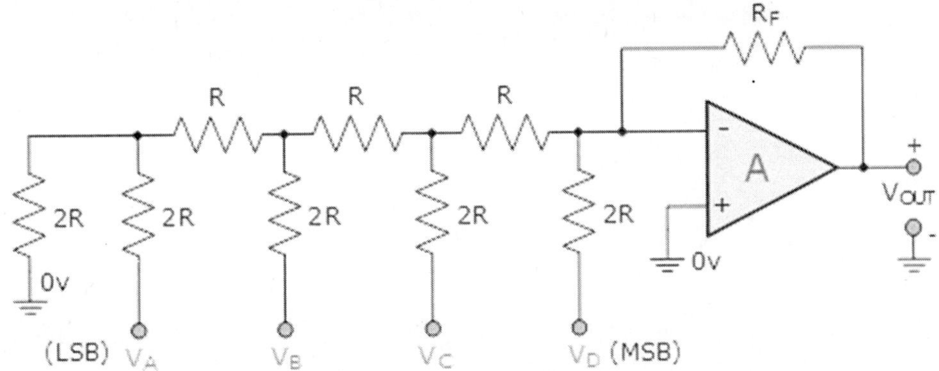

Fig. 6.17: R/2R ladder DAC.

The digital logic circuit used to drive the D/A converter can be generated by combinational or sequential logic circuits, data registers, counters or simply switches. The interfacing of a R-2R D/A converter of "n"-bits will depend upon its application. All-in-one boards such as the Arduino or Raspberry Pi have *digital-to-analogue converters* built-in so make interfacing and programming much easier. There are many popular DAC's available such as the 8-bit DAC0808.

Example 6.8: A 4-bit R-2R digital-to-analogue converter is constructed to control the speed of a small DC motor using the output from a digital logic circuit. If the logic circuit uses 10 volt CMOS devices, calculate:

1. analogue output voltage from the DAC when the input code is hexadecimal number "B".
2. Also determine the resolution of the DAC.

Solution:

1. The hexadecimal letter "B" is equal to the number eleven in decimal. The decimal number eleven is equal to the binary code "1011" in binary i.e.,

$$B_{16} = (11)_{10} = (1011)_2.$$

Thus, for our 4-bit binary number of 1011_2, input bit D = 1, bit C = 0, bit B = 1 and bit A = 1. If we assume that feedback resistor R_F is equal to "R", then our R-2R D/A converter circuit will look like as shown in Fig. 6.18:

The digital logic circuit uses 10-volt CMOS devices, so the input voltage to the R-2R network will be 10 volts. Also, being a 4-bit ladder DAC, there will be 2^4 possible input combinations, so using our equation from above, the output voltage for a binary code of 1011_2 is calculated as:

$$V_{OUT} = \frac{1V_A + 2V_B + 4V_C + 8V_D}{2^n} = \frac{1 \times 10 + 2 \times 10 + 4 \times 0 + 8 \times 10}{16} = \frac{110}{16}$$
$$= 6.875\ Volts$$

Therefore, the analogues output voltage used to control the DC motor when the input code is 1011_2 is calculated as, -6.875 volts.

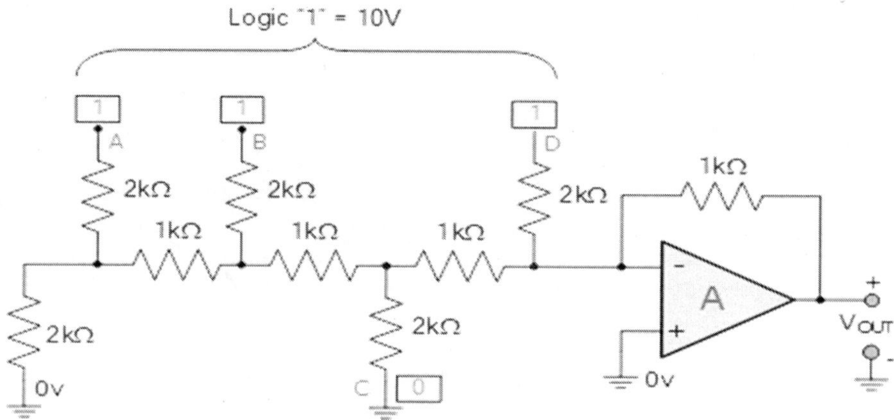

Fig. 6.18: Example 6.8:

Note that the output voltage is negative due to the inverting input of the operational amplifier.

2. The resolution of the converter will be equal to the value of the least significant bit (LSB) which is given as:

$$Resolution = (V)_{LSB} = \frac{(V)_{IN}}{2^n} = \frac{10}{16} = 0.625\ Volts$$

Then the smallest step change of the analogue output voltage, V_{OUT} for a 1-bit LSB change of the digital input of this 4-bit R-2R digital-to-analogue converter example is: 0.625 volts. That is the output voltage changes in steps or increments of 0.625 volts and not as a straight linear value.

6.3.7 4-bit Binary Counting R-2R DAC

Hopefully by now we understand that we can make a R-2R ladder DAC using just two resistor values, one the base value "R" and the other twice or double the value being "2R". In our simple example below, we have made a 4-bit R-2R DAC with four input data lines, A, B, C, and D giving us 16 (2^4) different input combinations from "0000" to "1111" (Refer Fig. 6.19).

The binary code for these four digital input lines can be generated in many different ways, using micro-controllers, digital circuits, mechanical or solid-state switches. But one interesting option is to use a 4-bit binary counter such as the 74LS93. The **74LS93** is a 4-bit J-K ripple counter which can be configured to count-up from 0000_2 to 1111_2 (MOD-16) and reset back to zero (0000) again by the application of a single external clock signal. The 74LS93 is an asynchronous counter commonly called a "ripple" counter because of the way that the internal J-K bistables respond to the clock or timing input producing a 4-bit binary output.

The frequency (or period) of this external clock or timing pulse is divided by a factor of 2, 4, 8, and 16 by the counters output lines as the clock pulse appears to ripple through the four J-K flip-flops producing the required 4-bit output count sequence from 0000_2 to 1111_2.

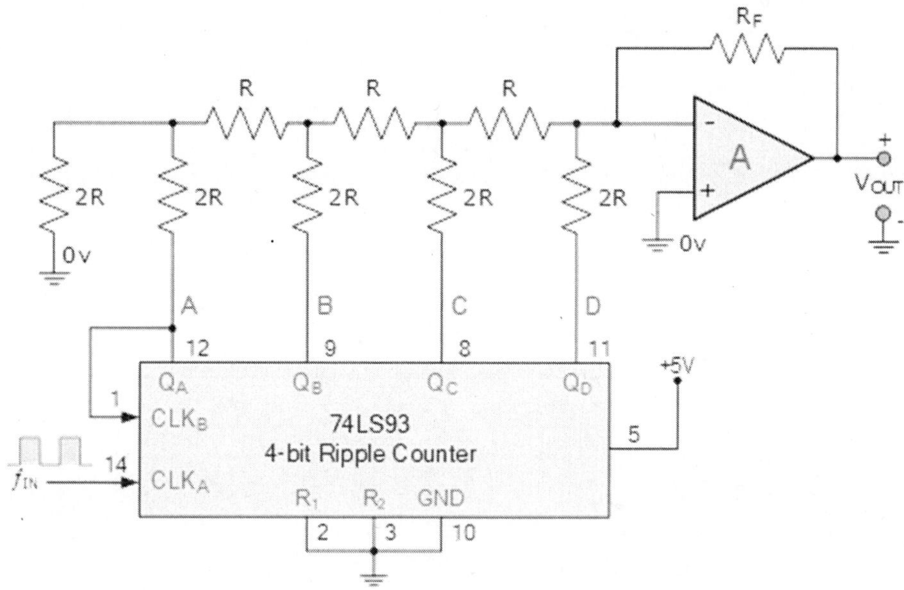

Fig. 6.19: 4-bit Binary Counting R-2R DAC

Note that to count upwards from 0000 to 1111, the external CLK_B input must be connected to the Q_A (pin-12) output and the input timing pulses are applied to input CLK_A (pin-14). This simple 4-bit asynchronous up counter built around the 74LS93 binary ripple counter as the same counting sequence given in the above table. On the application of a clock pulse the outputs: Q_A, Q_B, Q_C, and Q_D change by one step.

The input of the operational amplifier detects this step change and outputs a negative voltage (inverting op-amp) relative to the binary code at the R-2R ladder inputs. The output voltage value for each step will correspond to that given in the table above.

The ripple counter will count up in sequence with the four outputs producing an output sequence of binary values up to the 15th clock pulse where the outputs are set to 1111_2 (decimal 15) producing the maximum negative output voltage of the digital-to-analogue converter.

On the 16th pulse the counters output sequence is reset and the count returns back to 0000, which resets the op-amps output back to zero volts. The application of the next clock pulse begins a new counting cycle from zero to $V_{OUT(max)}$. We can show the output sequence for this simple 4-bit binary asynchronous counting R-2R D/A converter in the following timing diagram (Refer Fig. 6.20).

Clearly then, the output voltage of the operational amplifier varies from zero volts to its maximum negative voltage as the ripple counter counts from 0000_2 to 1111_2 respectively. This simple circuit could be used to vary the brightness of a lamp connected to the op-amps output, or continually vary the speed of a DC motor from slow to fast, and back to slow again at a rate determined by the clock period.

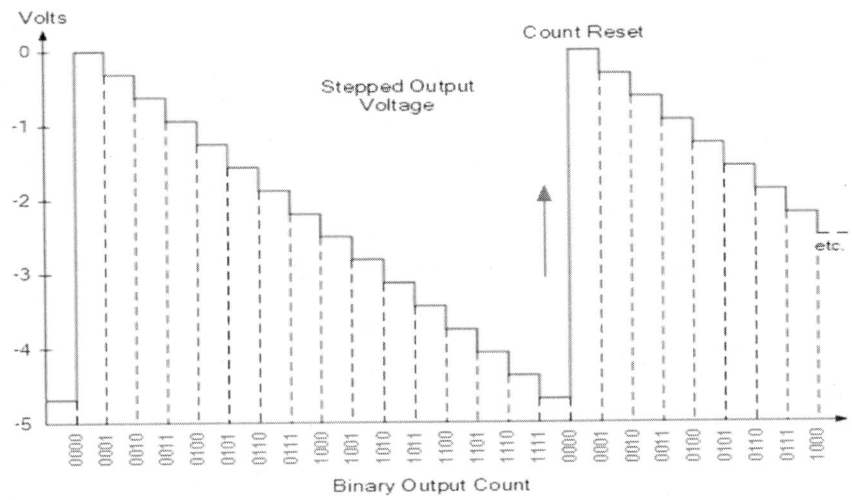

Fig. 6.20: 4-bit R-2R DAC Timing Diagram

Here the ripple counter and R-2R DAC are configured for 4-bit operation but using commonly available binary ripple counters such as the CMOS 4024 7-bit (÷128), the CMOS 4040 12-bit (÷4096) or the larger CMOS 4060 14-bit (÷16,384) counter and adding more input resistors to the R-2R ladder network such as those available from ***Bournes***, the resolution (LSB) of the circuit can be greatly lowered producing a smoother output signal from the *R-2R digital-to-analogue converter*.

6.3.8 DAC Applications

DACs are used whenever the output of a digital circuit has to provide an analog voltage or current to drive an analog device. Some of the most common applications are described in the following paragraphs.

- Control The digital output from a computer can be converted to an analog control signal to adjust the speed of a motor or the temperature of a furnace, to control almost any physical variable.

- Automatic Testing Computers can be programmed to generate the analog signals (through a DAC) needed to test analog circuitry. The test circuit's analog output response will normally be converted to a digital value by an ADC and fed into the computer to be stored, displayed, and sometimes analyzed.
- Signal Reconstruction In many applications, an analog signal is digitized, meaning that successive points on the signal are converted to their digital equivalent and stored in memory. This conversion is performed by an analog-to-digital converter (ADC).
- A DAC can then be used to convert the stored digitized data back to analog-one point at a time-thereby reconstructing the original signal. This combination of digitizing and reconstruction is used in digital storage oscilloscopes, audio compact disk systems, and digital audio and video recording.

6.4 Analog to Digital Convertors

An analog-to-digital converter takes an analog input voltage and after a certain amount of time produces a digital output code which represents the analog input. The A/D conversion process is generally more complex and time-consuming than the D/A process. The techniques that are used provide and insight into what factors determine an ADCs performance.

6.4.1 Introduction and Specifications of A/D converters

Fig. 6.2 represents a general block diagram for ADC. The timing for the operation is provided by the input clock signal. The control unit contains the logic circuitry for generating the proper sequence of operations. The START COMMAND, initiates the conversion process, the op-amp compactor has two analog inputs and a digital output that switches states, depending on which analog input is greater.

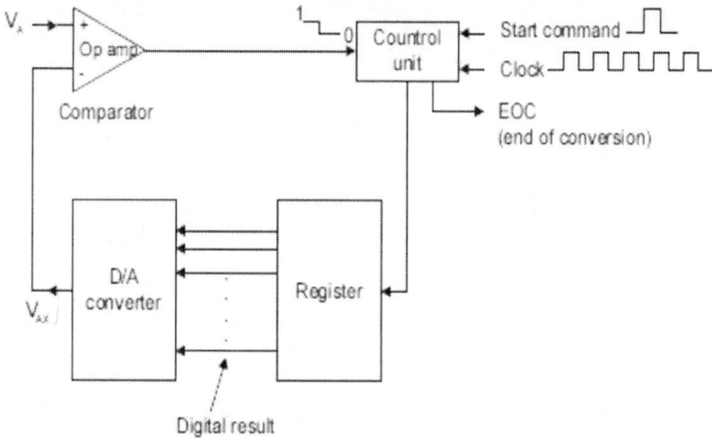

Fig. 6.21: General block diagram of ADC

The basic operation of ADCs of this type consists of the following steps:

- The START COMMAND pulse initiates the operation.

- At a rate determined by the clock, the control unit continually modifies the binary number that is stored in the register.
- The binary number in the register is converted to an analog voltage, VAX, by the DAC.
- The comparator compares VAX with the analog input VA. As long as VAX < VA1 the comparator output stays HIGH. When VAX exceeds VA by at least an amount = VT (threshold voltage), the comparator output goes LOW and stops the process of modifying the register number. At this point, VAX is a close approximation to VA. The digital number in the register, which is the digital equivalent of VAX, is also the approximate digital equivalent of VA1 within the resolution and accuracy of the system.
- The control logic activates the end-of-conversion signal, EOC, when the conversion is complete.

Specifications of A/D converters

Sample Rate: -

The rate at which the signals are converted from the analog domain to a stream of digital data is called the sample rate or *sampling* frequency. For example, barometric pressure changes very slowly over a period of minutes or hours, so you really don't need to sample it more than once per second. On the other hand, if you're trying to measure a RADAR signature, you need to sample hundreds of millions of times per second, or perhaps even into the billions of samples per second. it depends on the application.

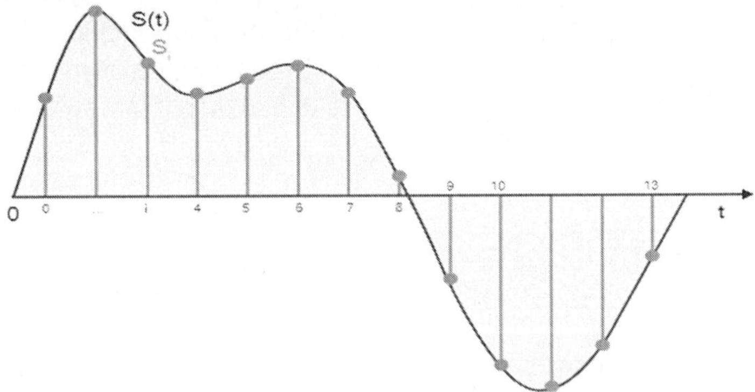

Fig. 6.22: Sampling representation with an interval of 1 sec.

In the world of data acquisition, we measure AC voltages and currents, shock and vibration, temperature, strain, pressure, and the like. These signals and sensors require sample rates in the range of DC to 200,000 samples per second (200 kS/s) on average, while a few applications require sampling up to 1,000,000 samples per second (1 MS/s). The sample rate is usually referred to as the T (time) or X-axis of measurement as shown in Fig. 6.22.

Moreover, the question arises Why Is the Sample Rate Important? The answer is, understanding signals and their highest possible frequencies is an important part of getting accurate measurements.

For example, let's say that we want to measure the output of an accelerometer. If we expect it to experience vibrations with a maximum frequency of 100 Hz, we must set the sample rate to at least double that (the Nyquist frequency), but in practice ten times oversampling is better in order to get a good quality representation of the signal shape. So, in this example, we set the sample rate to 1000 Hz and do the measurement.

In terms of digitizing voltage signals with our ADC, it is important that the sample rate is set appropriately. If we set it too high, we waste processing power and end up with data files that are unnecessarily large and hard to analyse. Moreover, if we set it too low, we could have missing vital dynamic signal components Ending up with false ("alias") signals (if the system lacks anti-aliasing filtering) as shown in Fig. 6.23.

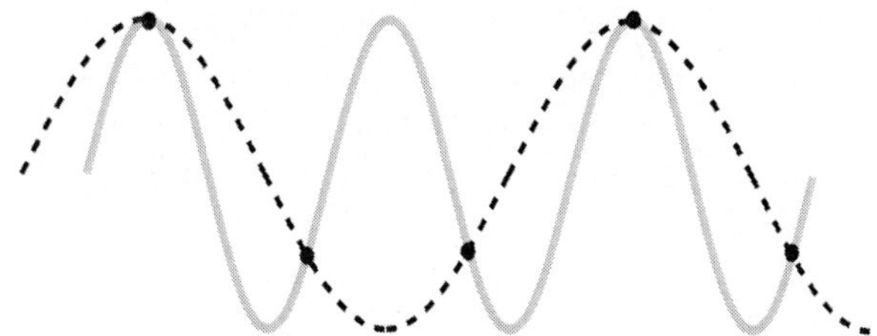

Fig.6.23: Demonstration of a false signal (alias) in black, caused by sampling too infrequently compared to the original signal. [Graphic is in the public domain]

Anti-Aliasing Filtering (AAF):

If we filter in the analog domain before the ADC, we can prevent the aliasing problem from ever occurring. Note that it is still important to set a high enough sample rate to capture the frequency range of interest, but at least with Anti-Aliasing Filters (AAF), we will avoid false signals from destroying the integrity of our measurements. The ideal AAF would have a very flat passband and very sharp cutoff at the Nyquist frequency (essentially half of the sample rate) as shown in Fig. 6.24.

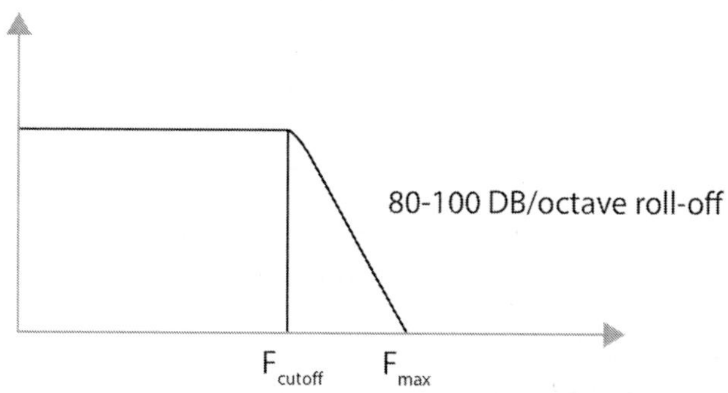

Fig. 6.24: Typical AAF configuration

Bit resolution: -

with how much precision can an ADC convert analog to digital? What is Bit Resolution and Why Does It Matter? While the sample rate as discussed in the previous section involves the time

(T or X) axis of our digital data stream, bit resolution, or a number of bits involves the amplitude (Y) axis.

In the early days of data acquisition (DAQ), 8-bit ADCs were common. As of this writing, in the world of DAQ systems, 24-bit ADCs are standard among most data acquisition systems designed to make dynamic measurements, and 16-bit ADCs are commonly considered the minimum resolution for signals in general. There are some low-end systems utilizing 12-bit ADCs.

Because each bit of resolution effectively doubles the possible resolution, systems with 24-bit ADCs provide 2^24 = 16,777,216. Thus, an incoming one-volt signal can be divided into more than 16 million steps on the Y-axis. 16,777,216 steps for a 24-bit ADCs are dramatically better than the maximum theoretical 65,656 steps of a 16-bit ADCs as shown in Fig. 6.25. Thus, the appearance of waveshapes is accordingly more accurate and has a lot more precision, the more resolution you have. This applies to the time axis, too.

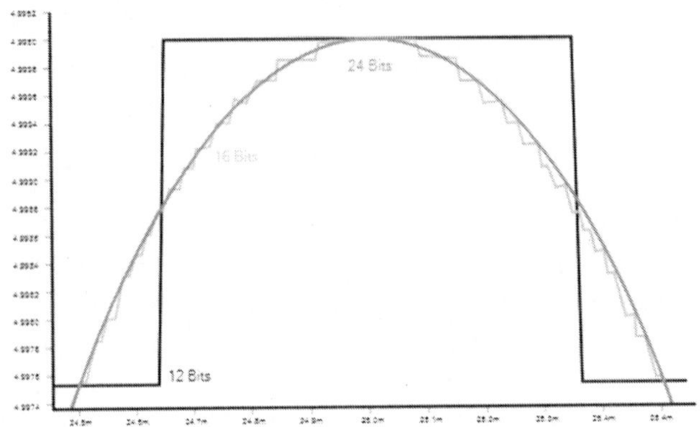

Fig. 6.25: 24-bit resolution (orange) vs. 16-bit resolution (gray)

Like DAC, ADCs are also having many important specifications. Some of them are Resolution, Quantization error, Conversion time, Analog error, Linearity error, DNL error, INL error & Input voltage range.

<u>Resolution:</u>

The resolution refers to the finest minimum change in the signal which is accepted for conversion, and it is decided with respect to number of bits. It is given as $1/2^n$, where 'n' is the number of bits in the digital output word. As it is clear, that the resolution can be improved by increasing the number of bits or the number of bits representing the given analog input voltage. Resolution can also be defined as the ratio of change in the value of input voltage Vi, needed to change the digital output by 1 LSB. It is given as

$$\text{Resolution} = \text{ViFS} / (2^n - 1)$$

Where,

ViFS = The full-scale input voltage.

n = The number of output bits.

Quantization Error (QE): If the binary output bit combination is such that for all the values of input voltage V_i between any two voltage levels, there is an unavoidable uncertainty about the exact value of V_i when the output is a particular binary combination. This uncertainty is termed as quantization error. Its value is ± (1/2) LSB. And it is given as,

$$QE = ViFS / 2(2^n - 1)$$

Where,

'ViFS' is the full-scale input voltage
'n' is the number of output bits.

Note: Maximum the number of bits selected, finer the resolution and smaller the quantization error.

Conversion Time: It is defined as the total time required for an A/D converter to convert an analog signal to digital output. It depends on the conversion technique and propagation delay of the circuit components.

Analog error: An error occurring due to the variations in DC switching point of the comparator, resistors, reference voltage source, ripples and noises introduced by the circuit components is termed as Analog error.

Linearity Error: It is defined as the measure of variation in voltage step size. It indicates the difference between the transitions for a minimum step of input voltage change. This is normally specified as fraction of LSB.

Differential Non-Linearity (DNL) Error: The analog input levels that trigger any two successive output codes should differ by 1 LSB. Any deviation from this 1 LSB value is called as DNL error.

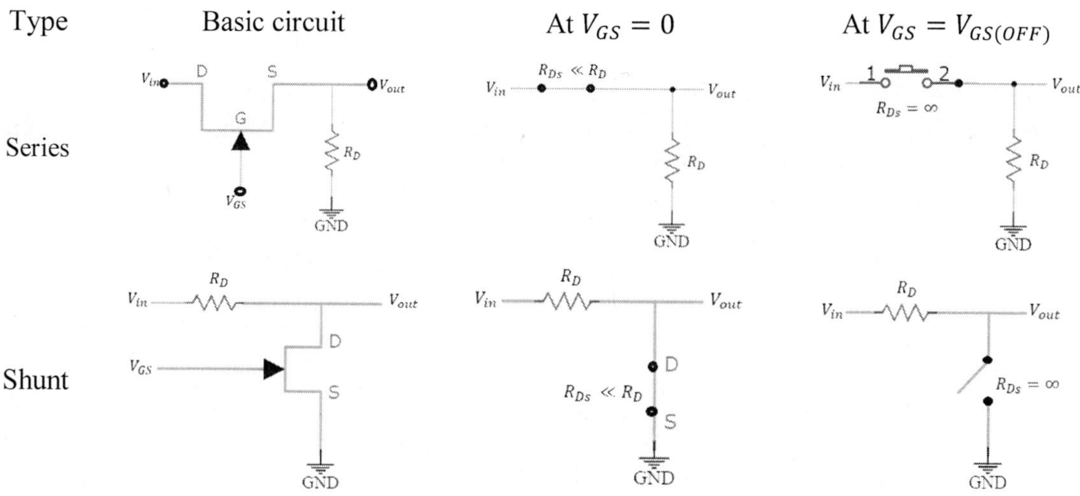

Fig. 6.26: Series and Shunt Analog switch

Integral Non-Linearity (INL) Error: The deviation of characteristics of an ADC due to missing codes causes INL error. The maximum deviation of the code from its ideal value after nulling the offset and gain errors is called as Integral Non-Linearity Error.

Input Voltage Range: It is the range of voltage that an A/D converter can accept as its input without causing any overflow in its digital output.

Analog Switches: There were two types of analog switches. Series and Shunt switch as shown in Fig. 6.26. The Switch operation is shown for both the cases VGS=0 VGS= VGs (off)

6.4.2 Quantization and encoding A/D converter

Sample and hold circuit:

A basic sample-and-hold circuit is shown in Fig. 6.27 (a) in this circuit the voltage across the capacitor follows the input signal voltages V_i during the time switch S is closed, the capacitor holds the instantaneous value of the signal voltage attained just before the switch is opened. Thus, for every T_S the switch is closed for a short duration of time and then opened. The D.C. voltage across the capacitor gives the value of the signal at the instant the switch is opened. This D.C. voltage represents a sample of the signal and is converted to digital signal using A/D converter circuit during the hold period.

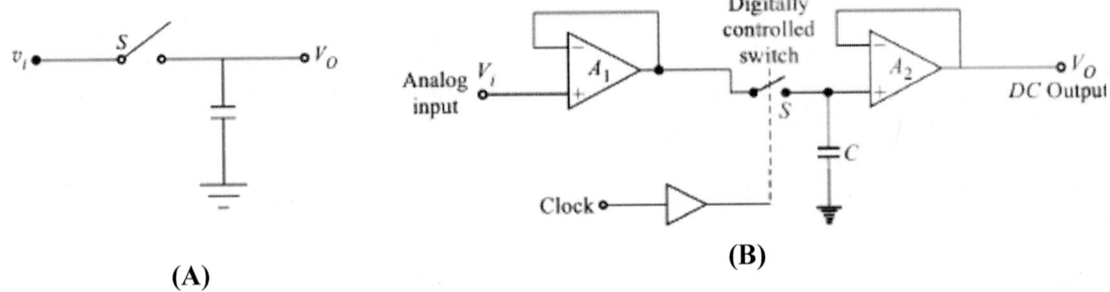

Fig. 6.27: (a) Basic Sample & Hold Circuit (b) A practical Sample & Hold Circuit

Fig. 6.27 (b) shows a practical S/H circuit in simplified form. It consists of two unity gain amplifiers A_1 and A_2 and a digitally controlled switch S. Analog input voltage is applied at the input of buffer amplifier A_1. The high input impedance of A_1 will prevent loading of the analog signal and its low output impedance will help in the fast charging of the capacitor C when the switch S is closed. The switch *is* controlled by a clock generator which makes the switch operate at regular interval as required according to the sampling rate. The voltage across the capacitor is applied at the input of the analog –to-digital converter through buffer amplifier A_2.

The high input impedance of A_2 avoids discharge of capacitor due to the loading effect of A/D converter. The accuracy of the circuit depends upon the holding of the charge in the capacitor, therefore, a capacitor with a very low leakage must be used. A capacitor with polycarbonate, polyethylene, or Teflon dielectric is preferred. Most of the other capacitors do not retain the stored charge for a sufficiently long duration due to polarization phenomenon.

There are five major types of ADCs in use.

1. Flash or parallel comparator A/D converter
2. successive approximation A/D converter
3. counting A/D converter

4. dual-slop converter
5. A/D converter using voltage to frequency and voltage to time conversion

6.4.3 Parallel or Flash Converter

Conceptually, flash conversion is very simple. It utilizes 2^N - 1 comparators to compare the input signal level with each of the 2^N - I possible quantization levels. The outputs of the comparators are processed by an encoding-logic block to provide the N bits of the output digital word. Note that a complete conversion can be obtained within one clock cycle. Although flash conversion is very fast, the price paid is a rather complex circuit implementation. Variations on the basic technique have been successfully employed in the design of IC converters. The fastest A/D conversion scheme is the simultaneous, parallel, or flash conversion process illustrated in Fig. 6.28.

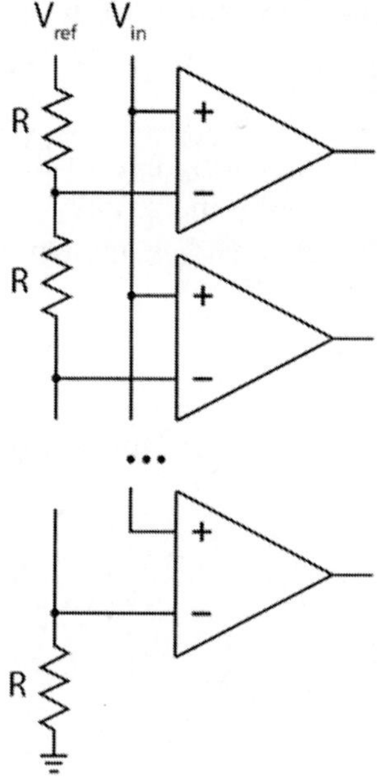

Fig. 6.28: Flash ADC diagram

Flash ADCs are fast and operate virtually without latency, which is why they are the architecture of choice when the highest possible sample rates are needed. They convert analog to a digital signal by comparing it with known reference values. The more known references that are used in the conversion process, the more accuracy can be achieved. For example, if we want a Flash ADC with a 10-bit resolution, we would need to compare the incoming analog signal against 1024 known values. The 8-bit resolution would require 256 known values, and so on.

The more resolution we want, the bigger and more power-hungry the Flash ADC becomes - and the sample rate has to be reduced. For that reason, the 8-bit resolution is generally the "sweet spot" for these ADCs. Flash ADCs can operate into the low GS/s and still provide an 8-bit resolution.

Pros

- Fastest ADC type
- Instant conversion without latency

Cons

- Circuit gets bigger and more power-consuming with each bit
- The resolution effectively limited to 8-bit

Applications

Applications for Flash ADCs include the fastest digital oscilloscopes, microwave measurements, fiber optics, RADAR detection, and wideband radio.

6.4.4 Successive Approximation ADCs (SAR)

The "bread and butter" ADC of the DAQ world is the SAR analog-to-digital converter (Successive Approximation Register) as shown in Fig. 6.29. It offers an excellent balance of speed and resolution and handles a wide variety of signals with excellent fidelity.

It's been around for a long time; therefore, SAR designs are stable and reliable, and the chips are relatively inexpensive. They can be configured for both low-end A/D cards, where a single ADC chip is "shared" by multiple input channels (multiplexed A/D boards), or in configurations where each input channel has its own ADC for true simultaneous sampling.

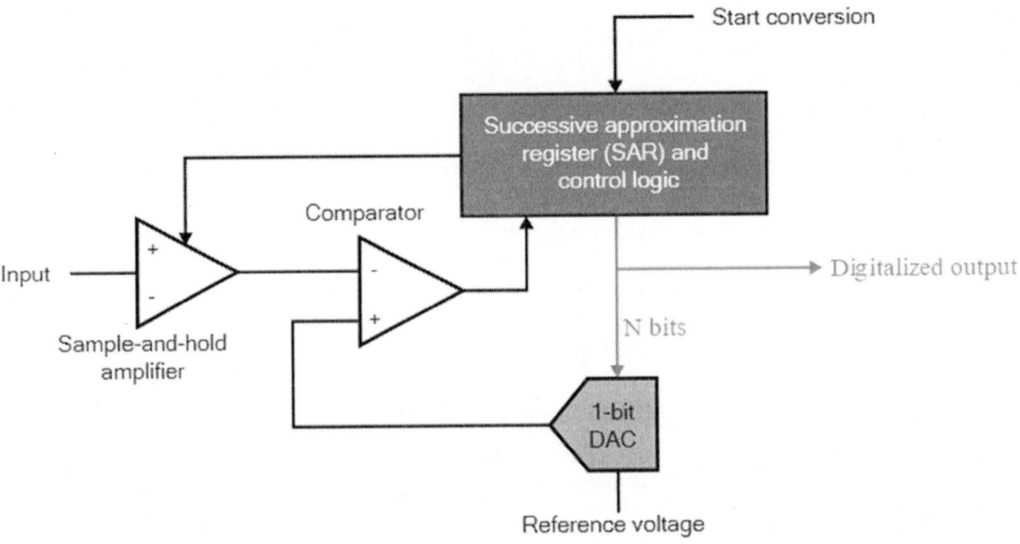

Fig. 6.29: SAR ADC converter block diagram

The analog input of most ADCs is 5V, which is why nearly all signal conditioning front-ends provide a conditioned output that is the same. The typical SAR ADC uses a sample-and-hold circuit that takes in the conditioned analog voltage from the signal conditioning front-end.

An on-board DAC creates an analog reference voltage equal to the digital code output of the sample and holds a circuit. Both of these are fed into a comparator which sends the result of the comparison to the SAR. This process continues for "n" successive times, with "n" being the bit resolution of the ADC itself, until the closest value to the actual signal is found.

SAR ADCs do not have any inherent anti-aliasing filtering (AAF), so unless this is added before the ADC by the DAQ system, if the engineer selects too low of a sample rate, false signals (aka "aliases") will be digitized by the SAR ADC. Aliasing is particularly problematic because it is impossible to correct it after digitization. There is no way to fix it with software. It must be prevented either by always sampling faster than the Nyquist frequency of all input signals or by filtering the signals before and within the ADC

Pros

- Simple circuit with only one comparator needed
- Higher sample rates possible compared to delta-sigma ADCs
- Handles natural and unnatural waveshapes well

Cons

- Anti-aliasing filtering must be added externally
- Bit resolution and dynamic range limited compared to delta-sigma ADCs

Applications

Applications for SAR ADCs include DAQ systems, from low-end multiplexed ADC systems to higher speed single ADC per channel systems, industrial control and measurement, CMOS imaging.

6.4.5 Counting A/D converter

The counter-type Analog-to-Digital Converter (ADC) is also known as the digital ramp ADC. It is because the output of the counter is fed to a Digital-to-Analog Converter (DAC), and while the counter increments its count, the output of the DAC increases in ramp fashion or staircase fashion.

Digital-Ramp ADC:

One of the simplest versions of the general ADC is shown in Fig. 6.30below.

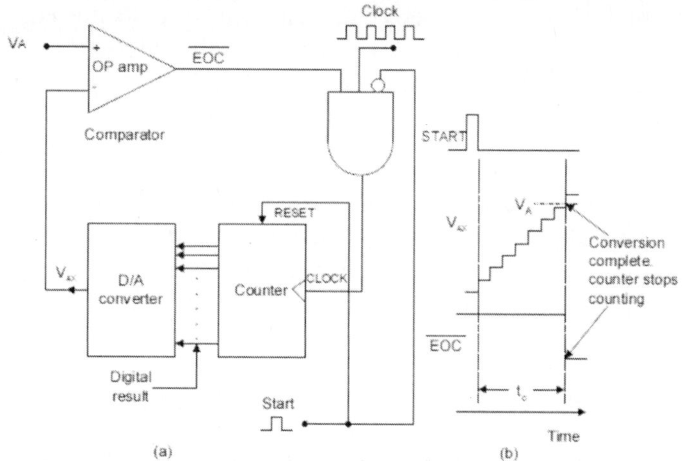

Fig. 6.30: Digital ramp ADC

It uses a binary counter as the register and allows the clock to increment the counter one step at a time until $V_{AX} \geq V_A$. It is called a digital-ramp ADC because the wave form at V_{AX} is a step-by-step ramp (actually a staircase) like the one shown in Fig. 6.30. It is also referred to as a counter-type ADC or a digital-ramp ADC. It contains a counter, a DAC, an analog comparator, and a control AND gate. The comparator output serves as the active-LOW end-of conversion signal $\overline{EOC}$. If we assume that V_A, the analog voltage to be converted, is positive, the operation proceeds as follows:

- A START pulse is applied to reset the counter to zero. The HIGH at START also inhibits clock pulse form passing through the AND gate into the counter.
- With all 0's at its input, the DAC's output will be VAX = 0V.
- Since VA > VAX, the comparator output, EOC, will be HIGH.
- When START returns LOW, the AND gate is enabled and clock pulses get through to the counter.
- As the counter advances, the DAC output, VAX, increases one step at a time as shown in Fig. 6.30. This continues unit VAX reaches a step that exceeds VA by an amount equal to or greater than VT (typically 10 to 100 μV). At this point, EOC will go LOW and inhibit the flow of pulses into the counter and the counter will stop counting.
- The conversion process is now complete as signaled by the HIGH-to-LOW transition at EOC, and the contents of the counter are the digital representation of VA.
- The counter will hold the digital value until the next START pulse initiates a new conversion.

Conversion Time (TC):

The conversion time is the time interval between the end of the START pulse and the activation of the ($\overline{EOC}$) output. The counter starts counting from zero and counts up until VAX exceeds VA, at which point ($\overline{EOC}$)goes LOW to end the conversion process. It should be clear that the value of conversion time, to, depends on VA.

A larger value will require more steps before the staircase voltage exceeds VA. The maximum conversion time will occur when VA is just below full scale so that VAX has to go to the last step to activate ($\overline{EOC}$) For an N-bit converter this will be

tc (max) = 2N - 1 clock cycles

Sometimes, average conversion time is specified; it is half of the maximum conversion time. The major disadvantage of the digital-ramp method is that conversion time essentially doubles for each bit that is added to the counter, so that resolution can be improved only at the cost of a longer tc. Applications, however, the relative simplicity of the digital-ramp converter is an advantage over the more complex, higher-speed ADCs.

A/D Resolution and Accuracy:

Resolution of the ADC is equal to the resolution of the DAC that it contains. The DAC output voltage V_{AX} is a staircase waveform that goes up in discrete steps until it exceeds V_A. Thus, V_{AX} approximates the value of V_A, and the best we can expect is that V_{AX} is within 10 mV of V_A if the resolution (step size) is 10 mV. We can think of the resolution as being a built-in error that is often referred to as quantization error. This quantization error, can be reduced by increasing the number of bits in the counter and DAC.

Example 6.9: Assume the following values for the ADC clock frequency = 1 MHz; VT = 0.1 mV; DAC has F.S. output = 10.23 V and a 10-bit input. Determine the following values.

a) The digital equivalent obtained for VA = 3.728 V.
b) The conversion time.
c) The resolution of this converter.

Solution:

a) The DAC has a 10-bit input and a 10.23V F.S. output. Thus, the number of total possible steps is 2^{10} - 1 = 1023, and so the step size is

$$\frac{10.23V}{1023} = 10mV$$

This means that V_{AX} increases in steps of 10 mV as the counter counts up from zero. Since V_A= 3.728 V and V_T = 0.1 mV, V_{AX} has to reach 3.7281 V or more before the comparator switches LOW. This will require.

$$\frac{3.728V}{10mV} = 372.81 = 373\ steps$$

At the end of the conversion, then, the counter will hold the binary equivalent of 373, which is 0101110101. This is the desired digital equivalent of V_A = 3.728 V, as produced by this ADC.

b) Three hundred seventy-three steps were required to complete the conversion. Thus, 373 clock pulses occurred at the rate of one per microsecond. This gives a total conversion time of 373 µs.

c) The resolution of this converter is equal to step size of the DAC, which is 10mV. In percent it is 1/1023 × 100% ≈ 0.1%.

Example 6.10: For the same ADC of problem 6.9 determine the approximate range of analog input voltages that will produce the same digital result of $0101110101_2 = 373_{10}$.

Solution:

Table below shows the ideal DAC output voltage, V_{AX}, for several of the steps on and around the 373rd. If V_A is slightly smaller than 3.72 V (by an amount < VT),

Step	V_{AX} (V)
371	3.71
372	3.72
373	3.73
374	3.74
375	3.75

Then $\overline{EOC}$ won't go LOW when V_{AX} reaches the 3.72-V step, but will go LOW on the 3.73-V step. If V_A is slightly smaller than 3.73 V (by an amount < V_T), then $\overline{EOC}$. Won't go LOW until V_{AX} reaches the 3.74-V step. Thus, as long as V_A is between approximately 3.72 V and 3.73-V, $\overline{EOC}$ will go LOW when V_{AX} reaches the 3.73-V step. The exact range of V_A values is

$$3.72\ V - V_T \text{ to } 3.73\ V - V_T$$

but since V_T is so small, we can simply say that the range is approximately 3.72 V to 3.73 V - a range equal to 10 mV, the DAC's resolution.

Example 6.11: A certain 8-bit ADC has a full-scale input of 2.55 V (i.e., $V_A = 2.55$ V produces a digital output of 11111111). It has a specified error of 0.1% F.S. Determine the maximum amount by which the V_{AX} output can differ from the analog input.

Solution:

The step size is 2.55 V/ (2^8 - 1), which is exactly 10 mV. This means that even if the DAC has no inaccuracies, the V_{AX} output could be off by as much as 10 mV because V_{AX} can change only in 10-mV steps; this is the quantization error. The specified error of 0.1% F.S. is 0.1% × 2.55 V = 2.55 mV. This means that the V_{AX} value can be off by as much as 2.55 mV because of component inaccuracies. Thus, the total possible error could be as much as 10 mV + 2.55 mV = 12.55 mV.

Applications:

Almost any measurable quantity present as a voltage can be digitized by an A/D converter and displayed. A/D converters are the heart of digital voltmeters and digital MultiMate's. Analog voice signals are converted to digital form for transmission over long distances. At their destination they are reconverted to analog. In digital audio record- the analog audio signal produced by a microphone is digitized (using an ADC), then stored on some medium such as magnetic tape, magnetic disk or optical disk. Later the stored data are played back by sending them to a DAC to reconstruct the analog signal, which is fed to the amplifier and speaker system to produce the recorded sound.

6.4.6 Dual slope A/D converter

A very popular high-resolution (12- to 14-bit) (but slow) A/D conversion scheme is illustrated in Fig. 6.31. To see how it operates, refer to the Fig. 6.31 and assume that the analog input signal V_A is negative. Prior to the start of the conversion cycle, switch S_2 is closed thus discharging capacitor C and setting $v_1 = 0$. The conversion cycle begins with opening S_2 and connecting the integrator input through switch S_1 to the analog input signal. Since V_A is negative, a current 1= v_A/R will flow through R in the direction away from the integrator. Thus, V_I rises linearly with a slope of I/C= V_A/RC, as indicated in Fig. 6.32.

Simultaneously; the counter is enabled and it counts the pulses from a fixed-frequency clock. This phase of the conversion process continues for a fixed duration T. It ends when the counter has accumulated a fixed count denoted n_{REF}' usually, for an N-bit converter, $n_{REF} = 2^N$. Denoting the peak voltage at the output of the integrator as V_{PEAK}, we can write with reference to Fig. 6.32 (b)

$$\frac{V_{PEAK}}{T_1} = \frac{v_A}{RC}$$

At the end of this phase, the counter is reset to zero. Phase II of the conversion begins at t= T_1 by connecting the integrator input through switch S_1 to the positive reference voltage V_{REF}. The current into the integrator reverses direction and is equal to V_{REF}/R. Thus, V_I decreases linearly with a slope of (V_{REF}/RC). Simultaneously the counter is enabled and it counts the pulses from the fixed-frequency clock.

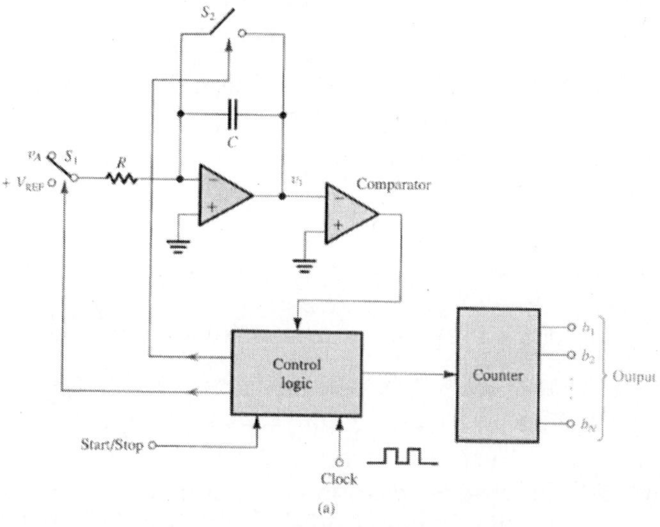

Fig. 6.31: Dual Slope A/D Convertor

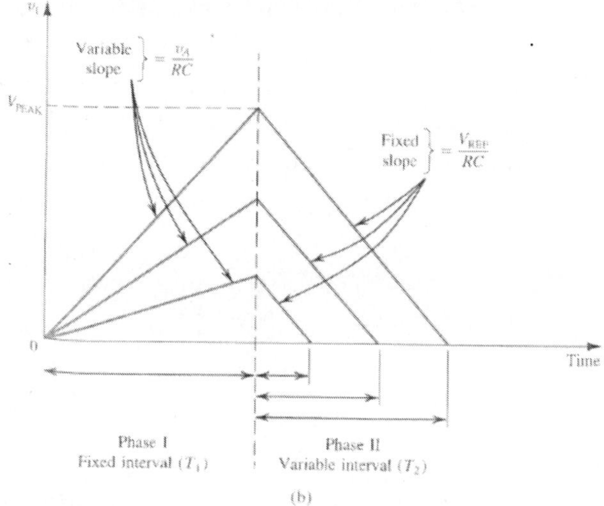

Fig. 6.32: The dual-slope A/D conversion method. Note that VA is assumed to be negative.

When V_I reaches zero volts, the comparator signals the control logic to stop the counter. Denoting the duration of phase II by T_2, we can write, by reference to Fig. 6.32 (b),

$$\frac{V_{PEAK}}{T_2} = \frac{v_{REF}}{RC}$$

Equations above can be combined to yield

$$T_2 = T_1\left(\frac{v_A}{V_{REF}}\right)$$

Hence, the counter reading, n_{REF}, can be given as

$$n = n_{REF}\left(\frac{v_A}{V_{REF}}\right)$$

Thus, at the end of the conversion process is the digital equivalent of V_A. The dual-slope converter features high accuracy, since its performance is independent of the exact values of R and C. There exist many commercial implementations of the dual slope method, some of which utilize CMOS technology.

Pros

- Very precise and accurate measurements

Cons

- Slow conversion time due to the ramp-up and ramp-down iteration

Applications: Applications for dual slope ADCs include handheld and benchtop mustimeters.

6.4.7 A/D converter using voltage to frequency conversion

A voltage to frequency converter (VCF) is a device which accepts at its input an analog voltage or current signal and provides at its output a train of pulses or square waves at a frequency which is proportional to the input value. A VFC converter can be used as a building block in an analog-to-digital (A/D) converter by using the VFC to clock a counter for a certain period of time (the "count time" or "gate time" or "conversion time") and reading the output digital data. This digital data will be proportional to the analog input.

In a voltage to frequency converter ADC generally voltage is integrated until it reaches some pre-set threshold, at which point a comparator trips, doing 2 things as shown in Fig. 6.33.

1) short circuit the capacitor, resetting the integrator to zero.
2) produce a pulse that is counted.

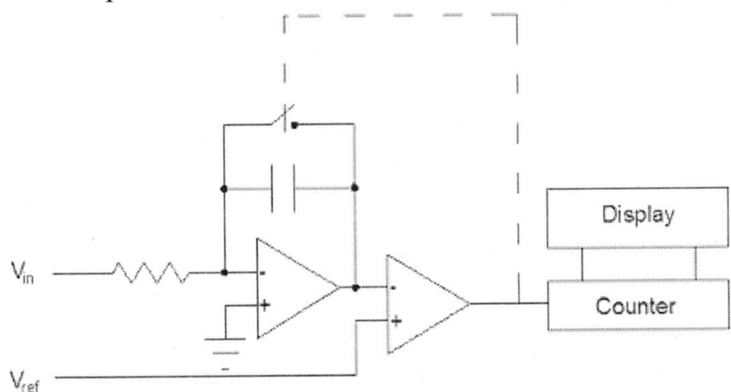

Fig. 6.33: Voltage to frequency converter ADC

A VFC integrate noise, and so are useful under circumstances similar to dual slope units. VFC based ADCs are precise, accurate, simple, and inexpensive. The precision is directly proportional to the time over which counting occurs, and inversely proportional to the time required to integrate a single count.

If we count pulses for a fixed period of time, then we know how often the comparator tripped during that time. Suppose we want to display 1.000 for a 1 V input, and suppose we look at the input for 1.000 s. Then we'd want the comparator to trip 1000 times per second. Further, suppose the comparator is set up to trip at a potential of 5 V, Then, a 1 V input should integrate to 5 V in 1 ms.

In the dual slope module, we explained how integrators work. From the argument there, the RC time constant of the integrator is set by

$$V_{out} = - V_{in}T/(RC).$$

Ignoring sign,

$$5\text{ V} = 1\text{ V} * 1\text{ ms}/RC.$$

$$RC = 1\text{ ms}/5 = 200\ \mu\text{s}.$$ *(For $R = 1\ k\Omega$, $C = 0.2\ \mu F$.)*

Changing the expected magnitude of the input voltage or the number of significant figures desired may mean a different integration time, R, or C value, but for a wide range of inputs the integrate, short, count, repeat cycle works well. The first very high resolution (up to 6 digits or ~ 20 bits) ADCs worked on this principle.

Practically VFCs are dependent on the design strategy. the capacitor; the time while the capacitor is discharging called "dead time." The bigger the resistance of the switch used to short the capacitor, the longer the switch must be closed. Alternatively, if the switch isn't closed for long enough, there will be residual potential on the integrator, and the time to integrate to threshold will be reduced. In practice, if one knows how long the switch is closed, one knows how long the dead time is, and a correction factor may be computed.

Suppose the integrator in a VFC is shorted for 1 μs every time the comparator "hits." In 1.0000 s, the V to F counts up 9997 counts, seemingly indicating 0.9997 V. However, that means that the integrator wasn't integrating for 9997 μs or 9.997 ms. Because the apparent count rate was 1 count every 1/9997 s or once per 0.1 ms, the counter missed integrating for nearly 100 counts. The corrected count is 9997 + 100 = 10097, giving a corrected output reading of 1.0097 V, about a 1% correction. While correction factors are easier to account for in software than in hardware, many systems account for the dead time in hardware.

However, to get rid of the dead time, one can integrate in one direction until we reach one threshold level, then integrate in the opposite sense until a second threshold (possibly zero, though usually a bit above zero to avoid noise in the zero level) is reached. We continue switching back and forth, with the error due to switching time typically small compared to dead time error. We get one count for each combined up/down cycle. One might think that needing two comparators would be expensive and complex.

Fortunately, early in the era of integrated circuits, the 555 timers were fabricated. It goes "High" when an input voltage is 2/3 of a reference, then switches "Low" when the input to a second input drops to 1/3 of the same reference. The reference can be the power supply potential (cheap, simple, and readily available) or a separate, carefully controlled reference (more precise) as shown in Fig. 6.34.

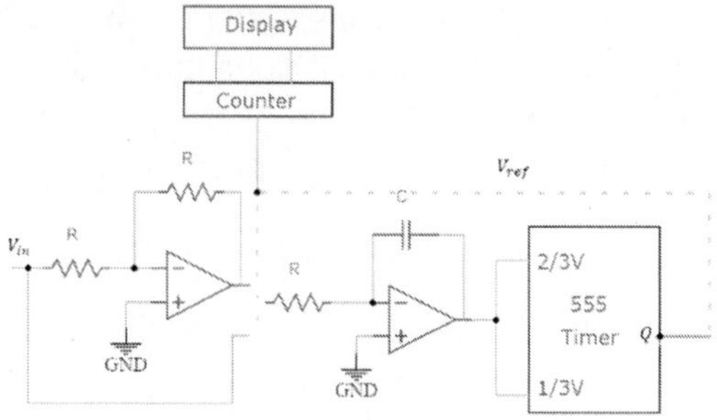

Fig. 6.34: An IC 555 based ADC using Voltage to frequency conversion

A 555 timer can toggle reliably at least 100 kHz, so 1s integrations can have 5 significant figures. 10 s integrations? 6 significant figures, provided only that there's no drift in the values of R, C, V_{ref}, or the mean of V_{in} during this time. We now come to a type of ADC that has become the tool of choice for audio digitization for computers, for inexpensive high-resolution digitizers, and for digital signal processors that need not digitize at high speeds: the sigma-delta (or, if the font display allows it, the Σ-Δ converter). Because the response frequency and signal averaging properties are as good as dual slope and V to F converters, the resolution better than many successive approximation converters and as good as many V to F systems, but the hardware is simpler to build and less subject to drift, in recent years the Σ-Δ converter has exploded in popularity. The only ADC whose hegemony is not threatened by the Σ-Δ converter is the flash converter.

6.4.8 A/D converter using voltage to time conversion

The Block diagram shows the basic voltage to time conversion type of A to D converter. Here the cycles of variable frequency source are counted for a fixed period. It is possible to make an A/D converter by counting the cycles of a fixed-frequency source for a variable period. For this, the analog voltage required to be converted to a proportional time period.

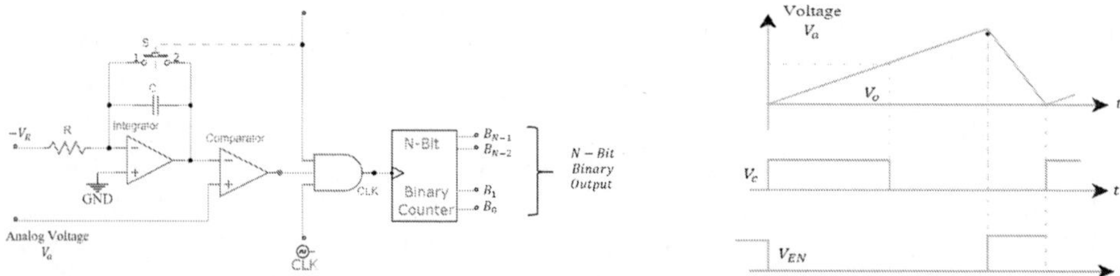

Fig. 6.35: A/D using Voltage to Time conversion

Fig. 6.36: Conversion Process

As shown in the diagram a negative reference voltage -VR is applied to an integrator, whose output is connected to the inverting input of the comparator. The output of the comparator is at 1 as long as the output of the integrator Vo is less than Va.

At t = T, Vc goes low and switch S remains open. When VEN goes high, the switch S is closed, thereby discharging the capacitor. Also, the NAND gate is disabled. The waveforms are shown here.

Comparison of Various ADC types

ADC TYPE	PROS	CONS	MAX RESOLUTION	MAX SAMPLE RATE	MAIN APPLICATIONS
Successive Approximation (SAR)	Good speed/resolution ratio	No inherent anti-aliasing protection	18 bits	10 MHz	Data Acquisition
Delta-sigma (ΔΣ)	High dynamic performance, inherent anti-aliasing protection	Hysteresis on unnatural signals	32 bits	1 MHz	Data Acquisition, Noise & Vibration, Audio
Dual Slope	Accurate, inexpensive	Low speed	20 bits	100 Hz	Voltmeters
Pipelined	Very fast	Limited resolution	16 bits	1 GHz	Oscilloscopes
Flash	Fastest	Low bit resolution	12 bits	10 GHz	Oscilloscopes

Unit Summary

The analog-to-digital(A/D) and digital-to-analog (D/A) converters form an important part of many digital system technique. Each ADC technology has its place. And because applications are so different, it is impossible to say one is better than another overall. However, it is absolutely possible to say one of them is better than another with respect to one or more of today's DAQ applications requirements. If you are measuring primarily static and quasi-static (slow) signals, you obviously don't need a super-high-speed system, but you probably want one with as much amplitude axis resolution as possible. For everyday DAQ systems, however, it's a bit more challenging since these systems are used in a variety of applications over time. The key is to select one which has the best overall performance and protections against noise, aliasing, and obsolescence.

Solved Examples

1. What is Data Acquisition (DAQ)?
   Data acquisition (commonly abbreviated as DAQ or DAS) is the process of sampling signals that measure real-world physical phenomena and converting them into a digital form that can be manipulated by a computer and software. Data Acquisition is generally accepted to be distinct from earlier forms of recording to tape recorders or paper charts. Unlike those methods, the signals are converted from the analog domain to the digital domain and then recorded to a digital medium such as ROM, flash media, or hard disk drives.
2. What are the components of Data Acquisition System

Modern digital data acquisition systems consist of four essential components that form the entire measurement chain of physics phenomena. It consist of sensors, Signal Conditioning, Analog-to-Digital Converter, and Computer with DAQ software for data logging and analysis

3. Name the essential parts of a DAC.
   Drive motors, Analog devices, Digitizer and Filter
4. Write down the drawback of weighted D/A converter.
   The main disadvantage of binary weighted D/A converter is the requirement of wide range of resister values. As the length of the binary word is increased. The range of resister values needed also increases.
5. List the broad classification of ADCs.
   Direct type ADC and Integrating type ADC.
6. List out the direct type ADCs.
   Flash (comparator) type converter
   Counter type converter
   Tracking or servo converter
   Successive approximation type converter
7. List out some integrating type converters.
   Charge balancing ADC and Dual slope ADC
8. What is integrating type converter?
   An ADC converter that perform conversion in an indirect manner by first changing the analog I/P signal to a linear function of time or frequency and then to a digital code is known as integrating type A/D converter.
9. Explain in brief the principle of operation of successive Approximation ADC.
   The circuit of successive approximation ADC consists of a successive approximation register (SAR), to find the required value of each bit by trial & error. With the arrival of START command, SAR sets the MSB bit to 1. The O/P is converted into an analog signal & it is compared with I/P signal. This O/P is low or High. This process continues until all bits are checked.
10. What are the main advantages of integrating type ADCs?
    The integrating type of ADC's do not need a sample/Hold circuit at the input.
    It is possible to transmit frequency even in noisy environment or in an isolated form.
11. Where are the successive approximation type ADC"s used?
    The Successive approximation ADCs are used in applications such as data loggers & instrumentation where conversion speed is important.
12. What is the main drawback of a dual-slop ADC?
    The dual slope ADC has long conversion time. This is the main drawback of dual slope ADC
13. State the advantages of dual slope ADC.
    It provides excellent noise rejection of ac signals whose periods are integral multiples of the integration time T.
14. Define conversion time

It is defined as the total time required to convert an analog signal into its digital output. It depends on the conversion technique used & the propagation delay of circuit components. The conversion time of a successive approximation type ADC is given by T(n+1) where T---clock period, Tc---conversion time, and n-no. of bits

15. Where are the successive approximation type ADCs used?
    The Successive approximation ADCs are used in applications such as data loggers & instrumentation where conversion speed is important.
16. What is the main drawback of a dual-slop ADC?
    The dual slope ADC has long conversion time. This is the main drawback of dual slope ADC
17. State the advantages of dual slope ADC
    It provides excellent noise rejection of ac signals whose periods are integral multiples of the integration time T.
18. Define resolution of a data converter.
    The resolution of a converter is the smallest change in voltage which may be produced at the output or input of the converter. Resolution (in volts)= VFS/2n-1=1 LSB increment. The resolution of an ADC is defined as the smallest change in analog input for a one bit change at the output.
19. Define accuracy of converter.
    Absolute accuracy: It is the maximum deviation between the actual converter output & the ideal converter output.
    Relative accuracy: It is the maximum deviation after gain & offset errors have been removed. The accuracy of a converter is also specified in form of LSB increments or % of full-scale voltage.
20. What is settling time?
    It represents the time it takes for the output to settle within a specified band ±½LSB of its final value following a code change at the input (usually a full-scale change). It depends upon the switching time of the logic circuitry due to internal parasitic capacitance & inductances. Settling time ranges from 100ns. 10Ws depending on word length & type circuit used.
21. Explain in brief stability of a converter:
    The performance of converter changes with temperature age & power supply variation. So all the relevant parameters such as offset, gain, linearity error & monotonicity must be specified over the full temperature & power supply ranges to have better stability performances.
22. What is meant by linearity?
    The linearity of an ADC/DAC is an important measure of its accuracy & tells us how close the converter output is to its ideal transfer characteristics. The linearity error is usually expressed as a fraction of LSB increment or percentage of full-scale voltage. A good converter exhibits a linearity error of less than ±½LSB.
23. What is monotonic DAC?
    A monotonic DAC is one whose analog output increases for an increase in digital input.
24. What are the specifications of D/A converter?

The specifications are accuracy, offset voltage, monotonicity, resolution, and settling time.

25. What is a sample and hold circuit? Where it is used?
    A sample and hold circuit is one which samples an input signal and holds on to its last sampled value until the input is sampled again. This circuit is mainly used in digital interfacing, analog to digital systems, and pulse code modulation systems.
26. Define sample period and hold period.
    The time during which the voltage across the capacitor in sample and hold circuit is equal to the input voltage is called sample period. The time period during which the voltage across the capacitor is held constant is called hold period.
27. Which is the fastest ADC and why?
    Simultaneous type A/D converter (flash type A/D converter) is the fastest because A/D conversion is performed simultaneously through a set of comparators.
28. What are the advantages and disadvantages of R-2R ladder DAC.
    Advantage: Easier to build, Number of bits can be expanded by adding more sections.
    Disadvantage: More power dissipation makes heating, which in turns develops non-linearties in DAC.
29. Give the disadvantages of flash type A/D converter.
    The simultaneous type A/D converter is not suitable for A/D conversion with more than 3 or 4 digital output bits. Then (2n- 1) comparators are required for an n-bit A/D converter and the number of comparators required doubles for each added bit.
30. Define quantization error.
    In A/D converter the smallest digital step is due to the LSB and it can be made smaller only by increasing the number of bits in the digital representation. This error is called quantization error.
31. Define Dither.
    It is a very small amount of random noise (white noise) which is added to the input before A/D conversion to improve the performance of A/D converter.
32. Define Delta modulation.
    Delta modulation is a technique in which derivative of the signal is Quantized. The delta modulation shows slope overload for fast input signals and their performance is dependent on input signal frequency.
33. Define slope overload noise and granular noise.
    Slope overload noise is introduced due to the use of a step size delta is too small to follow some portions of the waveform with a step size. Granular noise results from using a step size that is too large in parts of the Waveform having a small slope.
34. Define resolution of a data converter.
    Resolution of a converter is a smallest change in voltage which may be produced at the output.
35. Give the advantages of integrating type ADC.
    Integrating type ADC perform conversion in an indirect manner by first changing the analog input signal to a linear function of time or frequency and then to digital code. Here accuracy is more.
36. Compare and contrast binary ladder and R-2R ladder DAC?
    Binary ladder DAC: Requires wide range of resistor values.

R-2R ladder DAC: Only two resistor values are required.

37. Define conversion time of DAC.
    It is the total time required to convert digital signal into analog signal.
38. Define following performance parameters of D/A converters:
    Accuracy: It is the maximum deviation between the actual converter output and the ideal converter output.
    Monotonicity: Monotonic DAC is the one whose analog output increases for an increase in digital input.
39. Which is the fastest ADC? State the reason.
    Flash type ADC is the fastest ADC as the conversion takes place simultaneously rather than sequentially.
40. Define settling time of D/A converter.
    Time taken for the output to settle within specified band + ½ LSB of its final value.
41. What is the main drawback of dual slope ADC
    Long conversion time
42. Mention any two specifications of a D/A converter.
    Accuracy & Resolution.
43. For an n-bit flash type A/D converter, how many comparators are required? State the disadvantage of that type of converter.
    2n– 1.
44. State the principle of single slope A/D Converter.
    It uses a integrator to generate a sawtooth wafeform which is then compared against the analog input by a comparator
45. Give any two advantages of SA type ADC
    Efficient and Conversion speed is more
46. What is over sampling?
    Oversampling converters sample the analog signal at a rate much higher than the sampling rates normally required with nyquist converters.
47. What would be produced by a DAC whose output range is 0-10v and whose input binary number is 10111100 (for a 8 bit DAC)?
    Sol: V0 = 10V [1/21+0/22+1/23+1/24+1/25+1/26+0/27+0/28]= 10[47/64]= 7.34 V
48. What are advantages and disadvantages of R-2R ladder DAC?
    Adv:-Easier to build accurately as only two precision mental film resistors are required
    Number of bits can be expanded by adding more sections of same R/2R values.
    In inverted R/2R ladder DAC, node voltages remain constant with changing input binary This avoids any slo down effects by stray capacitances.
    Disadv:-
    With increasing output bits the circuit becomes larger
    The switches used are noted for the sources of errors.
49. What is a sample and hold circuit? Where it is used.
    A sample and hold circuit is one which samples an input s gnal and holds on to its last sampled value until the input is sampled again. This circuit is mainly used in digital interfacing, analog to digital systems, and pulse code modulation systems.

50. Define sample period and hold period.
    The time during which the voltage across the capacitor in sample and hold circuit is equal to the input voltage is called sample period. The time period during which the voltage across the capacitor is held constant is called hold period.
51. What are the specifications of data convertors?
    conversion time, settling time, accuracy, linearity, monotonic etc.

Review Questions:

1. Determine the number of comparators and resistors required for 8-bit flash type ADC.
2. Mention two advantages of R-2R ladder type Digital to Analog converter when compared to weighted resistor type Digital to Analog Converter.
3. Why is an interval R-2R ladder network DAC better than R-2R ladder DAC?
4. Which is the fastest ADC and Why?
5. Can you explain the process of binary weighted conversion used by successive approximation ADCs?
6. What causes nonlinearity errors in an ADC?
7. Is there a way to reduce distortion caused by mismatched impedances?
8. What is the effect of noise on an ADC?
9. How is ADC accuracy calculated?
10. What is the relationship between bit depth and resolution?
11. What factors affect the accuracy of an ADC?
12. What are the different types of inputs supported by an ADC?
13. How do you determine which type of ADC should be used for a given application?
14. What is the importance of voltage references in an ADC design?
15. What happens if an integrator circuit is connected to a constant input signal?
16. What are the benefits of using a dual slope converter over a single slope one?
17. What types of ADCs are available and what are their differences?
18. Why do we need to convert analog signals into digital signals?
19. What are the main components of an ADC?
20. How can the resolution of an ADC be improved?

Multiple Choice Questions (MCQs)

1. Find out the resolution of 8 bit DAC/ADC?
    a) 562
    b) 625
    c) 256
    d) 265
2. Non-linearity in the output of converter is expressed in
    a) None of the mentioned
    b) Percentage of reference voltage
    c) Percentage of resolution
    d) Percentage of full-scale voltage

3. A binary input 000 is fed to a 3bit DAC/ADC. The resultant output is 101. Find the type of error?
   a) Settling error
   b) Gain error
   c) Offset error
   d) Linearity error
4. How many equal intervals are present in a 14-bit D-A converter?
   a) 16383
   b) 4095
   c) 65535
   d) 1023
5. Resolution of a 6 bit DAC can be stated as
   a) Resolution of 1 part in 63
   b) 6-bit resolution
   c) Resolution of 1.568% of full scale
   d) All of the mentioned
6. Find the resolution of a 10-bit AD converter for an input range of 10v?
   a) 97.7mv
   b) 9.77mv
   c) 0.977mv
   d) 977mv
7. A good converter exhibits a linearity error
   a) Less than or equal to (1/2)LSB
   b) Greater than equal to (1/2)LSB
   c) Greater than or equal to (1/2)LSB
   d) None of the mentioned
8. The maximum deviation between actual and ideal converter output after the removal of error is
   a) Absolute accuracy
   b) Relative accuracy
   c) Relative /absolute accuracy
   d) Linearity
9. A monotonic DAC is one whose analog output increases for
   a) Decreases in digital input
   b) An increases in analog input
   c) An increases in digital input
   d) Decreases in analog input
10. All the commercially available DAC are
    a) Monotonic
    b) Non-monotonic
    c) Either monotonic or non-monotonic
    d) None of the mentioned
11. The time taken for the output to settle within a specified band of its final value is referred as
    a) Conversion time
    b) Settling time

c) Take off time
d) All of the mentioned

12. The primary disadvantage of the flash analog-to digital converter (ADC) is that:
    1. it requires the input voltage to be applied to the inputs simultaneously
    2. a long conversion time is required
    3. a large number of output lines is required to simultaneously decode the input voltage
    4. a large number of comparators is required to represent a reasonable sized binary number
13. What is the major advantage of the R/2R ladder digital-to-analog (DAC), as compared to a binary-weighted digital-to-analog DAC converter?
    a) The virtual ground is eliminated and the circuit is therefore easier to understand and troubleshoot.
    b) Its operation is much easier to analyze.
    c) It has fewer parts for the same number of inputs.
    d) It only uses two different resistor values.
14. The resolution of a 0–5 V 6-bit digital-to-analog converter (DAC) is:
    a) 63%
    b) 64%
    c) 1.56%
    d) 15.6%
15. The difference between analog voltage represented by two adjacent digital codes, or the analog step size, is the:
    a) monotonicity
    b) resolution
    c) accuracy
    d) quantization
16. The practical use of binary-weighted digital-to-analog converters is limited to:
    a) 4-bit D/A converters
    b) op-amp comparators
    c) R/2R ladder D/A converters
    d) 8-bit D/A converters
17. What is the resolution of a digital-to-analog converter (DAC)?
    a) It is the comparison between the actual output of the converter and its expected output.
    b) It is the deviation between the ideal straight-line output and the actual output of the converter.
    c) It is the smallest analog output change that can occur as a result of an increment in the digital input.
    d) It is its ability to resolve between forward and reverse steps when sequenced over its entire range.
18. A 4-bit R/2R digital-to-analog (DAC) converter has a reference of 5 volts. What is the analog output for the input code 0101.
    a) 0.3125 V
    b) 3.125 V
    c) 0.78125 V
    d) –3.125 V

19. Which of the following is a type of error associated with digital-to-analog converters (DACs)?
    a) offset error
    b) nonmonotonic and offset error
    c) incorrect output codes
    d) nonmonotonic error
20. Which is not an analog-to-digital (ADC) conversion error?
    a) differential nonlinearity
    b) missing code
    c) offset
    d) incorrect code
21. Sample-and-hold circuits in analog-to digital converters (ADCs) are designed to:
    a) sample and hold the D/A converter staircase waveform during the conversion process
    b) sample and hold the output of the binary counter during the conversion process
    c) stabilize the comparator's threshold voltage during the conversion process
    d) stabilize the input analog signal during the conversion process
22. In a flash analog-to-digital converter, the output of each comparator is connected to an input of a:
    a) multiplexer
    b) priority encoder
    c) · decoder
    d) demultiplexer
23. Which of following is not a type of ADC?
    (a) Flash ADC
    (b) Dual slope ADC
    (c) Recessive approximation ADC
    (d) sigma-delta ADC
24. The throughput of a flash ADC is measured in
    (a) displacement per second
    (b) distance per second
    (c) samples per minute
    (d) samples per second
25. A digital voltmeter uses a
    (a) flash ADC
    (b) successive approximation ADC
    (c) sigma-delta ADC
    (d) dual-slope ADC
26. The most common ADC seen in telecommunications based on audio signals is
    a) flash ADC
    b) successive approximation ADC
    c) sigma-delta ADC
    d) dual-slope ADC
27. In a binary weighted DAC, the lowest-value resistor corresponds to
    a) the highest binary weighted input
    b) the lowest binary weighted input
    c) the first input

   d) the last input
28. The resolution of a 12-bit Analog to Digital converter in percent is
   a) 0.01220
   b) 0.04882
   c) .02441
   d) 0.09760
29. An n-bit Analog to Digital converter is required to convert analog input in the range (0-5) V to an accuracy of 10 mV. The value of n should be
   a) 16
   b) 9
   c) 10
   d) 8
30. What is the value of LSB of an 8 bit digital to analog converter for 0- 12.8 V range?
   a) 1.6 V
   b) 50 mV
   c) 0.625 V
   d) 1.28 V

References and Suggested Readings

- *Floyd, T. L. (2005) Digital Fundamentals, Prentice-Hall Inc., USA.*
- *Tocci, R. J. and N. S. Widmer. 2004. Digital Systems Principles and Applications, 9th ed. Upper Saddle River, NJ: Prentice Hall.*
- *Victor, P. Nelson. DIGITAL LOGIC CIRCUIT ANALYSIS AND DESIGN. Prentice hall, 2020.*
- *Yarbrough, John M. Digital logic: Applications and design. Eagan: West Publishing Company, 1997.*

7 Semiconductor Memories

UNIT SPECIFICS

Through this unit we have discussed the following aspects:

- *To understand the terminology associated with memory systems*
- *To describe the difference between read/write memory and read-only memory &*
- *Different terms like: read, write, access time, nibble, byte, bus, word, word length, address, volatile, non-volatile etc.*
- *To distinguish among the various types of ROMs and cite comes common applications.*
- *To describe the structure and operation of SRAM and DRAM.*
- *To understand the construction and operation of Charge-Couple Devices, magnetic bubble memory and other memory devices*
- *How to implement combinational and sequential circuits using ROM.*

RATIONALE

The amount of memory required in a particular system depends on the type of application, but, in general, the number of transistors utilized for the information (data) storage function is much larger than the number of transistors used in logic operations and for other purposes.

Semiconductor memories are electronic data storage devices that use semiconductor technology. They are widely used in electronic devices such as computers, smartphones, digital cameras, and gaming consoles. There are two main types of semiconductor memories: volatile and non-volatile. Volatile memories require a continuous power supply to retain data, while non-volatile memories can retain data even when the power supply is turned off.

Some common types of semiconductor memories include:

1. Random Access Memory (RAM): A volatile memory that is used for temporary data storage.
2. Read-Only Memory (ROM): A non-volatile memory that is used to store permanent data.
3. Flash memory: A non-volatile memory which is used to store data that needs to be retained even when the power is turned off.
4. Electrically Erasable Programmable Read-Only Memory (EEPROM): This is a non-volatile memory that can be reprogrammed multiple times.

The ever-increasing demand for larger data storage capacity has driven the fabrication technology and memory development towards more compact design rules and, consequently, toward higher data storage densities. Here are some of the key trends and advances that are likely to shape the future of semiconductor memories:

1. Increased Density: This allows for more data to be stored on smaller chips, which can help to reduce costs and improve performance.
2. New Memory Technologies: New memory technologies such as resistive RAM (RRAM), phase-change memory (PCM), and magnetic RAM (MRAM) offers faster read and write

speeds, lower power consumption, and greater durability than traditional memory technologies.

3. AI and Machine Learning: This is driving the development of new memory architectures and technologies that are optimized for AI workloads.
4. Integration with Processing Units: This can help to reduce latency and improve performance by allowing data to be processed directly within the memory chip.

PRE-REQUISITES

Basic Digital fundamentals, Decoder, Encoder, Logic Circuits, A basic Knowledge of Semiconductor physics.

UNIT OUTCOMES

List of outcomes of this unit is as follows:

U7-O1: Determine the capacity of a memory device from its inputs and outputs.

U7-O2: Outline the steps that require reads from or writes to memory.

U7-O3: Compare EPROM, EEPROM, and flash memory .

U7-O4: To describe the read and write operation of random-access memory.

U7-O5: Combine memory ICs to form memory modules with larger word size and capacity.

Unit-7 Outcomes	***EXPECTED MAPPING WITH COURSE OUTCOMES*** *(1- Weak Correlation; 2- Medium correlation; 3- Strong Correlation)*					
	CO-1	***CO-2***	***CO-3***	***CO-4***	***CO-5***	***CO-6***
U7-O1	3	3	3	-	3	1
U7-O2	1	1	2	2	1	-
U7-O3	2	1	3	1	2	1
U7-O4	-	-	3	1	2	2
U7-O5	3	3	3	-	3	1

➢ Scan the below QR codes for more information on the topic written below the QR code.

Semiconductor Memories: RAM (Random Access Memory)

SRAM vs DRAM

Memory Terminology and applications

Unit outcomes

In this unit students will be able to understand the variety of memory, its organization those available in integrated circuit packages, the operations, size. Characteristics and various applications also explained.

7.1 Introduction

Semiconductor memory is used in any electronics assembly that uses computer processing technology. Often referred to as computer memory, these electronic components are essential for any computer or processor-based PCB assembly as shown in Fig. 7.1. This memory cards have become commonplace items for temporarily storing data - everything from the portable flash memory cards used for transferring files, to semiconductor memory cards used in cameras, mobile phones etc. The use of semiconductor memory has grown, and the size of these memory cards has increased as the need for larger and larger amounts of storage is needed.

Fig. 7.1: Printed circuit board containing computer memory

To meet the rapid growing needs for semiconductor memory, there are many types and technologies that are used. As the demand grows new computer memory technologies are being introduced and the existing types and technologies are being further developed.

A variety of different memory technologies are available - each one suited to different applications. Names such as ROM, RAM, EPROM, EEPROM, Flash memory, DRAM, SRAM, SDRAM, as well as F-RAM and MRAM are available, and new types are being developed to enable improved performance. Terms like DDR3, DDR4, DDR5 and many more are seen and these refer to different types of SDRAM computer memory.

7.1.1 Memory and its capacity:

Basically, memory is a means of storing data or information in the form of binary words. The information usually consists of program i.e. set of instruction that a computer executes to achieve a desired result. Memory card to store data is called data memory and memory used to store program is called program memory.

Computers which store programs in their memory are called stored-program type computers. All modern computers are of the stored- program type. In these computers, programs are stored as a set of machine language instructions, in binary codes. Each memory location is identified by an address. The number of storage locations can vary from a few in same memories to millions in others.

Each storage location can accommodate one or more bits. Generally, the total number of bits a memory can store is its capacity. Sometimes the capacity is specified in forms of bytes. Memories are made up of storage elements. (FFs or capacitors in semiconductor memories and magnetic domains in magnetic memories), each of which stores one bit of data. A storage element is called a ***cell***.

These memories have become very popular due to their small size, low cast, high speed, high reliability and ease of expansion of the memory size. Therefore, it is necessary for a designer of digital processors to know thoroughly the principles of operation and limitations of various semiconductor memory devices. So, a device for storing information that is fabricated by using integrated circuit technology is known as semiconductor memory. Also known as integrated-circuit memory, large-scale integrated memory, memory chip, semiconductor storage, transistor memory.

The block diagram of a memory device is shown in Fig. 7.2.

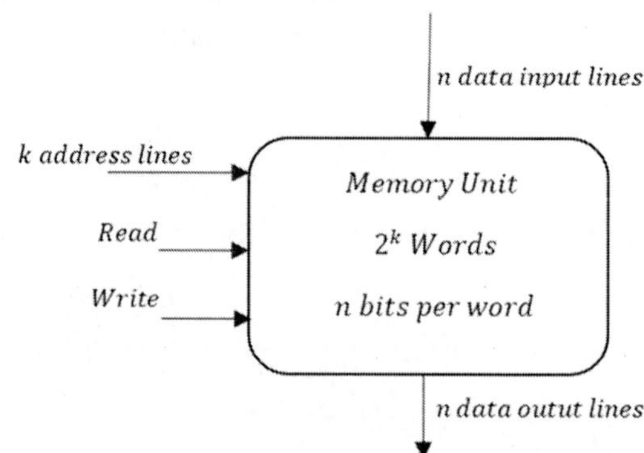

Fig. 7.2: Block diagram of memory unit

Each of the *M* locations of the memory is defined by a unique address and, therefore, for accessing any one of the *M* locations, *k* inputs are required, where $2^k = M$. This set of lines is referred to as address inputs or address bus. The address is specified in the binary form. For convenience, octal and hexadecimal representations are commonly employed.

In fact, the address input is applied to a *k-to-n* decoder circuit, which activates one of its *n* outputs depending on the address lines *k* and, thus, the desired memory location is selected. Semiconductor memory is the main memory element of a microcomputer-based system and is used to store program and data, code and information permanently. The semiconductor memory is directly accessible by the microprocessor, and the access time of the data present in the primary memory must be compatible with the operating time of the microprocessor.

Memory Sizes: Use power-of-2 multipliers

$$Kilo\ (K) = 2^{10} = 1024 \cong 10^3$$
$$Mega\ (M) = 2^{20} = 1{,}048{,}576 \cong 10^6$$
$$Giga\ (G) = 2^{30} = 1{,}073{,}741{,}824 \cong 10^9$$
$$Tera\ (T) = 2^{40} = 1{,}099{,}511{,}627{,}776 \cong 10^{12}$$

Example 7.1: How many address lines required to address a 32 K Byte memory capacity

Solution:

Represent 32K in the power of 2, i.e.,

$$32K = 2^x; \qquad Hence\ 32K = 32 \times 1K = 2^5 \times 2^{10} = 2^{15}$$

Hence, 32K memory requires 15 lines.

Note: Here Byte is used to represent the word length or size of one memory location i.e., each memory location can store 8-bit (1 Byte) data. As group of 8 bits is called a byte. Most computer memories use words that are multiples of 8-bit length. Thus, a 16-bit word contains two bytes, and a 32–bit word is made of four bytes. The capacity of a memory unit is usually stated as the total number of bytes that the unit can store. There are various units of Binary Data such as, Bits, Bytes, Nibbles and Words. However, as a rule, memories store data in units that have from one to eight bits. The Smallest unit of binary data is the bit. In many applications, data are handled in an 8-bit unit called a byte or in multiples of 8-bit units. The byte can be split into two 4-bitunits that are called nibbles. A complete unit of information is called a word and generally consists of one or more bytes. Some memories store data in 9-bit groups; a 9-bit group consists of a byte plus a parity bit.

A memory is just like a human brain. It is used to store data and instruction. Computer memory is the storage space in computer where data is to be processed and instructions required for processing are stored. The memory is divided into large number of small parts. Each part is called a cell. Each location or cell has a unique address which varies from zero to memory size minus one. For example, if computer has *64k* words, then this memory unit has *64 * 1024 = 65536* memory location. The address of these locations varies from *0 to 65535*.The specification for the memory size is possibly the most important key parameters to be specified. The way in which the memory is specified is standardised by JEDEC (JEDEC Standard 100B.01) and this format is used virtually universally for memory specifications.

Example 7.2: How Many address lines requires for a memory capacity of *64K X 8*.

Solution:

A *64K X 8* memory is organized into words of eight bits, or octets, and has a capacity of *64 K* octets. As each word requires an address line and as

$$1K = 2^{10} = 1024$$

implementing this memory requires a total *n*, number of address lines,

Hence

$$n = \log_2(64 \times 1024) = \log_2(2^{16}) = 16$$

Note: The set of wires through which the information to be stored or read is transmitted is called the data bus.

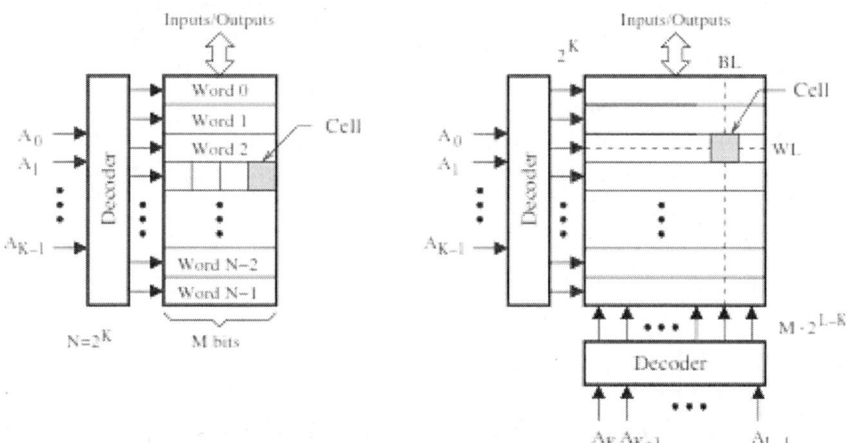

Fig. 7.3: (a) Column structure for a memory; (b) Matrix structure for a memory

A memory can be organized in a column structure, as shown in Fig. 7.3 (a). The decoder is used to select, from *k-bit* addresses, an address of *2K lines*, or an *M-bit word*. When the number of bits,

K, becomes very high, a regular structure is only obtained by adopting a matrix configuration, as illustrated in Fig. 7.3 (b).

The address of a word is divided into a line address $(A_0, A_1, A_2, \dots \dots \dots, A_{K-1},)$ and a column address $(A_0, A_1, A_2, \dots \dots \dots, A_{L-1},)$. Thus, in order to select a word, a horizontal line or a word line *(WL)* and a vertical line, or bit line *(BL),* must be activated.

7.1.2 Memory organization and operation:

Basic Semiconductor Memory Array: Each storage element in a memory can retain either a 1 or a 0 and is called a cell. Memories are made up of arrays of cells, as illustrated in Fig. 7.4 using 64 cells as an example. Each block in the memory array represents one storage cell, and its location can be identified by specifying a row and a column.

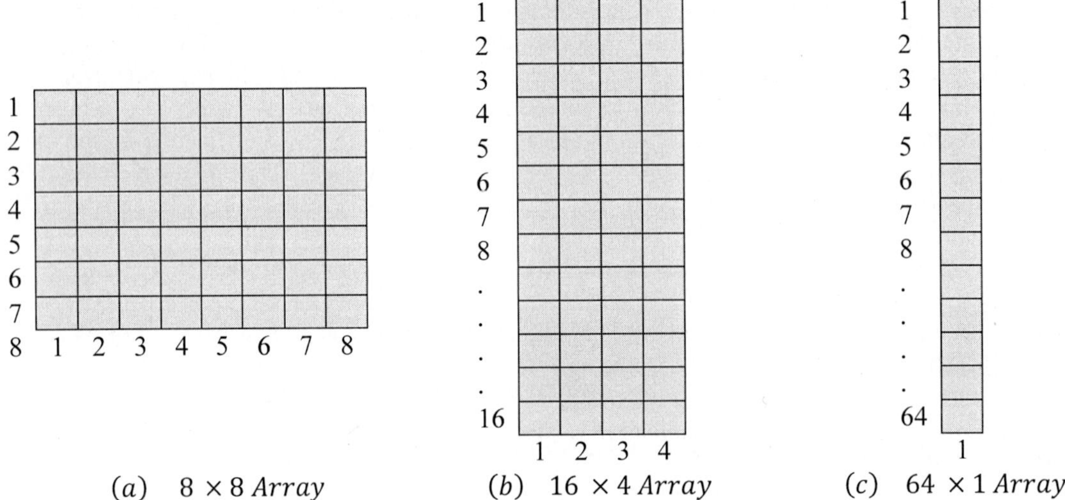

Fig. 7.4: A 64-cell memory array organized in three different ways

Memories are designed such that they can be connected to the same data bus. Each memory chip then possesses a selected input, $\overline{CS}$ (chip select) or $\overline{CE}$ (chip enable), that can be used to avoid conflicts when the bus is in use for other purposes (Refer Fig. 7.5). Each of these control inputs can serve to connect or disconnect a memory from the bus.

- if $\overline{CS}$ takes the logic state 0, the memory is selected and connected to the data bus;
- if $\overline{CS}$ assumes the logic state 1, the data buffer is set to the high impedance state in order to disconnect the memory.

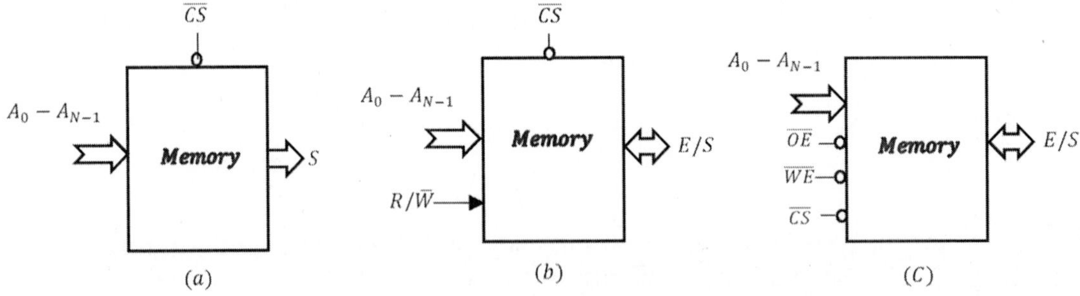

Fig. 7.6: a) Read only memory; b) and c) read/write memory

A memory has an output enable input that, often symbolized as $\overline{OE}$, can be used to enable or disable the output, and a write enable input (or read/write R/$\overline{W}$ input) that can be used to select which operation is to take place, read or write. The conditions for a read operation are:

- if $\overline{OE}$, takes the logic state 0, it is possible to read data:
- if $\overline{OE}$, takes the logic state 1, the data bus is in a high-impedance state and reading data are not possible.

The necessary conditions for a write operation are as follows:

- if $\overline{WE}$ takes the logic state 0, it is possible to write data:
- if $\overline{WE}$ takes the logic state 1, it is not possible to write data.

A signal may take the low level, high level or an intermediate level that corresponds to the high impedance state (Refer Fig. 7.6).

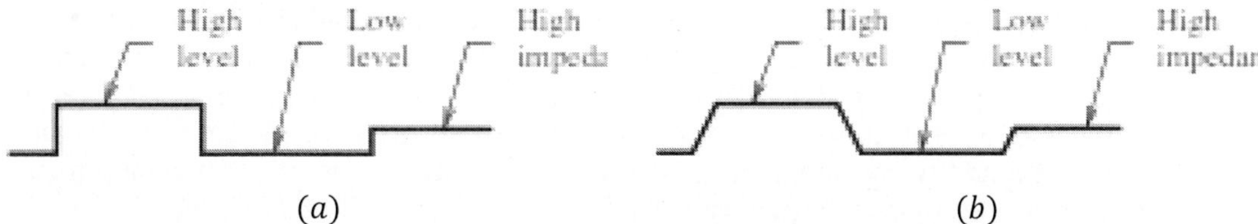

Fig. 7.6: Representation of a signal: a) ideal case; b) considering the rise and fall times

Data representation on a bus is illustrated in Fig. 7.7. Two states can be observed:

- a low-impedance state for which the logic level of each wire may be 0 or 1 (two horizontal lines; the hexadecimal value of the binary configuration of the wires is written between these lines);
- a high impedance state for which the logic level is not defined (a median line). In this case, the bus is disconnected.

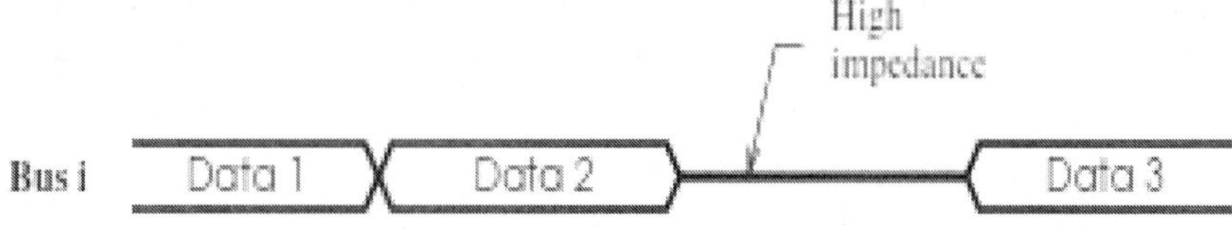

Fig. 7.7: Data representation on a bus

All memory regardless of its type or use, consists of location for storing binary information or bits. Each location is identified by an address. A word is the fundamental group of bits used to represent one entity of information such as one numerical value. The word size- the number of bits in a word- varies among computer system and may range from 4 to 64 or more bits. The word size is usually expressed as a certain number of bytes. For example, a 16- bit word is 2 bytes.

A memory location is thus a set of devices capable of storing one word. For example, each memory location in an 8-bit microcomputer one that uses 8-bit words might consists of eight latches. Each latch stores one bit of a word, and is referred to as a cell. The capacity or size of a memory is the total number of bits or bytes or words that it can store. For convenience, the size of a memory is expressed as a multiple of 2^{10}=1024, which is abbreviated as *'k'*. For example, a memory of size 2048 Byte is said to be *2k*, a memory of size 16,384 Byte is *16k* and a memory of size 65,536 Bytes is *64k*.

Each memory system requires several different types of input and output lines to perform the following functions.

- Select the address in memory that is to be accessed for a read or write operation.
- Select either a read or a write operation to be performed.
- Supply the input data to be stored in memory during a write operation.
- Hold the output data coming from the memory during a read operation.
- Enable or disable the memory, so that it will or will not respond to the address inputs and read/write command.

A *memory unit* is a collection of storage cells, together with associated circuits needed to transistor information into or out of a device. The architecture of memory is such that information can be selectively retrieved from any of its internal locations. A memory unit stores binary information in groups of bits called words. A word in memory is an entity of bits that move in and out of storage as a unit. A memory word is a group of 1's and 0's and may represent a number, an instruction, one or more alphanumeric character, or any binary coded information.

Example 7.3: Consider a memory of size 16 words. Find the binary address of each location.

Solution:

Table 7.1: Table for example 7.3

Word number	Binary address			
	A_3	A_2	A_1	A_0
0	0	0	0	0
1	0	0	0	1
2	0	0	1	0
.	.	.	.	.
.	.	.	.	.
.	.	.	.	.
.	.	.	.	.
12	1	1	0	0
13	1	1	0	1
14	1	1	1	0
15	1	1	1	1

Since Size of the Memory (M) = 16, therefore

$$M = 2^k; i.e., 16 = 2^k; \ gives, k = 4$$

So, for selecting one out of 16 words, a 4-bit address is required.

The address is specified as A_3, A_2, A_1, A_0, where A_3, is the Most-Significant Bit (MSB) and A_0, represents the Least Significant Bit (LSB) of the address (Refer Table 7.1). The address of each location is given in the table.

7.2 Memory Terminology:

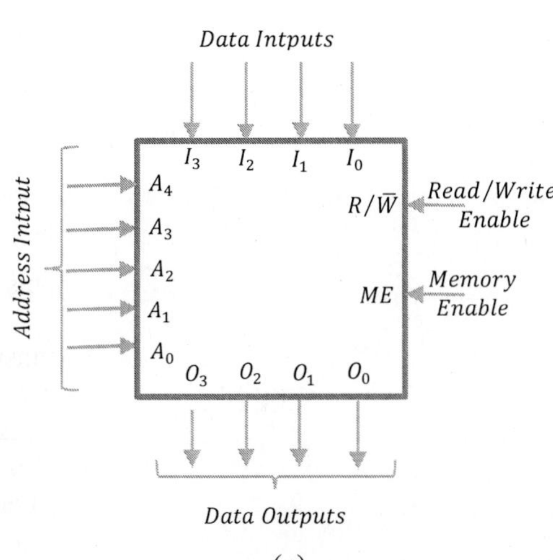

Memory Cells				Address
0	1	1	0	00000
1	0	0	1	00001
1	1	1	1	00010
1	0	0	0	00011
0	0	0	1	00100
0	0	1	0	00101
.	.	.	.	.
.	.	.	.	.
.	.	.	.	.
.	.	.	.	.
0	0	1	1	11101
1	1	0	1	11110
1	1	1	1	11111

(a) (b)

Fig. 7.8: Diagram of a 32 X4 Memory and arrangement of Memory Cells

Data Input and data Output Lines:

A *32 x 4* memory stores 32, 4-bit words. Since the word size of 4-bits, there are four *input data lines* I_0 to I_3 and four *output data lines* O_0 to O_3 (Refer Fig. 7.8 (a)). During a write operation the data to be stored in memory have to be applied to the data input lines. During a read operation a word being read from memory appears at the data output lines.

Address inputs lines:

Since, this memory stores 32 words, it has 32 different storage locations and therefore, 32 different binary addresses ranging from 00000 to 11111(0 to 31 decimal) (Refer Fig. 7.8 (b)). Thus, there are five *address inputs* A to A_4. To access one of the memory locations for a read or write operation, the 5-bit address code for that particular location is applied to the address inputs. In general, N address inputs are required for a memory that has a capacity of 2^N words. Each address location contains four memory cells that hold 1s and 0s to make up the data word stored at that location.

The $R/\overline{W}$ input:

Read/write ($R/\overline{W}$) input line determines the memory operation that would take place. Some memory systems use two separate inputs, one for read and one for write. When a single $R/\overline{W}$ input is used, the read operation take place for $R/\overline{W}= 1$ (Active High) and the write operation takes place for $R/\overline{W} = 0$ (Active Low).

Memory ENABLE:

An active-high input enables the memory to operate normally when it is kept High. A Low on this input disables the memory, preventing it to respond to address and $R/\overline{W}$ inputs. This type of input is useful when several memory modules are combined to form Larger memory.

Memory Cell:

A device or electrical circuit used to store a single bit (0 or 1). Example of memory cells are a flip-flop, a charged capacitor, and a single spot on magnetic or disk.

Memory Word:

A group of bits (cells) in a memory that represents instructions or data of some type. For example, a register consisting of eight FFs can be considered to be a memory that is storing an 8-bit word.

Byte:

A special term used for a group of 8-bits. A byte always consists of 8-bits. Word sizes can be expressed in byte as well as in bits. For example, a word size of 8-bits is also a word of one byte; a word size of 16-bits is two bytes word, and so on.

Write and Read operation:

The process of storing data in memory is called writing in memory. Retrieving data from memory is called reading memory.

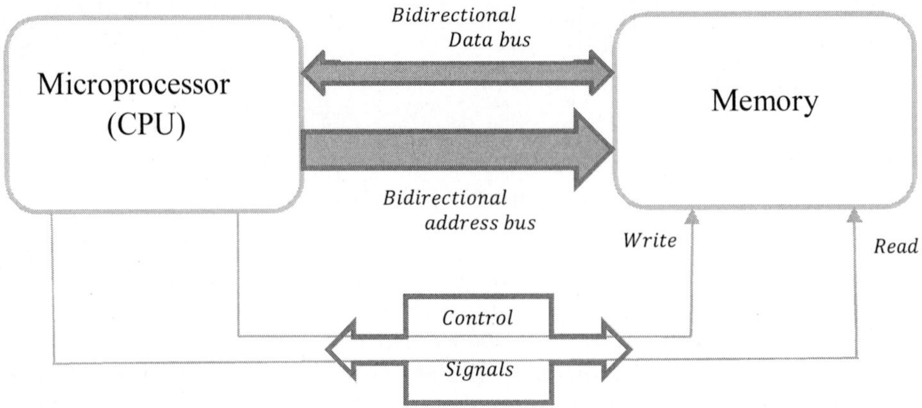

Fig. 7.9: Microcomputer System

The microprocessor serves as the central processing unit (CPU) for the computer. It contains an arithmetic /Logic unit (ALU), register array and control unit that is uses to perform read and write operations as well as to execute programs as shown in Fig. 7.9. The control signals are labelled as read and write. The CPU activates these when a read or a write operation is to be performed.

The 8-bit data bus consisting of eight lines on which data bits D0 through D7 are transmitted. It is called a bidirectional data bus, because words can be transmitted from the CPU to memory (write) or from memory to the CPU (read). The unidirectional address bus is the set of lines over which the CPU transmits the address bits corresponding to the memory address to be read or write into.

The address bus is a 16-bit bus (A_0 through A_{15}) meaning that the CPU can access (read or write into) up to $2^{16} = 65{,}536$ different memory addresses.

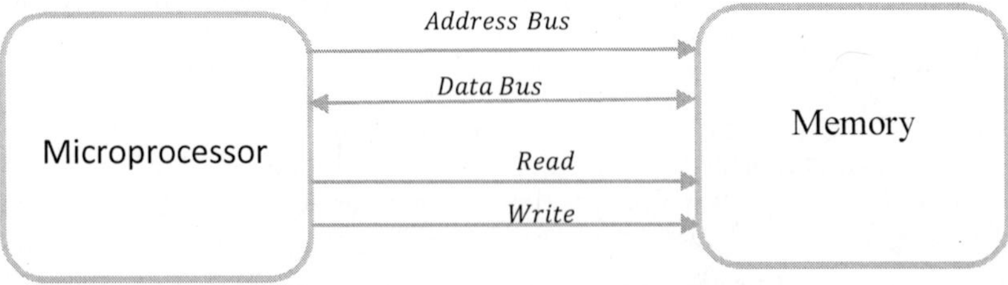

Fig. 7.10: Communication between Microprocessor and Memory

Memory cycle:

The memory unit of both RAM and ROM. It is connected to the microprocessor through the address and data buses and the read and write control. This is shown schematically in Fig. 7.10. A memory cycle is defined as the time required to access the memory to read or write a byte.

A number of control inputs are required to give commands to the device to perform the desired operation. For example, a command signal is required to tell the memory whether a read or write ($R/\bar{W}$) operation is desired.

Table 7.2: Control inputs to Memory chip

Memory Enable	Read/Write	Memory Operation
0	X	None
1	0	Memory Write
1	1	Memory Read

Form Table 7.2 it has been observed that, when $R/\bar{W}$ is HIGH, the data bus will be used for reading the memory whereas when $R/\bar{W}$ is LOW, the data on the bus will go into the memory. Other command includes inputs memory or chip enable, chip select (CS) etc.

In addition to the above-mentioned functional pins, a minimum of two pins are required for power supply and ground.

7.3 Memory reading and writing:

The internal organization of a 16 x 4 memory chip is illustrated in Fig. 7.11.

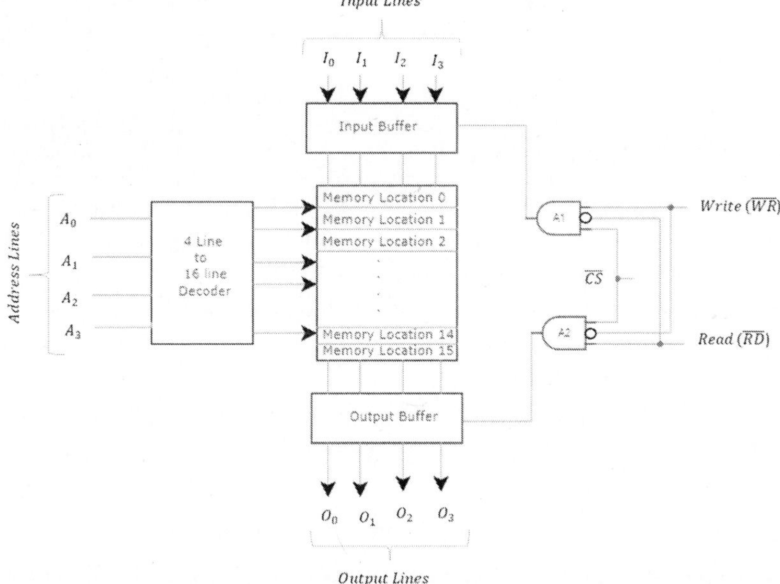

Fig.7.11: The internal organization of a 16 x 4 memory chip

Write operation:

To write a word into the selected memory Location requires Logic 1 voltage to be applied to $\overline{CS}$ (chip select) and write inputs and *Logic 0* voltage to *Read (RD)* input. This combination of inputs give output of AND gates A_1 and A_2 as 1 and 0 respectively. A Logic 1 at the output of A_1 Enables the input buffers so that the 4-bit word applied to the data inputs will be loaded (entered) into the selected memory Location. Whereas, Logic 0 at the output of A_2 disable the output buffers so that data outputs are not available.

For writing a word into a particular memory Location the following sequence of operation is to be performed.

1. The chip select signal is applied to the $\overline{CS}$ terminal.
2. The word to be stored is applied to the data input terminals.
3. The address of the desired memory Location is applied to the address-input terminals.
4. A write command is applied to the write-control input terminal with RD = 0.

The machine cycle diagram shown in Fig. 7.12 represents the various timing terms used in write operation

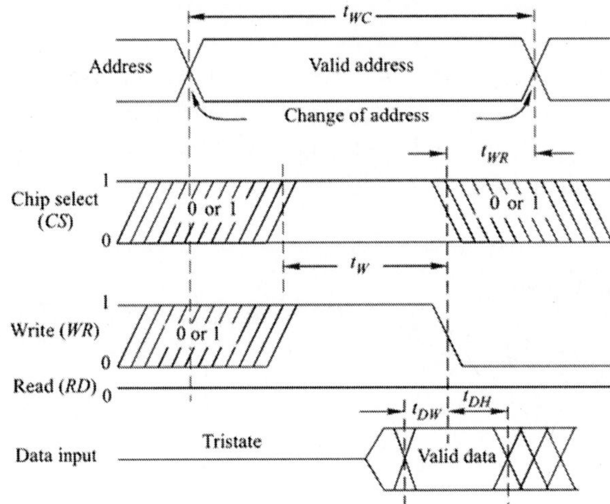

Fig. 7.12: Write operation waveform

- Write Cycle Time (t_{WC}): This is the minimum amount of time for which the valid address must be present for writing a word in the memory. In other words, it is the minimum time required between successive write operations.
- Write Pulse Time (t_W): This is the minimum length of the write pulse.
- Write Release Time (t_{WR}): This is the minimum amount of time for which the address must be valid after the write pulse ends.
- Data Set Up Time (t_{DW}): This is the minimum amount of time for which the data must be valid before the write pulse ends.
- Data Hold Time (t_{DH}): This is the minimum amount of time for which the data must be valid after the write pulse ends.

Read operation:

In order to read the contents of a selected memory Location, the Read (RD) and the chip select ($\overline{CS}$) inputs must be at Logic 1 level and WR at logic 0 level. This gives output of A_2 =1 which enables the output buffers so that the contents of the selected memory location will appear at the data output.

To read (or retrieve) a data word, known to be stored at a particular address, the following sequence of operations is required to be performed.

1. The chip select signal is applied to the $\overline{CS}$ terminal.
2. The address of the desired memory Location is applied to the address-input terminals.
3. A read-command signal is applied to the read control-input terminal.

In response to the above operations, the data word stored at the addressed location appears on the data-output terminals. Fig. 7.13 illustrates the various waveforms during the read operation.

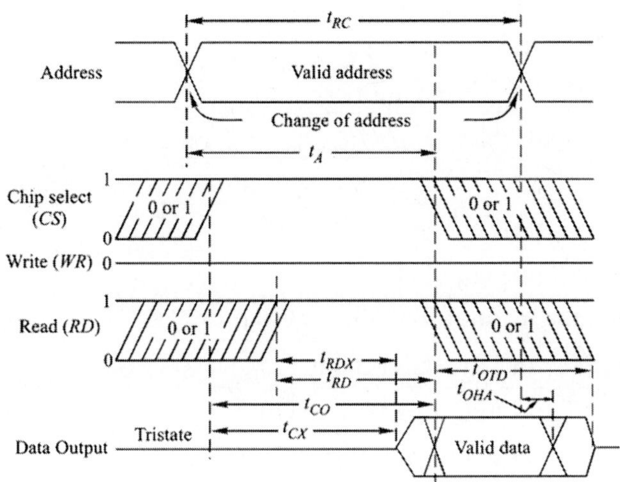

Fig. 7.13: Read operation waveform

The important timing characteristics of the read cycle are:

- Read Cycle Time (t_{RC}): This is the minimum amount of time for which the valid address must be present for reading a word from the memory. In other words, it is the minimum time required between successive read operations.
- Access Time (t_A): This is the maximum time from the start of the valid address of the read cycle to the time when the valid data is available at the data outputs. The access time is at most equal to the read-cycle time, i.e., $t_A \leq t_{RC}$. In other words, the data outputs might be ready before the memory is actually ready for the next read operation.
- Read to Output Valid Time (t_{RD}): This is the maximum time delay between the beginning of the read pulse and the availability of valid data at the data outputs.
- Read to Output Active Time (t_{RDX}):This is the minimum time delay between the beginning of the read pulse and the output buffers coming to active state (from the high-impedance state).
- Chip-Select to Output Valid Time (t_{CO}) :This is the maximum time delay between the beginning of the chip-select pulse and availability of valid data at the data outputs.
- Chip-Select to Output Active Time (t_{CX}): This is the minimum time delay between the beginning of the chip-select pulse and the output buffers coming to active state.
- Output Tristate From Read (t_{OTD}): This is the maximum time delay between the end of the read pulse and the output buffers going to high-impedance state.
- Data Hold Time (t_{OHA}): This is the minimum time for which the valid data is available at the data outputs after the address ends.

The write- and read-cycle timings of a typical memory chip are given in Table 7.3.

Table 7.3: Timing Parameter of a typical Memory Chip

Parameter	Time (ns)
Write Operation	
Write Cycle Time (t_{WC}):	200
Write Pulse Time (t_W):	120
Write Release Time (t_{WR}):	0
Data Set Up Time (t_{DW}):	120
Data Hold Time (t_{DH}):	0
Read Operation	
Read Cycle Time (t_{RC}):	200
Access Time (t_A):	200
Read to Output Valid Time (t_{RD}):	70
Read to Output Active Time (t_{RDX}):	20
Chip-Select to Output Valid Time (t_{CO})	70
Chip-Select to Output Active Time (t_{CX}):	20
Output Tristate From Read (t_{OTD}):	60
Data Hold Time (t_{OHA}):	50

Example 7.4: Show the memory cycle timing waveforms for the write and read operations. Assume a CPU clock of 150 MHz and a memory cycle time of 20 ns.

Solution:

Step 1:

From the problem description we know that CPU operates with a clock frequency of 150 MHz, hence the time period of one CPU cycle is:

$$T_{CPU} = \frac{1}{150\ MHZ} = \frac{1}{150\ .\ 10^6\ HZ} = 6.667ns$$

The memory cycle time is *20 ns*.

Step 2:

The write cycle shows three 20-ns cycle: T1, T2 and T3

- At the beginning of cycle T1 CPU provides the address and input data to the memory. After the signals in the address lines are stable the memory enable and read/write signal has been activated.
- For the write operation the memory enable signal switches to the high level and the read/write signal switches to the low level. The two control signals stay active for the 20 ns.
- At the completion of the clock cycle T3, the memory write operation is completed and the CPU can access the memory again with the next T1 cycle.

The memory write cycle timing waveforms for the write operation is shown in Fig. 7.14.

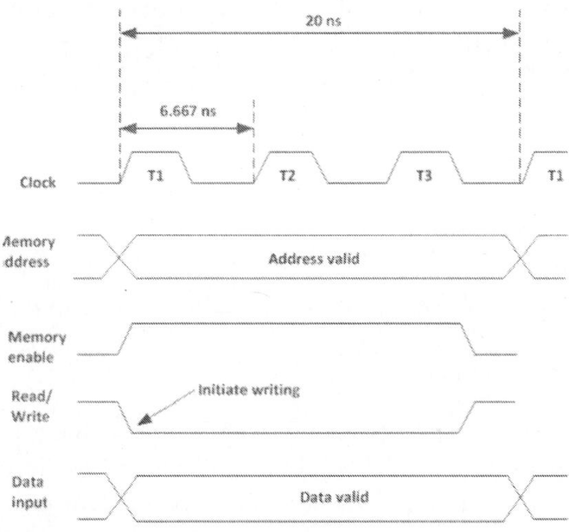

Fig. 7.14: Timing waveforms of write cycle

Step 3

- For a read operation the memory enables and read/write signals must be in their high level. After that, the memory places word content selected by the address into the output data lines within a *20 ns* interval (or less) from the time that the memory enable signal is activated.
- At the completion of the clock cycle T3, the memory read operation is completed and the CPU can access the memory again with the next T1 cycle.
- The memory Read cycle timing waveforms for the read operation is shown in Fig. 7.15.

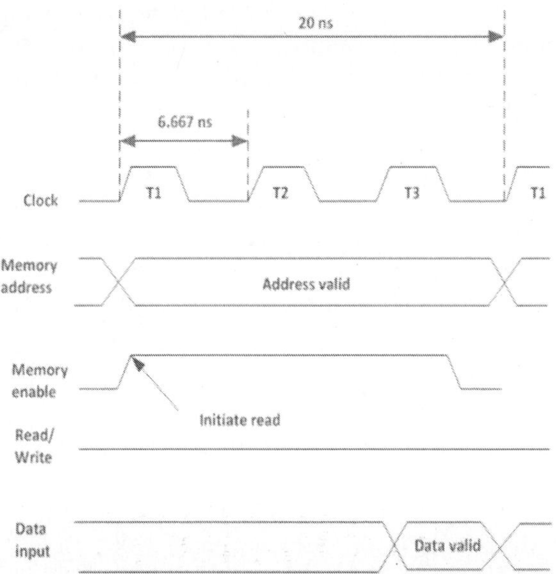

Fig. 7.15: Timing waveforms of read cycle

Example 11.2: For the memory timing of Table 7.3, find the maximum rate (Words/Second) at which

(a) Data can be stored

(b) Data can be Read

Solution:

(a) The maximum rate at which data can be stores is given as

$$\frac{1}{t_{WC}} = \frac{1}{200 \times 10^{-9}} = 5 \times 10^6 words/s$$

(b) The maximum rate at which data can be read is given as

$$\frac{1}{t_{RC}} = \frac{1}{200 \times 10^{-9}} = 5 \times 10^6 words/s$$

7.4 Expanding memory size

In many memory applications, the required memory capacity i.e. the number of words or word size cannot be satisfied by a single available memory IC chip. Therefore, several similar chips have to be combined suitably to provide the desired number of words or word size. Available memory can be expanded to increase the word length (number of bits in each address) or the word capacity (number of different addresses) or both.

Memory expansion is accomplished by adding an appropriate number of memory chips to the address, data, and control buses. However, it is also possible to expand word size. if it is required to have a memory of word size *n* and the word size of the available ICs is *N* i.e., *n* >*N*, then the number of IC chips required is an integer, next higher to the value *n/N*. These chips are to be connected in the following way.

- Concept the corresponding address lines of each chip individually i.e. A_0 of each chip is connected together and it becomes A_0 of the overall memory. Similarly, connect other address lines together.
- Connect the *RD* input of each IC together and it becomes the read input for the overall memory. Similarly, connect the *WR* and *CS* inputs.

So, the number of data-input/output lines will be equal to the product of the number of chips used and the word size of each chip.

Example 7.3: Obtain a 16 x 8 memory using 16 x 4 memory ICs.

Solution:

Since the word size required is *n =8* and the word size of available IC is *N= 4*, therefore *n/N =2* chips are required to obtain the desired memory. Fig. 7.16 shows the relevant connection of two chips. Here, we have assumed bidirectional input/output (I/O) lines which is a common in available memory chips. In this 16 x 8 memory,

- The higher order four bits (D_7, D_6, D_5, D_4) of each 8-bit word are located in memory M_1 and
- The lower order four bits (D_3, D_2, D_1, D_0)are located in memory M_2.

Memory chips can be combined together to produce a memory, with the desired number of memory locations.

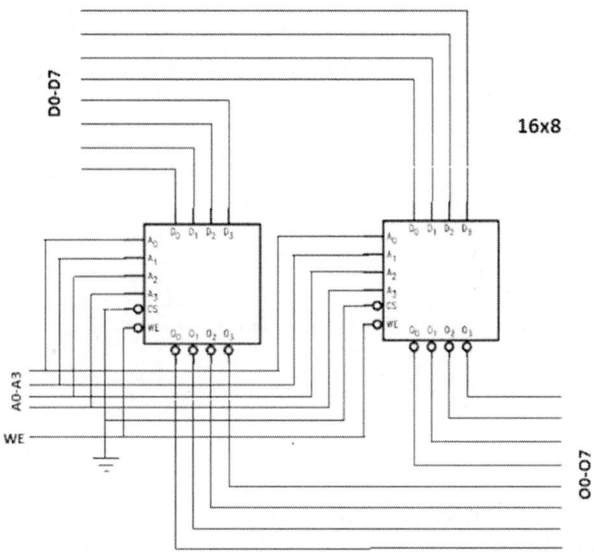

Fig. 7.16: **16** $\times$ **8** Memory obtained by two **16** $\times$ **4** Memory Chips.

Expanding word capacity: Memory chips can be combined together to produce a memory, with the desired number of locations. to obtain a memory of capacity m words, using the memory chips with M words each, the number of chips required is an integer next higher to the value m/M. these chips are connected in the following way.

- Connect the corresponding address lines of each chip respectively.
- Connect the RD input of each chip together. Similarly connect the WR inputs.
- Use a decoder of proper size and connect each of its output to one of the CS terminal of memory chips.

For example, if eight chips are to be connected, a 3-line-to-8-line decoder is required to select one out of eight chips at any one time.

Example 7.5: Suppose that a 2M x 16 main memory is built using 256K x 8 RAM chips and memory is word addressable. How many RAM chips are necessary?

Solution:

2M = 2^{21}, 256K = 2^{18}, 16 = 2^4, and 8 = 2^3.

The number is (2M•16) / (256K•8) = (2^{21}•2^4) / (2^{18}•2^3)

= 2^{25} / 2^{21} = 2^4 = **16**.

Example 7.6: Obtain a 2048 x 8 memory using 256 x 8 memory chips.

Solution:

The number of chips required is $\frac{\mathbf{2048}}{\mathbf{256}} = \mathbf{8}$.

At any one time, only one of the 2048 locations are to be accessed, which will be in one of the eight chips. That means only one of the eight chips must get selected at a time. For selecting one out of 2048 locations, the number of address lines required is 11 i.e., 2^{11} = 2048. The lower order eight bits of the address $A_7 - A_0$ will be same for each chip, and the higher order three bits of the address $A_{10} - A_8$ must select one out of the eight chips.

For this purpose, a 3-line-to-8-line decoder is required. Here, we have assumed a common terminal ($R/\overline{W}$) for read and write. For the read operation, logic 1 is to be applied to ($R/\overline{W}$), whereas logic 0 is to be applied for the write operation. The chip-select input is assumed to be active-low. The address range of the chips are given in Table 7.4.

Table 7.4: A 2048 x 8 memory obtained by combining Eight 256 x 8 Memory Chips

Memory chip	Addresses (Hex.)
M0	000-0FF
M1	100-1FF
M2	200-2FF
M3	300-3FF
M4	400-4FF
M5	500-5FF
M6	600-6FF
M7	700-7FF

7.5 Classification and Characteristics of Memories:

various memory devices can be classified on the basis of their principle of operation, physical characteristics, modes of access, technology used for fabrication etc. Principle of operation memories can be classified according to their principle of operation.

The classification of Memory is given in Fig. 7.17.

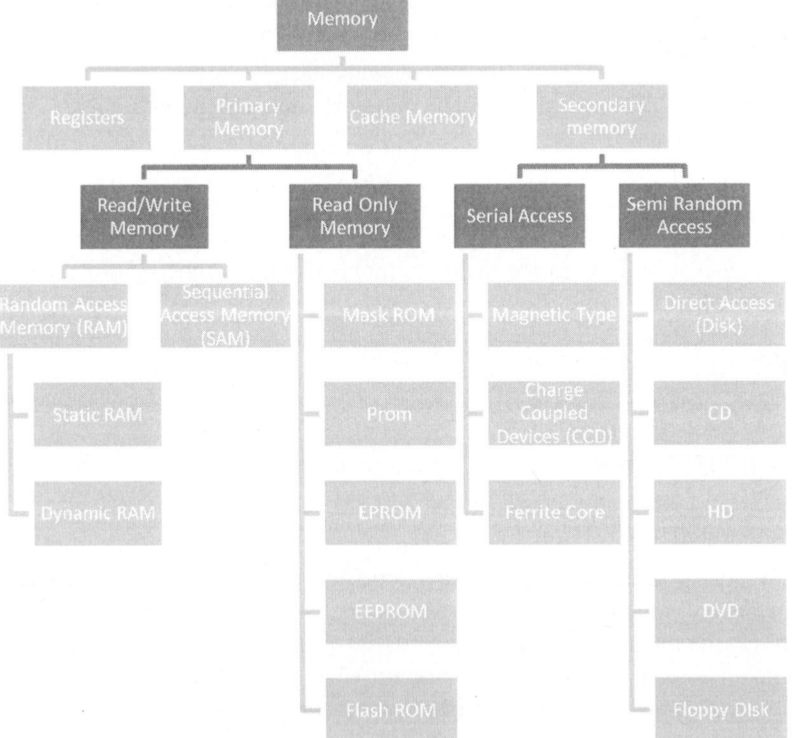

Fig. 7.17: Classification of memory

- *Registers:* Registers are memories located within the Central Processing Unit (CPU). Various types of registers are available within CPU. Registers are small but CPU can access it quickly. Some of the registers are instruction Register, ALU I/O registers, Status Register, Stack pointer register, Program counter, etc.
- *Primary Memory:* Primary memory, is volatile memory as main memory is the primary storage location in a computer system that stores data and instructions that are actively used by the CPU for processing. It is classified into two types, namely RAM and ROM. When a program is executed, it is loaded into primary memory from secondary memory. The CPU then accesses the data and instructions from primary memory to perform the required operations.
- *Cache Memory:* The Cache memory, also known as the CPU Memory, is a fast-operating static RAM (SRAM) that can be more easily accessed by the processing unit as compared to regular RAMs. Since it is designed using the SRAM, it is more expensive. It is also faster than other memories. The cache memory usually appears together on the same chip as the computers' microprocessors, with the chip area being too small. Thus, the cache size is just a fraction of the main memory, being comparatively small. The cache access time is as less as one clock cycle, whereas the main memory access requires several clock cycles.
- *Secondary Memory:* Secondary memory is a non-volatile memory used to store data and programs for long-term use. Examples of secondary memory devices include hard disk drives (HDD), solid-state drives (SSD), USB flash drives, CDs, DVDs, and magnetic tape. These devices have much larger storage capacity than RAM and are typically used to store large amounts of data that are not frequently accessed by the computer's processor.

7.5.1 Primary Memory

Random Access Memories (RWM or RAM):

In this type of memories, the memory Locations are organized in such a way so that any memory Location requires equal time for writing or reading. This type of memory is also known as read-and-write (RWM) or RAM. RAMs can be static or dynamic and one fabricated using bipolar or unipolar technologies.

The RAM is a form of semiconductor memory technology that is used for reading and writing data in any order. In other words, as it is required by the processor. It is used for such applications as the computer or processor memory where variables and other storage are required on a random basis. Data is stored and read many times to and from this type of memory.

The communication between a memory and its environment is achieved through data input and output lines, address selection lines, and control lines that specify the direction of transfer. The n data input lines provide the information to be stored in memory and the n data output lines supply the information coming out of memory.

The k address lines specify the particular word chosen among the many available. The two control inputs specify the direction of transfer desired. The write input causes binary data to be transferred into the memory, and the read input causes binary data to be transferred out of memory. The time taken to transfer information to or from any desired location is always same hence it is called Random Access Memory (RAM).

Memory size of RAM = $2^n \times m$, where,

- n: address line,
- m: data lines.

For example, the contents of a 1024 X 16 memory is given by Table 7.5

Table 7.5: Content of a 1024 x 16 memory

Memory Address		Memory Content
Binary	Decimal	
0000000000	0	1010111011111000
0000000001	1	1110111011111000
0000000010	2	1011111011111000
0000000011	3	1010111000111000
.	.	.
.	.	.
.	.	.
.	.	.
.	.	.
1111111110	1022	1010111000011000
1111111111	1023	1110111010101000

Write and Read Operations:

The two operations that a random-access memory can perform are the write and read operations as given by Table 7.6. The write signal specifies a transfer-in operation and the read signal specifies a transfer-out operation. On accepting one of these control signals, the internal circuits inside the memory provide the desired function. The steps that must be taken for the purpose of transferring a new word to be stored into memory are as follows:

1. Transfer the binary address of the desired word to the address lines.
2. Transfer the data bits that must be stored in memory to the data input lines.
3. Activate the write input.

Table 7.6: Control inputs to memory chip

Memory Enable	Read /Write	Memory Operation
0	X	*None*
1	0	*Write to selected word*
1	1	*Read to selected word*

The memory unit will then take the bits from the input data lines and store them in the word specified by the address lines. The steps that must be taken for the purpose of transferring a stored word out of memory are as follows:

1. Transfer the binary address of the desired word to the address lines.
2. Activate the read input.

The memory unit will then take the bits from the word that has been selected by the address and apply them to the output data lines. The content of the selected word does not change after reading.

Random access memory is used in huge quantities in computer applications as current day computing and processing technology requires large amounts of memory to enable them to handle the memory hungry applications used today. Many types of RAM including SDRAM with its DDR3, DDR4, and soon DDR5variants are used in huge quantities. RAM can be classified into two types, namely Static RAM and Dynamic RAM.

- Dynamic RAM (DRAM):

Dynamic RAM is a form of random-access memory. DRAM uses a capacitor to store each bit of data, and the level of charge on each capacitor determines whether that bit is a logical 1 or 0. However these capacitors do not hold their charge indefinitely, and therefore the data needs to be refreshed periodically. As a result of this dynamic refreshing it gains its name of being a dynamic RAM.

DRAM is the form of semiconductor memory that is often used in equipment including personal computers and workstations where it forms the main RAM for the computer. The semiconductor devices are normally available as integrated circuits for use in PCB assembly in the form of surface mount devices or less frequently now as leaded components. The DRAM is the main memory in computers and graphics cards. It is also used in many portable devices and video game consoles.

Advantages of DRAM

1. Data is stored in MOS capacity.
2. Only MOSFET is used for implementation.
3. It is Slow compared to Static RAM.
4. Dissipate less power.
5. The memory capacity of Dynamic RAM is more.
6. It can be used as Main memory.

Disadvantages of DRAM

1. Complex manufacturing process
2. Data requires refreshing
3. More complex external circuitry required (read and refresh periodically)
4. Volatile memory.
5. Relatively slow operational speed
6. Need to refresh the capacitor charge every once in two milliseconds.
7. In this type of RAM, data is stored on capacitors and requires a periodic refreshment.

- Static RAM (SRAM):

SRAM is stands for Static Random-Access Memory. Its symbol and circuit diagram is shown in Fig. 7.18.

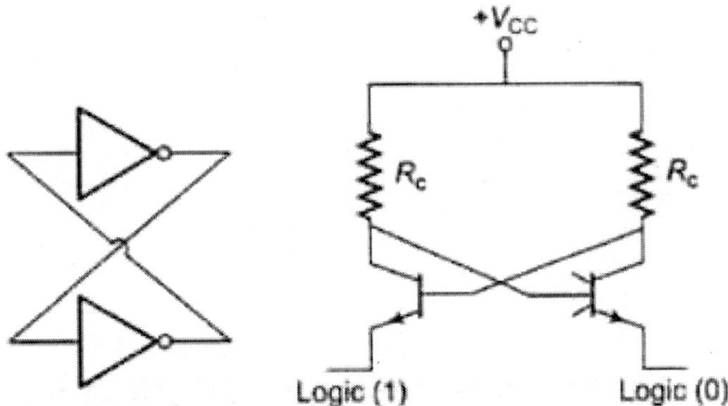

Fig. 7.18: Symbol and circuit diagram of SRAM

This form of semiconductor memory gains its name from the fact that, unlike DRAM, the data does not need to be refreshed dynamically. These semiconductor devices are able to support faster read and write times than DRAM (typically 10 ns against 60 ns for DRAM), and in addition its cycle time is much shorter because it does not need to pause between accesses. Due to the high speed of operation, SRAM is used for cache memory and as part of the digital-to-analog converter on video cards. It is also found in CDs, printers, routers, DVDs and digital cameras.

However, they consume more power, they are less dense and more expensive than DRAM. As a result of this SRAM is normally used for caches, while DRAM is used as the main semiconductor memory technology.

Advantage of SRAM

- Data is stored in flip flop like structure.
- It can be implemented with BJT or MOSFET.
- It is Faster
- No refreshing required.

Disadvantage of SRAM

- Dissipates more power.
- The memory capacity of Static RAM is less.
- It can be used as Cache memory.

Advantages of static RAM over Dynamic RAM:

- The access time of SRAM is less and thus these memories are faster memories.
- As SRAM is consists of flip-flops thus, refreshing is not required.
- Less number of memory cells are required in SRAM for a unit area.

- Synchronous DRAM (SDRAM)

The SDRAM "synchronizes" the speed of the memory along with the CPU clock speed. By doing so, the memory controller (which is a digital circuitry managing the flow of data from and to the main memory) is aware of the exact clock cycle by which the demanded data will be ready. Thus, the CPU's efficiency is improved, and it can do many more instructions at a given time. A typical SDRAM works at speeds of up to 133 MHz.

- Rambus DRAM (RDRAM)

The Rambus DRAM is named after the company that introduced it, Rambus. It was mainly used for video game devices and on-computer graphics cards, having transfer speeds running up to 1 GHz.

- Double Data Rate SDRAM (DDR SDRAM)

This memory has nearly double the bandwidth of a single data rate (SDR) SDRAM. It works on the principle of "double pumping" – this permits data to be transferred on both the rising & falling edges of the clock. This type of memory has been succeeded by the DDR2, DDR3, DDR4 and most recently, the DDR5 SDRAM.

- Magneto-resistive RAM (MRAM)

This is Magneto-resistive RAM, or Magnetic RAM. It is a non-volatile RAM memory technology that uses magnetic charges to store data instead of electric charges. Unlike technologies including DRAM, which require a constant flow of electricity to maintain the integrity of the data, MRAM retains data even when the power is removed. An additional advantage is that it only requires low power for active operation. As a result this technology could become a major player in the electronics industry now that production processes have been developed to enable it to be produced.

- Phase Change Random Access Memory (PCM):

This type of memory is known as Phase Change Random Access Memory, P-RAM or just Phase Change Memory, PCM. It is based around a phenomenon where a form of chalcogenide glass change is state or phase between an amorphous state (high resistance) and a polycrystalline state (low resistance). It is possible to detect the state of an individual cell and hence use this for data storage. Currently this type of memory has not been widely commercialized, but it is expected to be a competitor for flash memory.

Semiconductor memory technology is developing at a fast rate to meet the ever-growing needs of the electronics industry. Not only are the existing technologies themselves being developed, but considerable amounts of research are being invested in new types of semiconductor memory technology in terms of the memory technologies currently in use, SDRAM versions like DDR4 are being further developed to provide DDR5 which will offer significant performances improvements. In time, DDR5 will be developed to provide the next generation of SDRAM. Other forms of memory are seen around the home in the form of USB memory sticks, compact Flash, CF cards or SD memory cards for cameras and other applications as well as solid state hard drives for computers. The semiconductor devices are available in a wide range of formats to meet the differing PCB assembly and other needs.

- Content addressable memory (CAM)

The content addressable memory (CAM) is a special purpose random access memory device that can be accessed by searching for data content. For this purpose, it is addressed by associating the input data, referred to as key, simultaneously with all the stored words and produces output signals to indicate the match conditions between the key and the stored words. This operation is referred to as association or interrogation and this type of memory is also known as associative memory.

After identifying the locations whose contents match the key, read or write operations can be performed to these locations. The key to be used may either consist of the entire data word or only some specific bits of the data word, i.e. the other bits can be masked.

A CAM differs from the conventional memory organization in that the addressing of a location in the latter has no relation to the memory content. A CAM has the ability to search out or interrogate stored data on the basis of its contents and, therefore, can be a powerful asset in many applications. For example, consider a list containing the names of persons, their ages, professions, and nationalities stored in a CAM. If one is interested in finding out engineers in the list, the CAM is able to check every memory location simultaneously by using the coded form for engineer as the key. On the other hand, if it is required to find the engineers of Indian nationality, the key will consist of the combination of the codes corresponding to engineer and Indian nationality. All the memory locations with engineers of Indian nationality will be identified and the remaining data (name and age) can then be retrieved by using the read operation. To do the same search process with a conventional memory, each memory word is to be read out and compared with the key. This search is a serial process and hence time consuming. Thus, CAMs are better suited for information retrieval than the conventional memories.

CAMs are manufactured using MOS, CMOS or bipolar technologies. The most popular CAMs use ECL circuitry because of its high-speed operation

Operation of CAM: A CAM can perform three basic operations: read, write and associate. Fig. 7.19 shows a block diagram of a CAM. The CAM shown in Fig. 7.19 has the storage capacity is $M X N$ bits and is organized as M words of N bits each. It has M data input and N data output lines (one line for each bit of a word). The data input lines I_0 through I_{N-1} are used to input. Data to be written into the memory and for key word in case of associate operation. Data are read out of the CAM at the data output lines D_0 through D_{N-1}. The Y lines i.e., Y_0 to Y_{N-1} are bidirectional. During a read or write operation, these lines are used to select the storage location. There is one address input line for each word in the CAM. For example, Y_0 is the address line for memory location *0*, Y_1 for memory location *1* and so on. Notice that linear selection addressing is used in CAMs rather than coincident selection addressing.

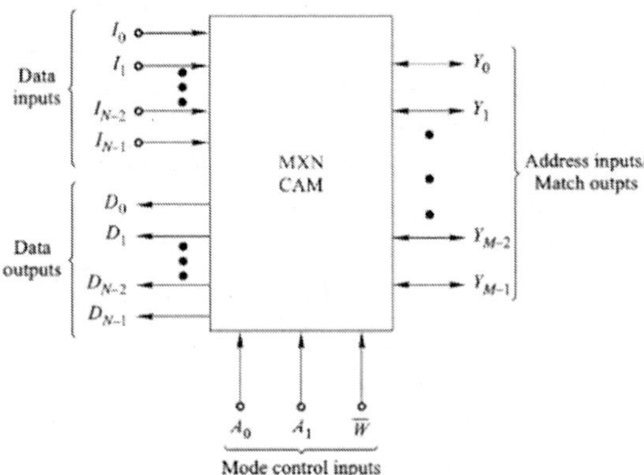

Fig. 7.19: Block diagram of a CAM

The Y lines serve as match output lines one for each memory location, when an association operation is performed. For example, if the keyword matches with the word stored in memory locations *5* and *8*, lines Y_5 and Y_8 will become HIGH to indicate the match condition.

The mode control inputs are used to select the required operation. The read and write operations are performed in a manner similar to that used for RAM. However, during the write operation, the input data also appear at the data outputs. The reading of the data is non-destructive. The internal architecture of a typical *8 X 2* ECL CAM is shown in Fig. 7.20.

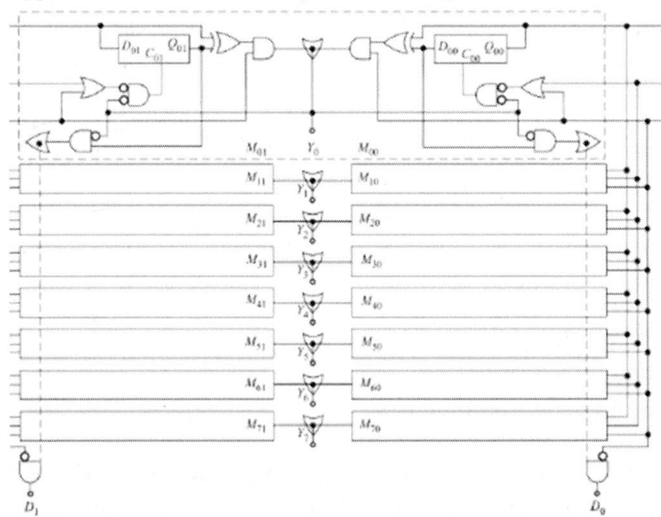

Fig. 7.20: Internal Architecture of an ECL 8 X 2 CAM

It has eight 2-bit storage locations M_0, through M_7, each consisting of two parts, one for the higher order bit (M_{01} through M_{71}) and the other, for the lower order bit (M_{00}, through M_{70}). The detailed circuitry of one of the locations M_0 is shown in the Fig. 7.20 and all other locations have similar circuitry.

Data to be written into the memory and for key word in case of associate operation. Data are read out of the CAM at the data output lines D_0 through D_{N-1}.The Y lines (Y_0 through Y_{M-1}) are bidirectional. During a read or write operation, these lines are used to select the storage location. There is one address input line for each word in the CAM.

Each memory location consists of two D-type FLIP-FLOPs, one for each bit, and some logic gates to control the function of the CAM for read, write, and associate operations. The outputs Y_0 through Y_7 and the data outputs D_1 and D_0 are produced using *wired-OR* connections as shown in Fig. 7.20. The operation of this CAM is summarized in Table 7.7.

Table 7.7: Operations of the 8 X 2 CAM

Operation	Control Inputs			Data Inputs		Data Inputs		Y_n	
	A_1	A_0	$\overline{W}$	I_1	I_0	D_1	D_0	Input /Output	Operation
Associate		1	X	1/0	1/0	0	0	Output	1-Mismatch 0-Match
Associate (Higher Bit Masked)		1	1	X	1/0	D_1	0	Output	1-Lower bit Mismatch 0-Lower Bit Match
Associate (Lower Bit Masked)		0	1	1/0	X	0	D_0	Output	1-Higher bit Mismatch)- Higher Bit Match
Read		0	1	X	X	D_1	D_0	Input	0- Selected Address
Write		0	0	1/0	1/0	I_1	I_0	Input	0- Selected Address
Associate and write at Match Address		1	0	1/0	1/0	I_1	0	Output	1-Lower bit Mismatch 0-Lower Bit Match
		0	0	1/0	1/0	0	I_0	Output	1-Higher bit Mismatch)- Higher Bit Match

The word length and/or word size can be expanded by suitably connecting the available CAM chips in a manner similar to the one used for RAMs and ROMs expansion. Mainly CAM has been utilized in network routing and switching devices.

Advantages of RAM:

1. Fast operating speed (< 150 nsec),
2. Low power dissipation (< 1mW),
3. Economy,
4. Compatibility,
5. Non-destructive read-out.

Read-only Memories (ROM):

Read-only Memories (ROM), as the name suggest, is meant only for reading the information from it. This does not mean that information is not written into it, because unless some information is stored into it, there cannot be anything to read from. Actually, the process of entering information into this type of memory is much more complicated than for RAM and is done outside the system where it is used. Therefore, it is called as read-only memory. It is used to store information which is fixed, such as tables for various functions, fixed data and instructions. The ROMs are also organized so that every memory Location required equal time for reading the data already stored in that Location. It consists of n input lines and m output lines. Each bit combination of the input variables is called an address. Each bit combination that comes out of the output lines is called a word.

As shown in Fig. 7.21, in a ROM the number of bits per word is equal to the number of output lines, m. An address is essentially a binary number that denotes one of the min terms of n variables. The number of distinct addresses possible with n input variables is 2^n. An output word can be selected by a unique address, and since there are 2^n distinct addresses in a ROM, there are 2^n distinct words that are said to be stored in the unit. The word available on the output lines at any given time depends on the address value applied to the input lines. A ROM is characterized by the number of words 2^n and the number of bits per word m.

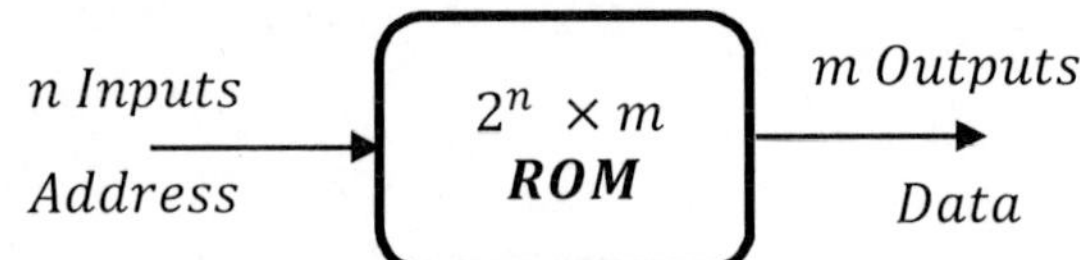

Fig. 7.21: Block diagram of a ROM

A read-only memory (ROM) includes both the decoder and the OR gates within a single IC package as shown in Fig. 7.22. The connections between the outputs of the decoder and the inputs of the OR gates can be specified for each particular configuration.

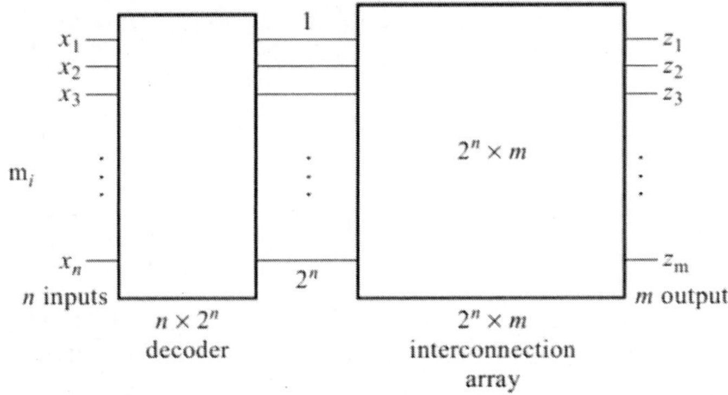

Fig. 7.22: Basic structure of ROM

A ROM is essentially a memory (or storage) device in which permanent binary information is stored. The binary information must be specified by the designer and is then embedded in the unit to form the required interconnection pattern. ROMs come with special internal electronic fuses that can be "programmed" for a specific configuration. Once the pattern is established, it stays within the unit even when power is turned off and on again. In general, a $2^n \times m$ ROM will have an internal n *to* 2^n decoder and m OR gates. Each OR gate has 2^n inputs, which are connected to each of the outputs of the decoder. For example, consider a 32 x 5 ROM as shown in Fig. 7.23

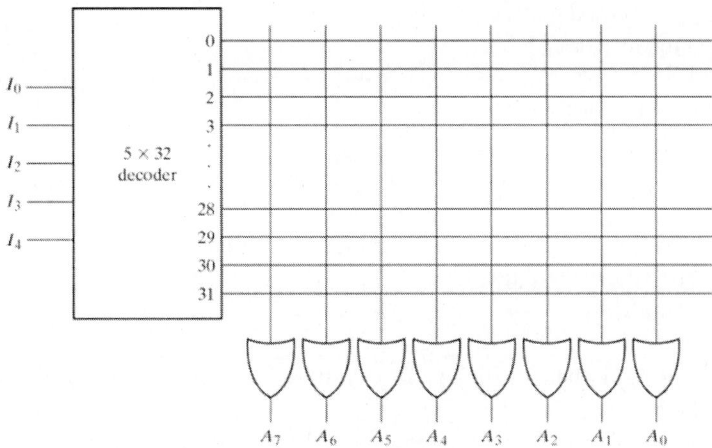

Fig. 7.23: 32 x 8 ROM

The unit consists of 32 words of 8 bits each. This means that there are eight output lines and that there are 32, distinct words stored in the unit, each of which may be applied to the output lines. The particular word selected that is presently available on the output lines is determined from the five input lines. There are only five inputs in a *32 X 8* ROM because $2^5 = 32$, and with five variables, we can specify 32 addresses or min terms. For each address input, there is a unique selected word. Thus, if the input address is 00000, word number 0 is selected and it appears on the output lines. If the input address is 11111, word number 31 is selected and applied to the output lines. In between, there are 30 other addresses that can select the other 30 words.

Applications of ROM:

- Pre-programmed toy circuit,
- Pre-programmed robot circuit,
- Standard look up table,
- Arithmetic function table generator,
- User defined code generator,
- Character generator,
- Printable or displayable fonts table

The ROM chips are also crucial for the basic input/output system (BIOS), reading-writing to peripheral devices, data management, etc. and are programmed before they are included in the computer system. It is also known as masked memory. ROM is read-only and cannot have the data altered easily; thus, it is mainly used for firmware purposes.

A firmware is a software program or instruction sets that are embedded into hardware. It has instructions on how a device works and cooperates with the other hardware components linked to the system. Since this information need not be continuously meddled with, the amount of memory a ROM holds is small. A ROM stores data that are used repeatedly in system applications, such as tables, conversions, or programmed instructions for system initialization and operation. ROMs retain stored data when the power is OFF and are therefore non-volatile memories.

The ROM (Read-Only Memory) is a type of non-volatile memory that aids in storing and retrieving information on a computer. As the name goes, this type of memory allows memory to be read-only. Since it is non-volatile, it does not require a constant power supply to retain the data stored in it. These are essential instructions for a system, such as loading the OS into the RAM, running hardware diagnostics, etc.

Advantages of ROM:

- Ease and speed of design,
- Faster than MSI devices (PLD and FPGA)
- The program that generates the ROM contents can easily be structured to handle unusual or undefined cases,
- A ROM's function is easily modified just by changing the stored pattern, usually without changing any external connections,
- More economical.

Disadvantages of ROM:

- For functions more than 20 inputs, a ROM based circuit is impractical because of the limit on ROM sizes that are available.
- For simple to moderately complex functions, ROM based circuit may be costly: consume more power; run slower.

The various types of read-only memories are further sub-classified on the basis of technique employed for storing information into memory (referred to as programming) or their eras ability properties these are:

1. Masked ROM (ROM)
2. Programmed ROM (PROM)
3. Erasable PROM (EPROM)
4. Electrically Erasable PROM (EEPROM)

- Masked ROM:

it is programmed at the time of manufacturing, as the Last process of fabrication, according to the information specified by user. It is referred to as custom programmed or mask programmed. The data stored cannot be changed after fabrication. Since this process is quite costly, therefore this type of ROM is suitable only for bulk requirement in order of millions of chips. The mask ROM is usually referred to simply as a ROM.

It is permanently programmed during the manufacturing process to provide widely used standard functions, such as popular conversions, or to provide user-specified functions. Once the memory is programmed, it cannot be changed. Most IC ROMs utilize the presence or absence of a transistor connection at a row/column junction to represent a 1 or a 0. The presence of a connection from a at row line to the gate of a transistor represents a 1 at that location because when the row line is taken HIGH; all transistors with a gate connection to that row line turn on and connect the HIGH (1) to the associated column lines. At row/column junctions where there are no gate connections, the column lines remain LOW (0) when the row is addressed.

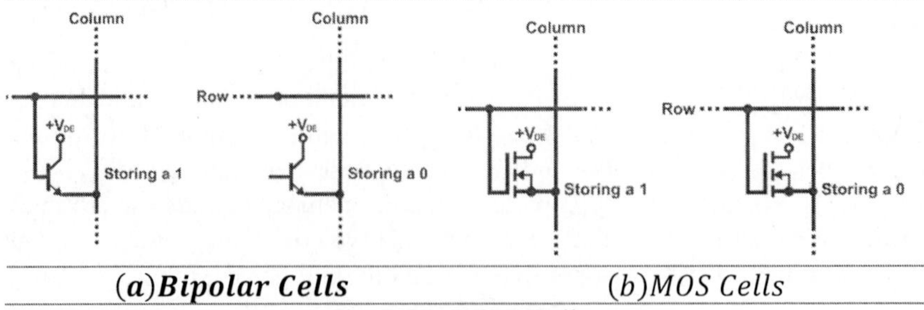

(a) ***Bipolar Cells*** (b) *MOS Cells*

Fig. 7.24: ROM Cells

- Programmable Read-Only Memory (PROM):

This type of ROM is programmed by the user using a special circuit a PROM programmer. A PROM can be programmed only once after which its contents are permanently fixed as in a ROM. This type of ROM is suitable for storage of data which is of permanent nature. The chip is avilable without any data stored from the vendor. The PROM (Programmable Read-only memory), comes from the manufacturer un-programmed and are custom programmed in the field to meet the user's needs.

A PROM uses some type of fusing process to store bits, in which a memory link is burned open or left intact to represent a o or a 1. The fusing process is irreversible; once a PROM is programmed, it cannot be changed. The fusible links are manufactured into the PROM between the source of each cell's transistor and its column line. In the programming process, a sufficient current is injected through the fusible link to bum it open to create a stored O. The link is left intact for a stored 1. All drains are commonly connected to VDD.

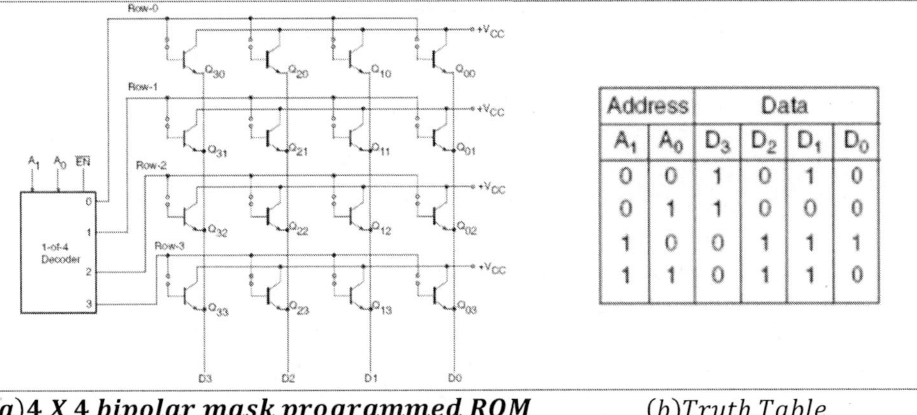

Address		Data			
A_1	A_0	D_3	D_2	D_1	D_0
0	0	1	0	1	0
0	1	1	0	0	0
1	0	0	1	1	1
1	1	0	1	1	0

(a)4 X 4 bipolar mask programmed ROM *(b)Truth Table*

Fig. 7.25: PROM array with fusible links and its Truth Table

Three basic fuse technologies used in PROMs are metal links, silicon links, and *pn* junctions. A brief description of each of these follows.

1. Metal links are made of a material such as nichrome. Each bit in the memory array is represented by a separate link. During programming, the link is either "blown" open or left intact. This is done basically by first addressing a given cell and then forcing a sufficient amount of current through the link to cause it to open. When the fuse is intact, the memory cell is configured as a logic 1 and when fuse is blown (open circuit) the memory cell is logic o.
2. Silicon links are formed by narrow, notched strips of polycrystalline silicon. Programming of these fuses requires melting of the links by passing a sufficient amount of current through them. This amount of current causes a high temperature at the fuse location that oxidizes the silicon and forms insulation around the now-open link.
3. Shorted junction, or avalanche-induced migration, technology consists basically of two pn junctions arranged back-to-back. During programming, one of the diode junctions is avalanched, and the resulting voltage and heat cause aluminium ions to migrate and short the junction. The remaining junction is then used as a forward- biased diode to represent a data bit.

- Erasable and Programmable ROM:

This type of ROMs is reprogrammable i.e. it can be programmed again and again. It is referred to as erasable and programmable ROM. There are two techniques used for erasing; in one technique the chip is exposed to ultraviolet radiation and in the other technique the contents are altered electrically.

The erasable programmable ROM using ultraviolet radiation for erasing is known as EPROM, whereas the device using electrical voltage for erasing is known as Electrically Alterable ROM (EAROM).

An EPROM is an erasable PROM. Unlike an ordinary PROM, an EPROM can be reprogrammed if an existing program in the memory array is erased first. An EPROM uses an NMOSFET array with an isolated-gate structure. The isolated transistor gate has no electrical connections and can store an electrical charge for indefinite periods of time. The data bits in this type of array are represented by the presence or absence of a stored gate charge. Erasure of a data bit is a process that removes the gate charge. This basic type of erasable PROMs is the UltraViolet Erasable PROM (UVEPROM). UV EPROM devices has the transparent quartz lid on the package, as shown in Figure below.

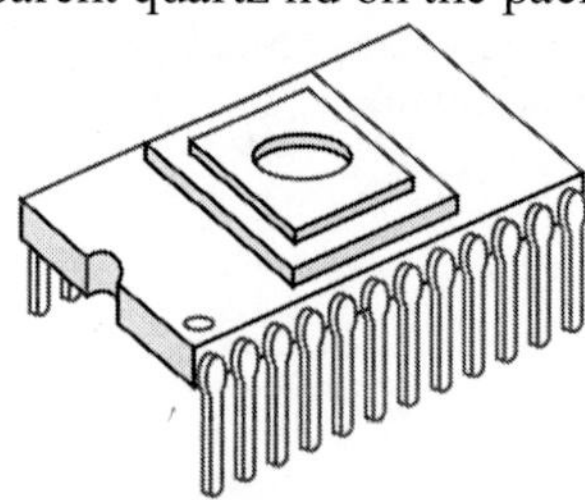

Fig. 7.26: Ultraviolet Erasable PROM

The isolated gate in the FET of an ultraviolet EPROM is "floating" within an oxide insulating material. The programming process causes electrons to be removed from the floating gate. Erasure is done by exposure of the memory array chip to high-intensity ultraviolet radiation through the quartz window on top of the package. The positive charge stored on the gate is neutralized after several minutes to an hour of exposure time. In EPROM's, it is not possible to erase selective information, when erased the entire information is lost. The chip can be reprogrammed. It is ideally suited for product development, college laboratories, etc.

During programming, address and data are applied to address and data pins of the EPROM. The program pulse is applied to the program input of the EPROM. The program pulse duration is around 50msec and its amplitude depends on EPROM IC. It is typically 11.5V to 25V.In EPROM, it is possible to program any location at any time- either individually, sequentially or at random.

- <u>EEPROM (Electrically Erasable PROM):</u>

The EEPROM (Electrically Erasable PROM), also uses MOS circuitry. Data is stored as charge or no charge on an insulating layer, which is made very thin (<200A). Therefore, a voltage as low as 20- 25V can be used to move charges across the thin barrier in either direction for programming or erasing ROM. An electrically erasable PROM can be both erased and programmed with electrical pulses. Since it can be both electrically written into and electrically erased, the EEPROM can be rapidly programmed and erased in-circuit for reprogramming. It allows selective erasing at the register level rather than erasing all the information, since the information can be changed by using electrical signals.

At it has chip erase mode by which the entire chip can be erased in 10 msec. Hence EEPROM's are most expensive. In EEPROM data can be written to it and it can be erased using an electrical voltage, like other types of PROM, EEPROM retains the contents of the memory even when the power is turned off. Also, like other types of ROM, EEPROM is not as fast as RAM. EEPROM memory cells are made from floating-gate MOSFETs (known as FGMOS). It is also called as

Flash Memory. The content can be re-programmed by applying suitable voltages to the EEPROM pins.

- Flash Memory:

Flash memory may be considered as a development of EEPROM technology. Data can be written to it and it can be erased, although only in blocks, but data can be read on an individual cell basis. To erase and re-program areas of the chip, programming voltages at levels that are available within electronic equivalent are used. It is also non-volatile, and this makes it particularly useful. As a result flash memory is widely used in many applications including memory cards for digital cameras, mobile phones, computer memory sticks and many other applications.

Flash memory stores data can be in an array of memory cells. The memory cells are made from floating-gate MOSFET (known as FGMOS). Theses FG MOSFETs (or FGMOS in short) have the ability to store an electrically charge for extended periods of time (2 to 10 years) even without a connecting to a power supply. The Flash Memories are very important data storage devices for mobile applications.

Disadvantages of Flash Memory:

1. Higher cost per bit than hard derives
2. Slower than other forms of memory
3. Limited number of write/erase cycles
4. Data must be erased before new data can be written
5. Data typically erased and written in blocks.

- Content Addressable Memories (CAM):

it is a special purpose random access memory which performs association operation in addition to read/write operations.

7.5.2 Secondary Memory

Examples of secondary memory devices include hard disk drives (HDD), solid-state drives (SSD), USB flash drives, CDs, DVDs, and magnetic tape. These devices have much larger storage capacity than RAM and are typically used to store large amounts of data that are not frequently accessed by the computer's processor. Further Secondary Memory may be classified as

- Serial Access
- Semi random Access

In the *serial access memories*, the memory locations are organized in sequence i.e. one after other, Such as a magnetic tape audio/ video cassette. Writing into or reading from it is in sequential manner. Therefore, the time required for accessing a memory Location for writing or reading is different for different Locations. There are two types of semiconductor sequential memories. These are

- Shift registers and
- Charge coupled devices (CCD)

Shift registers can be either static or dynamic. In a static memory, the contents of the memory Location do not change with time as long as power is on. On the other hand, in dynamic memories the information is stored in MOS capacitors which changes with time and therefore, it has to be refreshed at regular intervals. The dynamic memories are simpler, less expensive, require less power, have high packing density in comparison to static memories and are therefore widely used in digital systems. However, the cost of the additional circuitry for refreshing may increase the system cost. The charge coupled devices are implemented using MOS technology. These devices have high density and Low cost.

- **Sequential access Memory (SAM):**

Sequential access means the system must search the storage device from the beginning of the memory address until it finds the required piece of data. Memory device which supports such access is called a Sequential Access Memory or Serial Access Memory. Magnetic tape is an example of serial access memory.

In computing, Sequential Access Memory (SAM) is a class of data storage devices that read their data in sequence. This is in contrast to random access memory (RAM) where data can be accessed in any order. Sequential access devices are usually a form of magnetic memory. While sequential access memory is read in Sequence, accesses can still be made to arbitrary locations by "seeking" to the requested location. This operation, however, is often relatively inefficient (See seek time, rotational latency).

Magnetic sequential access memory is typically used for secondary storage in general-purpose computers due to their higher density at lower cost compared to RAM, as well as resistance to wear and non-volatility. Examples of SAM devices still in use include hard disks, CD-ROMs and magnetic tapes. Historically, drum memory has also been used.

In a sequential memory, the words are stored-in and read-out in sequence. For example, if the Pth location is being accessed at a particular time, then the (p+q)th location is not accessible , unless all the intermediate locations have been accessed one-by-one in sequence. In other words, the access to a particular location is obtained by waiting until the desired location is reached. This means that the access time is not same for all the locations. Shift registers are examples of sequential memories. Another kind of shift registers gaining popularity in recent years are charge-coupled devices (CCDs) and bubble memories.

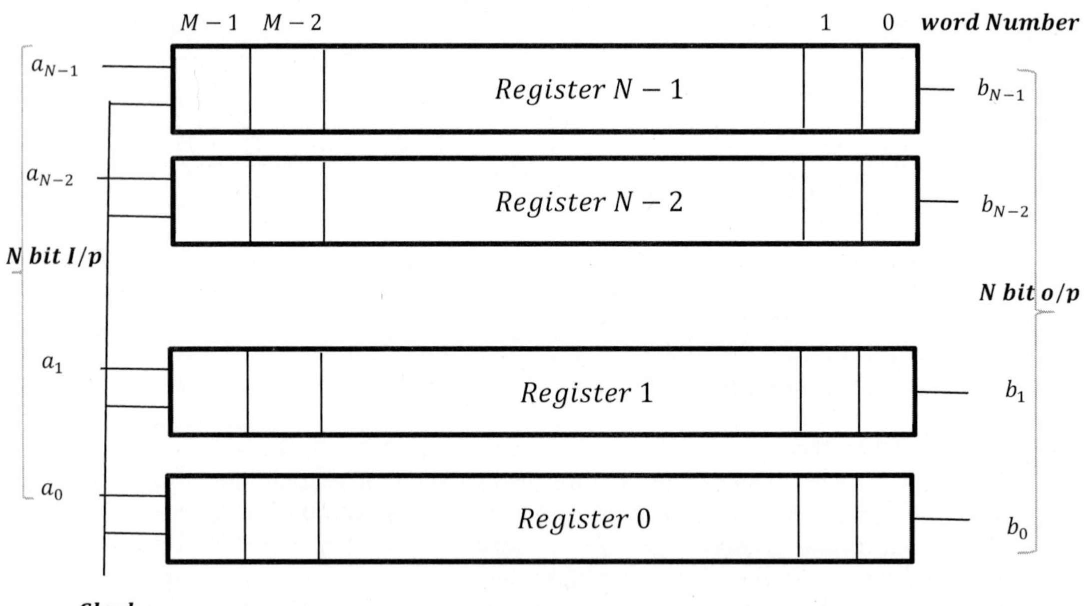

Fig. 7.27: M X N Sequential memory

A sequential memory of size $M X N$ is shown in Fig. 7.27 it requires N shift registers, each of M stages. Each register holds one of the N bits of each of the M words with each clock cycle, the bits in each register will advance towards the right by one position and stored words will appear sequentially at the outputs of the registers. This configuration is referred to as first-in-first-out (FIFO) sequential-memory system. Since the word which is entered first will be the one read out first. In contrast, if the outputs are taken from the $(M-1)^{th}$ stage instead of 0th stage, the resulting configuration is called as last-in-first-out (LIFO). The shift registers used in this configuration must have a provision of shifting the bits in either direction. In the FIFO sequential memory system, the word being read may be transferred back to the left-most register position. This is known as circulating shift register.

- **Charge coupled device (CCD):**

It is a device used for the movement of electrical charge within it for the charge manipulation, which is done by changing the signals through stages within the device one at a time. It can be treated as CCD sensor, which is used in the digital and video cameras for taking images and recording videos through photoelectric effect. It is used for converting the captured light into digital data, which is recorded by the camera. It can be defined as a light-sensitive integrated circuit imprinted on a silicon surface to form light-sensitive elements called pixels, and each pixel is converted into an electrical charge. It is an array of MOS capacitors operating as a dynamic shift register. CCDs are simple, versatile, and low-cost devices and can be used wherever a serially accessed memory is required. There are different CCDs such as electron multiplying CCDs, intensified CCD, frame-transfer CCD and buried-channel CCD. A CCD can be simply defined as Charge Transfer Device. The inventors of the CCD, Smith and Boyle also discovered a CCD with greatly enriched performance than a general Surface Channel CCD and other CCDs; it is known as Buried channel CCD and is majorly used for practical applications.

The operation of CCDs involves the following steps as shown in 7.28.:

1. Conversion of digital input signal into charge,
2. Transfer of charge through various stages in sequential manner, and
3. Conversion of charge at the output into digital signal.

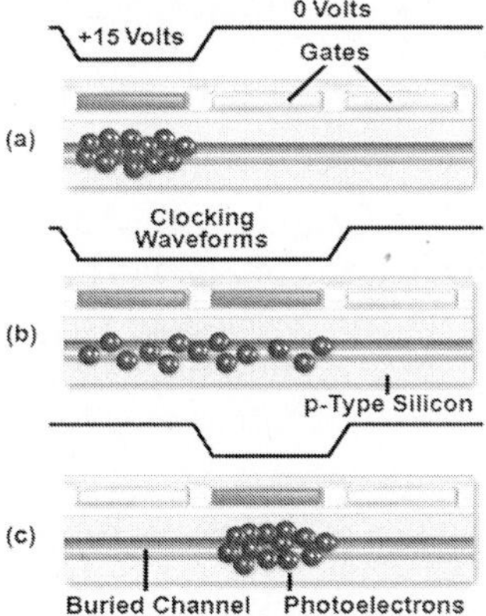

Fig. 7.28: Three Phase CCD Clocking Systems

During each charge transfer step, a small amount of charge is lost. Also, due to thermal effects, undesirable charge may be generated which is known as dark current. To overcome these defects, the charge is recirculated around the shift register for refreshing.

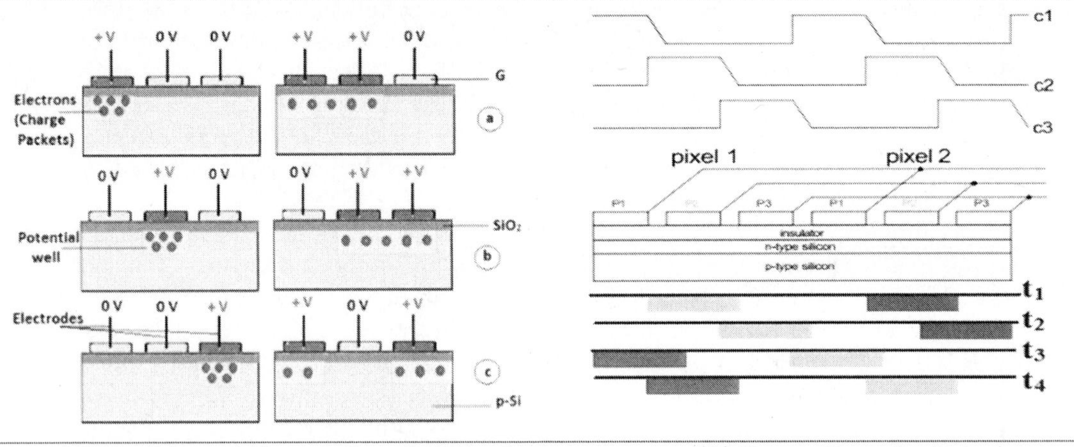

(A)**Working of Charge Coupled Device** (B)Charge Transfer Process

Fig. 7.29: Charge Coupled Devices (CCD) (A) Working (B) Charge Transfer Process

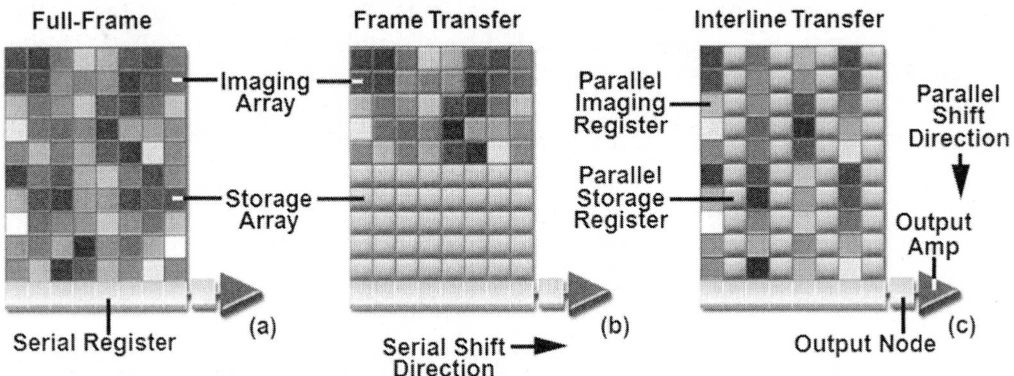

Fig. 7.30: Charge coupled device principle

The silicon epitaxial layer acting as a photoactive region and a shift- register-transmission region are used for capturing images using a CCD. Through the lens image is projected onto the photo active region consisting of capacitor array. Thus, the electric charge proportional to the light intensity of the image pixel colour in the colour spectrum at that location is accumulated at each capacitor. If the image gets detected by this capacitor array, then the electrical charge accumulated in each capacitor is transferred to its neighbour capacitor by performing as a shift register controlled by the control circuit.

In Fig. 7.29 from a, b and c, the transfer of charge packets is shown according to the voltage applied to the gate terminals. At last, in the array electrical charge of last capacitor is transferred into the charge amplifier in which the electric charge is converted into a voltage. Thus, from the continuous operation of these tasks, entire charges of the capacitor array in the semiconductor are converted into a sequence of voltages.

The charge packets can be moved from cell to cell by using many schemes in Bucket Brigade style. There are various techniques such as two phase, three phase, four phases, and so on. Every cell consists of *n-wires* passing through it in n-phase scheme. The height of the potential wells is controlled by using each wire connected to transfer clock. Charge packets can be pushed and pulled along the line of the CCD by varying the height of the potential well.

Consider a three-phase charge transfer, in the above figure, the three clocks (C1, C2 and C3) which are identical in shape but in different phases are shown. If gate B goes high and gate A goes low, then the charge will move from space A to space B.

The pixels can be transferred through the parallel vertical registers or vertical CCD (V-CCD) and parallel horizontal registers or horizontal CCD (H-CCD). The charge or image can be transferred using different scanning architectures such as full frame readout, frame transfer and interline transfer as shown in Fig. 7.30. The charge coupled device principle can be easily understandable with the following transfer schemes:

- *Full-Frame Readout:* It is the simplest scanning architecture which requires a shutter in a number of applications to cut off the light input and to avoid smearing during the passage of charges through parallel-vertical registers or vertical CCD and parallel-horizontal registers or horizontal CCD and then transferred to output in serial.
- *Frame Transfer:* By using the bucket brigade process the image can be transferred from image array to opaque frame storage array. As it does not use any serial register, it is a fast process compared to other processes.
- *Interline transfer:* Each pixel consists of a photodiode and opaque charge storage cell. As shown in the figure, the image charge is first transferred from light sensitive PD to the opaque V-CCD. This transfer, as the image is hidden, in one transfer cycle produces a minimum image smear; hence, the fastest optical shuttering can be achieved.

The final process on the CCD is the reading of each pixel so that the size of the associated charge cloud can be measured. At the end of the readout register is an amplifier which measure the value of each charge cloud and converts it into a voltage, a typical conversion factor being around 5-10μV per electron with "typical" full well values being about 100,000 electrons or so.

CCD characteristics:

1. Quantum Efficiency: The percentage of photons that are actually detected is known as the Quantum Efficiency (QE). For example, the human eye only has a QE of about 20%, photographic film has a QE of around 10%, and the best CCDs can achieve a QE of over 80%. Quantum efficiency will vary with wavelength.
2. Wavelength range: CCDs can have a wide wavelength range ranging from about 400nm (blue) to about 1050nm (Infra-red) with a peak sensitivity at around 700nm. However, using a process known as back thinning, it is possible to extend the wavelength range of a CCD down into shorter wavelengths such as the Extreme Ultraviolet and X-ray.
3. Dynamic Range: The difference between a brightest possible source and the faintest possible source that the detector can accurately see in the same image is known as the dynamic range. When light falls onto a CCD the photons are converted into electrons. Consequently, the dynamic range of a CCD is usually discussed in terms of the minimum and maximum number of electrons that can be imaged. For a typical scientific CCD this may occur at around 150,000 electrons or so. The minimum signal that can around 2-4 electrons for each pixel.

4. Linearity: if it detects 10000 photons, it will convert these to 10000 electrons. In such a situation, we say that the detector has a linear response. Such a response is obviously very useful to determine intensity of an image.
5. Noise: One of the most important aspects of CCD performance is its noise response. There are two types of the noise of a CCD, these are:
    a. Dark current – It is thermally generated noise. At room temperature, the noise performance of a CCD can be as much as thousands of electrons per pixel per second. Consequently, the full well capacity of each pixel will be reached in a few seconds and the CCD will be saturated. Dark current can be massively reduced by cooling. For example, the noise performance of the CCD could be reduced from thousands of electrons at room temperature to only tens of electrons per pixel per second at -40 degrees C. By cooling down to temperatures below about -70 degrees C dark current can be virtually eliminated (substantially below one electron per pixel per second). A second way of reducing noise is to slightly alter the CCD processing technique to produce a Multi-Pinned -Phase (MPP) CCD. This technique can reduce the dark current to very low levels (a few hundred electrons per pixel per second at room temperature).
    b. Readout noise - The readout noise is the conversion of the electrons in each pixel to a voltage on the CCD output node (a typical value would be around 4μV per electron). The magnitude of this noise depends on the size of the output node. Noise values of 2-3 electrons rms (root mean square) are now typical for many CCDs but some companies have.

7.6 Classification based on physical characteristics:

Memories can be classified according to their physical characteristics, as

- Erasable and Non-erasable
- Volatile and Non-volatile

Erasable or non-erasable: A memory in which the information stored can be erased and new information stored is called erasable memory. On the other hand, the information stored in the non- erasable memory cannot be erased, for example ROM is non-erasable.

Volatile or non- Volatile: If the information stored in a memory is lost when electrical power is switched off, the memory is referred to as a volatile memory. For example, the RAM is a volatile memory. On the other hand in a non-volatile memory, the information once stored remains in fact until changed. All the type of ROMs are non-volatile memories.

7.7 Classification based on Mode of Access:

Mode of access refers to the manner in which a memory Location is accessed for reading or writing. In case of ROMS only reading is possible. There are two modes of access. These are:

- Sequential access and
- Random access.

Sequential memories are referred to as sequentially accessed memories, whereas RAM, ROM, and CAM are random access memories. In *random-access* memories any memory location requires equal time for accessing (access time), whereas the access time is different for different Locations in case of sequentially accessed memory.

7.8 Classification based on Fabrication Technology:

Memories can be classified on the basis of the fabrication technology used. The two broad categories of memories based on fabrication technology used are:

- Bipolar and
- Unipolar (MOS)

Static RAM, ROM and PROM can be fabricated using either bipolar technology (TTL, ECL etc.) or MOS technology, whereas dynamic RAM, EPROM and EAROM can be fabricated only using unipolar devices (MOSFETs).

7.9 Word Length Expansion

To increase the word length of a memory, the number of bits in the data bus must be increased. An 8-bit word length can be achieved by using two memories each with 4-bit words as illustrated in Figure below. The 16-bit address bus is commonly connected to both memories so that the combination memory still has the same number of addresses (2^{16} = 65,536) as each individual memory. The 4-bit data buses from the two memories are combined to form an 8-bit data bus. Now when an address is selected, eight bits are produced on the data bus-four from each ROM.

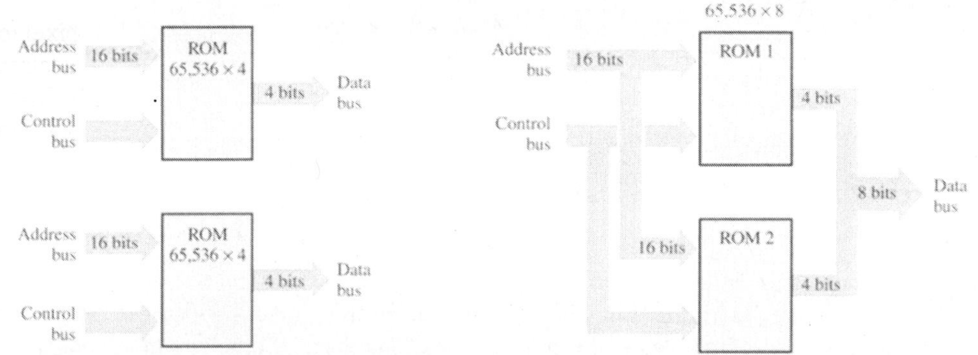

(a) Two separate 65, 536 x 4 ROMs (b) One 65,536 x 8 ROM by Using65,536 x 4 ROMs

Fig. 7.31: Illustrate of word-length expansion

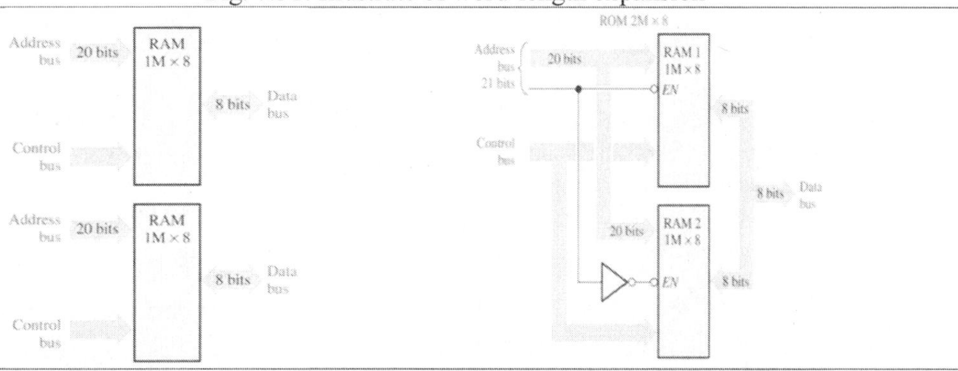

(a) Individual memories each store 1M X 8 require 20 bit address bus

(a) Memories expanded to form a 2z × 8 RAM requires 21 bit address bus

Fig. 7.32: Word capacity Expansion

Increase the word capacity, the number of addresses is increased. Two 1M x 8 RAMs are expanded to form a 2M x 8 memory is shown in Fig. 7.31. Each individual memory has 20 address bits to select its 1,048,576 addresses. The expanded memory has 2,097,152 addresses and therefore requires 21 address bits, as shown in part (b). The twenty-first address bit is used to enable the at appropriate memory chip. The data bus for the expanded memory remains eight bits wide.

Example7.8: Use 512k x 4 RAMs to implement a 1M x 4 memory.

Solution: The expanded addressing is achieved by connecting the chip enable (E0') input to the twentieth address bit (A19). Input (E1') is used as an enable input common to both memories. When the twentieth address bit (A19) is LOW, RAM 1 is selected (RAM 2 is disabled), and the nineteen lower-order address bits (A0 —A18) access each of the addresses in RAM 1. When the twentieth address bit (A19) is HIGH, RAM 2 is enabled by a LOW on the inverter output (RAM 1 is disabled), and the nineteen lower-order address bits A0 —A18) access each of the RAM 2 addresses.

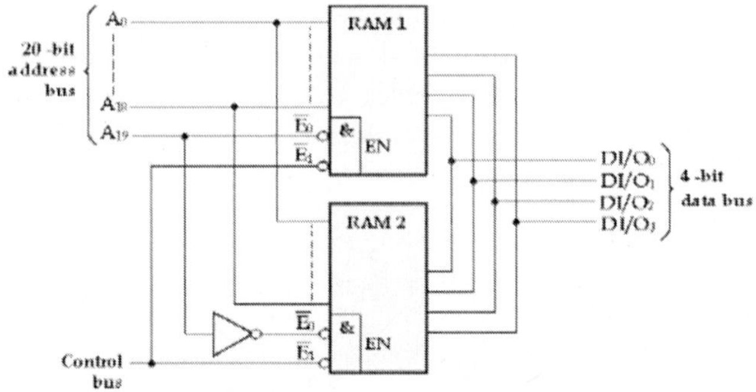

Fig. 7.33: 1M X 4 RAM using 512K X 4 RAM

Applications of Semiconductor Memories:

Semiconductor memories are widely used in various digital electronic devices due to their fast access time, high storage density, low power consumption, and small size. Some common applications of semiconductor memories are:

1. Personal computers and laptops: Random Access Memory (RAM) and Read-Only Memory (ROM) are used to store data and program code in personal computers and laptops.
2. Mobile phones and tablets: Flash memory is commonly used in mobile phones and tablets for storage of operating system, applications, and user data.
3. Digital cameras: Flash memory is used to store pictures and videos in digital cameras.
4. Solid State Drives (SSDs): SSDs use NAND flash memory to store data in place of traditional hard disk drives (HDDs) in computers.
5. Gaming consoles: RAM and flash memory are used in gaming consoles for game storage and quick access to game data.
6. Automotive electronics: EEPROM (Electrically Erasable Programmable Read-Only Memory) and flash memory are used in automotive electronics for storing data such as engine control parameters, fault codes, and user preferences.
7. Consumer electronics: Semiconductor memories are used in various consumer electronics such as televisions, digital music players, and GPS devices.
8. Medical devices: EEPROM and flash memory are used in medical devices for storing data such as patient information, diagnostic results, and treatment history.

9. Industrial automation: Semiconductor memories are used in industrial automation systems for storing program code, data, and configuration settings.

These are just a few examples of the many applications of semiconductor memories.

Unit summary

Semiconductor memories play a critical role in digital systems as they provide a fast, reliable, and efficient means of storing digital information. Here are some ways in which semiconductor memories are used in digital systems:

1. Random Access Memory (RAM): RAM is used as the main working memory in digital systems, storing data temporarily during program execution. RAM is volatile memory, meaning that its contents are lost when the power is turned off.
2. Read-Only Memory (ROM): ROM is used to store program code, configuration data, and other data that needs to be retained even when the power is turned off. ROM is non-volatile memory, meaning that its contents remain intact even when power is removed.
3. Flash Memory: Flash memory is a type of non-volatile memory that can be programmed and erased in blocks. It is commonly used in digital systems for mass storage, such as in Solid State Drives (SSDs) and memory cards.
4. Electrically Erasable Programmable Read-Only Memory (EEPROM): EEPROM is a type of non-volatile memory that can be reprogrammed electrically. It is commonly used in digital systems for storing configuration data and other types of non-volatile data.
5. Cache Memory: Cache memory is a small amount of high-speed memory that is used to temporarily store frequently accessed data, reducing the time required to access it from slower memory such as RAM.
6. Register Files: Register files are small arrays of memory used to store data during program execution, allowing for fast access and manipulation of data by the CPU.

Overall, semiconductor memories are an essential component of digital systems, providing fast and efficient access to data and program code.

Solved Problems

Q.1 What is the difference between RAM and ROM?

Solution:

Factors	RAM	ROM
Nature of working	Volatile memory – needs power supply	Non-volatile memory – doesn't need power
Speed	Fast	Not as fast as RAM
Storage Capacity	High Capacity, ranging from 1 to 256 GB	Low Capacity of about 4-8MB
Space it occupies	RAM data takes a lot of space	RAM takes up less space
Cost	More expensive	Affordable
Data	Can be altered anytime	Cannot be modified/limited changes permitted but not done easily
Application	Used to store program codes which CPU needs immediately	Used to store booting instructions or firmware coding

Q.2: What is the difference between SRAM and DRAM?
Solution:

Factors	SRAM	DRAM
Number of transistors	6 Transistors	1 Transistor
Charge Leakage	Not evident	A lot of discharge happens, thus there is a refresh circuitry
Speed	Quite fast	Comparatively slower
Power Consumption	Low	High
Space it occupies	Less space	More space
Cost	Expensive	Cheap
Density	Less dense	Denser
Application	Cache Memory	Main memory

Q.3: What is the difference between EPROM and EEPROM?
Solution:

Factors	EPROM	EEPROM
The medium of erasure of data	UV Light is used to erase data	An electrical signal is used to delete data
Updating is done by	Need to eject the EPROM chip to update the data	No ejection is required for reading or erasing the data
Timeline of technology	This is an older technology.	This is a newer technology.
Design on the case	Has a transparent quartz crystal window on the top	The memory is entirely enclosed in an opaque case.
Preceded by	EPROM is the updated version of the PROM.	EEPROM is the updated version of the EPROM.
Time for erasing	Erasing the contents takes about 15-20 minutes.	Erasing the contents takes about 5ms only.
Programming Technique	A hot electron injection technique is used for reprogramming the EPROM.	The tunnel effect is used for programming the EEPROM.

Q.4: Describe the three features of the RAM and ROM.
Solution: Feature of RAM

- It is volatile.
- High speed of operation.
- All the unsaved data lost if the sudden power cut off.
- There are basic two types of RAM: Static RAM and Dynamic RAM.

Features of ROM

- It is non – volatile.
- It cannot be used as read / write memory.
- It stores data permanently.
- There are following types of ROM: PROM, EPROM, EEPROM and MROM

Q.5: Describe the three features of PROM and EPROM.
Solution: Features of PROM

- The programming in the PROM is done by the programmer.

- Once the programming is done in the PROM, it cannot be altered.

Features of EPROM

- The contents of the EPROM can be erased and also new contents can be stored in it.
- It is non – volatile.
- The contents of the EPROM can be erased either by certain electrical voltages or ultraviolet light exposing on them
- Same device is used repeatedly

Q.6: What is meaning of 32 × 8 ROM?

Solution: 32 × 8 ROM represent a ROM that can store 32 bytes of data and it contains 32 possible addresses. As there are 8 data outputs, each word contains 8 bits

Q.7: Define Access time of memory? What is its significance?

Solution: The propagation delay time between ROM input and appearance of data output is called as access time of memory. The access time measured the operating speed of the ROM

Q.8: What Is Static Ram and Dynamic Ram?

Answer:

- SRAM: No refreshing, 6 to 8 MOS transistors are required to form one memory cell, Information stored as voltage level in a flip flop.
- Dynamic RAM: Refreshed periodically, 3 to 4 transistors are required to form one memory cell, Information is stored as a charge in the gate to substrate capacitance.

Q.9: Explain What Is Virtual Memory?

Answer: This is a method of extending the available physical memory on a computer. In a virtual memory system, the operating system creates a page file, or swap file, and divides memory into units called pages. Recently referenced pages are located in physical memory, or RAM. If a page of memory is not referenced for a while, it is written to the page file. This is called "swapping" or "paging out" memory. If that piece of memory is then later referenced by a program, the operating system reads the memory page back from the page file into physical memory, also called "swapping" or "paging in" memory. The total amount of memory that is available to programs is the amount of physical memory in the computer in addition to the size of the pagefile.

Q.10: Explain the difference between RDRAM and SDRAM?

Answer: RDRAM stands for Rambus Dynamic Random-Access Memory. SDRAM stands for Synchronous Dynamic Random-Access Memory. The two memories are completely different memory technologies and are not compatible with each other. RDRAM is a unique design developed by a company called Rambus, Inc. RDRAM is extremely fast and uses a narrow, high-bandwidth "channel" to transmit data at speeds much faster than SDRAM.

Q.11: List the benefits of upgrading computers memory?

Answer: Upgrading your memory is typically the easiest and least expensive way to upgrade your computer for a significant boost in performance. The computer's RAM memory is its workspace, or where all of the instructions it needs to act on are stored temporarily. Think of the RAM as the desk you use to sort through your work. If the size of that desk is small, your efficiency is limited in comparison to a larger desk that allows you to work more effectively and efficiently. Similarly, a computer with more RAM can work more efficiently because it does not need to retrieve information from the hard disk drive as often. A memory upgrade is particularly helpful for users who work with large files, have more than one program open at one time, or use memory-intensive applications such as games or graphics and video editing software.

Q.11: How to know its time for a memory upgrade?

Answer: There are several signs indicating it may be time to upgrade your memory. If you see your mouse pointer turn into an hourglass for significant periods of time, if you hear your hard drive working, or if your computer seems to work more slowly than you expect, the reason is probably insufficient memory. When physical memory is insufficient, the system uses Hard Disk Space as memory. This is called "Virtual Memory". Since access time of Physical memory is in tens of Nanoseconds and Access time of Hard Disk is in Milliseconds, the system slows down considerably.

Q.12: Explain on What basis Microprocessor Speed Depend?

Answer: The processing speed depends on

- Data Bus width
- Processor clock speed
- Processor pipelining
- Width of the data and address bus, i.e. Max Data that can be fetched at one stretch.
- Support for floating point operations for faster floating-point operations.

Q.13: Can adding more RAM make internet browsing faster?

Answer: Maybe. Internet browsing speed depends on a huge number of factors, including your connection speed, traffic on the site you're visiting, and the other components in your system. You will probably notice the biggest improvement from additional RAM if are viewing or working with large files (such as photos and digital audio and video) or if you switch between your browser and other applications often.

Q.14: Explain the difference between 72-Bit And 64-Bit Memory?

Answer: 72-bit memory is commonly known as ECC memory. It has an additional 8 bits for Error Correction Check 64-bit memory is non-ECC. 72 bit or 64-bit configuration are typically found in 168 pin DIMMs.

Q.15: Explain the concept of Virtual Memory?

Answer: In computing, virtual memory is a memory management technique developed for multitasking kernels. This technique virtualizes a computer architecture's various forms of computer data storage such as random-access memory and disk storage, allowing a program to be designed as though there is only one kind of memory, "virtual" memory, which behaves like directly and contiguous addressable read/write memory.

Q.16: What is RAM parity?

Answer: RAM parity checking is the storing of a redundant parity bit representing the parity odd or even of a small amount of computer data typically one byte stored in random access memory, and the subsequent comparison of the stored and the computed parity to detect whether a data error has occurred. The parity bit was originally stored in additional individual memory chips; with the introduction of plug-in DIMM, SIMM, etc. modules, they became available in non-parity and parity (with an extra bit per byte, storing 9 bits for every 8 bits of actual data) versions.

Q.17: What is shadow RAM?

Answer: Shadow RAM is a copy of Basic Input/output Operating System (BIOS) routines from read-only memory (ROM) into a special area of random-access memory (RAM) so that they can be accessed more quickly. Access in shadow RAM is typically in the 60-100 nanosecond range whereas ROM access is in the 125-250 ns range. In some operating systems such as DOS, certain BIOS routines are not only used during the boot or startup of the system, but also during normal operation, especially to drive the video display terminal. In Windows and OS/2, however, these routines are not used and the use of shadow RAM is not necessary. In some systems, the user can turn the use of shadow RAM off or on.

Q.18: What is t_{RAS}?

Answer: $t_{RAS}$is the minimum number of clock cycles needed to access a certain row of data in RAM between the data request and the precharge command. It's known as active to precharge delay.

Q.19: What is t_{RP}?

Answer: t_{RP} is the number of clock cycles needed to terminate access to an open row of memory, and open access to the next row. It stands for row precharge time.

Q.20: What is t_{CAS}?

Answer: t_{CAS} is the number of clock cycles needed to access a certain column of data in SDRAM. CAS latency is the column address strobe time, sometimes referred to as tCL.

Q.21: What is t_{RCD} timing?

Answer: t_{RCD}tRCD is the number of clock cycles delay required between an active command row address strobe (RAS) and a CAS. It is the time required between the memory controller asserting a row address, and then asserting a column address during the subsequent read or write command. t_{RCD} stands for row address to column address delay time.

Q. 22: What is ECC memory?

Answer: Error-Correcting Code memory (ECC memory) is a type of computer data storage that can detect and correct the more common kinds of internal data corruption. ECC memory is used in most computers where data corruption cannot be tolerated under any circumstances, such as for scientific or financial computing.

ECC memory maintains a memory system effectively free from single-bit errors: the data that is read from each word is always the same as the data that had been written to it, even if a single bit actually stored, or more in some cases, has been flipped to the wrong state. Some non-ECC memory with parity support allows errors to be detected, but not corrected; otherwise errors that may occur are not detected.

Q.23: What is FeRAM?

Answer: Ferroelectric RAM is a random-access memory similar in construction to DRAM but uses a ferroelectric layer instead of a dielectric layer to achieve non-volatility. FeRAM is one of a growing number of alternative non-volatile random-access memory technologies that offer the same functionality as flash memory. FeRAM advantages over flash include: lower power usage, faster write performance and a much greater maximum number of write-erase cycles. Disadvantages of FeRAM are much lower storage densities than flash devices, storage capacity limitations, and higher cost.

Q.24: What is RDRAM?

Answer: Direct Rambus DRAM or DRDRAM (sometimes just called Rambus DRAM or RDRAM) is a type of synchronous dynamic RAM. RDRAM was developed by Rambus inc., in the mid-1990s as a replacement for then-prevalent DIMM SDRAM memory architecture. RDRAM was initially expected to become the standard in PC memory, especially after Intel agreed to license the Rambus technology for use with its future chipsets. Further, RDRAM was expected to become a standard for VRAM. However, RDRAM got embroiled in a standards war with an alternative technology - DDR SDRAM, quickly losing out on grounds of price, and, later on, performance. By the early 2000s, RDRAM was no longer supported by any mainstream computing architecture.

Q.25: What is SDRAM?

Answer: Synchronous dynamic Random-Access Memory (SDRAM) is dynamic random-access memory (DRAM) that is synchronized with the system bus. Classic DRAM has an asynchronous interface, which means that it responds as quickly as possible to changes in control inputs. SDRAM has a synchronous interface, meaning that it waits for a clock signal before responding to control inputs and is therefore synchronized with the computer's system bus. The clock is used to drive an internal finite state machine that pipelines incoming commands. The data storage area is divided into several banks, allowing the chip to work on several memory access commands at a time, interleaved among the separate banks. This allows higher data access rates than an asynchronous DRAM.

Review Questions

1. Write a short-note on semi-conductor memories
2. With neat diagram explain the working of MOS DRAM Cell.
3. Discuss various types of read-only memories.
4. Static and Dynamic RAM
5. With neat circuit diagram explain the operation of bipolar static RAM cell.
6. Discuss various types of ROMs and their applications in brief.
7. Write short notes on:
    i. Static and Dynamic MOS RAM
    ii. ROM, EPROM, EEPROM
8. Distinguish between volatile and non-volatile memories.
9. Explain the differences between SRAM and DRAM.
10. Write a short note on static bi-polar RAM cell.
11. With the help of neat circuit diagram give the working of static MOS RAM cell.
12. Write a short note on dynamic RAM.
13. A certain memory has a capacity of 8K X 8
    i. How many data inputs and data outputs does it have
    ii. How many address lines does it have
    iii. What is its capacity in bytes
14. Implement the following Boolean expression using ROM

$$F_1(A_1, A_0) = \sum m(1,2)$$

$$F_2(A_1, A_0) = \sum m(0,1,3)$$

15. Design a combinational circuit using ROM. The circuit accepts 3-bit binary number and generates its equivalent Excess-3 Code.

Multiple Choice Questions

Choose the Correct answer:

1. ROM consist of ___________
    a. NOR and OR arrays
    b. NAND and NOR arrays
    c. NAND and OR arrays
    d. NOR and AND arrays

2. For programmability, PLDs use __________
   a) PROM b) EPROM c) CDROM d) PLA
3. Which of the following is a reprogrammable gate array?
   a) EPROM b) FPGA c) Both EPROM and FPGA d) ROM
4. The difference between FPGA and PLD is that __________
   a. FPGA is slower than PLD
   b. FPGA has high power dissipation
   c. FPGA incorporates logic blocks
   d. All of the Mentioned
5. How many storage locations are available when a memory device has twelve address lines?
   a. 144
   b. 512
   c. 2048
   d. 4096
6. The access time (t_{acc}) of a memory IC is governed by the IC's:
   a. internal address buffer
   b. internal address decoder
   c. volatility
   d. internal address decoder and volatility
7. What is the principal advantage of using address multiplexing with DRAM memory?
   a. reduced memory access time
   b. reduced requirement for constant refreshing of the memory contents
   c. reduced pin count and decrease in package size
   d. no requirement for a chip-select input line, thereby reducing the pin count
8. PLA is used to implement ____________
   a. A complex sequential circuit
   b. A simple sequential circuit
   c. A complex combinational circuit
   d. A simple combinational circuit
9. A PLA is similar to a ROM in concept except that ____________
   a. It hasn't capability to read only
   b. It hasn't capability to read or write operation
   c. It doesn't provide full decoding to the variables
   d. It hasn't capability to write only
10. For programmable logic functions, which type of PLD should be used
   a) PLA b) PAL c) CPLD d) SLD
11. A sequential access memory is one in which __________
   a. A particular memory location is accessed rapidly
   b. A particular memory location is accessed sequentially
   c. A particular memory location is accessed serially
   d. A particular memory location is accessed parallelly
12. An example of sequential access memory is __________
   a) Floppy disk b) Hard disk c) Magnetic tape memory d) RAM
13. A Random-Access Memory is one in which __________
   a. Any location can be accessed sequentially

b. Any location can be accessed randomly
c. Any location can be accessed serially
d. Any location can be accessed parallel

14. As the storage capacity of main memory is inadequate, which memory is used to enhance?
    a. Secondary Memory
    b. Auxiliary Memory
    c. Static Memory
    d. Both Secondary Memory and Auxiliary Memory
15. A dynamic memory is one in which __________
    a. Content changes with time
    b. Content doesn't change with time
    c. Memory is static always
    d. Memory is dynamic always
16. Static memory holds data as long as __________
    a. AC power is applied
    b. DC power is applied
    c. Capacitor is fully charged
    d. High Conductivity
17. The example of dynamic memory is __________
    a. CCD
    b. Semiconductor dynamic RAM
    c. Both CCD and semiconductor dynamic RAM
    d. Floppy-Disk
18. In dynamic memory, CCD stands for __________
    a. Charged Count Devices
    b. Change Coupled Devices
    c. Charge Coupled Devices
    d. Charged Compact Disk
19. The example of non-volatile memory device is __________
    a. Magnetic Core Memory
    b. Read Only Memory
    c. Random Access Memory
    d. Both Magnetic Core Memory and Read Only Memory
20. By which technology, semiconductor memories are constructed?
    a) PLD b) LSI c) VLSI d) Both LSI and VLSI
21. The capacity of a memory unit is ____________
    a) The number of binary input stored b) The number of words stored
    b) The number of bytes stored d) All of the Mentioned
22. In ROM, each bit combination that comes out of the output lines is called ____________
    a) Memory unit b) Storage class c) Data word d) Address
23. Fill in the Blanks:
    1. PLDs with programmable AND and fixed OR arrays are called __________
    2. A large memory is compressed into a small one by using __________
    3. Data stored in an electronic memory cell can be accessed at random and on demand using____
    4. The evolution of PLD began with__________

5. The inputs in the PLD is given through ___________
6. Based on method of access, memory devices are classified into ___________ categories.
7. A memory is a collection of ___________
8. Each word stored in a memory location is represented by ___________
9. MOS ROM is constructed using _________
10. The MOS technology-based semiconductor ROMs are classified into _____ categories.
11. If a RAM chip has 'n' address input lines then it can access memory locations up to __________

References and Suggested Readings

- *A. Anand Kumar, Fundamentals of digital circuits Second Edition, PHI Learning Private Limited, ISBN (978-81-203-3679-7)*
- *Cook, N. P. (2003) Practical Digital Electronics, Prentice-Hall, NJ, USA,*
- *Floyd, T. L. (2005) Digital Fundamentals, Prentice-Hall Inc., USA.*
- *Tocci, R. J. and N. S. Widmer. 2004. Digital Systems Principles and Applications, 9th ed. Upper Saddle River, NJ: Prentice Hall.*
- *Tokheim, R. L. (1994) Schaum's Outline Series of Digital Principles, McGraw-Hill Companies Inc., USA.*
- *Yarbrough, John M. Digital logic: Applications and design. Eagan: West Publishing Company, 1997.*

8 Programmable Logic Devices

UNIT SPECIFICS

Through this unit we have discussed the following aspects:

- *To understand the terminology associated with programmable logic device (PLD)*
- *To describe the difference between several type of PLD's.*
- *To understand different type of devices like PAL, PLA and their uses.*
- *To distinguish among the various types of ROMs and cite comes common applications.*
- *To describe the structure and operation of CPLD.*
- *To understand the construction and operation of FPGA.*

RATIONALE

A programmable logic device (PLD) is an electronic component that can be programmed to perform various logical functions. PLDs are used to create digital circuits that can be customized to meet specific design requirements. There are several types of PLDs, including Programmable Logic Arrays (PLAs), Programmable Array Logic (PALs), and Complex Programmable Logic Devices (CPLDs).

PLAs are the simplest type of PLD, consisting of an array of AND gates followed by an array of OR gates. PALs are similar to PLAs, but they have additional programmable connections between the inputs and outputs. CPLDs are more complex, and they typically consist of multiple PLDs connected together with a programmable interconnect matrix. The programming of a PLD involves defining the logical functions that the device should perform, which is typically done using a Hardware Description Language (HDL) such as VHDL or Verilog. Once programmed, the PLD can be used in a wide range of applications, including digital signal processing, data processing, and control systems.

One of the advantages of using PLDs is their flexibility, as they can be reprogrammed multiple times to perform different functions, which makes them cost-effective compared to traditional application-specific integrated circuits (ASICs). However, they can be more expensive than discrete logic circuits for small-scale applications.

PRE-REQUISITES

Basic Digital fundamentals, Decoder, Encoder, Logic Circuits, Semiconductor circuits.

UNIT OUTCOMES

List of outcomes of this unit is as follows:

U8-O1: To understand the basic principles in the construction of the programmable devices.

U8-O2: To understand and implement various designs with PAL, PLA and CPLD

U8-O3: To identify the designing part of various Programmable Logic Devices.

U8-O4: To describe the FPGA in association with XILINX package.

U8-O5: To understand Xilinx XC9500 CPLD Family and its operation.

Unit-8 Outcomes	EXPECTED MAPPING WITH COURSE OUTCOMES (1- Weak Correlation; 2- Medium correlation; 3- Strong Correlation)					
	CO-1	**CO-2**	**CO-3**	**CO-4**	**CO-5**	**CO-6**
U8-O1	3	3	3	-	3	1
U8-O2	1	1	2	2	1	-
U8-O3	2	1	3	1	2	1
U8-O4	-	-	3	1	2	2
U8-O5	3	3	3	-	3	1

➢ Scan the below QR codes for more information on the topic written below the QR code.

Introduction of PLDs

Programmable Logic Array (PLA)

Programmable Array Logic (PAL)

Difference b/w PAL and PLA

Xilinx XC9500 CPLD Family

VHDL Basics

Unit outcomes

In this unit students will be able to understand different types of PLDs available, such as Programmable Logic Arrays (PLAs), Programmable Array Logic (PALs), Field Programmable Gate Arrays (FPGAs), and Complex Programmable Logic Devices (CPLDs). Also, the students will be able to understand that by using a programmable logic device, the same hardware can be used to implement different functions simply by changing the programming. This can reduce costs and improve *flexibility* in design.

8.1 Introduction

A Programmable Logic Device (PLD) is an Integrated Circuit (IC) with several gates, flip-flops, and other components that can be programmed by the user to carry out various tasks. A programmable logic device is, in general, is an IC that can be configured by the user to perform logic operations. It enables the user to implement complex digital logic designs on a single chip.

PLDs contain embedded logic gates and/or connections that may be modified or modified through programming. With a 'regular' structure that enables the designer to customize it for any particular application, PLD's may be programmed by the user to execute a specific function as necessary for the given application. Programmable logic devices make the construction of complicated logic simpler than fixed logic devices and may perform better. A PLD may be programmed for a particular purpose by altering the connections between its internal gates.

Programmable Logic Devices (PLDs) contain an array of AND gates & another array of OR gates. A general structure of PLD is shown in Fig. 8.1. The ***pro****cess* of entering the information into these devices is known as ***pro****gramming*. Basically, users can program these devices or ICs electrically in order to implement the Boolean functions based on the requirement. Here, the term programming is used in a context that refers to *hardware programming* but not software programming.

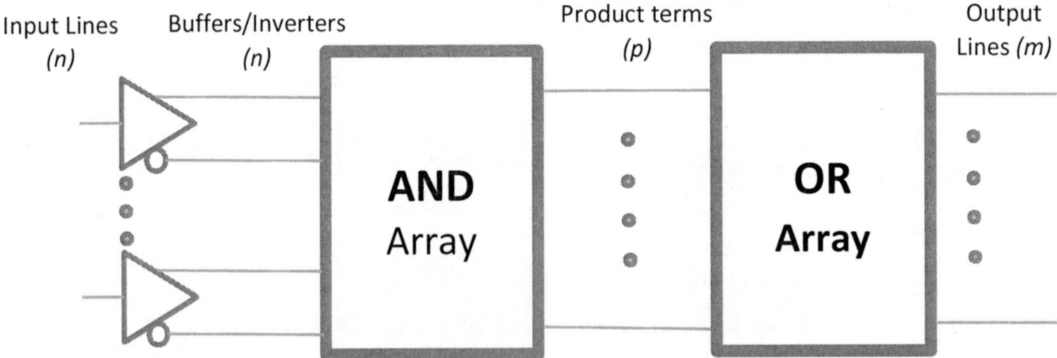

Fig. 8.1: General structure of PLDs.

8.1.1 Fixed logic versus Programmable Logic

There are two main categories of logic devices i.e. fixed logic devices and programmable logic devices. A comparison between these two is mentioned in Table 8.1.

Table 8.1: Comparison between Fixed Logic Devices and Programmable Logic Devices

S.No.	Fixed logic devices	Programmable logic devices.
1.	These are fixed functions ICs/Devices like Multiplexers, Demultiplexers, adders, Counters, etc.	These are Application Specific integrated Circuits (ASICs) and the outcome can change accordingly.
2.	The circuit configurations in the fixed logic system are permanent. They have designed to carry out a predetermined set of actions. The logic cannot be altered after it has been constructed and coded. For routine activities, this method is a great advantage.	Application Specific Integrated Circuits (ASICs) are produced by an IC manufacturer according to the user's specifications. ASICs are being designed and created by the designer to fulfil the unique requirements of a user. Hence, the designer aims to give an adaptable, practical, and economical solution for the user.
3.	One tiny mistake in the manufacturing process like uploading the wrong code in the device, and the entire system is discarded, and a new design is developed. That's quite some risk that companies aren't willing to take unless necessary.	It is easy-to-program, affordable and equipped with better features. Inexpensive software is used to develop, code and test the required design.
4.	Fixed logic does not allow the users to expand or build on their existing functionalities.	A programmable logic seems more feasible and beneficial.
5.	Fixed function ICs have the following advantages over PLDs • Compact circuitry • Higher switching speed	ASICs have many advantages like • Lower quantity production costs • Design security • Reduced power requirement • Higher density • Reduced space requirement

8.1.2 Implementing Boolean functions

Every Boolean logic can be decomposed into product-of-sum (POS) or sum-of-product (SOP) by Karnaugh map (k-map),

$$S = A \oplus B \oplus C = \bar{A}\bar{B}C + \bar{A}B\bar{C} + \bar{A}\bar{B}\bar{C} + ABC$$
$$= (A + B + \bar{C})(A + \bar{B} + C)(\bar{A} + B + \bar{C})(\bar{A} + \bar{B} + \bar{C})$$

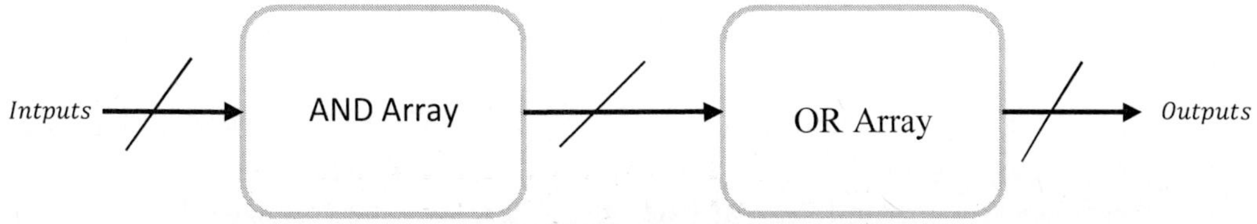

Fig. 8.2: PLD with AND Array and OR Array

PLDs are typically built with an array of AND gates (AND-array) and an array of OR gates (OR-array) to implement the sum-of-products as shown in figure above (Refer Fig. 8.2).

In order to show the internal logic diagram for such technologies in a concise form, it is necessary to have special symbols for array logic. Figure shows the conventional and array logic symbols for a multiple input AND gate and a multiple input OR gate as shown in Fig. 8.3.

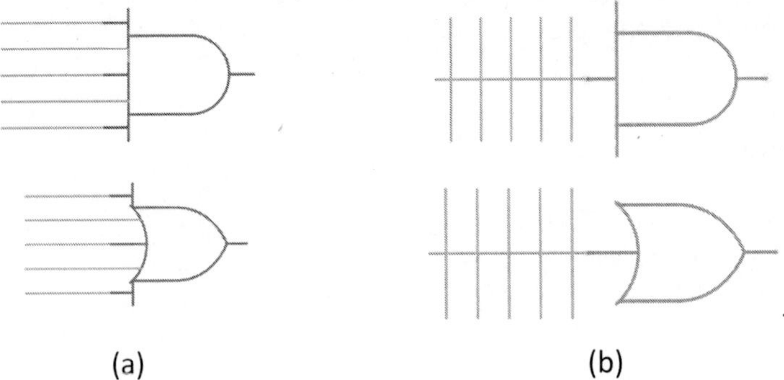

Fig. 8.3: logic symbol for multiple input AND gate and OR gate (a) Conventional Symbol (b)array Logic Symbol

One of the simplest programming technologies is to use fuses. In the original state of the device, all the fuses are intact. Programming the device involves blowing those fuses along the paths that must be removed in order to obtain the particular configuration of the desired logic function. Anti-fuse employs a thin barrier of non-conducting amorphous silicon between two metal conductors. Usually in mesh structure. When a sufficiently high voltage is applied across the amorphous silicon it is turned into a polycrystalline silicon-metal alloy with a low resistance, which is conductive as shown in Fig. 8.4.

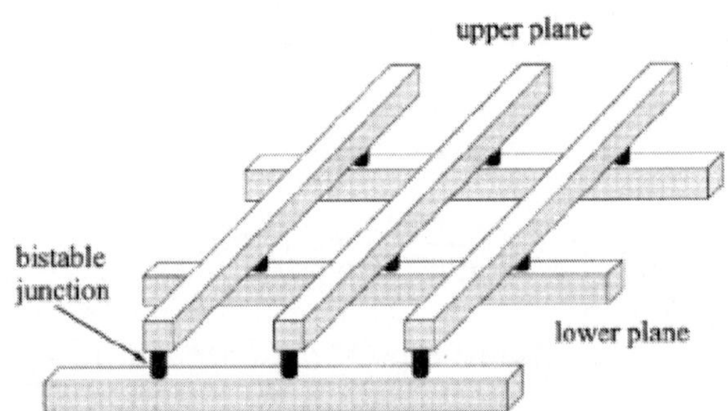

Fig. 8.4: Polycrystalline silicon-metal alloy

Problems of using standard ICs: Problems of using standard ICs in logic design are that they require hundreds or thousands of these ICs, considerable amount of circuit board space, a great deal of time and cost in inserting, soldering, and testing. Also require keeping a significant inventory of ICs.

Advantages of using PLDs: Advantages of using PLDs are less board space, faster, lower power requirements (i.e., smaller power supplies), less costly assembly processes, higher reliability (fewer ICs and circuit connections means easier troubleshooting), and availability of design software.

8.1.3 Array Logic Symbols

PLDs have hundreds of gated interconnected through hundreds of electronic fuses. It is sometime convenient to draw the internal logic of such device in a compact form referred to as array logic. Array logic symbols for AND gate as well as OR gate is shown in Fig. 8.6. However, Fig. 8.7 represents OR and AND PLD notation for before and after programming.

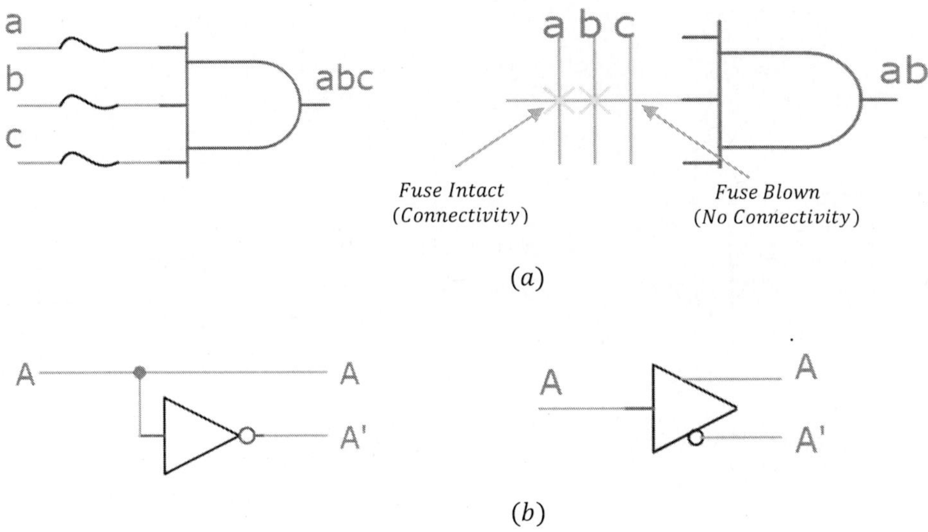

(b)

Fig. 8.5: Array Logic Symbols

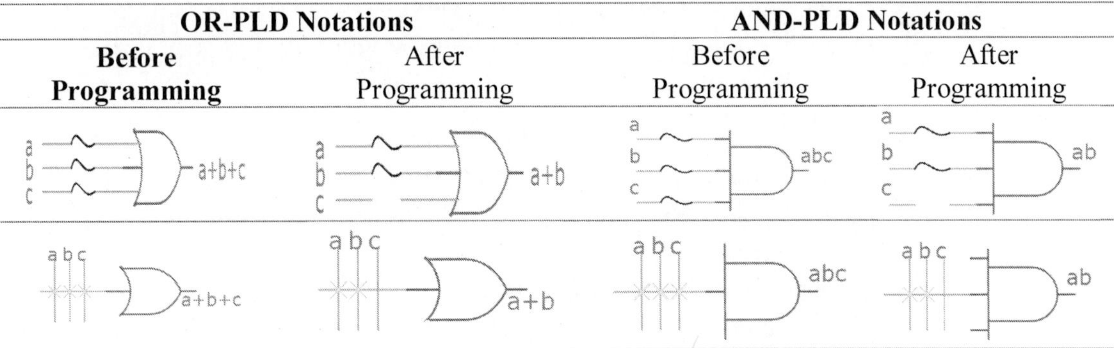

Fig. 8.6: Programming by blowing fuses

8.1.4 Types of PLDs

As per the increasing order of complexity PLDs can broadly be categorized into as, Simple Programmable Logic Devices (SPLDs), programmable logic array and generic array logic; Complex Programmable Logic Devices (CPLDs) and Field-Programmable Gate Arrays (FPGAs). Fig. 8.7 shows the classification and major categories of PLDs.

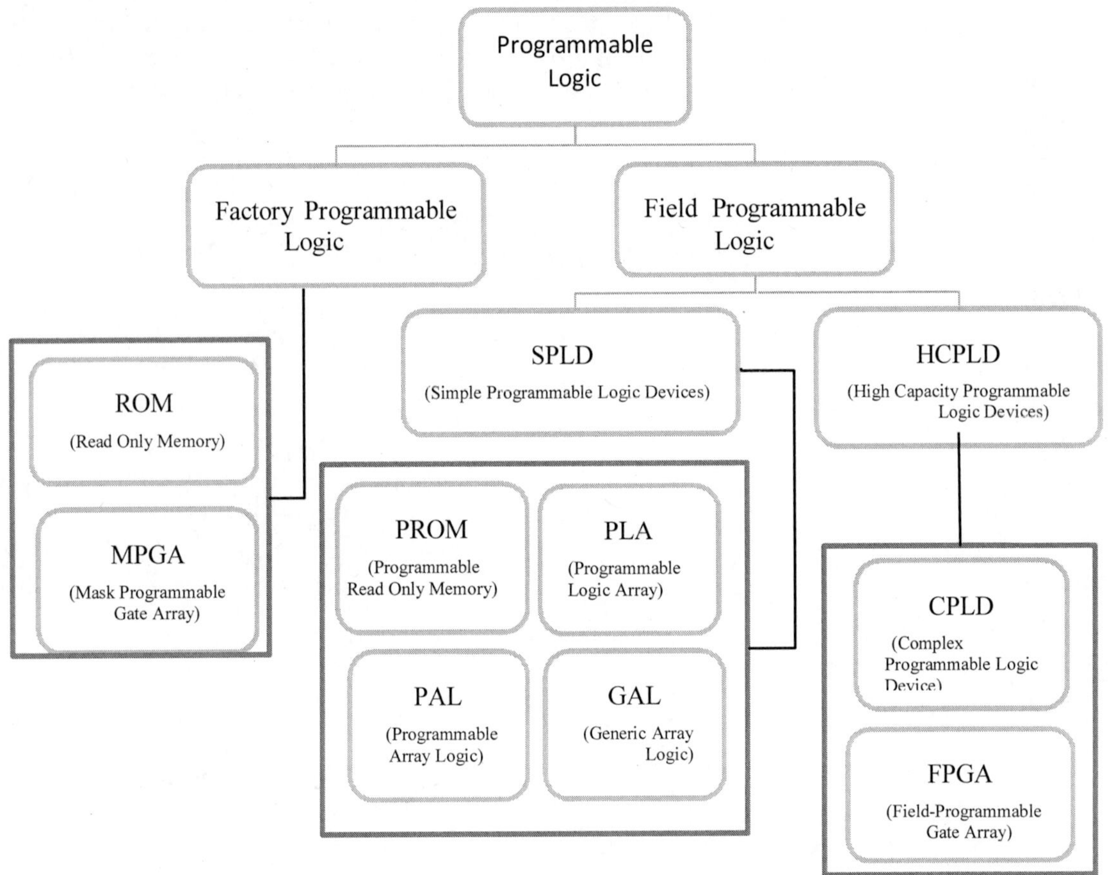

Fig. 8.7: Major Categories of Programmable Logic Devices

PLDs are broadly classified into simple and complex programmable logic devices further, this is grouped as,

- SPLDs (Simple Programmable Logic Devices)
  - PROM (Programmable Read-Only Memory)
  - PLA (Programmable Logic Array)
  - PAL (Programmable Array Logic)
  - GAL (Generic Array Logic)
- HCPLD (High Capacity Programmable Logic Device)
  - CPLD (Complex Programmable Logic Device)
  - FPGA (Field-Programmable Gate Array)

Simple Programmable Logic Devices (SPLDs): Programmable Logic Devices (PLDs) are used in many applications to replace fixed-function circuits. They save space and reduce the actual number and cost of devices in a given design. An SPLD consists of an array of AND gates and OR gates that can be programmed to achieve specified logic functions. Four types of devices that are classified as SPLDs are the programmable array Logic (PAL), the Generic Array Logic (GAL), the Programmable Logic Array (PLA) and the programmable read only memory (PROM).

CPLDs (complex programmable logic devices): CPLDs have a much higher capacity than SPLDs. Permit more complex logic circuits to be programmed into them. A typical CPLD is the equivalent of function two to sixty-four SPLDs. There are several forms of CPLD, which vary in complexity and programming capability. CPLD typically come in 44-pin to 160-pin packages depending on the complexity.

FPGAs (field-programmable gate arrays): FPGAs are different from SPLDs and CPLDs. In their internal organization and have the greatest Logic capacity. FPGAs consists of an array of anywhere from sixty-four to thousands of logic-gate groups that are sometimes called logic blocks. FPGAs come in packages ranging up to 1000 pins or more.

PLD Programming: a logic circuit design for a PLD is entered using one of the two basic methods: schematic entry or text-based entry. Sometimes a combination of both methods is used.

- In the schematic entry method, the software allows the user to enter a logic design using logic components (e.g. logic gates, flip-flops) and to interconnect them on the computer screen to form a schematic diagram.
- In the text Based entry method also known as language-based entry, the software allows the user to enter a logic design in the form of text using a hardware description Language (HDL). Several HDLs are available, such as VHDL and Verilog HDL developed for programming. PLDs, are widely used in various application like automation, healthcare, agriculture, utility etc.
- An HDL that is becoming widely used, especially for programming CPLDs and FPGAs is VHDL, a standard developed by the department of Defense and adopted by the IEEE (Institute of Electrical and Electronics Engineering). The Latest version of VHDL is IEEE std.1076-1993. Verilog is another popular HDL for programming CPLDs and FPGAs.

8.1.5 Programmable Connections in PLDs

The programmable connections of AND-OR arrays for different types of PLDs are described here. Figure shows the locations of the programmable connections for the three types or Basic configuration of three PLDs are shown in figure below.

The PROM (Programmable Read Only Memory) has a fixed AND array (constructed as a decoder) and programmable connections for the output OR gates array. The PROM implements Boolean functions in sum-of-minterms form. The PAL (Programmable Array Logic) device has a programmable AND array and fixed connections for the OR array. The PLA (Programmable Logic Array) has programmable connections for both AND and OR arrays. So, it is the most flexible type of PLD.

8.1.6 Applications of Programmable Logic Devices

- Glue Logic
- State Machines
- Counters
- Synchronization
- Decoders
- Bus Interfaces
- Parallel-to-Serial
- Serial-to-Parallel

8.2 Simple Programmable Logic Devices

A combinational PLD is an integrated circuit with programmable gates divided into an AND array and an OR array to provide an AND-OR sum of product implementation.

- PROM: Fixed AND array constructed as a decoder and programmable OR array. {Refer Fig. 8.8 (a)}.

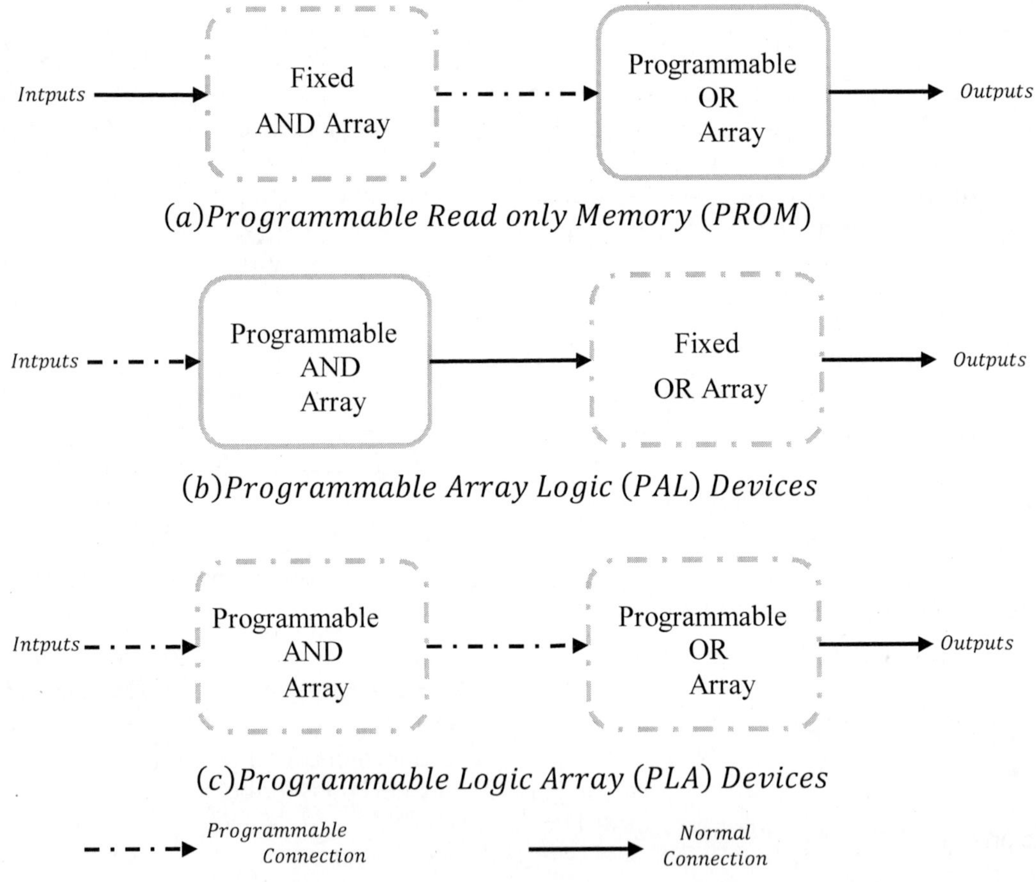

Fig. 8.8: programmable connections of AND-OR arrays

- PAL: Programmable AND array and fixed OR array. {Refer Fig. 8.8 (b)}
- PLA: Both AND array and OR arrays can be programmed. {Refer Fig. 8.8(c)}

The required paths in a ROM may be programmed in four different ways.

1. Mask programming: fabrication process
2. Read-only memory or PROM: blown fuse /fuse intact
3. Erasable PROM or EPROM: placed under a special ultraviolet light for a given period of time will erase the pattern in ROM.
4. Electrically-erasable PROM (EEPROM): erased with an electrical signal instead of ultraviolet light.

8.2.1 Programmable Read Only Memory (PROM)

Read Only Memory or ROM is a memory device, which stores the binary information permanently. If the ROM has programmable feature, then it is called as *Programmable ROM* or *PROM.* The user has the flexibility to program the binary information electrically once by using PROM programmer.

PROM is a programmable logic device that has fixed AND array & Programmable OR array. The **block diagram** of PROM is shown in Fig. 8.9.

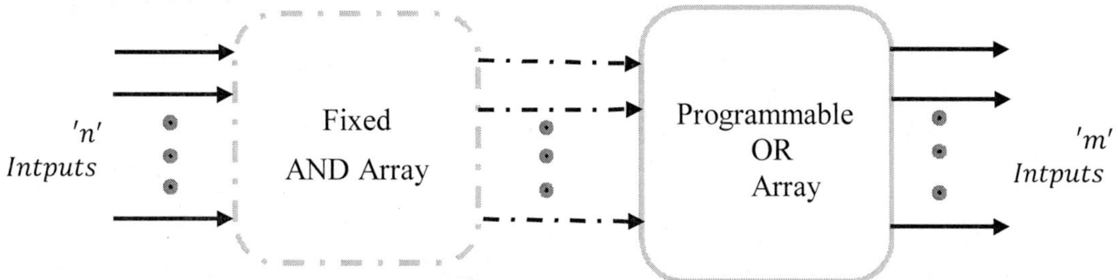

Fig. 8.9: Block diagram of PROM

Here, the inputs of AND gates are not of programmable type. So, we have to generate 2^n product terms by using 2^n AND gates having n inputs each. We can implement these product terms by using $n \times 2^n$ decoder. So, this decoder generates *'n' minterms*.

Here, the inputs of OR gates are programmable. That means, we can program any number of required product terms, since all the outputs of AND gates are applied as inputs to each OR gate. Therefore, the outputs of PROM will be in the form of *sum of minterms*.

Thus, we generate 2^n product terms using 2^n AND gates having n inputs each, using $n \times 2^n$ decoder. This decoder generates *'n'* min-terms. Each AND gate generate one of the possible AND products (i.e., min-terms). Given a $2^k \times n$ ROM, we can implement any combinational circuit with at most k inputs and at most n outputs. Because,

- k-to-2^k decoder will generate all 2^k possible min-terms
- Each of the OR gates must implement a $\sum m()$
- Each $\sum m()$ can be programmed

The procedure for implementing a ROM-based circuit is as follows,

For example, it has been given that

$$f(a,b,c) = a'b' + abc; \quad g(a,b,c) = a'b'c' + ab + bc; \quad h(a,b,c) = a'b' + c$$

Its solution can be obtained as,

- Express $f(a.b.c);\ g(a,b,c);\ h(a,b,c)\ in\ \sum m(a,b,c)\ format\ by\ using\ Truth\ Table$
- Program the ROM based on the $\sum m(a,b,c\ ...)'s$ combinations of a, b, c.

This can be better understood by the example below.

Example 8.1: There are 3 inputs and 3 outputs, thus we need a 8x3 ROM block. Given, f = Σm (0, 1, 7); g = Σm (0, 3, 6, 7); and h = Σm (0, 1, 3, 5, 7)

Solution:

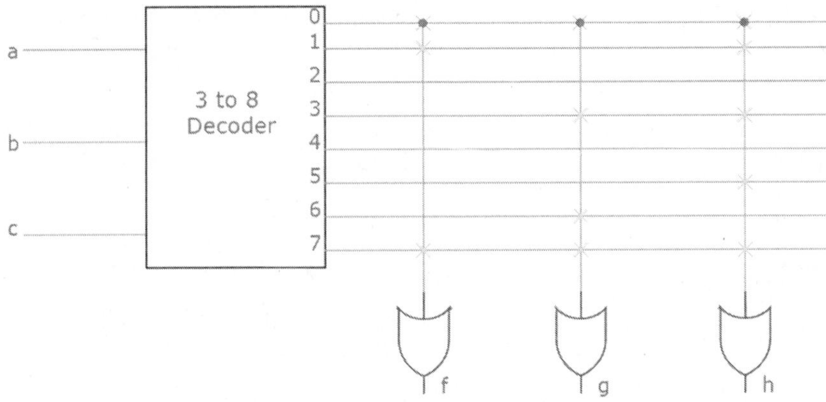

Fig. 8.10: Solution for Example 8.1

Another practical application of PROM device is BCD to 7 Segment Display Controller and the corresponding input and output relationship are shown in Fig. 8.11.

Numerical Value	Binary Value				Input to 7-segment display						
	A	B	C	D	C_0	C_1	C_2	C_3	C_4	C_5	C_6
0	0	0	0	0	1	1	1	1	1	1	0
1	0	0	0	1	0	1	1	0	0	0	0
2	0	0	1	0	1	1	0	1	1	0	1
3	0	0	1	1	1	1	1	1	0	0	1
4	0	1	0	0	0	1	1	0	0	1	1
5	0	1	0	1	1	0	1	1	0	1	1
6	0	1	1	0	1	0	1	1	1	1	1
7	0	1	1	1	1	1	1	0	0	0	0
8	1	0	0	0	1	1	1	1	1	1	1
9	1	0	0	1	1	1	1	0	0	1	1

Fig. 8.11: Truth table for BCD to 7 Segment Display converter

Example 8.2: Implement the following **Boolean functions** using PROM.

$$A(X,Y,Z) = \sum m(5,6,7)$$

$$B(X,Y,Z) = \sum m(3,5,6,7)$$

Solution: The given two functions are in sum of min terms form and each function is having three variables X, Y & Z. So, we require a 3 to 8 decoder and two programmable OR gates for producing these two functions. The corresponding PROM is shown in Fig. 8.12.

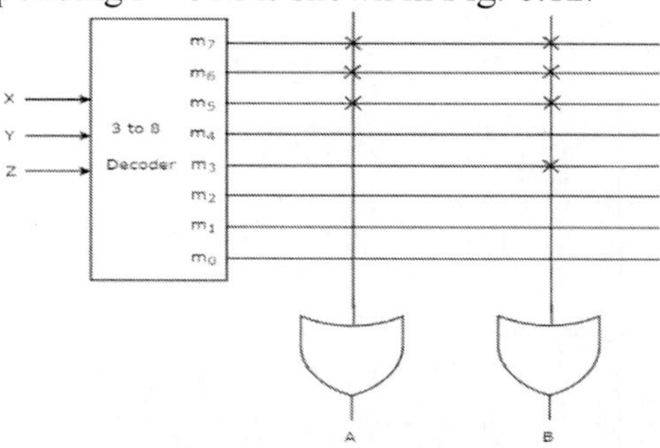

Fig. 8.12: Example 8.2

Here, 3 to 8 decoder generates eight min terms. The two programmable OR gates have the access of all these min terms. But, only the required min terms are programmed in order to produce the respective Boolean functions by each OR gate. The symbol 'X' is used for programmable connections.

8.2.2 Programmable logic array (PLA)

In PLAs, instead of using a decoder as in PROMs, a number of AND gates (say *k*) is used where *k* < *2 n*, (*n* is the number of inputs). Each of the AND gates can be programmed to generate a product term of the input variables and does not generate all the minterms as in the ROM. The AND and OR gates inside the PLA are initially fabricated with the links (fuses) among them. The specific Boolean functions are implemented in sum of products form by opening appropriate links and leaving the desired connections.

A block diagram of the PLA is shown in Fig. 8.13. It consists of *n* inputs, *m* outputs, and *k* product terms. The product terms constitute a group of *k* AND gates each of *2n* inputs. Links are inserted between all *n* inputs and their complement values to each of the AND gates. Links are also provided between the outputs of the AND gates and the inputs of the OR gates.

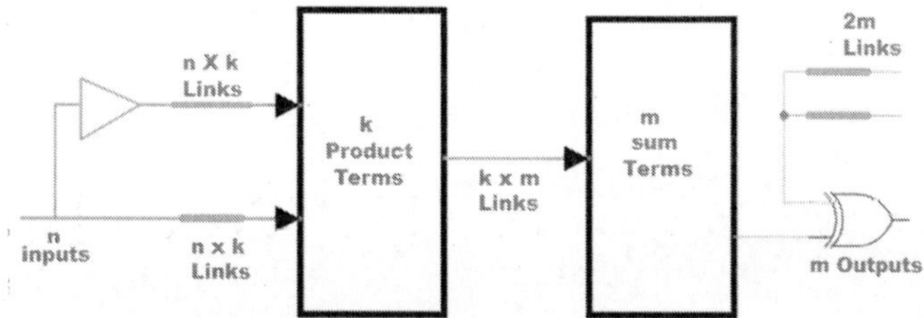

Fig. 8.13: Block diagram of PLA

Since PLA has m-outputs, the number of OR gates are m. The output of each OR gate goes to an XOR gate, where the other input has two sets of links, one connected to logic 0 and other to logic 1. It allows the output function to be generated either in the true form or in the complement form. The output is inverted when the XOR input is connected to 1 (since X ⊕ 1 = X'). The output does not change when the XOR input is connected to 0 (since X ⊕ 0 = X). Thus, the total number of programmable links is $2n \times k + k \times m + 2m$. The size of the PLA is specified by the number of inputs (n), the number of product terms (k), and the number of outputs (m), (the number of sum terms is equal to the number of outputs).

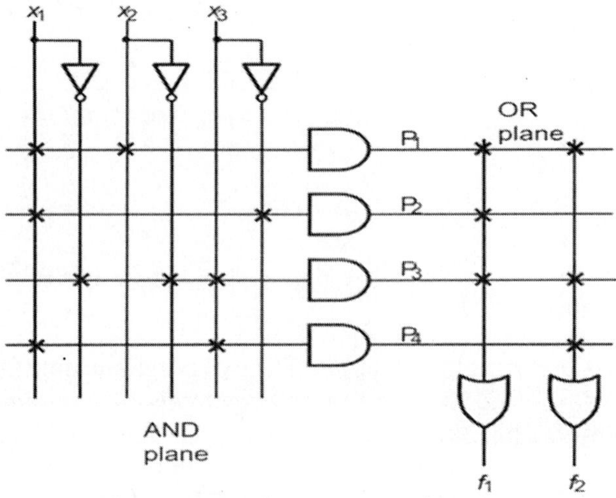

Fig. 8.14: Basic Structure of PLA

As shown in Fig. 8.14, here, the inputs of AND gates are programmable. That means each AND gate has both normal and complemented inputs of variables. So, based on the requirement, we can program any of those inputs. We can generate only the required product terms by using these AND gates. However, the inputs of OR gates are also programmable. So, we can program any number of required product terms, since all the outputs of AND gates are applied as inputs to each OR gate. Therefore, the outputs of PAL will be in the form of sum of products form. It can be better explained by using example below

Example 8.3: Implement the combinational circuit having the shown truth table (8.2), using PLA.

Table 8.: Example 8.3

A	B	C	F_1	F_2
0	0	0	1	1
0	0	1	1	0
0	1	0	1	0
0	1	1	0	0
1	0	0	1	0
1	0	1	0	1
1	1	0	0	1
1	1	1	0	1

Solution: Each product term in the expression requires an AND gate. To minimize the cost, it is necessary to simplify the function to a minimum number of product terms using K-map as shown in Fig. 8.15.

$A \backslash BC$	00	01	11	10
0	1	1	0	1
1	1	0	0	0

(a)

$A \backslash BC$	00	01	11	10
0	1	0	0	0
1	0	1	1	1

(b)

$$F_1 = \bar{A}\bar{B} + \bar{A}\bar{C} + \bar{B}\bar{C}$$
$$\overline{F_1} = AB + AC + BC$$

$$F_2 = AB + AC + \bar{A}\bar{B}\bar{C}$$
$$F_2 = \bar{A}C + \bar{A}B + A\bar{B}\bar{C}$$

Fig. 8.15: K-Map for example 8.3

Designing using a PLA, a careful investigation must be taken in order to reduce the distinct product terms. Both the true and complement forms of each function should be simplified to see which one can be expressed with fewer product terms and which one provides product terms that are common to other functions. The equations above give only 4 distinct product terms: AB, AC, BC, and A'B'C'. So, the PLA table (Table 8.3) and circuit (Fig. 8.16) can be designed as follows,

Table 8.3: PLA programmable table

S.No.	Product Term	Inputs			Outputs	
		A	B	C	$C\ (F_1)$	$T(F_2)$
1	AB	1	1	-	1	1
2	AC	1	-	1	1	1
3	BC	-	1	1	1	-
4	$\bar{A}\bar{B}\bar{C}$	0	0	0	-	1

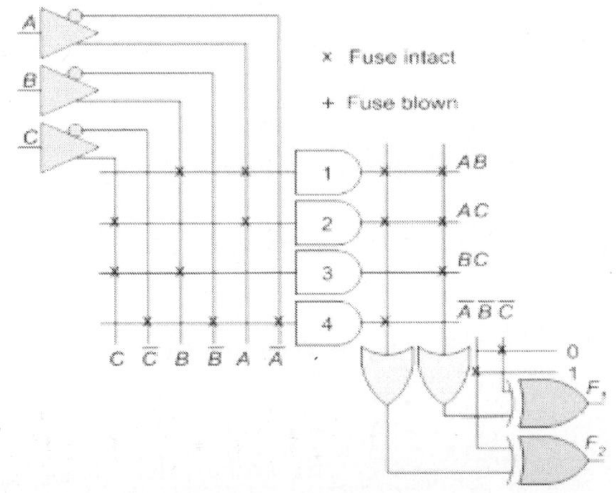

Fig. 8.16: PLA implementation for example 8.3

For each product term, the inputs are marked with *1, 0,* or – (dash).

- If a variable in the product term appears in its normal form (unprimed), the corresponding input variable is marked with a 1. A 1 in the Inputs column specifies a path from the corresponding input to the input of the AND gate that forms the product term.
- A 0 in the Inputs column specifies a path from the corresponding complemented input to the input of the AND gate.
- A dash (-) specifies no connection.

The appropriate fuses are blown and the ones left intact form the desired paths. It is assumed that the open terminals in the AND gate behave like a 1 input. In the Outputs column, a T (true) specifies that the other input of the corresponding XOR gate can be connected to 0, and a C (complement) specifies a connection to 1. Note that output F1 is the normal (or true) output even though a C (for complement) is marked over it. This is because F1' is generated with AND-OR circuit prior to the output XOR. The output XOR complements the function F1' to produce the true F1 output as its second input is connected to logic 1.

Limitations of PLAs

PLAs come in various sizes. Typical size is 16 inputs, 32 product terms, 8 outputs. Each AND gate have large fan-in. This limits the number of inputs that can be provided in a PLA o 16 inputs forms 216, possible input combinations; only 32 permitted (since 32 AND gates) in a typical PLA 32 AND terms permitted large fan-in for OR gates as well This makes PLAs slower and slightly more expensive.

Applications of PLA:

- PLA may be used to provide control over data path.
- PLA may be used as a counter.
- PLA may be used as a decoders.
- PLA may be used as a BUS interface in programmed I/O.

Example 8.4: Implement the combinational circuit with a PLA having 3 inputs, 4 product terms and 2 outputs for the functions.

$$F_1(A,B,C) = \sum m\ (3,5,6,7) \text{ and } F_2(A,B,C) = \sum m\ (0,2,4,7)$$

Solution:

Step 1: Truth table for the given functions (Refer Table 8.4)

Table 8.4: Example 8.4

Decimal Value	**Binary Value**			F_1	F_2
	A	B	C		
0	0	0	0	0	1
2	0	1	0	0	1
3	0	1	1	1	0
4	1	0	0	0	1
5	1	0	1	1	0
6	1	1	0	1	0
7	1	1	1	1	1

For the above truth table K-map can be drawn as shown below

Step 2: K-map Simplification

A \ BC	00	01	11	10
0	0	0	1	0
1	0	1	1	1

(a)

A \ BC	00	01	11	10
0	1	0	0	1
1	1	0	1	0

(b)

$$F_1 = AC + AB + AC \qquad F_2 = \bar{B}\bar{C} + \bar{A}\bar{C} + ABC$$

With this simplification, total number of product term is 6. But we require only 4 product terms. Therefore, find out F1' and F2'.

$$\overline{F_1} = \bar{B}\bar{C} + \bar{A}\bar{C} + \bar{A}\bar{B} \qquad \overline{F_2} = \bar{A}C + \bar{B}C + AB\bar{C}$$

Now select, F1' and F2, the product terms are B'C', A'C', A'B' and ABC

Step 3: Design PLA Program table (Refer table 8.5)

Table 8.5: PLA Program table for example 8.2

S.No.	**Product Term**	**Inputs** A	B	C	**Outputs** $F_1(C)$	$F_2(T)$
1	$\bar{B}\bar{C}$	-	0	0	1	1
2	$\bar{A}\bar{C}$	0	-	0	1	1
3	$\bar{A}\bar{B}$	0	0	-	1	-
4	ABC	1	1	1	-	1

Step 4: Draw the PLA circuit Diagram (Refer Fig. 8.17)

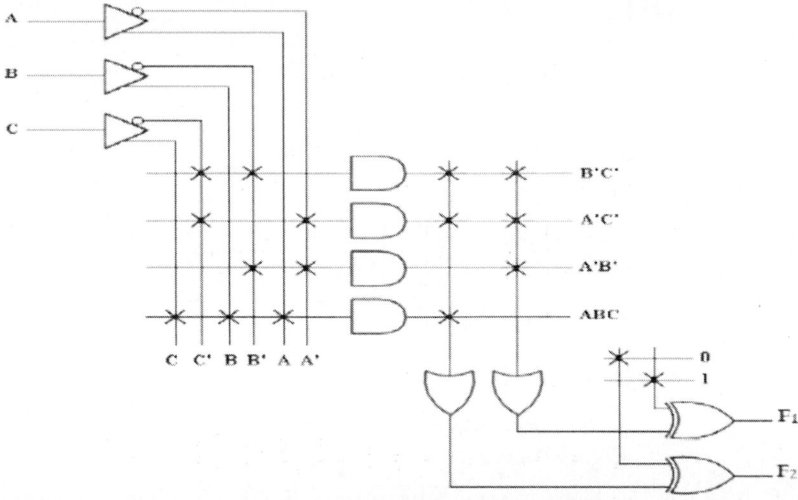

Fig. 8.17: PLA Circuit Diagram

8.2.3 Programmable array logic (PAL)

PAL is a programmable logic device that has Programmable AND array & fixed OR array. The advantage of PAL is that we can generate only the required product terms of Boolean function instead of generating all the min terms by using programmable AND gates. The block diagram of PAL is shown in Fig. 8.18.

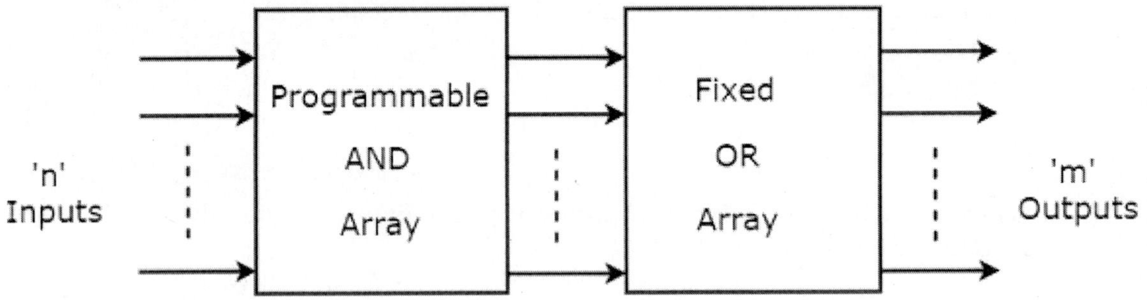

Fig. 8.18: block diagram of PAL

Here, the inputs of AND gates are programmable. That means each AND gate has both normal and complemented inputs of variables. So, based on the requirement, we can program any of those inputs. So, we can generate only the required product terms by using these AND gates.

Here, the inputs of OR gates are not of programmable type. So, the number of inputs to each OR gate will be of fixed type. Hence, apply those required product terms to each OR gate as inputs. Therefore, the outputs of PAL will be in the form of sum of products form (Refer Fig. 8.19).

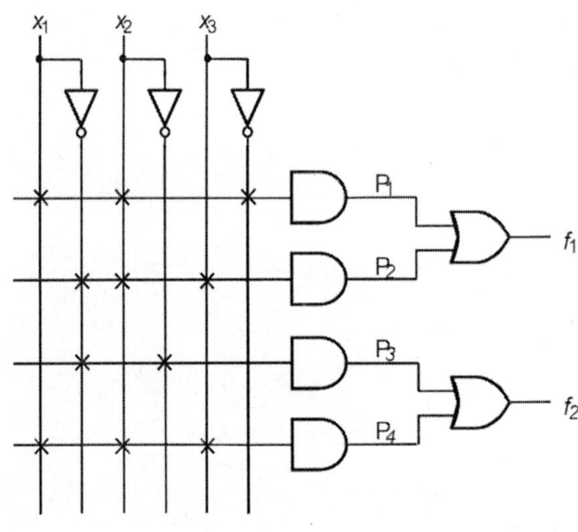

Fig. 8.19: Basic Structure of PAL

As shown in Fig. 8.20, the device has 4 inputs and 4 outputs. Each input has a buffer-inverter gate, and each output is generated by a fixed OR gate. The device has 4 sections, each composed of a 3-wide AND-OR array, meaning that there are 3 programmable AND gates in each section.

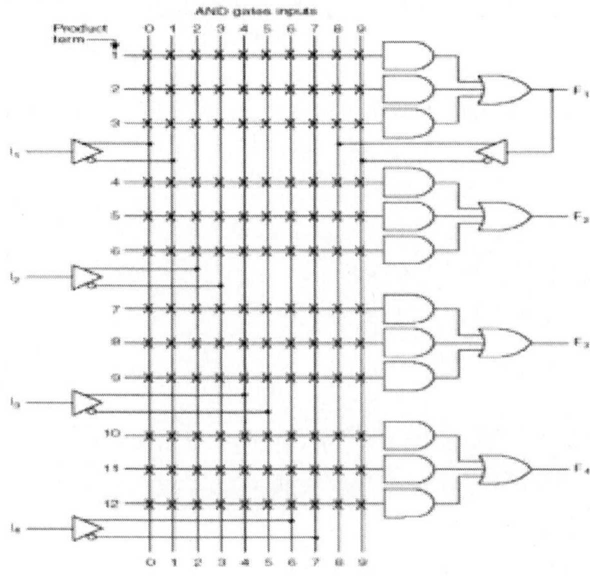

Fig. 8.20: 4 inputs and 4 outputs PAL

Each AND gate have 10 programmable input connections indicating by 10 vertical lines intersecting each horizontal line. The horizontal line symbolizes the multiple input configuration of an AND gate. One of the outputs F1 is connected to a buffer-inverter gate and is fed back into the inputs of the AND gates through programmed connections.

Designing using a PAL device, the Boolean functions must be simplified to fit into each section. The number of product terms in each section is fixed and if the number of terms in the function is too large, it may be necessary to use two or more sections to implement one Boolean function.

Example 8.4: For the given PAL circuit derive the output equation.

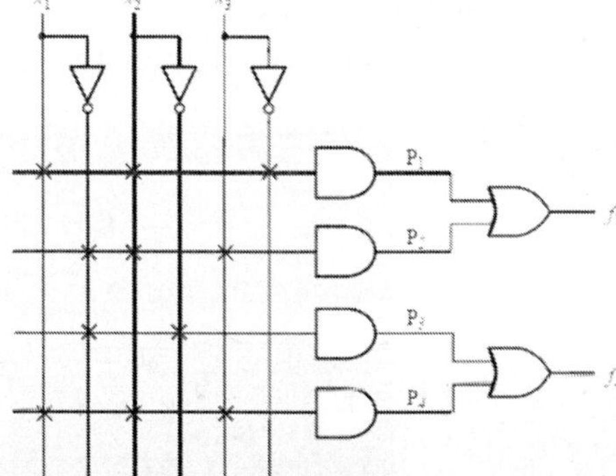

Fig. 8.21: Circuit Diagram for $\boldsymbol{f_1}$ and $\boldsymbol{f_2}$

Solution:

$f_1 = x_1.x_2.\overline{x_3} + \overline{x_1}.x_2.x_3$ $\qquad$ $f_2 = \overline{x_1}.\overline{x_2}.x_3 + x_1.x_2.x_3$

Example 8.5: Implement the following Boolean functions using PAL.

$$A = XY + XZ'$$
$$A = XY' + YZ'$$

Solution: The given two functions are in sum of products form. There are two product terms present in each Boolean function. So, we require four programmable AND gates & two fixed OR gates for producing those two functions. The corresponding PAL is shown in Fig. 8.22.

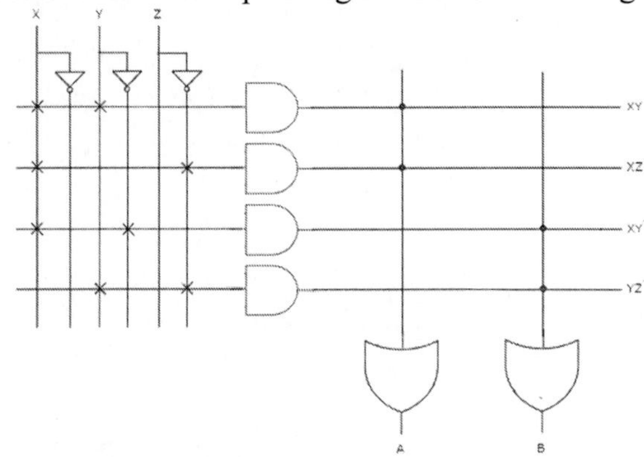

Fig. 8.22: Solution for example 8.5

The programmable AND gates have the access of both normal and complemented inputs of variables. In the above figure, the inputs X, X', Y, Y' Z & Z', are available at the inputs of each AND gate. So, program only the required literals in order to generate one product term by each AND gate. The symbol 'X' is used for programmable connections.

Here, the inputs of OR gates are of fixed type. So, the necessary product terms are connected to inputs of each OR gate. So that the OR gates produce the respective Boolean functions. The symbol '.' is used for fixed connections.

Example 8.6: Implement the following Boolean functions using the PAL device.

$$W(A,B,C,D) = \sum m\,(2,12,13)$$
$$X(A,B,C,D) = \sum m\,(7,8,9,10,11,12,13,14,15)$$
$$Y(A,B,C,D) = \sum m\,(0,2,3,4,5,6,7,8,10,11,15)$$
$$Z(A,B,C,D) = \sum m\,(1,2,8,12,13)$$

Solution: Simplifying the 4 functions to a minimum number of terms results in the following Boolean functions:

$$W = AB\bar{C} + \bar{A}\bar{B}C\bar{D}$$
$$X = A + BCD$$
$$Y = \bar{A}B + CD + \bar{B}\bar{D}$$
$$Z = AB\bar{C} + \bar{A}\bar{B}CD + A\bar{C}\bar{D} + \bar{A}\bar{B}\bar{C}\bar{D} = W + A\bar{C}\bar{D} + \bar{A}\bar{B}\bar{C}\bar{D}$$

Note that the function for Z has four product terms. The logical sum of two of these terms is equal to W. Thus, by using W, it is possible to reduce the number of terms for Z from four to three, so that the function can fit into the given PAL device. Hence prepare a PAL as given by Table 8.6.

Table 8.6: Solution for example 8.6

S. No.	Product Term	AND Inputs					Output equation
		A	B	C	D	W	
1	$AB\bar{C}$	1	1	-	0	-	$W = AB\bar{C} + \bar{A}\bar{B}C\bar{D}$
2	$\bar{A}\bar{B}C\bar{D}$	0	0	1	0	-	
3	A	1	-	-	-	-	$X = A + BCD$
4	BCD	-	1	1	1	-	
5	$\bar{A}B$	0	1	-	-	-	
6	CD	-	-	1	1	-	$\bar{A}B + CD + \bar{B}\bar{D}$
7	$\bar{B}\bar{D}$	-	0	-	0	-	
8	$A\bar{C}\bar{D}$	1	-	0	0	-	
9	$\bar{A}\bar{B}\bar{C}\bar{D}$	0	0	0	0	-	$Z = W + A\bar{C}\bar{D} + \bar{A}\bar{B}\bar{C}\bar{D}$
10	W	-	-	-	-	1	

The PAL programming table is similar to the table used for the PLA, except that only the inputs of the AND gates need to be programmed. Fig. 8.23 shows the connection map for the PAL device, as specified in the programming table.

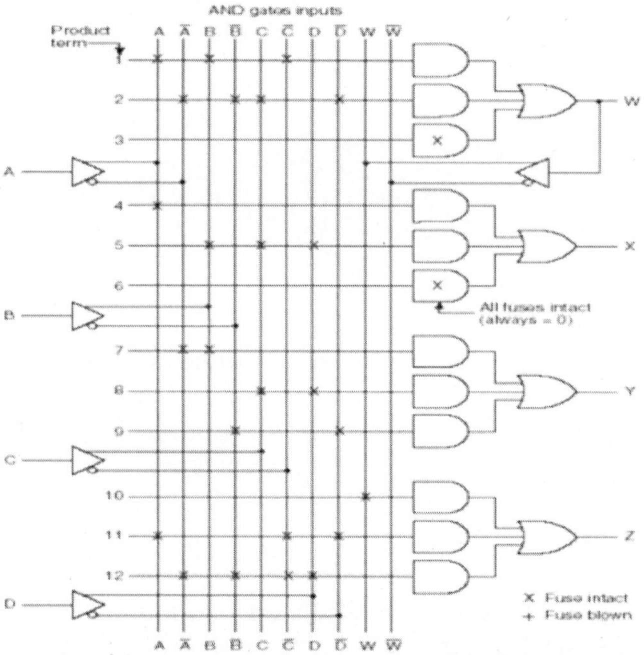

Fig. 8.23: Connection map for the PAL device example 8.6

Since both W and X have two product terms, third AND gate is not used. If all the inputs to this AND gate left intact, then its output will always be 0, because it receives both the true and complement of each input variable i.e., AA' =0 Inferences: If an I/O pin's output-control gate produces a constant 1, the output is always enabled, but the pin may still be used as an input too. Outputs can be used to generate first-pass "helper terms" for logic functions that cannot be performed in a single pass with the limited number of AND terms available for a single output.

Concept of Macro-cell

PALs have the same limitations as PLAs (small number of allowed AND terms) plus they have a fixed OR plane i.e., less flexibility than PLAs. PALs are simpler to manufacture, cheaper, and faster (better performance). PALs also often have extra circuitry connected to the output of each OR gate. The OR gate plus this circuitry is called a macro-cell as shown in **Fig. 8.24.**

- Enable=0; can be used to allow pin for F1 to be used as an additional input pin to the PAL
- Enable=1, Select=0; is the normal for typical PAL operation.
- Enable=1, Select=1; allows the PAL to synchronize the output change with a clock pulse.

The feedback of the AND plane provides for multi-level design

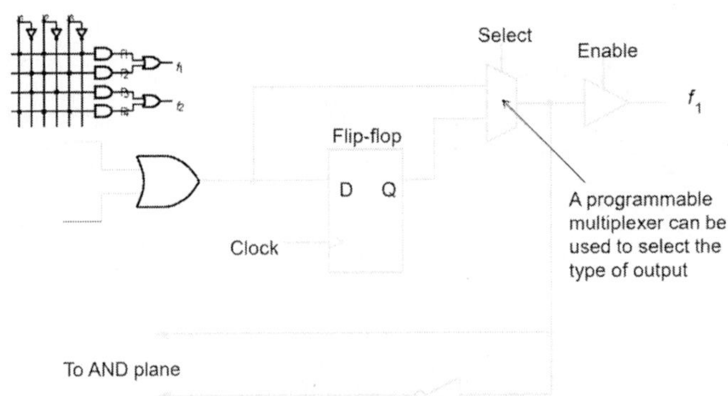

Fig. 8.24: Macrocell in a CPLD

Table 8.7: Comparison of PAL Vs PLA

Programmable Array Logic (PAL)	**Programmable Logic Array (PLA)**
The construction of PAL can be done using the programmable collection of AND & OR gates	The construction of PLA can be done using the programmable collection of AND & fixed collection of OR gates.
The availability of PAL is less prolific	The availability of PLA is more
The flexibility of PAL programming is more	The flexibility of PLA is less
The cost of a PAL is expensive	The cost of PLA is middle range
The number of functions implemented in PAL is large	The number of functions implemented in PLA is limited
The speed of PAL is slow	The speed of PLA is high

Table 8.8: Comparison of ROM, PAL and PLA

Parameter	Rom	PAL	PLA
Structure	Full AND plane and general OR plane	Programmable AND plane and fixed OR plane	Programmable AND plane as well as OR plane
Cost	Cheap (High Volume)	Intermediate Cost	Most expensive (Most Complex in design)
Speed	Medium	High Speed (Only one programmable plane that is much smaller than ROM's decoder)	Slow (Two Programmable planes)
Implementation	Can implement any function of n inputs.	Can implement functions limited by number of terms	Can implement any function upto a product term limit.

8.3 Programming SPLDs:

SPLDs must be programmed so that the switches are in the correct places, CAD tools are usually used to do this. A fuse map is created by the CAD tool and then that map is downloaded to the device via a special programming unit. There are two basic types of programming techniques,

1. Removable sockets on a PCB
2. In system programming (ISP) on a PCB (This approach is not very common for PLAs and PALs but it is quite common for more complex PLDs.)

Removable SPLD Socket Package: The SPLD is removed from the PCB, placed into the unit and programmed there. A SPLD socket package is shown in **Fig. 8.25.**

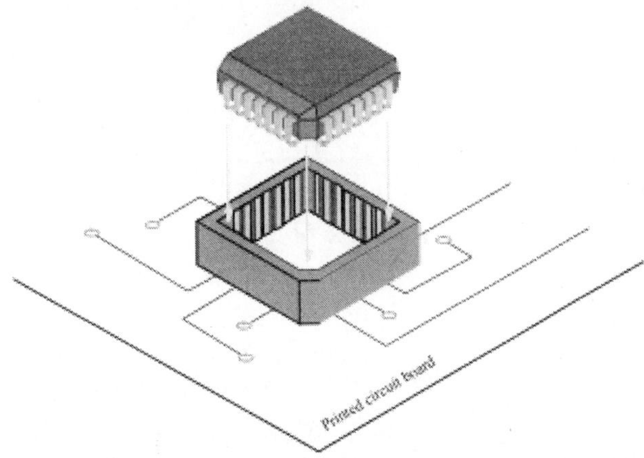

Fig. 8.25: SPLD Socket Package

In-System Programming (ISP): Used when the SPLD cannot be removed from the PCB. A special cable and PCB connection are required to program the SPLD from an attached computer. Very common approach to programming more complex PLDs like CPLDs, FPGAs, etc.

8.3.1 Complex Programmable logic devices (CPLDS)

A CPLD contains a bunch of PLD blocks whose inputs and outputs are connected together by a global interconnection matrix as shown in **Fig. 8.26**. A CPLD is an arrangement of many SPLD-like blocks on a single chip. These circuit blocks might be either PAL-like or PLA-like blocks. Thus, a CPLD has two levels of programmability, in first stage each PLD block can be programmed, further the interconnections between the PLDs can be programmed.

Characteristics of CPLDs:

- They have a higher input to logic gate ratio.
- These devices are denser than SPLDs but have better functional abilities.
- CPLDs are based on EPROM or EEPROM technology.

If you require a larger number of macro-cells for a given application, ranging anywhere between 32 to 1000 macro-cells, then a Complex Programmable Logic Device is the solution. Thus, we use CPLD in applications involving larger I/Os, but data processing is relatively low.

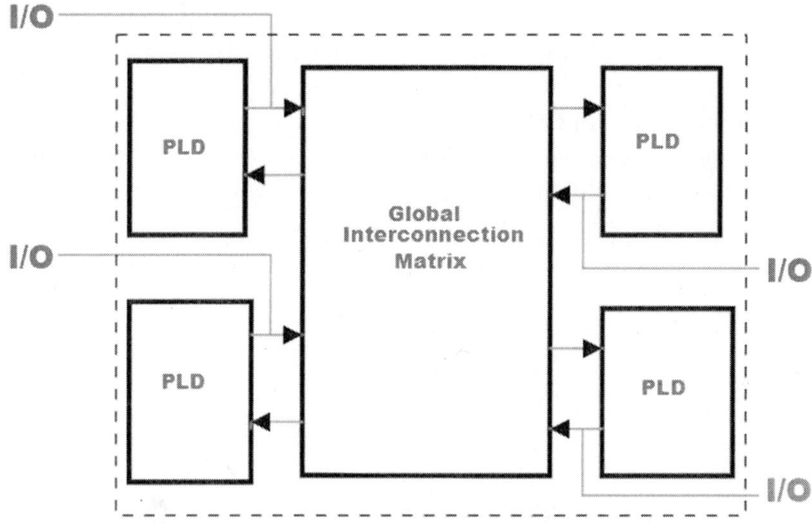

Fig. 8.26: structure of CPLD

8.3.2 Xilinx XC9500 CPLD Family

The Xilinx XC9500 series is a family of CPLDs with a similar architecture but varying numbers of external input/output (I/O) pins and internal PLDs which is called as function blocks (FBs). Its architecture is shown in **Fig. 8.27**. Each internal PLD has 36 inputs and 18 macro-cells according to the number of chip family. Macro-cells whose outputs are usable only internally are sometimes called buried macro-cells.

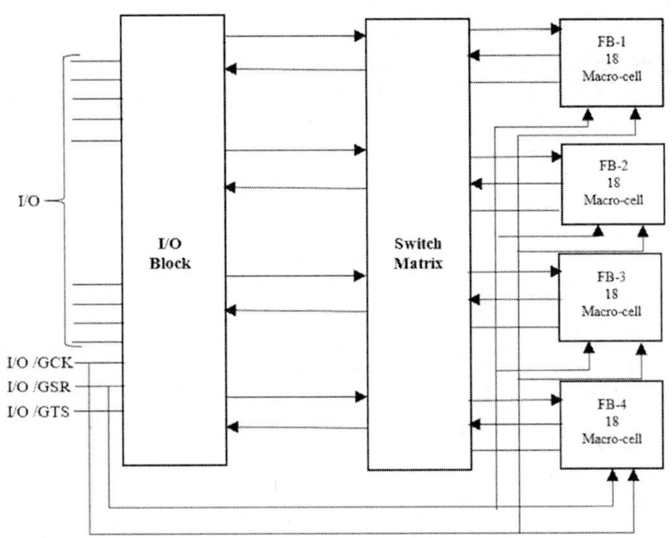

Fig. 8.27: Block Diagram of Xilinx XC9500 CPLD Family (Courtesy: Xilinx), FB-Function Block

Fig. 8.27 shows the block diagram of the internal architecture of a typical XC9500-family CPLD. The external I/O pins can be used as input, output or bi-directional pins according to device programming. The pins marked I/O/GCK, I/O/GSR and I/O/GTS are special purpose pins. Any of these pins can be used as global clocks (GCK). The same pin can be used as an asynchronous preset or clear. Two or four pins can be used as global three state controls (GTS).

Architectural Description:

- Each External I/O pin can be used as an input, and output or a bidirectional pin according to device programming.
- Any of the 3 pins at the bottom can be used as 'Global Clocks'(GCK). One pin can be used as a 'Global set/reset (GSR). 2 or 4 pins depending on the devices can be used as 'Global Three State Controls (GTS).
- Each macro cell can be programmed to use a selected clock input. Each macro cell can use this signal as an asynchronous Preset or Clear. One of the signals can be selected in each macrocell's output is hooked to an external I/O pin.
- The internal PLDs in Xilinx are called as function blocks (FBs). Each internal PLD has 36 inputs and 18 macrocells and outputs. Hence, it can be called as a "36v18"
- Only four functional blocks (FB) are shown but XC 9500 scales to accommodate 16 FBs in the XC 95288.
- Regardless to the specific family member each FB programmable receives 36 signals from the switch matrix.
- The inputs to the Switch matrix are the 18 macro cell outputs from each of the functional blocks and the external inputs from the I/O pins.

Function Block (FB) Architecture

The basic structure of an XC9500 FB is shown in Fig. 8.28. The programmable AND array has just 90 product terms. It has fewer AND terms per microcell. Each FB will receive 36 signals from the switch matrix, the macro-cell outputs from each of the FB and the external inputs from the I/O pins are applied to the switching matrix. Each FB has 18 outputs which run "under" the switch matrix and connect to the I/O blocks. These signals are only the output enable signals for the I/O block output drivers

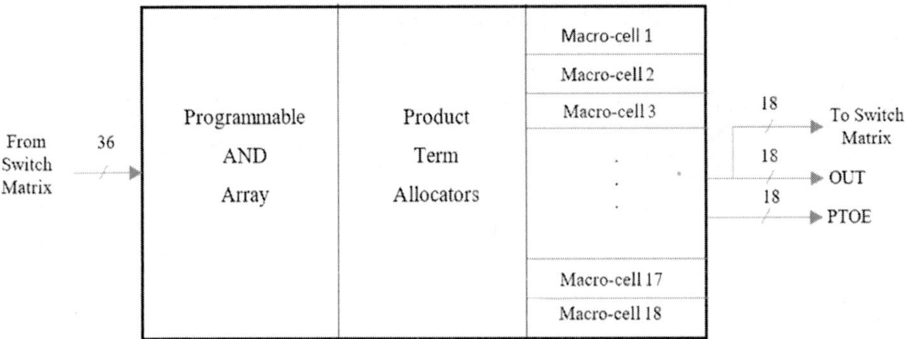

Fig. 8.28: The basic structure of an XC9500 (Courtesy: Xilinx)

Macro-cell in XC9500:

The XC9500 and other CPLDs have product-term allocators that allow a macro-cell's unused product terms to be used by other nearby macro-cells in the same FB. **Fig. 8.29** shows the logic diagram of the XC9500 product-term allocator and macro-cell. In **Fig. 8.29**,

- The rectangular boxes labelled S1–S8 are programmable signal-steering elements that connect their input to one of their two or three outputs.
- The trapezoidal boxes labeled M1–M5 are programmable multiplexers that connect one of their two to four inputs to their output.
- The five AND gates associated with the macro-cell appear on the left-hand side of the figure. Each one is connected to a signal-steering box whose top output connects the product term to the macro-cell's main OR gate G4.

Considering just this, only five product terms are available per macro-cell. However, the top, sixth input of G4 connects to another OR gate G3 that receives product terms from the macro-cells above and below the current one. Any of the macro-cell's product terms that are not otherwise used can be steered through S1–S5 to be combined in an OR gate G1 whose output can eventually be steered to the macro-cell above or below by S8. Before steering, product-term allocator these product terms may be combined with product terms from below or above through S6, S7, and G2 (Refer Fig. 8.29).

Thus, product terms can be *"daisy-chained"* through successive macro-cells to create larger sums of products. In principle, all 90 product terms in the FB could be combined and steered to one macro-cell, although that would leave 17 out of the FB's 18 macro-cells with no product terms at all.

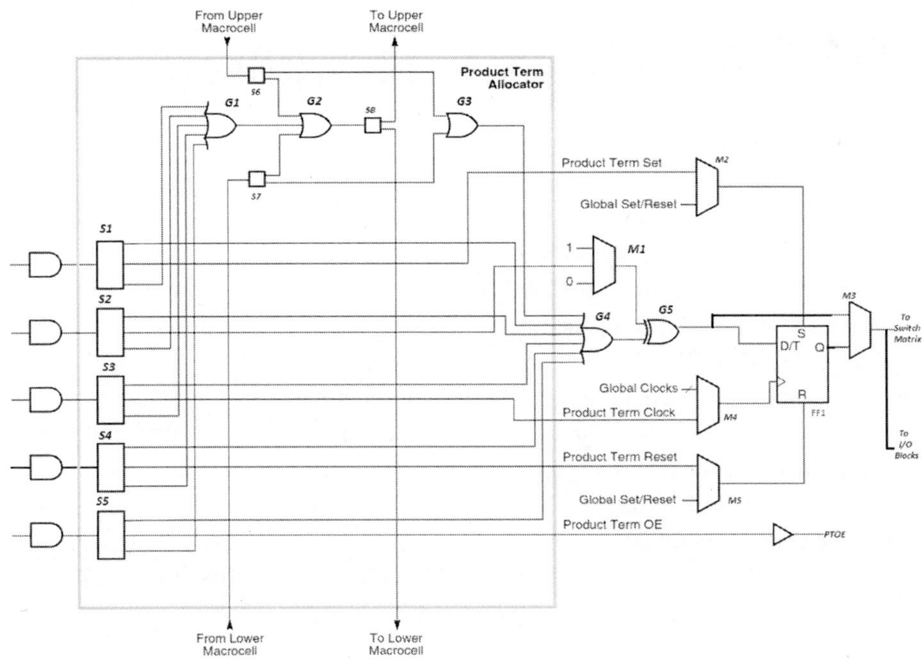

Fig. 8.29: Product term Allocator Logic (Courtesy: Xilinx)

Switch Matrix:

The Fast CONNECT switch matrix connects signals to the FB inputs. All IOB outputs (corresponding to user pin inputs) and all FB outputs drive the Fast CONNECT matrix. Any of these (up to a FB fan-in limit of 36) may be selected, through user programming, to drive each FB with a uniform delay. The switch matrix is capable of combining multiple internal connections into a single wired-AND output before driving the destination FB. This provides additional logic capability and increases the effective logic fan-in of the destination FB without any additional timing delay.

I/O Block:

The I/O Block (IOB) interfaces between the internal logic and the device user I/O pins. Each IOB includes an input buffer, output driver, output enable selection multiplexer, and user programmable ground control. The input buffer is compatible with standard 5V CMOS, 5V TTL, and 3.3V signal levels. The input buffer uses the internal 5V voltage supply (VCCINT) to ensure that the input thresholds are constant and do not vary with the VCCIO voltage.

The output enable may be generated from one of four options: a product term signal from the macrocell, any of the global OE signals, always 1, or always 0. There are two global output enables for devices with up to 144 macrocells, and four global output enables for the rest of the devices. Both polarities of any of the global 3-state control (GTS) pins may be used within the device.

Features:

- High-performance
- Large density range: 36 to 288 macrocells with 800 to6,400 usable gates
- 5V in-system programmable
- Endurance of 10,000 program/erase cycles
- Program/erase over full commercial voltage and temperature range
- Enhanced pin-locking architecture
- Flexible 36V18 Function Block
- 90 product terms drive any or all of 18 macrocells within Function Block
- Global and product term clocks, output enables, set and reset signals
- Extensive IEEE Std 1149.1 boundary-scan (JTAG) support
- Programmable power reduction mode in each macrocell
- Slew rate control on individual outputs - User programmable ground pin capability
- Extended pattern security features for design protection
- High-drive 24 mA outputs
- 3.3V or 5V I/O capability

Example 8.7: Use a CPLD to implement the function, from interconnection wires

$$f = x_1x_3x_6' + x_1x_4x_5x_6' + x_2x_3x_7 + x_2x_4x_5x_7$$

Solution:

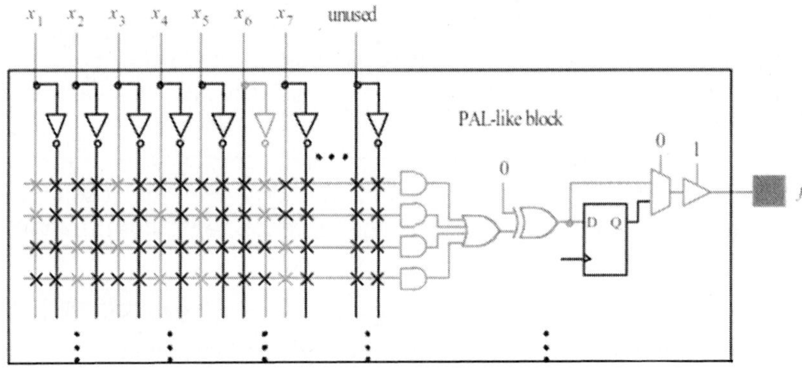

Fig. 8.30: example 8.7

Applications of CPLD

- Complex programmable logic devices are ideal for high performance, critical control applications.
- CPLD can be used in digital designs to perform the functions of boot loader
- CPLD is used for loading the configuration data of a field programmable gate array from non-volatile memory.
- Generally, these are used in small design applications like address decoding
- CPLDs are frequently used many applications like in cost sensitive, battery operated portable devices due to its low size and usage of low power.

8.4 Field Programmable Gate Array (FPGA)

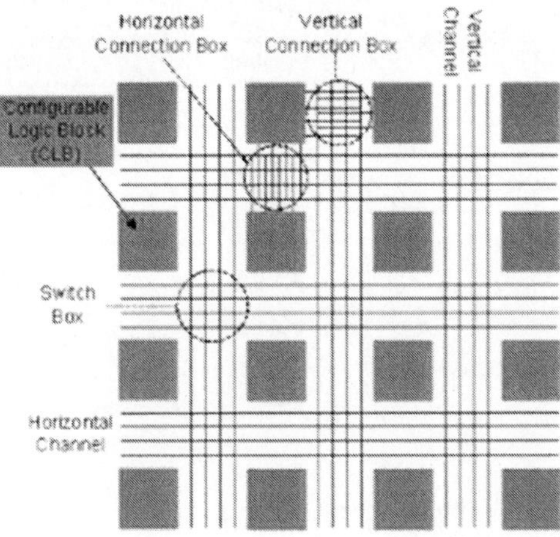

Fig. 8.31: FPGA Architecture

A Field Programmable Gate Array has an entire logic system integrated on a single chip. The basic FPGA architecture is shown in **Fig. 8.31**. It offers excellent flexibility for reprogramming to the system designers. Logic circuitry involving more than a thousand gates use FPGAs. Compared to a normal custom system chip, the FPGA has ten times better integration density. The FPGA consists of 3 main structures:

- Programmable logic structure,
- Programmable routing structure, and
- Programmable Input/output i.e., I/O.

8.4.1 Programmable Logic Structure

The programmable logic structure FPGA consists of a 2-dimensional array of configurable logic blocks (CLBs). A logic block in an FPGA often consists of a LUT, a flip-flop and a multiplexer to select register output as shown in **Fig. 8.32.** The structure has been shown in **Fig. 8.33**.

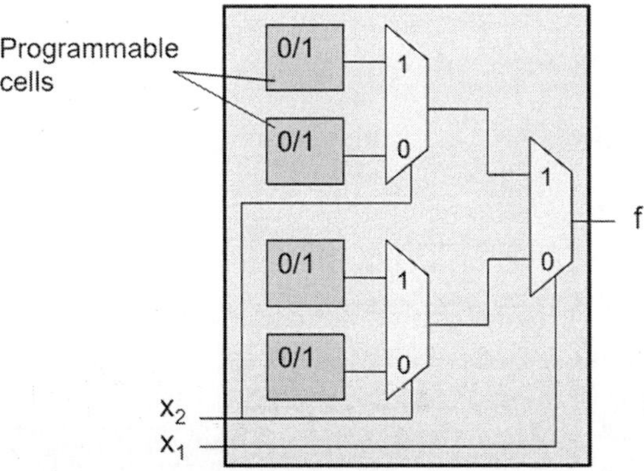

Fig. 8.32: Two input LUT

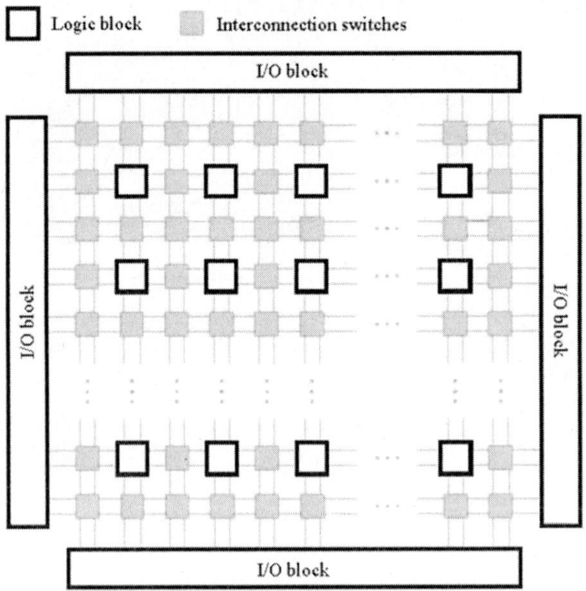

Fig. 8.33: Structure of an FPGA

A LUT with *n* inputs can realize all combinational functions with up to *n* inputs as shown in **Fig. 8.34**. The usual size of LUT in an FPGA is $n = 4$. These logic blocks have a lookup table (LUT) in which the sequential circuitry is implemented. Each CLB can be configured (programmed) to implement any Boolean function of its input variables. Typically, CLBs have between 4-6 input variables.

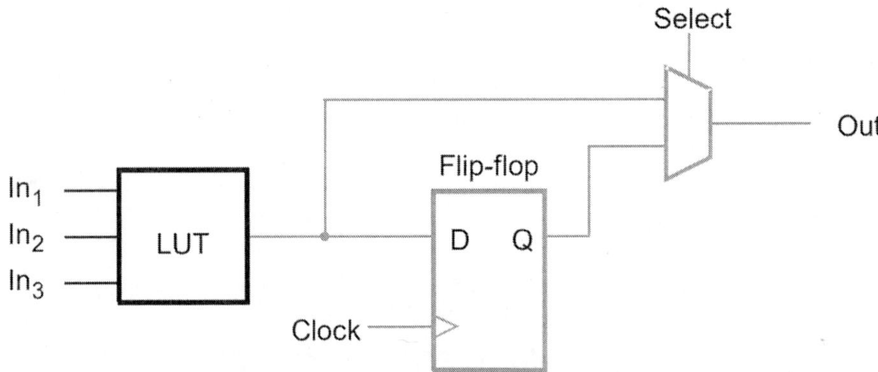

Fig. 8.34: Logic Block in a FPGA

Functions of larger number of variables are implemented using more than one CLB. In addition, each CLB typically contains 1 or 2 FFs to allow implementation of sequential logic. Large designs are partitioned and mapped to a number of CLBs with each CLB configured (programmed) to perform a particular function. These CLBs are then connected together to fully implement the target design. Connecting the CLBs is done using the FPGA programmable routing structure as shown in **Fig. 8.33**.

8.4.2 Configurable Logic Blocks (CLBs):

Look-up Table (LUT)-Based CLB: The basic unit of look-up table-based FPGAs is the configurable logic block. The configurable logic block implements the logic functions. The look-up table-based FPGA is the Xilinx 4000-series FPGA. Further, configurable logic block implements functions.

PLA-Based CLB: PLA-based FPGA devices are based on conventional PLDs. The important advantage of this structure is the logic circuits are implemented using only a few level logics. To improve integration density logic expander is used.

Multiplexer-Based CLB: In Multiplexer-based FPGAs to implement the logic circuits the multiplexers are used. The main advantage of multiplexer-based FPGA is to provide more functionality by using minimum transistors. Due to large number of inputs, multiplexer-based FPGAs place high demands on routing.

The logic circuit design procedure using FPGA involves the following steps:

1. Capture the logic circuit to be implemented with a suitable software package, using a library of logic elements which are various configurations of basic modules available in the FPGA. In addition, many FPGA libraries also contain predesigned circuits for multiplexers, encoders, adders and so on. Predesigned circuits make design much easier.
2. Functional simulation: It simulates the circuits to determine whether it is functioning properly.
3. Configure and interconnect the modules of the FPGA to produce the desired logic circuit. This may be done automatically by routing software called router. Once the routing is over, it is now possible to determine the actual circuit delays which can now be introduced into the simulation model. Now, an accurate simulation of the circuit can be available.
4. Programming: it is a completely automated step in which FPGA interconnections are done. The routing of the devices determines in the previous step in now made into a fuse map. Then, this fuse map is used in conjunction with a device programmer to make the internal device connections.
5. Testing: After programming, it must be tested. If the designed function is not fulfilled, it must be reprogrammed, with careful simulation, reprogramming can be minimized.

8.4.3 Xilinx 4000 FPGA Family:

The principle CLB elements are shown in following figure. Each CLB contains a pair of flip-flops and two independent 4-input function generators. These function generators have a good deal of flexibility as most combinatorial logic functions need less than four inputs. Thirteen CLB inputs and four CLB outputs provide access to the functional flip-flops. Configurable Logic Blocks implement most of the logic in an FPGA.

Two 4-input function generators (F and G) offer unrestricted versatility. Most combinatorial logic functions need four or fewer inputs. However, a third function generator (H) is provided. The H function generator has three inputs. One or both of these inputs can be the outputs of F and G; the other input(s) are from outside the CLB. The CLB can therefore implement certain functions of up to nine variables, like parity check or expandable-identity comparison of two sets of four inputs.

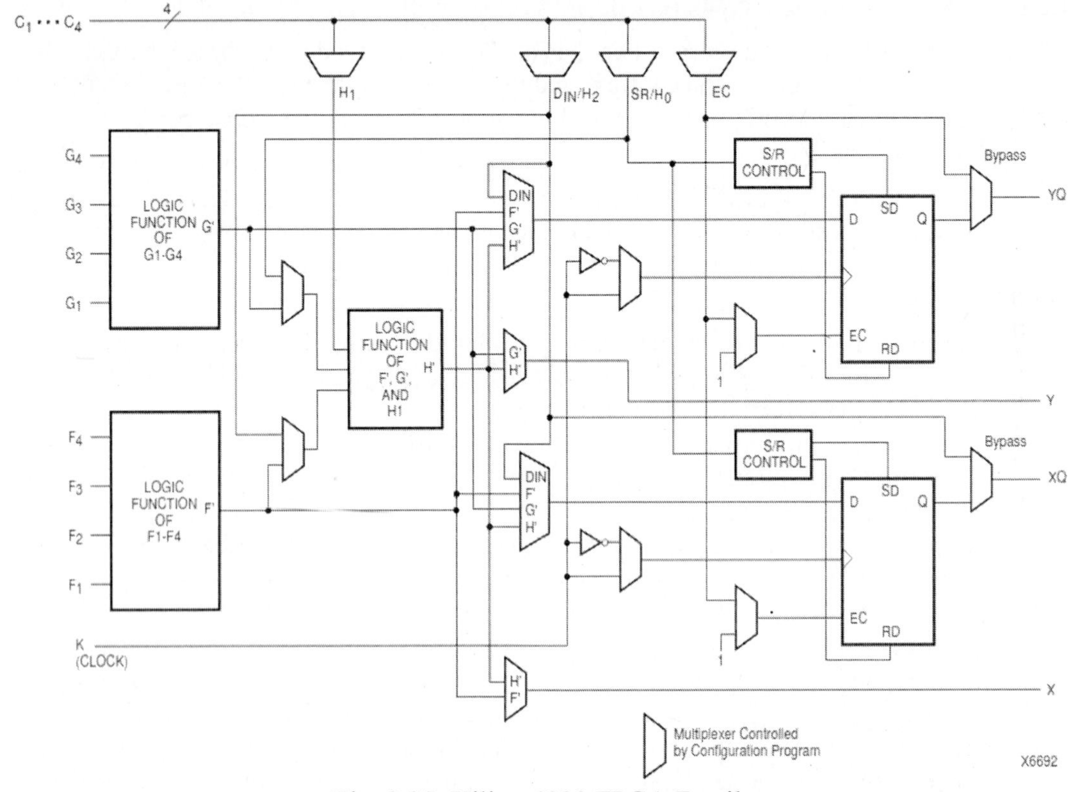

Fig. 8.35: Xilinx 4000 FPGA Family

Each CLB contains two flip-flops that can be used to store the function generator outputs. However, the flip-flops and function generators can also be used independently. DIN can be used as a direct input to either of the two flip-flops. H1 can drive the other flip-flop through the H function generator. Function generator outputs can also be accessed from outside the CLB, using two outputs independent of the flip-flop outputs. This versatility increases logic density and simplifies routing. Thirteen CLB inputs and four CLB outputs provide access to the function generators and flip-flops. These inputs and outputs connect to the programmable interconnect resources outside the block.

Four independent inputs are provided to each of two function generators (F1 - F4 and G1 - G4). These function generators, whose outputs are labelled F' and G', are each capable of implementing any arbitrarily defined Boolean function of four inputs. The function generators are implemented as memory look-up tables. The propagation delay is therefore independent of the function implemented. A third function generator, labelled H', can implement any Boolean function of its three inputs. Two of these inputs can optionally be the F' and G' functional generator out-puts. Alternatively, one or both of these inputs can come from outside the CLB (H2, H0). The third input must come from outside the block (H1). Signals from the function generators can exit the CLB on two outputs. F' or H' can be connected to the X output. G' or H' can be connected to the Y output.

A CLB can be used to implement any of the following functions:

- Any function of up to four variables, plus any second function of up to four unrelated variables, plus any third function of up to three unrelated variables.
- Any single function of five variables
- Any function of four variables together with some functions of six variables
- Some functions of up to nine variables Implementing wide functions in a single block reduces both the number of blocks required and the delay in the signal path, achieving both increased density and speed.

The versatility of the CLB function generators significantly improves system speed. In addition, the design software tools can deal with each function generator independently. This flexibility improves cell usage. The flexibility and symmetry of the CLB architecture facilitates the placement and routing of a given application. Since the function generators and flip-flops have independent inputs and outputs, each can be treated as a separate entity during placement to achieve high packing density. Inputs, outputs and the functions themselves can freely swap positions within thew CLB to avoid routing congestion during the placement and routing operation.

8.4.4 Programmable Routing Structure

To allow for flexible interconnection of CLBs, FPGAs have 3 programmable routing resources:

1. *Vertical and horizontal routing channels* which consist of different length wires that can be connected together if needed. These channels run vertically and horizontally between columns and rows of CLBs as shown in the Figure.
2. *Connection boxes*, which are a set of programmable links that can connect input and output pins of the CLBs to wires of the vertical or the horizontal routing channels.
3. *Switch boxes,* located at the intersection of the vertical and horizontal channels. These are a set of programmable links that can connect wire segments in the horizontal and vertical channels.

8.4.5 Programmable I/O

These are buffers, that can be configured either as input buffers, output buffers or input/output buffers (IOBs). They allow the pins of the FPGA chip to function either as input pins, output pins or input/output pins. The IOBs provide a simple interface between the internal user logic and the package pins.

Input Signals:

Two paths, labelled I1 and I2, bring input signals into the array. Inputs also connect to an input register that can be programmed as either an edge-triggered flip-flop or a level sensitive transparent-Low latch. The choice is made by placing the appropriate primitive from the symbol library. The inputs can be globally configured for either TTL (1.2V) or CMOS (2.5V) thresholds.

The two global adjustments of input threshold and output level are independent of each other. There is a slight hysteresis of about 300mV.Seperate clock signals are provided for the input and output registers; these clocks can be inverted, generating either falling-edge or rising edge triggered flip-flops. As is the case with the CLB registers, a global set/reset signal can be used to set or clear the input and output registers whenever the RESET net is alive.

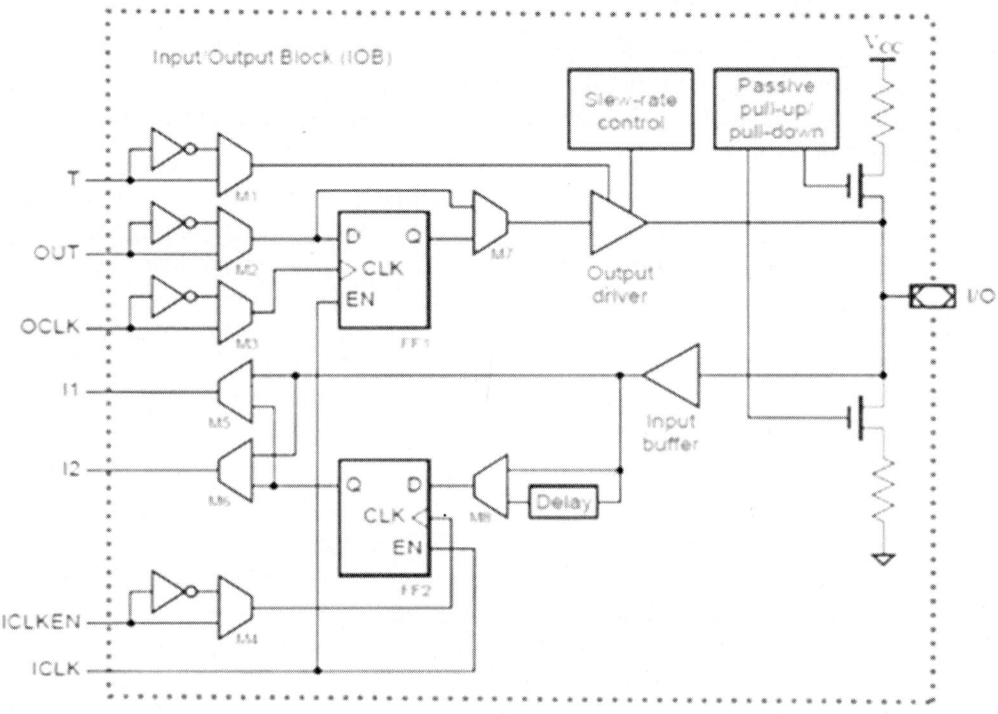

Fig. 8.36: Programmable I/O

Registered Inputs:

The I1 and I2 signals that exit the block can each carry either the direct or registered input signal. The input and output storage elements in each IOB have a common clock enable input, which through configuration can be activated individually for the input or output flip- flop or both. This clock enable operates exactly like the EC pin on the XC4000E CLB. It cannot be inverted within the IOB.

Applications of FPGAs

- Implementation of random logic and easy to changes at system-level.
- Can eliminate need for full-custom chips that helps in prototyping to have more/better/faster debugging done than possible with simulation
- Ensemble of gate arrays used to emulate a circuit to be manufactured
- Reconfigurable hardware i.e., one hardware block used to implement more than one functions must be mutually-exclusive in time
- Can greatly reduce cost while enhancing flexibility
- Special-purpose computation engines i.e., hardware dedicated to solving one problem (or class of problems)
- Accelerators attached to general-purpose computers

Table 8.9: Comparison between PROM, PLA and PAL

S. No.	PROM	PLA	PAL
1	It consists of a programmable OR array followed by a fixed AND array.	It comprises both programmable OR array and AND array.	It comprises a programmable AND array followed by a fixed OR array.
2	It is cheaper and simpler to use.	It is costliest amongst all and is also complex to use.	It is cheaper and simpler to use.
3	In this, all min-terms are decoded.	AND array can be programmed to get the desired min-term.	AND array can be programmed to get the desired min-term.
3	If a Boolean function is in standard SOP form, then only it can be implemented using PROM.	Any Boolean function needs not to be in standard SOP form, it can still be implemented using PLA.	Any Boolean function needs not to be in standard SOP form, it can still be implemented using PLA.

Example 8.8: Use an FPGA with 2 input LUTS to implement the function,

$$f = x_1x_3x_6' + x_1x_4x_5x_6' + x_2x_3x_7 + x_2x_4x_5x_7$$

Solution:

Fan-in of expression is too large for FPGA. This was simple to do in a CPLD Factor f to get sub-expressions with max *fan-in = 2*. The implementation of expression above is shown in Fig. 8.37. The goal is to build expressions out of 2-input LUTs

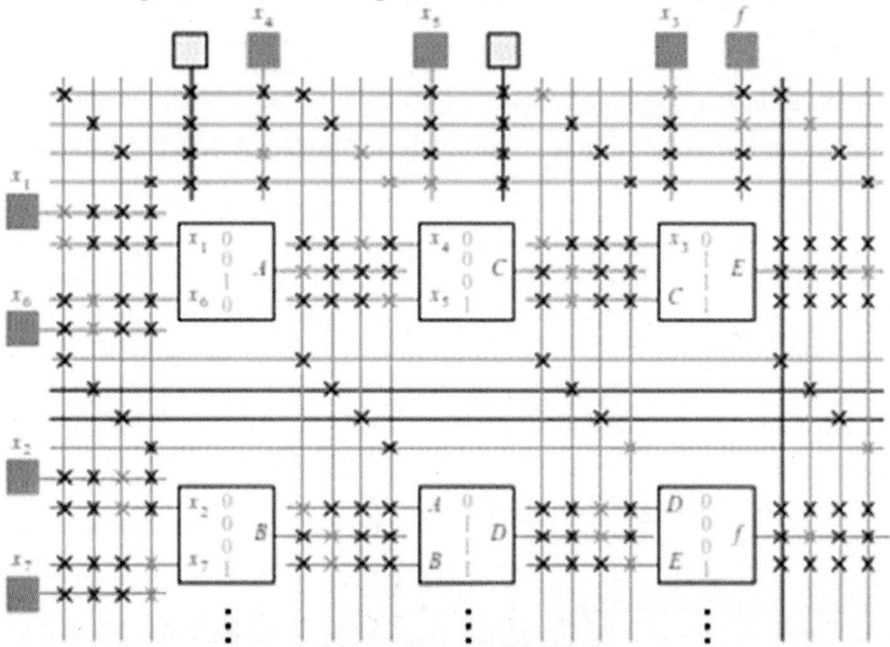

Fig. 8.37: Example 8.8

Example for four and three input functions:

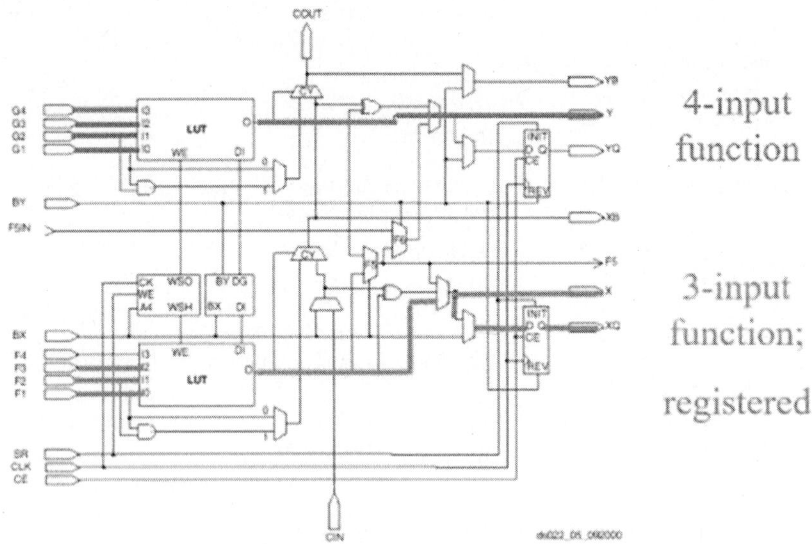

Fig. 8.38: For four and three input functions

Unit summary:

Over the last few years, programmable logic suppliers have made such phenomenal technical advances that PLDs are now seen as the logical solution of choice from many designers. PLD (Programmable Logic Device) and FPGA (Field-Programmable Gate Array) are two types of digital logic devices that allow designers to create custom digital circuits without the need for custom ASICs (Application-Specific Integrated Circuits).

A PLD is a general term that refers to any type of programmable logic device that can be programmed by a user to perform a specific function. PLDs typically include simple programmable logic devices (SPLDs) and complex programmable logic devices (CPLDs). An FPGA, on the other hand, is a type of PLD that is based on a matrix of programmable logic blocks (PLBs) interconnected by programmable interconnects. The PLBs can be configured to perform any logic function, and the interconnects can be programmed to connect the PLBs in any desired configuration.

FPGAs are more flexible and powerful than PLDs, as they can be programmed to perform complex logic functions and are capable of implementing complete digital systems. However, they are also more expensive and require more design effort than PLDs.

Here are some of the specific benefits and applications of PLDs:

1. Flexibility: PLDs are highly flexible and can be programmed to perform any digital logic function. This means that designers can quickly and easily create custom digital circuits that meet their specific needs.
2. Low cost: PLDs are less expensive than ASICs, which require custom design and manufacturing. This makes them a cost-effective solution for many digital design applications.

3. Fast turnaround time: Because PLDs can be programmed rather than manufactured, they can be produced more quickly than ASICs. This can be important for applications where time-to-market is critical.
4. Small form factor: PLDs are available in a variety of small form factors, including surface-mount packages and chip-scale packages. This makes them well-suited for applications where space is limited.
5. Reduced power consumption: PLDs typically consume less power than ASICs, which can be important in battery-powered applications or other applications where power consumption is a concern.

Both PLDs and FPGAs have a wide range of applications in digital design, including in telecommunications, control systems, image processing, and many others.

Solved examples

Example 8.9: Let us implement the following Boolean functions using PROM.

$$A(X,Y,Z) = \sum m(5,6,7)$$
$$B(X,Y,Z) = \sum m(3,5,6,7)$$

Solution:

The given two functions are in sum of min terms form and each function is having three variables X, Y & Z. So, we require a 3 to 8 decoder and two programmable OR gates for producing these two functions. The corresponding PROM is shown in **Fig. 8.39.**

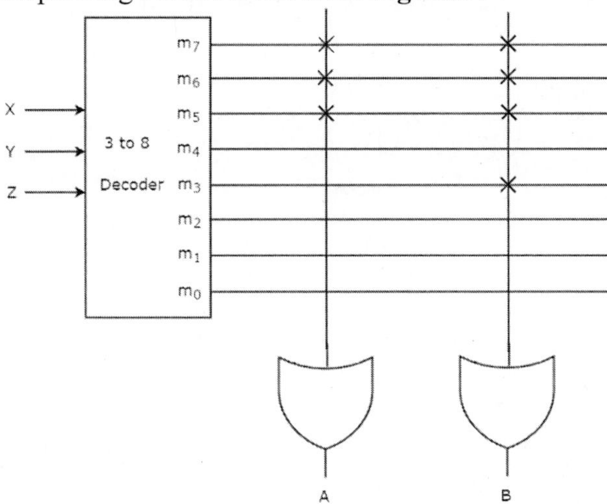

Fig. 8.39: Solution for example 8.9

Here, 3 to 8 decoder generates eight min terms. The two programmable OR gates have the access of all these min terms. But, only the required min terms are programmed in order to produce the respective Boolean functions by each OR gate. The symbol 'X' is used for programmable connections.

Example 8.10: Design a combinational circuit using ROM. The circuit accepts a 3-bit number and generate an output binary equal to the square of the number.
Solution: Derive truth table and Block diagram as shown in **Fig. 8.40**

A2	A1	A0	F1 / B5	F2 / B4	F3 / B3	F4 / B2	B1	B0
0	0	0	0	0	0	0	0	0
0	0	1	0	0	0	0	0	1
0	1	0	0	0	0	1	0	0
0	1	1	0	0	1	0	0	1
1	0	0	0	1	0	0	0	0
1	0	1	0	1	1	0	0	1
1	1	0	1	0	0	1	0	0
1	1	1	1	1	0	0	0	1

ROM Truth Table Block Diagram

Fig. 8.40: Truth Table and Block Diagram for example 8.10

B1 is ALWAYS *0*, no need to generate it using the ROM. Whereas, *B0* is equal to *A0*, no need to generate it using the ROM. Therefore, the minimum size of ROM needed is *23 X 4 or 8 X 4*.

Example 8.11: Tabulate the truth for an *8 X 4* ROM that implements the following four Boolean functions:

$A(X,Y,Z) = \sum m(3,6,7)$; $B(X,Y,Z) = \sum m(0,1,4,5,6)$; $C(X,Y,Z) = \sum m(2,3,4)$; and $D(X,Y,Z) = \sum m(2,3,4,7)$

Solution: Derive truth table and Block diagram as shown in **Fig. 8.41**

Decimal	Inputs			Outputs			
	X	Y	Z	A	B	C	D
0	0	0	0	0	1	0	0
1	0	0	1	0	1	0	0
2	0	1	0	0	0	1	1
3	0	1	1	1	0	1	1
4	1	0	0	0	1	1	1
5	1	0	1	0	1	0	0
6	1	1	0	1	1	0	0
7	1	1	1	1	0	0	1

Fig. 8.41: Truth Table and Block Diagram for example

Example 8.12: Implement the following Boolean functions using PAL.

$$A = XY + X\bar{Z} \quad and \quad B = X\bar{Y} + Y\bar{Z}$$

Solution:

The given two functions are in sum of products form. There are two product terms present in each Boolean function. So, we require four programmable AND gates & two fixed OR gates for producing those two functions. The corresponding PAL is shown in the Fig. 8.42.

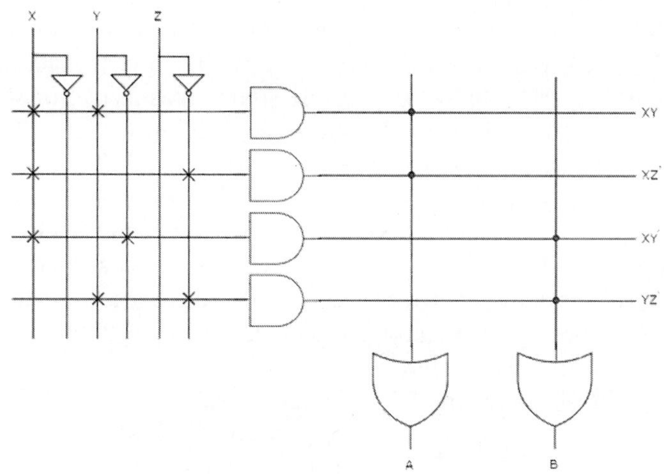

Fig. 8.42: PAL for example 8.12

The programmable AND gates have the access of both normal and complemented inputs of variables. In the above figure, the inputs X, X′, Y, Y′, Z & Z′, are available at the inputs of each AND gate. So, program only the required literals in order to generate one product term by each AND gate. The symbol 'X' is used for programmable connections.

Here, the inputs of OR gates are of fixed type. So, the necessary product terms are connected to inputs of each OR gate. So that the OR gates produce the respective Boolean functions. The symbol '.' is used for fixed connections.

Example 8.13: Implement the following **Boolean functions** using PLA.

$$A = XY + XZ'$$
$$B = XY' + YZ + XZ'$$

Solution:

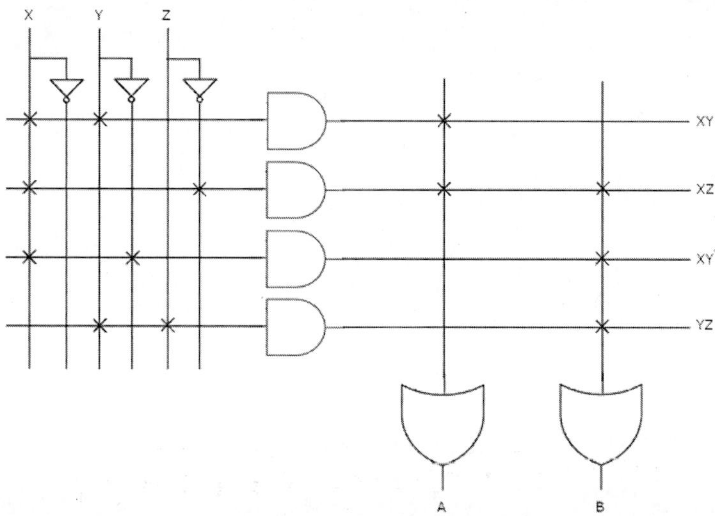

Fig. 8.43: example 8.13

The given two functions are in sum of products form. The number of product terms present in the given Boolean functions A & B are two and three respectively. One product term, Z'X is common in each function. So, we require four programmable AND gates & two programmable OR gates for producing those two functions. The corresponding **PLA** is shown in Fig. 8.43.

The **programmable AND gates** have the access of both normal and complemented inputs of variables. In the above figure, the inputs X, X′, Y, Y′, Z & Z′, are available at the inputs of each AND gate. So, program only the required literals in order to generate one product term by each AND gate. All these product terms are available at the inputs of each **programmable OR gate.** But, only program the required product terms in order to produce the respective Boolean functions by each OR gate. The symbol 'X' is used for programmable connections.

Example 8.14: Using PROM realize the following expression

$$F_1(A,B,C) = \sum m(0,1,3,5,7)$$

$$F_2(A,B,C) = \sum m(1,2,5,6)$$

Solution:

Truth table (Refer Table 8.10) and PROM diagram (Refer **Fig. 8.44)** for the given function is shown below.

Table 8.10: Truth table

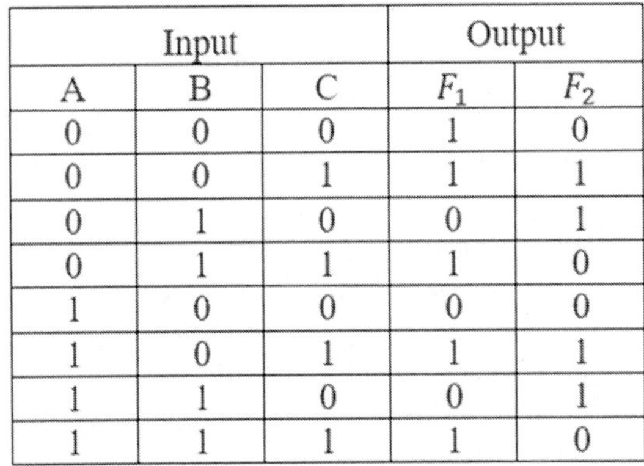

Input			Output	
A	B	C	F_1	F_2
0	0	0	1	0
0	0	1	1	1
0	1	0	0	1
0	1	1	1	0
1	0	0	0	0
1	0	1	1	1
1	1	0	0	1
1	1	1	1	0

Fig. 8.44: PROM Diagram

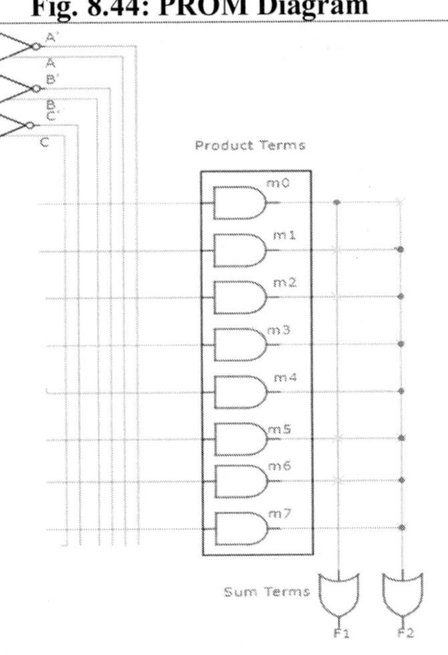

Review Questions

1. Design a combinational circuit using PROM. The circuit accepts 3-bit binary and generates its equivalent Excess-3 code.
2. Design a BCD to Excess-3 code converter and implement using suitable PLA.
3. Implement the following Boolean expression using PAL,

$$F_1(A,B,C) = \sum m(3,5,7)$$

$$F_2(A,B,C) = \sum m(4,5,7)$$

4. Implement the following Boolean expressions using a suitable PLA.

$$A(x,y,z) = \sum m(1,2,4,6)$$

$$B(x,y,z) = \sum m(0,1,2,6,7)$$

$$C(x,y,z) = \sum m(2,6)$$

$$D(x,y,z) = \sum m(1,2,3,5,7)$$

5. Implement the following two Boolean functions with a PLA:

$$F_1(A,B,C) = \sum m(0,1,2,4)$$

$$F_2(A,B,C) = \sum m(0,5,6,7)$$

6. How does a programmable logic device differ from a fixed logic device? What are the primary advantages of using programmable logic devices?
7. Distinguish between a programmable logic array (PLA) device and a programmable array logic (PAL) device in terms of architecture and capability to implement Boolean functions.
8. How does a generic array logic (GAL) device differ from its PAL counterpart? Do they differ in their internal architecture? If yes, then how?
9. What are complex programmable logic devices (CPLDs)? Briefly outline salient features of these devices and application areas where these devices fit the best.
10. How does the architecture of a typical FPGA device differ from that of a CPLD? In what way does the architecture affect the timing performance in the two cases?
11. What are the various interconnect technologies used for the purpose of programming PLDs? Briefly describe each one of them.
12. What is a hardware description language? What are the requirements of a good HDL? Briefly describe the salient features of VHDL and Verilog.
13. What do you understand by the following as regards programmable logic devices?
    a. combinational and registered outputs;
    b. configurable output logic cell;
    c. reprogrammable PLD;
    d. in-system programmability.
14. Determine the size of PROM required for implementing the following logic circuits.
    a. 16-to-1 multiplexer;
    b. four-bit binary adder.

15. Determine the number of programmable interconnections in the following programmable logic devices.
    a. 1K × 4 PROM;
    b. PLA device with four input variables, 32 AND gates and four OR gates;
    c. PAL device with eight input variables, 16 AND gates and four OR gates.
16. Figure below shows a portion of the internal logic diagram of a certain PAL device that uses anti-fuse interconnect technology. In the diagram shown, a cross (×) represents an unprogrammed interconnect and the absence of a cross (×) at an intersection of input and product lines represents programmed interconnects; a dot (•) represents a hard-wired interconnect. Write
    a. The Boolean expression for Y and
    b. The Boolean expression for Y if the interconnect technology were fuse based.

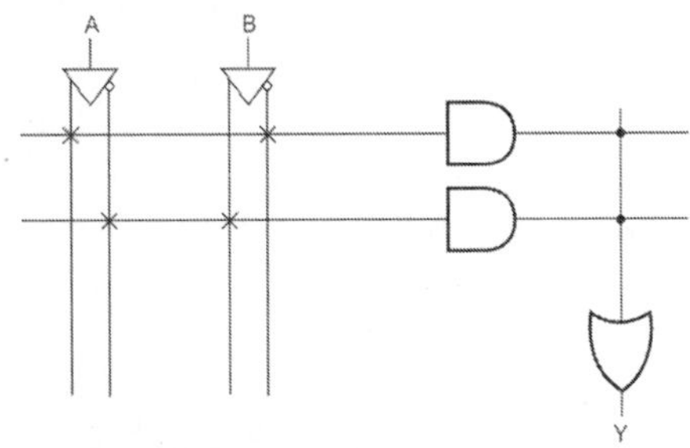

Fig. 8.45: Problem no. 16

17. A and B are two binary variables. The objective is to design a magnitude comparator to produce A =B, A <B and A >B outputs. Design a suitable PLD with a PAL-like architecture using anti-fuse-based interconnects.
18. Figure 9.45 shows a programmed PAL device using fuse-based interconnects. Examine the logic diagram and determine the logic block implemented by the PLD. A cross (×) represents an unprogrammed interconnection and a dot (•) represents a hard-wired interconnection.

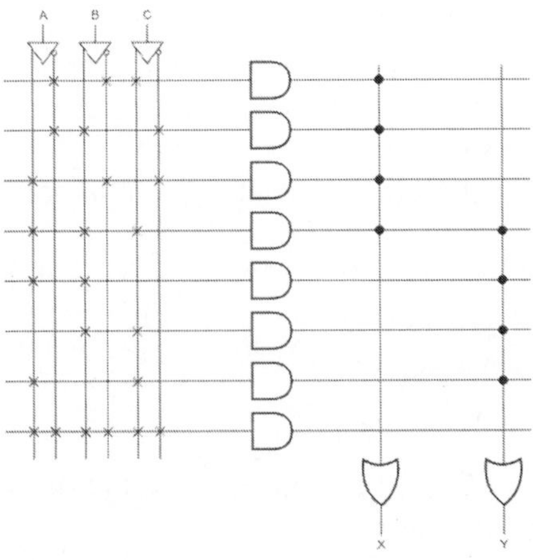

Fig. 8.46: Problem No. 18

Multiple Choice Questions (MCQs)

1. The inputs in the PLD is given through ____________
   a) NAND gates
   b) OR gates
   c) NOR gates
   d) AND gates
2. PAL refers to ____________
   a) Programmable Array Loaded
   b) Programmable Logic Array
   c) Programmable Array Logic
   d) Programmable AND Logic
3. Outputs of the AND gate in PLD is known as ____________
   a) Input lines
   b) Output lines
   c) Strobe lines
   d) Control lines
4. PLA contains ____________
   a) AND and OR arrays
   b) NAND and OR arrays
   c) NOT and AND arrays
   d) NOR and OR arrays
5. PLA is used to implement ____________
   a) A complex sequential circuit
   b) A simple sequential circuit
   c) A complex combinational circuit
   d) A simple combinational circuit
6. A PLA is similar to a ROM in concept except that ____________
   a) It hasn't capability to read only
   b) It hasn't capability to read or write operation
   c) It doesn't provide full decoding to the variables
   d) It hasn't capability to write only
7. For programmable logic functions, which type of PLD should be used?
   a) PLA
   b) PAL
   c) CPLD
   d) SLD
8. The complex programmable logic device contains several PLD blocks and __________
   a) A language compiler
   b) AND/OR arrays
   c) Global interconnection matrix
   d) Field-programmable switches
9. Which type of device FPGA are?

a) SLD
b) SROM
c) EPROM
d) PLD

10. The difference between a PAL & a PLA is ____________
a) PALs and PLAs are the same thing
b) The PLA has a programmable OR plane and a programmable AND plane, while the PAL only has a programmable AND plane
c) The PAL has a programmable OR plane and a programmable AND plane, while the PLA only has a programmable AND plane
d) The PAL has more possible product terms than the PLA
11. If a PAL has been programmed once ____________
a) Its logic capacity is lost
b) Its outputs are only active HIGH
c) Its outputs are only active LOW
d) It cannot be reprogrammed
12. The FPGA refers to ____________
a) First programmable Gate Array
b) Field Programmable Gate Array
c) First Program Gate Array
d) Field Program Gate Array
13. The full form of VLSI is ____________
a) Very Long Single Integration
b) Very Least Scale Integration
c) Very Large Scale Integration
d) Very Long Scale Integration
14. In FPGA, vertical and horizontal directions are separated by ____________
a) A line
b) A channel
c) A strobe
d) A flip-flop
15. Applications of PLAs are ____________
a) Registered PALs
b) Configurable PALs
c) PAL programming
d) All of the Mentioned
16. Which type of PLD should be used to program basic logic functions?
a) PAL
b) PLA
c) CPLD
d)SLD
17. The content of a simple programmable logic device (PLD) consists of:
a) fuse-link arrays
b) thousands of basic logic gates

c) advanced sequential logic functions
d) thousands of basic logic gates and advanced sequential logic functions

18. Once a PAL has been programmed:
a) it cannot be reprogrammed.
b) its outputs are only active HIGHs
c) its outputs are only active LOWs
d) its logic capacity is lost

19. The complex programmable logic device (CPLD) contains several PLD blocks and:
a) field-programmable switches
b) AND/OR arrays
c) a global interconnection matrix
d) a language compiler

20. The OR array in a PAL is ________.
a) fixed
b) programmable
c) nonexistent; there is no OR array in a PAL
d) floating

References and Suggested Readings

- *Brown, S., Francis, R., Rose, J. and Vranesic, Z. (1992) Field Programmable Gate Arrays, Kluwer Academic Publishers, MA, USA.*
- *Chan, P., and Mourad, S. Digital Design Using Field Programmable Gate Arrays. Upper Saddle River, N.J.: Prentice Hall, 1994.*
- *Chartrand, L. (2003) Digital Fundamentals: Experiments and Concepts with CPLD, Thomson Delmar Learning, New York, USA.*
- *Dewey, Allen. Analysis and Design of Digital Systems with VHDL. Toronto, Ontario, Canada: Thomson Engineering, 1997.*
- *The Programmable Logic Data Book, 2nd ed. 1994. San Jose, CA: Xilinx, Inc.*
- *https://docs.xilinx.com/v/u/en-US/DS063*
- *https://www.xilinx.com/support/documents/data_sheets/4000.pdf*
- *IEEE Standard Hardware Description Language Based on the Verilog Hardware Description Language (IEEE Std 1364-2005). 2005. New York: Institute of Electrical and Electronics Engineers.*

Appendix-A

ASCII and Extended ASCII Characters

ASCII

ASCII (*which stands for American Standard Code for Information Interchange*) is a character encoding standard for text files in computers and other devices. ASCII is a subset of Unicode and is made up of 128 symbols in the character set. These symbols consist of letters (both uppercase and lowercase), numbers, punctuation marks, special characters and control characters. Each symbol in the character set can be represented by a Decimal value ranging from 0 to 127, as well as equivalent Hexadecimal and Octal values. The following is a listing of ASCII values displaying the Decimal, Hexadecimal, Octal and Character values for each ASCII character.

Dec	Hex	Oct	Char	Description
0	00	000	^@	Null (NUL)
1	01	001	^A	Start of heading (SOH)
2	02	002	^B	Start of text (STX)
3	03	003	^C	End of text (ETX)
4	04	004	^D	End of transmission (EOT)
5	05	005	^E	Enquiry (ENQ)
6	06	006	^F	Acknowledge (ACK)
7	07	007	^G	Bell (BEL)
8	08	010	^H	Backspace (BS)
9	09	011	^I	Horizontal tab (HT)
10	0A	012	^J	Line feed (LF)
11	0B	013	^K	Vertical tab (VT)
12	0C	014	^L	New page/form feed (FF)
13	0D	015	^M	Carriage return (CR)
14	0E	016	^N	Shift out (SO)
15	0F	017	^O	Shift in (SI)
16	10	020	^P	Data link escape (DLE)
17	11	021	^Q	Device control 1 (DC1)
18	12	022	^R	Device control 2 (DC2)
19	13	023	^S	Device control 3 (DC3)
20	14	024	^T	Device control 4 (DC4)
.	.	.	.	.
.	.	.	.	.
.	.	.	.	.
.	.	.	.	.
60	3C	074	<	Less-than sign
61	3D	075	=	Equal/Equality sign

Dec	Hex	Oct	Char	Description
62	3E	076	>	Greater-than sign
63	3F	077	?	Question mark
.	.	.	.	.
.	.	.	.	.

In the ASCII character set, the Decimal values 0 to 31 as well as the Decimal value 127 represent symbols that are non-printable. It is possible to generate these non-printable characters using a key sequence where ^ represents the control key on your keyboard. For example, you could generate a carriage return (Decimal value 13) by pressing the control key followed by the letter M on your keyboard (^M).

All other symbols in the character set can printed or represented on the screen. These printable character values can be seen in the Char field in the table above.

Appendix-B
Extended ASCII Characters

There are several different variations of the 8-bit ASCII table. The table below is according to Windows-1252 (CP-1252) which is a superset of ISO 8859-1, also called ISO Latin-1, in terms of printable characters, but differs from the IANA's ISO-8859-1 by using displayable characters rather than control characters in the 128 to 159 range.

Dec	Oct	Hex	BIN	Char	Description
128	200	80	10000000	€	Euro sign
129	201	81	10000001		*Unused*
130	202	82	10000010	‚	Single low-9 quotation mark
131	203	83	10000011	ƒ	Latin small letter f with hook
132	204	84	10000100	„	Double low-9 quotation mark
133	205	85	10000101	…	Horizontal ellipsis
134	206	86	10000110	†	Dagger
135	207	87	10000111	‡	Double dagger
136	210	88	10001000	ˆ	Modifier letter circumflex accent
137	211	89	10001001	‰	Per mille sign
138	212	8A	10001010	Š	Latin capital letter S with caron
153	231	99	10011001	™	Trade mark sign
154	232	9A	10011010	š	Latin small letter S with caron
155	233	9B	10011011	›	Single right-pointing angle quotation mark
156	234	9C	10011100	œ	Latin small ligature oe
157	235	9D	10011101		*Unused*
158	236	9E	10011110	ž	Latin small letter z with caron
159	237	9F	10011111	Ÿ	Latin capital letter Y with diaeresis
160	240	A0	10100000	NBSP	Non-breaking space

Dec	Oct	Hex	BIN	Char	Description
161	241	A1	10100001	¡	Inverted exclamation mark
162	242	A2	10100010	¢	Cent sign
163	243	A3	10100011	£	Pound sign
164	244	A4	10100100	¤	Currency sign
165	245	A5	10100101	¥	Yen sign
166	246	A6	10100110	¦	Pipe, broken vertical bar
167	247	A7	10100111	§	Section sign
168	250	A8	10101000	¨	Spacing diaeresis - umlaut
169	251	A9	10101001	©	Copyright sign
170	252	AA	10101010	ª	Feminine ordinal indicator
.	.	.	.	.	.
.	.	.	.	.	.
.	.	.	.	.	.
.	.	.	.	.	.
.	.	.	.	.	.
.	.	.	.	.	.
	.	.	.	.	.

(for more details please do visit)
https://www.ascii-code.com
OR
Scan the QR code

Appendix-C
EBCDIC Code

EBCDIC stands for ***Extended Binary Coded Decimal Interchange Code***. IBM invented this code to extend the Binary Coded Decimal which existed at that time. All the IBM computers and peripherals use this code. It is an 8-bit code and therefore can accommodate 256 characters.

Dec	Hex	Code	Dec	Hex	Code	Dec	Hex	Code	Dec	Hex	Code
0	00	NUL	10	0A		20	14		30	1E	IRS
1	01	SOH	11	0B	VT	21	15		31	1F	IUS
2	02	STX	12	0C	FF	22	16	BS	**32**	20	
3	03	ETX	13	0D	CR	23	17		**33**	21	
4	04		14	0E	SO	24	18	CAN	**34**	22	
5	05	HT	15	0F	SI	25	19	EM	**35**	23	
6	06		16	10	DLE	26	1A		**36**	24	

Dec	Hex	Code	Dec	Hex	Code	Dec	Hex	Code	Dec	Hex	Code
7	07	DEL	17	11		27	1B		**37**	25	LF
8	08		18	12		28	1C	IFS	**38**	26	ETB
9	09		19	13		29	1D	IGS	**39**	27	ESC
.	.	.	.	.	.	.	.	.	.	.	.
.	.	.	.	.	.	.	.	.	.	.	.
.	.	.	.	.	.	.	.	.	.	.	.
.	.	.	.	.	.	.	.	.	.	.	.
.	.	.	.	.	.	.	.	.	.	.	.
.	.	.	.	.	.	.	.	.	.	.	.

(for more details please do visit)
http://www.astrodigital.org/digital/ebcdic.html
OR
Scan the QR code

Appendix-D
Unicode

Unicode is an international character encoding standard that includes different languages, scripts and symbols. Each letter, digit or symbol has its own unique Unicode value. Unicode is an extension of ASCII that allows many more characters to be represented.

Since there are so many Unicode characters available, we have divided our list into ranges of 64 characters. Also, we do not represent every possible Unicode character in this table. You may notice that the first 4 *Unicode Ranges* include the same characters as the ASCII standard and extended ASCII. This is because ASCII is a subset of Unicode.

The following is a listing of Unicode characters and their corresponding Unicode, Decimal, Hexadecimal, Octal, HTML Code/HTML Entity, and UTF-8 values.

Unicode Characters (U+0000 - U+003F)
Unicode Characters (U+0040 - U+007F)
Unicode Characters (U+0080 - U+00BF)
Unicode Characters (U+00C0 - U+00FF)
Unicode Characters (U+0100 - U+013F)
Unicode Characters (U+0140 - U+017F)
Unicode Characters (U+0180 - U+01BF)
Unicode Characters (U+01C0 - U+01FF)
Unicode Characters (U+0200 - U+023F)
...and so on

Unicode Characters (U+0000 - U+003F)

Unicode	Char	Dec	Hex	Oct	HTML Code	Escaped Unicode	UTF-8	Description
U+0030	0	48	30	060	0	\u0030	30	Digit zero
U+0031	1	49	31	061	1	\u0031	31	Digit one
U+0032	2	50	32	062	2	\u0032	32	Digit two
U+0033	3	51	33	063	3	\u0033	33	Digit three
U+0034	4	52	34	064	4	\u0034	34	Digit four
U+0035	5	53	35	065	5	\u0035	35	Digit five
U+0036	6	54	36	066	6	\u0036	36	Digit six
U+0037	7	55	37	067	7	\u0037	37	Digit seven
U+0038	8	56	38	070	8	\u0038	38	Digit eight
U+0039	9	57	39	071	9	\u0039	39	Digit nine
U+003A	:	58	3A	072	:	\u003A	3A	Colon
U+003B	;	59	3B	073	;	\u003B	3B	Semicolon
U+003C	<	60	3C	074	<	\u003C	3C	Less-than sign
U+003D	=	61	3D	075	=	\u003D	3D	Equal/Equality sign
U+003E	>	62	3E	076	>	\u003E	3E	Greater-than sign
U+003F	?	63	3F	077	?	\u003F	3F	Question mark
.	.	.	.	.	.	.	.	.
.	.	.	.	.	.	.	.	.
.	.	.	.	.	.	.	.	.

In the same way other symbols are also have been defined.

(for more details please do visit)
https://www.techonthenet.com/unicode/chart.php
OR
Scan the QR code

Appendix-E
Basics of VHDL

Practical Significance: VHDL stands for very high-speed integrated circuit hardware description language which is one of the programming languages used to model/ describe the hardware of a digital system by dataflow, behavioural and structural style of modelling. This practical will help the students to model the logic gates using different modelling techniques.

Relevant Program Outcomes (POs)

- Discipline knowledge: Apply Electronics and Telecommunication engineering knowledge to solve broad-based Electronics and Telecommunications engineering related problems.
- Experiments and practice: Plan to perform experiments and practices to use the results to solve broad-based Electronics and Telecommunication engineering problems.

- Engineering tools: Apply relevant Electronics and Telecommunications technologies and tools with an understanding of the limitations.
- Lifelong learning: Engage in independent and life-long learning activities in the context of technological changes also in the Electronics and Telecommunication engineering and allied industry

Competency and Practical Skills

This practical is expected to develop the following skills for the industry-identified competency: 'Maintain VLSI based electronic circuits and equipments."

- Use relevant VHDL model for the given applications.
- Debug VHDL programme for the given application.

Relevant Course Outcome

- Debug VHDL programme for the given application.

Practical Outcome

- Implement any two logic gates using Data flow and Behavior Model.

Relevant Affective domain related Outcome(s)

- Follow safety practices.
- Maintain tools and equipment.
- Follow ethical practices.

Minimum Theoretical Background

VHDL is used for describing hardware. There are four styles of modelling/ description using VHDL.

- Dataflow Description: A data flow description describes the transfer of data from input to output and between signals. The view of data as flowing through a design, from input to output. An operation is defined in terms of collection of data transformations, expressed as concurrent statements.
- Behavioural Description: A Behavioural modelling describes the design in terms of circuit or system behaviour using algorithms. This high-level language uses language constructs that resemble a high-level software programming language.
- Structural Description: A structural description describes the circuit structure in terms of the logic gates used and the interconnect wiring between the logic gates to form a circuit netlist. This view is expressed by component instantiations.
- Mixed Style of Modelling: Mixed style of modelling is combination of different modelling styles to describe an entire hardware.

VHDL Flow Elements: The VHDL provides five different types of primary constructs called design units:

- Entity
- Architecture
- Configuration
- Package
- Package body

Entity declaration: It defines the names, input output signals and modes of a hardware module.

Syntax:

```
entity entity_name is
      Port declaration;
   end entity_name;
```

An entity declaration should starts with 'entity' and ends with 'end' keywords. Ports are interfaces through which an entity can communicate with its environment. Each port must have a name, direction and a type. An entity may have no port declaration also. The direction will be input, output or inout.

In	Port can be read
Out	Port can be written
Inout	Port can be read and written
Buffer	Port can be read and written, it can have only one source.

Architecture: It describes the internal description of design or it tells what is there inside design. Each entity has at least one architecture and an entity can have many architectures. Architecture can be described using structural, dataflow, behavioural or mixed style. Architecture can be used to describe a design at different levels of abstraction like gate level, register transfer level (RTL) or behaviour level.

Syntax:

```
architecture architecture_name of entity_name
architecture_declarative_part;
begin
    Statements;
end architecture_name;
```

Here we should specify the entity name for which we are writing the architecture body. The architecture statements should be inside the begin and end keyword. Architecture declarative part may contain variables, constants, or component declaration.

Question: Implement AND gate using Data flow and Behavior Model.

Solution:

Resources Required:

Sr. No	Instrument / Components	Specification	Quantity
1	Desktop PC	Loaded with open source EDA tool (VHDL)	1 No.

Precautions to be followed:

Check the syntax / rules of VHDL Programming.

Procedure:

1. Create the Xilinx ISE project for your top-level FPGA design, by doing the following in ISE:
    - In the ISE software, select File > New Project.
    - In the Project name and Project location fields, enter the project name and location, respectively.
    - Select HDL or Schematic as the Top-level source type, and click Next.
2. Create New Source file. Select Source Type" select the Source type and give the name to the source then click "Next"
3. Define Module: Enter the entity used in design and then click "Next".
4. Develop VHDL code for given problem and Synthesize the *.vhd* file and view RTL schematic.
5. Create Test Bench file- A test bench is HDL code that allows you to provide a documented, repeatable set of stimuli that is portable across different simulators. A test bench can be as simple as a file with clock and input data or a more complicated file that includes error checking, file input and output, and conditional testing.
6. Go to implementation to simulation tab, right click on main source file and create Test Bench file for simulation extend with *abc_tb*.

7.In order to view test files, select the box of "Simulation" in the "View Panel" of the "Design" panel. In the "Process Panel," double click on the "Behavioral Check Syntax" to make sure that you didn't make any syntax errors while making changes.

8.Double click on "Simulate Behavioral Model" in the "Process Pane", which will open the ISim software with your test bench loaded.

9.ISim simulator window will open with your simulation executed, where one can simulate designs and check errors if any.

VHDL code for AND gate

Data Flow Model:
VHDL Code

```
library IEEE;
use IEEE.STD_LOGIC_1164.ALL;
use IEEE.STD_LOGIC_ARITH.ALL;
use IEEE.STD_LOGIC_UNSIGNED.ALL;

Entity AND-GATE is
Port ( A,B : in  STD_LOGIC;
       P : out  STD_LOGIC);
end AND-GATE;

Architecture dataflow of AND_GATE is
begin
P < = A and B;
end dataflow;
```

Behavior Model:
VHDL Code

```
library IEEE;
use IEEE.STD_LOGIC_1164.ALL;
use IEEE.STD_LOGIC_ARITH.ALL;
use IEEE.STD_LOGIC_UNSIGNED.ALL

Entity AND is
port(
   A,B: in std_logic;
   C: out std_logic
);
end AND;

Architecture arch of AND is
   begin
   process(a, b)
   begin
   if a='1' and b='1' then
      c <= '1';
   else

c <= '0';
   end if;
   end process;
 end arch;
```

Observations Result:

Truth Table for AND gate has been verified

Practical Related Questions

1. Explain the term model. List different types of Model.
2. Compare different types of model.
3. Give the syntax of Entity and Architecture with example.
4. Implement NAND gate using Behavioral Model.
5. Explain the Process statement with an example.

more such questions so as to ensure the achievement of identified CO

Question: Implement Half/Full adder/ subtractor using FPGA.

Solution

Practical Significance

Field Programmable Gate Arrays (FPGAs) are semiconductor devices that are based around a matrix of configurable logic blocks (CLBs) connected via programmable interconnects. FPGAs can be reprogrammed to desired application or functionality requirements after manufacturing. This practical will help the students to implement the various combinational circuits like Half or Full adder/ subs tractor using FPGA.

Relevant Program Outcomes (POs)

- Discipline knowledge: Apply Electronics and Telecommunication engineering knowledge to solve broad-based Electronics and Telecommunications engineering related problems.
- Experiments and practice: Plan to perform experiments and practices to use the results to solve broad-based Electronics and Telecommunication engineering problems.
- Engineering tools: Apply relevant Electronics and Telecommunications technologies and tools with an understanding of the limitations.
- Lifelong learning: Engage in independent and life-long learning activities in the context of technological changes also in the Electronics and Telecommunication engineering and allied industry

Competency and Practical Skills

This practical is expected to develop the practical skills for the industry-identified competency: 'Maintain VLSI based electronic circuits and equipment." and debug VHDL programme for the given application.

Relevant Course Outcome

Debug VHDL programme for the given application.

Practical Outcome

Implement Half/Full adder/ subtractor using FPGA.

Relevant Affective domain related Outcome(s)

- Follow safety practices.
- Maintain tools and equipment.
- Follow ethical practices.

Minimum Theoretical Background

- Half Adder
- Full Adder:
- Half Subtractor:

Resources Required:

Sr. No.	Instrument /Components	Specification	Quantity
1.	FPGA Development kit	Device: Xilinx FPGA (XC3S400 PQ208), On board +5V, +3.3V, +2.5V supply to FPGA and other hardware circuit. On board, 2 Crystal 8MHz and 25MHz.JTAG Interface (Boundary Scan),PROM Interface (XCF02S),40 pin, 4 header connector for external I/O's	1 No.
2.	Desktop PC	Loaded with open source IDE, simulation and program downloading software.	1 No.

Precautions to be followed:

1. Check the syntax / rules of VHDL Programming.
2. Do not power up the board before completing connections.

Procedure:

1. Create the Xilinx ISE project for top-level FPGA design, by doing the following in ISE:
   - In the ISE software, select File > New Project.
   - In the Project name and Project location fields, enter the project name and location, respectively.
   - Select HDL or Schematic as the Top-level source type, and click Next.
2. Create New Source file. Select Source Type" select the Source type and give the name to the source then click "Next".
3. Define Module". Enter the entity used in design and then click "Next".
4. Develop VHDL code for given problem and Synthesize the .vhd file and view RTL schematic.
5. Create Test Bench file- A test bench is HDL code that allows documentation, repeatable set of stimuli that is portable across different simulators. A test bench can be as simple as a file with clock and input data or a more complicated file that includes error checking, file input and output, and conditional testing.
6. Go to implementation to simulation tab , right click on main source file and create Test Bench file for simulation.
7. In order to view test files, select the box of "Simulation" in the "View Panel" of the "Design" panel. In the "Process Panel," double click on the "Behavioral Check Syntax" to make sure that there are no syntax errors, while making changes.
8. Double click on "Simulate Behavioral Model" in the "Process Pane", which will open the ISim software with test bench loaded.
9. ISim simulator window will open with simulation executed, where one can simulate designs and check for errors.
10. After simulation implement 2 as 4 decoder Using FPGA development board.
11. Go to User Constraints – select Floorplan Area/IO/Logic (PlanAhead) – Windows – Properties – now assign pin configuration and save.
12. Go to Implement Design – Run all.
13. Go to Generate Programming File and Run
14. Go to Config. Target Device and Run – Impact wizard opened – select Device 2 (as per practical Kit) -- open Bit file -- and initialize Chain – Right click on Xilinx IC And select Program.

Sample Program:

Implement Half Adder and Full Adder using data flow model.

Data Flow Model for Half Adder

VHDL Code

```
library IEEE;
use IEEE.STD_LOGIC_1164.ALL;
use IEEE.STD_LOG IC_ARITH.ALL;
use IEEE.STD_LOGIC_UNSIGNED.ALL;
entity HA is
Port ( A, B : in std_logic;
S, C : out std_logic);
end HA;
architecture dataflow of HA is
```

Data Flow Model for Full Adder

VHDL Code

```
library IEEE;
use IEEE.STD_LOGIC_1164.ALL;
use IEEE.STD_LOG IC_ARITH.ALL;
useIEEE.STD_LOGIC_UNSIGNED.ALL;
entity full_adder is
port(A, B, Cin :in bit;
S, C:out bit);
end full_adder;
  architecture data of full_adder is
```

```
begin
S<= A xor B;
C<= A and B;
end dataflow;
```

```
begin
S<= A xor B xor Cin;
C <= ((A and B) or (B and Cin) or (A and
Cin));
end data;
```

Observations/Result:

Truth Table for Half Adder and Full Adder has been verified

Practical Related Questions

1. Write the VHDL code for half adder using Behavioral Model.
2. Write the VHDL code for full subtractor using if-then-else statement.
3. Write VHDL code for full adder using mixed style of model.

more such questions so as to ensure the achievement of identified CO

Question: Implement 8:1 Multiplexer using FPGA.

Solution

Practical Significance

In electronics, a multiplexer (or mux) is a device that selects between several analog or digital input signals and forwards it to a single output line. Multiplexers are used to implement Boolean functions of multiple variables. This practical will help the students to implement the 8:1 using FPGA.

Relevant Program Outcomes (POs)

- Discipline knowledge: Apply Electronics and Telecommunication engineering knowledge to solve broad-based Electronics and Telecommunications engineering related problems.
- Experiments and practice: Plan to perform experiments and practices to use the results to solve broad-based Electronics and Telecommunication engineering problems.
- Engineering tools: Apply relevant Electronics and Telecommunications technologies and tools with an understanding of the limitations.
- Lifelong learning: Engage in independent and life-long learning activities in the context of technological changes also in the Electronics and Telecommunication engineering and allied industry

Competency and Practical Skills

This practical is expected to develop the following skills for the industry-identifiedcompetency: 'Maintain VLSI based electronic circuits and equipments."

1. Use relevant VHDL model for the given applications.
2. Debug VHDL programme for the given application.

Relevant Course Outcome

Debug VHDL programme for the given application.

Practical Outcome

Implement 8:1 Multiplexer using FPGA.

Relevant Affective domain related Outcome(s)

- Follow safety practices.
- Maintain tools and equipment.
- Follow ethical practices.

Minimum Theoretical Background

1. Basic Idea behind Multiplexer
2. Working of Xilinx

Resources Required:

Sr. No.	Instrument /Components	Specification	Quantity
1.	FPGA Development kit	Device: Xilinx FPGA (XC3S400 PQ208), On board +5V, +3.3V, +2.5V supply to FPGA and other hardware circuit. On board, 2 Crystal 8MHz and 25MHz.JTAG Interface (Boundary Scan),PROM Interface (XCF02S),40 pin, 4 header connector for external I/O's	1 No.
2.	Desktop PC	Loaded with open source IDE, simulation and program downloading software.	1 No.

Precautions to be followed:

1. Check the syntax / rules of VHDL Programming.
2. Do not power up the board before completing connections.

Procedure:

1. Create the Xilinx ISE project for your top-level FPGA design, by doing the following in ISE:
   - In the ISE software, select File > New Project.
   - In the Project name and Project location fields, enter the project name and location, respectively.
   - Select HDL or Schematic as the Top-level source type, and click Next.
2. Create New Source file. Select Source Type" select the Source type and give the name to the source then click "Next".
3. Define Module enter the entity used in design and then click "Next".
4. Develop VHDL code for given problem and Synthesize the .vhd file and view RTL schematic.
5. Create Test Bench file- A test bench is HDL code that allows documentation, repeatable set of stimuli that is portable across different simulators. A test bench can be as simple as a file with clock and input data or a more complicated file that includes error checking, file input and output, and conditional testing.
6. Go to implementation to simulation tab , right click on main source file and create Test Bench file for simulation.
7. In order to view test files, select the box of "Simulation" in the "View Panel" of the "Design" panel. In the "Process Panel," double click on the "Behavioral Check Syntax" to make sure that there are on syntax errors.
8. Double click on "Simulate Behavioral Model" in the "Process Pane", which will open the ISim software with test bench loaded.
9. ISim simulator window will open with simulation executed, where can simulate designs and check for errors.
10. After simulation implement 8:1 multiplexer Using FPGA development board.
11. Go to User Constraints – select Floorplan Area/IO/Logic (PlanAhead) – Windows– Properties – now assign pin configuration and save.
12. Go to Implement Design – Run all.
13. Go to Generate Programming File and Run
14. Go to Config. Target Device and Run – Impact wizard opened – select Device 2 (as per practical Kit) -- open Bit file -- and initialize Chain – Right click on Xilinx IC And select Program.

Sample Program:

Implement 8:1 Multiplexer.

```
library IEEE;
use IEEE.STD_LOGIC_1164.ALL;
use IEEE.STD_LOG IC_ARITH.ALL;
use IEEE.STD_LOGIC_UNSIGNED.ALL;
entity mux8_1 is
port(
din : in STD_LOGIC_VECTOR(7 downto 0);
sel : in STD_LOGIC_VECTOR(2 downto 0)
Y : out STD_LOGIC
      );
end mux8_1;
architecture multiplexer8_1_arc of mux8_1 is
begin
     Y <= din(7) when (sel="000") else
        din(6) when (sel="001") else
        din(5) when (sel="010") else
        din(4) when (sel="011") else
        din(3) when (sel="100") else
        din(2) when (sel="101") else
        din(1) when (sel="110") else
        din(0);
end multiplexer8_1_arc
```

Observations/Result:

Truth Table for Half Adder and Full Adder has been verified

Practical Related Questions

1. Give the syntax of if statement.
2. Write the VHDL code for 4:1 multiplexer using the "case" statement.

more such questions so as to ensure the achievement of identified CO

CO AND PO ATTAINMENT TABLE

Course outcomes (COs) for this course can be mapped with the programme outcomes (POs) after the completion of the course and a correlation can be made for the attainment of POs to analyze the gap. After proper analysis of the gap in the attainment of POs necessary measures can be taken to vercome the gaps.

Table for CO and PO attainment

Course Outcomes	Expected Mapping with Programme Outcomes *(1- Weak Correlation; 2- Medium correlation; 3- Strong Correlation)*											
	PO-1	PO-2	PO-3	PO-4	PO-5	PO-6	PO-7	PO-8	PO-9	PO-10	PO-11	PO-12
CO-1												
CO-2												
CO-3												
CO-4												
CO-5												
CO-6												

The data filled in the above table can be used for gap analysis.

INDEX